Principles of
HEAT TRANSFER

Principles of
HEAT TRANSFER

D. Srinivasan
Formerly Professor
Anna University

NEW AGE

NEW AGE INTERNATIONAL (P) LIMITED, PUBLISHERS
New Delhi • Bangalore • Chennai • Cochin • Guwahati • Hyderabad
Jalandhar • Kolkata • Lucknow • Mumbai • Ranchi

Published by New Age International (P) Ltd., Publishers
First Edition: 2003
Reprint : 2007

ISBN : 81-224-1502-4

Rs. 205.00

C-07-04-1427

3 4 5 6 7 8 9 10

Printed in India at Mehra Offset, Delhi.

PUBLISHING FOR ONE WORLD

NEW AGE INTERNATIONAL (P) LIMITED, PUBLISHERS
(formerly Wiley Eastern Limited)
4835/24, Ansari Road, Daryaganj, New Delhi - 110002
Visit us at **www.newagepublishers.com**

To

All my family members

Preface

In the past two or three decades, a large number of text books on the subject of 'Heat Transfer' have been published. In each case the author or authors have undoubtedly felt that their new entry into the field offered something new or unique to the students engaged in the study of this important field. As a matter of fact, the field of heat transfer itself has advanced manifold during this time span.

Every teacher at the university level develops approaches to the teaching of certain material and a methodology of organizing the subject matter, which in his or her experience works best.

While preparing this book it was felt that the contents of this book are to be confined to the fundamental facts and leave all somewhat more complicated subject matter to the postgraduate study. It is believed that a student completing first or second level course in heat transfer should have a good understanding of the physical phenomena involved and certainly a capability to formulate and solve typical problems encountered in this field, viz. heat transfer. This textbook lays emphasis on fundamental aspects of heat transfer. It is needless to add here that no attempt has been made to be exhaustive concerning the advanced developments. The writing of any textbook requires some selection of material and some compromise concerning the level of treatment. The choices made in the preparation of this book are those thought to be most reasonable by this author.

The presentation adopted in the preparation endeavors to convey to the student in a simple manner, a physical understanding of the processes by which heat is transmitted and provide him or her with the tools necessary to get quantitative solution to engineering problems involving one or more of the basic modes of heat flow. Sufficient material has been included in the text to cater to the requirements of the undergraduate curriculum. Illustrations pertaining to the different modes of heat transfer and the design calculations of heat exchangers have been liberally included in the text.

The purpose of this book is to present a basic introduction to the field of engineering heat transfer. The book begins with a brief presentation of the importance of heat transfer in chemical and processing industry and the modes of heat transfer. Chapter 2, dealing with conduction, includes a few aspects of conduction phenomenon, analogy between heat flow and electricity flow, critical thickness and conduction with internal generation of heat. In Chapter 3, the concept of film coefficients is presented and the relationship between the individual and overall heat transfer coefficients are dealt with. The phenomenon of unsteady state heat transfer and the methods of solving one dimensional transient heat conduction problems have been discussed in Chapter 4, which is on unsteady state heat conduction. Also the application of molecular transport theory to the unsteady state heat conduction is included. Chapter 5, which is on convection, a general basic concept, the application of dimensional analysis in the case of forced and free convection, the heat transfer from fins, the heat transfer to fluids in laminar flow inside tubes, heat transfer from condensed vapours and boiling heat transfer are included. The various types of heat exchangers, the concept of capacity ratios, the effectiveness of heat exchanger, the log mean temperature difference, the number of transfer units (NTU) and calculations pertaining to heat exchanger design and the effectiveness–NTU relationship have been discussed in Chapter 6, which bears the title 'Industrial Heat Exchange Equipment'. In Chapter 7, which is on thermal energy transfer by radiation, the basic concepts and theory of radiation are presented. Chapter 8, deals with evaporation, the basic concepts and definitions, boiling point elevation, types of evaporators, single and multiple effect evaporation, the occurrence of heat transfer in evaporators and the analysis of performance calculations of multiple effect evaporators are discussed at some length. Chapter 9, the final chapter, presents a brief review of heat transfer principles.

D. SRINIVASAN

Nomenclature

A	Area of cross-section
a	Temperature coefficients of thermal conductivity
b	Width
c_p	Specific heat at constant pressure
d	Diameter
E	Emissive power, Heat Exchanger Effectiveness
F	Correction factor for LMTD
Gr	Grashof number
Gz	Graetz number
g	Acceleration due to gravity
h	Convection heat transfer coefficient
j	Colburn j factor
k	Thermal conductivity
LMTD	Log Mean Temperature Difference
P	Pressure
Pr	Prandtl number
Q	Quantity of heat, rate of heat transfer
R	Thermal resistance
Re	Reynold number
r	Radius
st	Stanton number
T, t	Temperature
U	Overall heat transfer coefficient
v	Velocity, volume (molal)
x	x-coordinate

Greek Letters

α	Thermal diffusivity, Absorptivity
β	Coefficient of volume expansion
ε	Emissivity
μ	Viscosity (dynamic)
ρ	Density
σ	Stefen— Boltzmann constant

Subscripts

c	Cold fluid
c	Critical
f	Film
h	Hot fluid
i	Inner surface
s	Saturation
v	vapour

Conversion Factors

Multiply	By	To obtain
acre	0.404	ha (hectares)
acre ft	1233	m^3
atm	14.7	lb/in^2
Btu	252	cal
Btu	1.054×10^3	J
Btu/ft^3	8905	cal/m^3
Btu/lb	2.32	J/g
Btu/lb	0.555	cal/g
Btu/sec	1.05	*kw*
Btu/ton	278	cal/tonne
cal	4.18	joules
cal	3.9×10^{-3}	Btu
cal/gm	1.80	Btu/lb
cal/m^3	1.12×10^{-4}	Btu/ft^3
cal/tonne	3.60×10^{-3}	Btu/ton
cm	0.393	in
ft	0.305	m
ft/min	0.00508	m/sec

ft/sec	0.305	m/sec
ft^2	0.0929	m^2
ft^3	0.0283	m^3
ft^3	28.3	liters
ft^3/sec	0.0283	m^3/sec
ft^3/sec	449	gal/min
ft lb (force)	1.357	joules
ft lb (force)	1.357	newton meters
gal	3.78×10^{-3}	m^3
gal	3.78	liters
gal/day/ft^2	0.0407	m^3/day/m^2
gal/min	2.23×10^{-3}	ft^3/sec
gal/min	0.0631	liters/sec
gal/min	0.227	m^3/hr
gal/min	6.31×10^{-5}	m^2/sec
gal/min/ft^2	2.42	m^2/hr/m^2
million gals/day (mgd)	43.8	liters/sec
mgd	3785	m^3/day
mgd	0.0438	m^2/sec
gms	2.2×10^{-3}	lb
hectares	2.47	acre
h.p.	0.745	kw
inches	2.54	cm
inches of Hg	0.49	lb/m^2
inches of Hg	3.38×10^3	newtons/m^2
inches of water	249	newtons/m^2
joules (J)	0.239	calorie
joules	9.48×10^{-4}	Btu
joule	0.738	ft lb
joule	2.78×10^{-7}	kwh
joule	1	newton meter
J/g	0.430	Btu/lb
J/sec	1	walt
kilogram (kg)	2.2	lb (mass)
kg	1.1×10^{-3}	tons
kg/ha	0.893	lb/acre
kg/hr	2.2	lb/hr
kg/m^3	0.0624	lb/ft^3

kg/m^3	1.68	lb/yd^3
kilometers (km)	0.622	mile
km/hr	0.622	m/hr
kilowatt (kw)	1.341	h.p.
kwh	3600	kilojoule
liters	0.264	gal
liters/sec	15.8	gal/min
meters (m)	3.28	ft
meters	1.094	yd
m/sec	3.28	ft/sec
m/sec	196.8	ft/min
m^2	10.74	ft^2
m^2	1.196	yd^2
m^3	35.3	ft^3
m^3	26.4	gal
m^3	1.31	yd^3
m/day	264	gal/day
m^3/hr	4.4	gpm
m^3/hr	6.38×10^{-3}	gpm
m^3/sec	35.31	ft^3/sec
m^3/sec	15,850	gpm
m^3/sec	22.8	mgd
miles	1.61	km
$miles^2$	2.59	km^2
mph	0.447	m/sec
milligram/liter	0.001	kg/m^3
mgd	43.8	liter/sec
mgd	157	m^3/sec
mgd	0.0438	m^3/sec
newton (N)	0.225	lb (force)
$newton/m^2$	2.94×10^{-4}	inches of Hg
$newton/m^2$	1.4×10^{-4}	lb/in^2
newton meters	1	joule
newton sec/m^2	10	poise
pounds (force)	4.45	newton
pounds $(force)/in^2$	6895	N/m^2
pounds (mass)	454	gm
pounds (mass)	0.454	$kg/m^2/kg$

pounds (mass)/ft^2/yr	4.89	kg/m^2/yr
pounds (mass)/yr/ft^3	16.0	kg/yr/m^3
pounds/acre	1.12	kg/ha
pounds/ft^3	16.04	kg/m^3
pounds/in^2	0.068	atmospheres
pounds/in^2	2.04	inches of Hg
pounds/in^2	7140	Newtons/m^2
tons	0.907	tonne
(2000 lbs)		(1000 kgs)
tons	907	kg
ton/acre	2.24	tonne/ha
tonne (1000 kgs)	1.10	tons (2000 lbs)
tonne/ha	0.446	tons/acre
yd	0.914	m
yd^3	0.765	m^3
watt	1	J/sec

Contents

1

Introduction

Heat transfer is one of the unit operations involving the transfer of thermal energy in the form of heat.

The laws of heat transmission are of controlling importance in the design and operation of many forms of stream generators, furnaces, preheaters, exchangers, coolers, evaporators and condensers in any processing industry.

1.1 THE OBJECTIVES

- To obtain the maximum heat transfer rate per unit surface compatible with economic factors.
- Salvaging of heat as in exchangers, recuperators and regenerators.
- The purpose may also be to minimize the heat flux by making use of insulation.

Heat is transferred in most cases by more than one of the modes and hence it is preferable to use the term transmission to describe the overall process and the respective names of the modes of transmission in the individual cases.

The three modes of heat transfer are (1) conduction (2) convection and (3) radiation.

Heat transfer can be accomplished by each one of the above three and/or combination of all the three. These are three different mechanisms and each one of it has a rate equation. Depending on the situation the three modes would come into operation.

As chemical engineers we are interested in studying/knowing.

(i) the basic mechanism of heat flow

(ii) the fundamental quantitative methods of calculation with special relevance to chemical engineering and

(iii) the practical utility viz., the application of the principles to the design of heating and cooling equipment.

1.2 CLASSIFICATION OF HEAT FLOW PROCESSES

Conduction

Conduction in a homogeneous opaque solid is the transfer of heat from one part to another under the influence of a temperature gradient without appreciable displacement of the particles. Conduction involves the transfer of the kinetic energy (transference of momentum of individual molecules without mixing) from one molecule to an adjacent molecule; it is the only mechanism of flow of heat in an opaque solid. In case of certain transparent solids such as glass and quartz the thermal energy is transmitted by radiation as well as conduction. In the case of gases and liquids conduction may be accompanied by convection and radiation. Within a fluid flowing in streamline (laminar) motion heat transfer by conduction occurs at right angles to the direction of fluid flow, e.g. the flow of heat through the brick wall of a furnace or the metal shell of a heat exchanger or boiler.

Convection

This involves the transfer of heat, by mixing one parcel of fluid with another. Convection occurs only in fluids. In practice heat flow by conduction through fluids is restricted to certain cases of laminar flow or heat flow through thin films. Heat flow by convection only unaccompanied by conduction is never encountered because of the laminar layer (present near the boundary i.e. tube wall) through which the heat must pass by conduction. As they occur simultaneously it is rather difficult to separate them nor is it useful. It is better to consider the total contribution of the two mechanisms to the transfer.

The motion of the fluid may be entirely the result of difference of density resulting from the temperature difference as in natural convection or the motion may be induced by mechanical means, by the use of a blower, agitator, stirrer as in forced convection. As was pointed out earlier, energy is also transferred by conduction and in transparent media by radiation. Examples include (i) the heating of a room by means of a steam radiator and the heating of water by a hot surface; (ii) heat transfer to the body of the fluid flowing through a pipe from the hot pipe wall or vice versa.

Radiation

Radiation is a term given to the transfer of energy through space by means of electro magnetic waves. If radiation is passing through an empty space, it is not transformed to heat or any other form of energy nor is it diverted from its path. When this energy strikes another body part may be reflected, part may be transmitted through the body in which case the body said to be diathermanous. The rest of the energy is absorbed and quantitatively transformed into heat. It is only the absorbed energy that manifests as heat.

It two bodies, one hotter than the other are placed within an enclosure there is a continuous interchange of energy between them. The hotter body radiates more energy than it absorbs; the cooler body absorbs more than it radiates. Even after equilibrium of temperature is attained the process continues, with each body radiating and absorbing energy. Certain gases as well as solids are capable of radiating and absorbing thermal energy.

Fused quartz transmits practically all of the radiation that strikes it: a polished opaque surface or mirror will reflect most of the radiation impinging on it. Whereas a black or matte surface will absorb most of the radiation received by it and will hence transform such absorbed energy quantitatively into heat.

Some gases may be transferent to radiation and heat flow through a fluid may occur by radiation–conduction–convection, e.g. loss of heat from a radiator or unlagged steam pipe to the surrounding air of the room, heat transfer in furnaces.

The term radiation is probably the most generally misused of the three; such expressions as radiation through furnaces walls being common.

Generally radiation becomes important at high temperatures and is independent of the circumstances of the flow of the fluid while conduction–convection is sensitive to the flow pattern (conditions) and is relatively unaffected by the temperature level.

2

Conduction

2.1 CONCEPT OF CONDUCTION

This concept is easily understood by studying this phenomenon in the case of solids because convection is not present in this situation.

The basic law is

$$\text{Rate} = \frac{\text{driving force}}{\text{resistance}}$$

The driving force is the temperature drop across the solid because heat flow can occur only if there is inequality of temperature.

2.2 FOURIER'S LAW OF HEAT CONDUCTION

Consider a flat wall of thickness and an area A.

Let the temperature drop be Δt

The law states,

$$\text{Rate of flow} \propto \text{temperature drop, area}$$

$$\propto \frac{1}{\text{thickness of the wall}}.$$

If it does not vary with time, then the rate of flow is constant and the process is one of steady state.

If Q is the heat flow in time Q, then Q/θ is the rate of heat flow and Fourier's law takes the following mathematical form

$$\frac{Q}{\theta} = q = \frac{k\,A\,\Delta t}{x} = \frac{\Delta t}{R}$$

where $\quad R = \dfrac{x}{k\,A}$

k is the thermal conductivity of the material in question—it is a physical property.

More generally the instantaneous rate of heat flow may be written as

$$\frac{dQ}{d\theta} = -k A \frac{dt}{dx}$$

where A = area of the section taken at right angle to the direction of heat flow

$-dt/dx$ = temperature gradient i.e. the rate of change of temperature 't' with respect to the length of the path x.

The above form is general for unidirectional conduction and can be applied to cases in which the temperature gradient $-dt/dx$ varies with time as well as the location of the point considered.

The process of heat flow in a case in which temperature varies with time and position is called unsteady state or transient heat conduction.

If the temperature gradient dt/dx does not vary with time, then the rate of heat flow is constant with time and

$$\frac{dQ}{d\theta} = \text{Constant} = q = -\frac{k A\, dt}{dx}$$

Since normally we know only the temperatures at the boundaries i.e. at the two faces of the wall and not the intermediate temperatures along the path of heat transfer, the ordinary use of Fourier's law requires that the differential equation be integrated over a specified path, viz. $x = 0$ to x = total length.

k may be a function of temperature. Similarly A may vary with x but is independent of the temperature.

By separating the variables

$$\frac{q\, dx}{A} = -k\, dt$$

Applying the boundary conditions

$$q \int_0^x \frac{dx}{A} = -\int_{t_1}^{t_2} k\, dt = \int_{t_2}^{t_1} k\, dt$$

where $t_1 > t_2$

q is constant

If A does not vary with x (flat wall is an example if this situation) and if k varies with temperature (normally the variation is taken as linear) so that k_m the arithmetic mean value of k may be considered constant

$$\frac{q\, x}{A} = k_m (t_1 - t_2) = k_m\, \Delta t \text{ or } q = \frac{k_m\, A\, \Delta t}{x}$$

2.3 THERMAL CONDUCTIVITY

It is a physical characteristic or property of the material from which the wall is made.

k is expressed in F.P.S. system as Btu per hour per square foot taken at right angles to the direction of heat flow, per unit temperature gradient °F per foot length of path.

Often it is improperly abbreviated as Btu per hr per sq. ft per °F per ft. It should be remembered that the units of k are

$$\frac{\text{Btu / hr sq ft}}{{}^\circ\text{F / ft}} = \frac{\text{Btu}}{(\text{hr})\,(\text{ft})\,({}^\circ\text{F})}$$

In other system of units

$$\text{C.G.S.}: \frac{\text{cal}}{(\text{sec})\,(\text{cm})\,(\text{K})}$$

$$\text{S.I. units}: \frac{\text{Joules}}{(\text{sec})\,(\text{m})\,{}^\circ\text{C} \text{ or } (\text{hr})\,(\text{m})\,{}^\circ\text{C}}$$

$$\frac{\text{watts}}{\text{m}\,{}^\circ\text{K or }{}^\circ\text{C}}$$

The numerical value of k depends on the material of which the body is made and upon its temperature. The variation with temperature is not much and for small values of Δt, k may be considered constant.

If it is linear

$$k = a + bt$$

a and b are constants

t—temperature

One can use an average value of $\bar{k}$ for k, which can be found either by averaging the individual values of k for the two end temperatures or by computing t_{av} (arithmetic) and using the value of k at that temperature.

k for liquids and gases are small and for solids it is very high. Its values for a few common materials are

k for *cu*	220 Btu/hr ft °F
for corkboard	0.025 Btu/hr ft °F
silver	240 Btu/hr ft °F
building bricks	0.4 Btu/hr ft °F
water	0.35 Btu/hr ft °F
air	0.017 Btu/hr ft °F

For insulating materials k is very low.

The temperature dependence of k may also be written as

$k =$ k_0 (1 + at)

k – at t °F

k_0 – at 0 °F

For non-metallic solids as t is increased, k increases for amorphous substances and k decreases for crystalline substances. The k of an alloy of two metals may be less than the value of k of either constituent, e.g.

constantan 60% cu

40% Ni

k of the alloy is 13 Btu/hr ft °F

whereas k of Cu = 222 Btu/hr ft °F

k of Ni = 36 Btu/hr ft °F

We know $q = -kA\, dt/dx$

For integrating this expression, variation of k with t and of A with x must be known.

If k is linearly changing with temperature

$k = k_0\,(1 + at)$

We can now write

$$-k\,dt = -k_0\,(1 + at)\,dt = \frac{q\,dx}{A}$$

$$-\int_{t_1}^{t_2} k\,dt = (t_1 - t_2)\,k_0\left(1 + a\,\frac{t_1 + t_2}{2}\right)$$

$$= q\int_1^2 \frac{dx}{A}$$

Hence, where k is linear in temperature, one can use the equation

$$k_a\,(t_1 - t_2) = q\int_1^2 \frac{dx}{A}$$

where $k_a = \dfrac{k_1 + k_2}{2}$

or

evaluated at the arithmetic mean of t_1 and t_2.

If k is not linear in temperature, some mean value k_m will apply. Irrespective of the relation between A and x

$$k_m = \frac{1}{t_1 - t_2} \int_{t_1}^{t_2} k \, dt$$

2.4 CONDUCTION OF HEAT THROUGH A SOLID OF CONSTANT CROSS-SECTIONAL AREA

A is constant and the integration of the rate equation between the limits of x_1 and x_2 gives (using k_m for the mean value of k)

$$q = \frac{k_m \, A \, (t_2 - t_1)}{(x_2 - x_1)}$$

$$= \frac{k_m \, A \, \Delta t}{x}$$

The heat transfer rate equations are summarized in the following table

Table 2.1 *Summary of heat transfer rate processes*

Mode of Heat Transfer	*Mechanism*	*Rate Equation*	*Transport Property or Coefficient*
Conduction	Diffusion of energy due to random molecular motion	$q = -k \dfrac{dT}{dx}$	k w/m °k
Convection	Diffusion of energy due to random molecular motion plus energy transfer due to bulk motion	$q = h\,(T_s - T_\infty)$	h w/m² °k
Radiation	Energy transfer by	$q = \varepsilon\,\sigma\left(T_S^4 - T_{sur}^4\right)$	ε
	electromagnetic waves	$q = h_r\, A\left(T_S - T_{sur}\right)$	h_r w/m² °k

The range of thermal conductivity for various states of matter at normal temperatures and pressure is shown in Fig. 2.1.

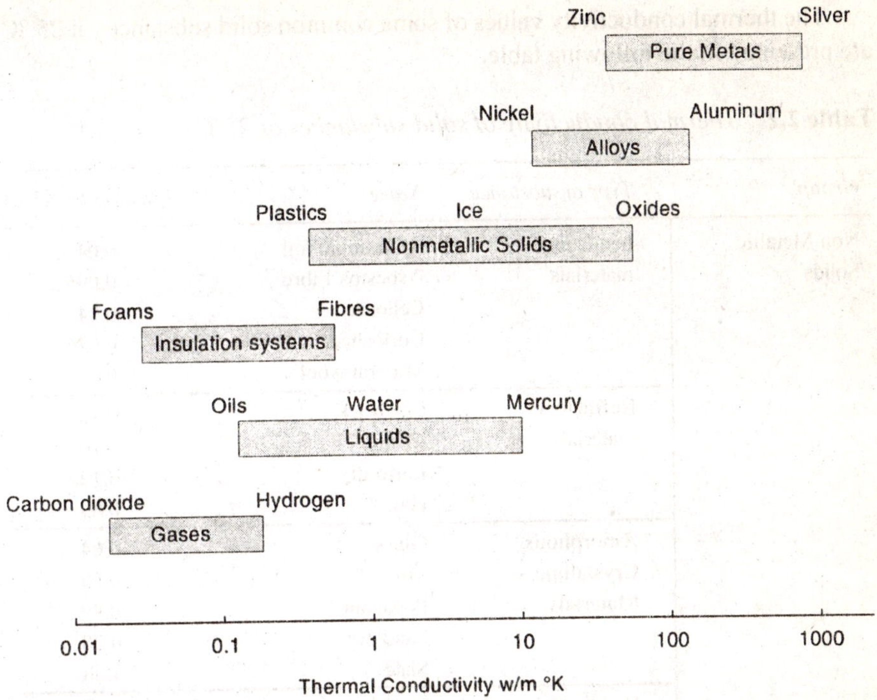

Fig. 2.1 *The range of thermal conductivity for various states of matter*

Thermal conductivity ranges for various classes of substances at moderate pressures and temperatures

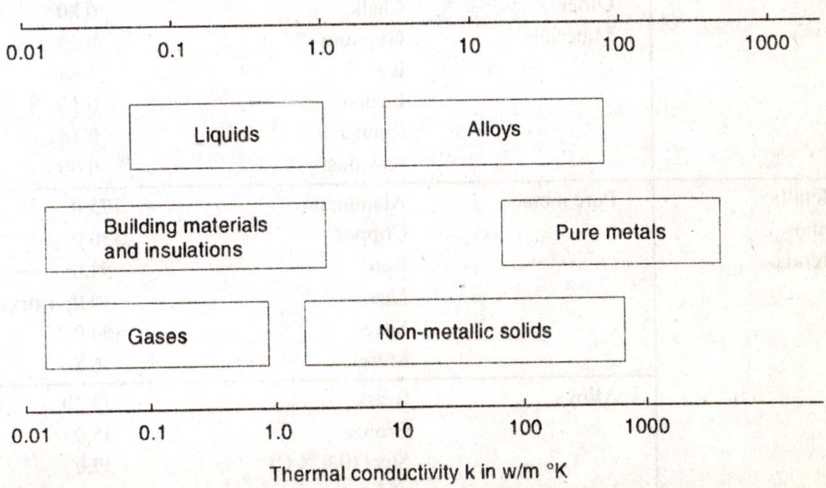

Fig. 2.2 *Thermal conductivity ranges for various classes of substances*

The thermal conductivity values of some common solid substances at 25 °C are presented in the following table.

Table 2.2 *Thermal conductivity of solid substances at 25°C*

Group	*Type of substance*	*Name*	*k kcal/m hr °C*
Non Metallic Solids	Insulating materials	Aluminium foil	0.04
		Asbestos Fibre	0.095
		Cellotex	0.04
		Cork sheet	0.036
		Mineral wool	0.04
	Refractory materials	Fire brick	0.12
		Coal	0.16
		Earth dry	0.12
		clay	0.89
	Amorphous Crystalline Materials	Glass	0.64
		Mica	0.50
		Porcelain	0.89
		Sand dry	0.28
		Slate	1.28
	Building Materials	Brick	0.20
		Concrete	1.10
		Gravel	0.31
		Linoleum	0.16
		Marble	1.12
		Plaster	0.67
		Cement	0.26
	Other Materials	Chalk	0.80
		Gypsum	0.25
		Ice	1.94
		Leather	0.13
		Rubber	0.14
		Saw dust	0.06
Metallic Solids Metals	Pure metals	Aluminium	175.0
		Copper	330.0
		Iron	54.0
		Lead	30.0
		Silver	360.0
		Mercury	6.8
	Alloys	Brass	73.50
		Bronze	55.0
		Steel (0.8 % C)	39.0

Illustration

1. Calculate the heat loss through a 9″ brick and mortar wall ($k_m = 0.4$) 10 ft high and 6 ft wide when the inner and outer surface temperatures are 330°F and 130°F respectively.

Solution

$$q = \frac{k_m\, A\, \Delta t}{x}$$

$$= \frac{(0 \cdot 4)\,(60)\,(330 - 130)}{9/12} = 6400 \text{ Btu / hr}$$

2. A layer of pulverised cork 6″ thick is used as a layer of insulation in a flat wall. The temperature of the cold side of the cork is 40°F and that of the warm side is 180°F The k of the cork material at 32°F is 0.021 Btu/hr ft °F and that at 200°F is 0.032. The area of the wall is 25 ft². Calculate the heat flow through the wall in Btu/hr

Solution The k_m has to be computed first; the arithmetic average temperature in the cork is $\dfrac{40 + 180}{2} = 110\ °\text{F}$

By linear interpolation k at 110°F is

$$k_m = 0.021 + \frac{(110 - 32)\,(0.032 - 0.021)}{(200 - 32)}$$

$$= 0.021 + 0.005$$

$$= 0.026$$

$$A = 25 \text{ ft}^2 \quad \Delta t = 140\ °\text{F}$$

$$x = 6/12 = 0.5 \text{ ft}$$

$$q = \frac{(0.026)\,(25)\,(140)}{(0.5)} = 182 \text{ Btu / hr}$$

2.5 COMPOUND RESISTANCE IN SERIES—SLAB GEOMETRY

Let us consider say three layers of different thicknesses and different materials; the k's of the three materials are k_A k_B and k_C respectively.

The area does not change with distance in the direction of heat transfer i.e. A (transport area) is constant.

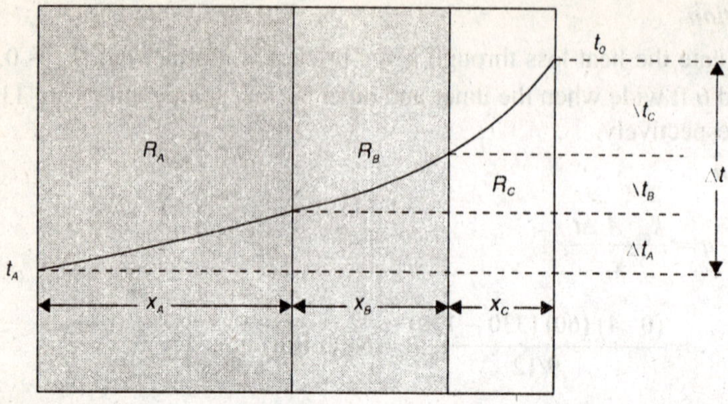

Fig. 2.3 *Compound resistances in series*

Δt = temperature drop in all the three layers

$$\Delta t = \Delta t_A + \Delta t_B + \Delta t_C$$

The temperature drops in the three layers are

$$\Delta t_A = q_A \frac{x_A}{k_A\ A} \qquad \Delta t_B = q_B \frac{x_B}{k_B\ A}$$

and $$\Delta t_C = q_C \frac{x_C}{k_C\ A}$$

The temperature drops when added together

$$\Delta t_A + \Delta t_B + \Delta t_C = \frac{q_A\ x_A}{A\ k_A} + \frac{q_B\ x_B}{A\ k_B} + \frac{q_C\ x_C}{A\ k_C}$$

$$= \Delta t$$

Since all the heat that passes through the first resistance, must pass through the second and then through the third, $q_A\, q_B$ and q_C must be equal and hence can be represented by a common term viz. q

q from the above expression is

$$q = \frac{\Delta t}{\dfrac{x_A}{k_A\ A} + \dfrac{x_B}{k_B\ A} + \dfrac{x_C}{k_C\ A}} = \frac{\Delta t}{R_A + R_B + R_C}$$

where R_A, R_B and R_c are the respective resistances.

Problem

A flat furnace wall consists of three layers: 4 inches of kaolin insulating fire brick on the inside, 5 inches of kaolin insulating brick in the middle and 5 inches of magnesite brick on the outside: If the inner surface is at 1375°F and the outer surface is at 215°F, what is the rate of heat loss per unit area through the wall? Use linear interpolatron of the thermal conductivity data given below ; what are the temperatures at the interfaces?

	°F	*k, Btu/hr ft °F*
Kaolin Insulating Brick	93.2	0.15
	2102	0.26
Kaolin Insulating Firebrick	392	0.05
	1900	0.113
Magnesite brick	400	2.2
(MgO, Fe_2O_3, CaO and $Si\ O_2$)	1200	1.6
	2190	1.1

Illustration

1. An industrial furnace wall is constructed of 0.7 ft thick fireclay brick having $k = 0.6$ Btu/hr ft °F. This is covered on the outer surface with 0.1 ft thick layer of insulating material having a $k = 0.04$ Btu/hr ft °F. The inner most surface is at 1800°F and the outer most is at 100°F . Calculate the steady state heat transfer per sq ft.

Solution

$$q = \frac{\Delta t}{\dfrac{x_A}{k_A\ A} + \dfrac{x_B}{k_B\ A}} \qquad A = 1\,\text{sq. ft}$$

x_A – thickness of brick wall

x_B = thickness of the insulation

$$= \frac{1800 - 100}{\dfrac{0.7}{0.6} + \dfrac{0.1}{0.04}}$$

= 464 Btu/hr ft^2

2. A frequently encountered engineering problem is the determination of the thickness of insulation that will result in a specified heat flux. If the maximum allowable heat transfer rate for the furnace of the previous problem is 300 Btu/hr ft^2, the brick wall is unchanged and the same insulation material is to be used, how thick must the insulation be?

Solution

$$q = \frac{\Delta t}{\dfrac{x_A}{k_A\ A} + \dfrac{x_B}{k_B\ A}} \qquad 300 = \frac{1700}{\dfrac{0.7}{0.6} + \dfrac{x_B}{0.04}}$$

$x_B = 0.18$ ft

3. A composite three- layered wall is formed of a 0.5 cm thick Aluminium plate, 0.25 cm thick layer of sheet asbestos and 2.0 cm thick layer of rock wool (density 64 kg/m^3). The asbestos is the centre layer. The inner Aluminium surface is at 500°C and the outer rock wool surface is at 50°C. Determine the flow of heat per unit area. The average values of the thermal conductivities for the temperature range in question of the three materials are

$k_{\text{Aluminium}} = 268.1$ w/m °K

$k_{\text{Asbestos}} = 0.1660$ w/m °K

$k_{\text{rockwool}} = 0.055$ w/m °K

Answer $q = 1184.0$ w/m^2

2.6 ANALOGY BETWEEN FLOW OF HEAT AND FLOW OF ELECTRICITY

The analogy between the flow of heat and the flow of electricity through a conductor can be made as follows

$$\text{(heat)} \quad \text{Rate} = \frac{\text{temperature drop}}{\text{resistance}}$$

$$\text{(electricity)} \quad \text{Amperes} = \frac{\text{volts}}{\text{ohms}}$$

rate of heat flow (Btu/ hr) is analogous to ampere

temperature drop—voltage

thermal resistance—electrical resistance (measured as ohms)

The rate of flow of heat through several resistances in series has been shown to be analogous to the current flowing through several electrical resistances in series.

In any electrical circuit the ratio of the potential drop over any one of several resistances to the total potential drops in the circuit is equal to the ratio of individual resistances to the total resistance.

In the same way, the ratio of the temperature differences in a thermal circuit to the total temperature drop is equal to the ratio of the individual thermal resistances to the total thermal resistance

Mathematically

$$\frac{\Delta t}{R} = \frac{\Delta t_A}{R_A} = \frac{\Delta t_B}{R_B} = \frac{\Delta t_C}{R_C}$$

Illustration on Computation of Temperature of the Interfaces

A flat furnace wall is constructed of a 4.5 inch layer of sil-o-cel brick with k of 0.08 backed by a 9 inch layer of common brick ($k = 0.8$). The temperature of the inner face of the wall is 1400°F and that of the outer face is 170°F, what is the temperature of the interface between the refractory brick and the common brick?

Solution

The thermal resistance $R = \dfrac{x}{k \cdot A}$

Considering 1 sq.ft of wall, $A = 1$

$$R_A \text{ (for Sil-o-cel brick)} = \frac{\frac{4.5}{12}}{(0.08)\,(1)} = 4.687$$

$$R_B \text{ (for commonl brick)} = \frac{\frac{9}{12}}{(0.8)\,(1)} = 0.938$$

Total resistance $R = 5.625$

We know $\dfrac{\Delta t}{R} = \dfrac{\Delta t_A}{R_A}$

i.e. $$\frac{\Delta t_A}{4.687} = \frac{1400 - 170}{5.625}$$

$$\Delta t_A = 1025\ ^\circ\text{F}$$

and hence the temperature of the interface is 1400 – 1025 = 375°F.

2.7 HEAT FLOW THROUGH A CYLINDER

L = Length of the cylinder; we are interested in calculating the rate of heat flow for this case.

t_i—temperature of the inside surface

t_0—temperature of the outside surface

$t_i > t_0$ is heat flows from inside to the outside in the radial direction.

This is a case where the transfer (heat)area varies with distance in the direction of transfer of heat; that is the c.s. of path is proportional to radius.

Consider a cylindrical shell, concentric with the main cylinder, shell radius being r.

$r_i < r < r_o$

The thickness of the wall is dr (of the shell)

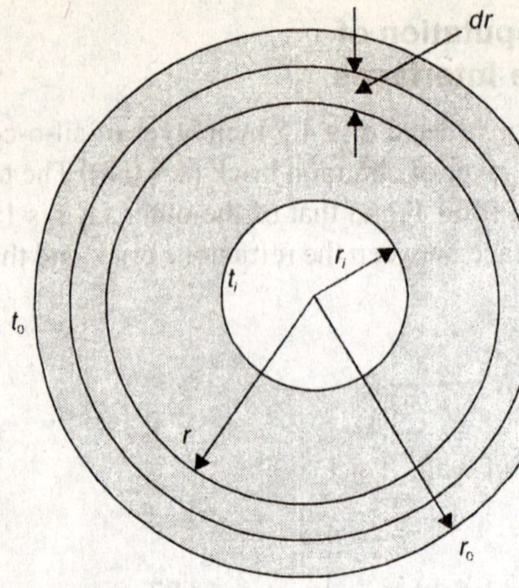

Fig. 2.4 *Heat flow through a cylinder*

Therefore the rate equation for heat transfer becomes

$$q = -k\,A\,\frac{dt}{dr}$$

temperature gradient is –ve

dr is very small compared to *r*.

$$A = 2\pi\, r\, L$$

$$\therefore q = -k\,\frac{dt}{dr}\,2\pi\, r\, L$$

Separating the variables

$$\int \frac{dr}{r} = -\frac{2\pi\, L\, k}{q}\int dt$$

The limit of the integration are r_i and r_0 and t_i and t_0.

$$\int_{r_i}^{r_0} \frac{dr}{r} = -\frac{2\pi\, L k}{q}\int_{t_i}^{t_0} dt$$

$$= \frac{2\pi\, L k}{q}\int_{t_0}^{t_i} dt$$

$$ln\,\frac{r_0}{r_i} = \frac{2\pi\, L k}{q}\,(t_i - t_0)$$

$$q = \frac{k(2\pi L)(t_i - t_0)}{ln\left(\frac{r_0}{r_i}\right)}$$

It may be desirable to express the heat flow in the following form

$$q = \frac{k_m A_m (t_i - t_0)}{(r_0 - r_i)}$$

Which is the general form of the equation for heat flow through a flat wall with the exception of A_m which must be so chosen that the equation is correct.

To determine A_m, we may equate the right sides of the above two equations

$$\frac{k_m (2\pi L)(t_i - t_o)}{ln\frac{r_o}{r_i}} = \frac{k_m A_m (t_i - t_o)}{r_o - r_i}$$

from which it is found that

$$A_m = \frac{2\pi L(r_o - r_i)}{ln\left(\frac{r_o}{r_i}\right)}$$

A_m is the area of a cylinder of length L and radius r_m where $r_m = \frac{r_o - r_i}{ln\frac{r_o}{r_i}}$

$$= \frac{r_o - r_i}{2.303\log\left(\frac{r_o}{r_i}\right)}$$

r_m = Logarithmic mean radius

Log mean value is less convenient than the arithmetic mean value. The arithematic mean value is sufficiently accurate if the tube is thin-walled.

If $\frac{r_o}{r_i} > 1.5$, it is better to use log mean radius.

Problem

A standard 2″ steel pipe carrying superheated steam is insulated with 1.25 inches of a moulded high temperature covering made of diatomaceous earth and asbestos ($k = 0.058$) This covering is insulated with 2.5 inches of laminated asbestos felt ($k = 0.042$). In a test run the temperature of the surroundings was 86°F, the average temperature of the steam side of the pipe was 900°F and the temperature of the outer surface of the lagging was 122°F.

(a) Calculate the heat loss, expressed in Btu/hr per ft length of the pipe.

(b) Calculate the temperature at the interface between the layers of insulation.

Solution

$$q = \frac{\Delta t}{\dfrac{x_A}{k_A\ A_A} + \dfrac{x_B}{k_B\ A_B} + \dfrac{x_C}{k_C\ A_C}}$$

i.e.
$$= \frac{\Delta t}{R_A + R_B + R_C}$$

Data needed

k of the steel is taken as 23.5 Btu/hr ft °F wall thickness of the pipe of 2″ standard diameter (*A*)

inner diameter of the pipe = 2.07″

outer diameter of the pipe = 2.37″

mean diameter = 2.22″

outer diameter of the first covering (*B*) = 4.87 that is (2.37 + 2.50)

log mean diameter = 3.48″

outer diameter of the second covering (*C*) = 9.87″

log mean diameter = 7.07″

$$\therefore\ q = \frac{900 - 122}{\dfrac{\left(\dfrac{0.154}{12}\right)}{23.5\left(\dfrac{2.22\pi}{12}\right)} + \dfrac{\left(\dfrac{1.25}{12}\right)}{(0.058)\left(\dfrac{3.48\pi}{12}\right)} + \dfrac{\left(\dfrac{2.5}{12}\right)}{(0.048)\left(\dfrac{7.07\pi}{12}\right)}}$$

$$= \frac{778}{0.00094 + 1.97 + 2.68} = 167 \text{ Btu / hr}$$

The temperature drop $\propto$ resistance

$$\frac{\Delta t}{R} = \frac{\Delta t_A}{R_A} = \frac{\Delta t_B}{R_B} = \frac{\Delta t_C}{R_C}$$

$$\frac{\Delta t}{R} = \frac{\Delta t_B}{R_B}$$

$R = 4.65$ $\qquad R_B = 1.97$

$\Delta t = 778$ $\qquad \Delta t_B = ?$

$$\Delta t_B = \frac{(1.97)(778)}{4.65}$$

$$= 330°\text{F}$$

$$\Delta t_B = 900 - t_{\text{interface}}$$

$$t_{B-C} = 570\ °\text{F}$$

(temperature at the interface of *B* and *C*)

Integral of –*k dt* when *k* is Dependent on Temperature

$$k = ko\,(1 + at)$$

$$-\int k\,dt = -\,ko \int (1 + at)$$

$$= -\left[k_o\left(1 + \frac{at^2}{2}\right)\right]_{t_1}^{t_2}$$

$$= -k_o\left[t_2 - t_1 + \frac{a\,t_2^2}{2} - \frac{a\,t_1^2}{2}\right]$$

$$= +k_o\left[t_1 - t_2 - \frac{a\,t_2^2}{2} + \frac{a\,t_1^2}{2}\right]$$

$$= k_o\left[(t_1 - t_2) + \frac{a\,(t_1 - t_2)\,(t_1 + t_2)}{2}\right]$$

$$= k_o\left\{[t_1 - t_2] + \left[1 + \frac{a\,(t_1 + t_2)}{2}\right]\right\}$$

Problem (on varying *k*)

A steam pipe 5 cm in outside diameter has an outside temperature of 175°C. The pipe is covered with a coating material 5 cm thick. The *k* of coating varies with temperature such that,

$k = -\,3.834 + (0.0212)T$, where T is in °*K* and *k* is in cal/sec cm °*K*

The outside surface of the coating is at 40°C. Calculate the heat loss per cm of the pipe length.

Solution

Let subscript 1, designate the inside of the insulation.

Let 2, designate the outside of the insulation

$r_1 = 2.5$ cm $\quad r_2 = 2.5 + 5.0 = 7.5$ cm

$T_1 = 175$ °C $\quad = 448$ °K

$T_2 = 40$ °C $\quad = 313$ °K

The rate equation is

$$q \int_{r_1}^{r_2} {dr}/{A} = -\int_{T_1}^{T_2} k\ dt$$

$$A = 2\pi\ r\ L \text{ and } k = -3.834 + (0.0212)\ T$$

$$q \int_{r_1}^{r_2} \frac{dr}{2\pi\ r\ L} = -\int_{T_1}^{T_2} \left[-3.834 + (0.0212)\ T\right] dT$$

$$\frac{q}{2\pi\ L}\ ln\left({r_2}/{r_1}\right) = -\left[-3.834\ (T_2 - T_1) + \frac{0.0212}{.2}\left(T_2^2 - T_1^2\right)\right]$$

$L = 1$ cm

from the above equation, q can be evaluated.

Problem

A standard 2-inch steel pipe containing steam in insulated with a 3-inch lagging of glass wool. The outside surface of the pipe is at 400°F and the outside surface of the covering is at 150°F. Estimate the rate of heat loss from the pipe, expressed as Btu/hr per 100 ft of the pipe.

2.8 HEAT FLOW THROUGH A SPHERE

To determine the mean area of transfer through a wall of a thick-walled hollow sphere of radii r_1 and r_2

We know from the rate equation

$$q \int_{r_1}^{r_2} \frac{dr}{A} = -k \int_{T_1}^{T_2} dT$$

Let us denote $\overline{A}$ as the mean transfer area of transfer in the case of this sphere

then $\quad q \cdot \dfrac{r_2 - r_1}{\overline{A}} = -k\ (T_2 - T_1)$

from which we get

$$\int_{r_1}^{r_2} \frac{dr}{A} = \frac{r_2 - r_1}{\overline{A}}$$

The heat transfer area for a sphere of radius r is $A = 4\pi\ r^2$

$$\frac{r_2 - r_1}{\bar{A}} = \int_{r_1}^{r_2} \frac{dr}{4\pi r^2}$$

$$= \frac{1}{4\pi}\left(\frac{-1}{r}\right)_{r_1}^{r_2}$$

$$= \frac{1}{4\pi}\left[\frac{-1}{r_2} + \frac{1}{r_1}\right]$$

$$\frac{r_2 - r_1}{\bar{A}} = \frac{1}{4\pi}\left(\frac{r_2 - r_1}{r_1\, r_2}\right) \quad \text{or} \quad \bar{A} = 4\pi\, r_1\, r_2$$

Illustration on Determining the Temperature Profile in a Sphere

A heated sphere of radius R is suspended in a large motionless body of fluid. It is desired to study the steady-state heat conduction in the fluid surrounding the sphere. It is assumed that the free convection effects can be neglected.

(i) Set up the differential equation describing the temperature T in the surrounding fluid as a function of r, the distance from the centre of the sphere. The k can be assumed to be constant.

(ii) Integrate the differential equation and use the boundary conditions

at $\quad r = R, \quad T = T_R$

$\quad r = \infty \quad T = T_\infty,$

to determine the constants of integration.

Determine the temperature profile and obtain an expression for the heat flux at the surface.

Solution

This is a case of steady-state heat transfer

Rate of heat transfer is constant. If does not change with distance in the direction of transfer—radial direction in the present case

$$q = \text{rate of heat transfer} \quad = \frac{Q}{\theta}$$

$$\frac{q}{A} = q' = \text{heat flux}$$

Since the rate of heat transfer is constant, change of the rate of heat transfer w.r.t. radial distance is zero.

i.e. $$\frac{d}{dr}\left(\frac{q}{A}\cdot A\right) = 0$$

$$\frac{d}{dr}(q'\ A) = 0$$

$$A = 4\pi\ r^2$$

$$\frac{d}{dr}\left(q'\ 4\pi\ r^2\right) = 0$$

$$\frac{d}{dr}(q'\ r) = 0$$

Integrating $q'\ r^2 = c_1$

$$q' = \frac{c_1}{r^2}$$

Substituting the Fourier's law of heat conduction

$$q' = -k\ \frac{dt}{dr} = \frac{c_1}{r^2}$$

$$\text{or } -k\ dT = \frac{c_1}{r^2}\ dr$$

Integrating we get $-kT = -\dfrac{c_1}{r} + c_2$

B.C's—Boundary conditions:

at $r = R,\ T = T_R$

at $r = \infty \quad T = T_\infty$ (far away from the sphere)

$$-k\ T_R = -\frac{c_1}{R} + C_2 \qquad \text{...(1)}$$

$$-k\ T_\infty = -\frac{c_1}{\infty} + C_2 \qquad \text{...(2)}$$

from (2) $c_2 = -k\ T_\infty$

Substituting in equation (1)

$$-k\ T_R = -\ \frac{c_1}{R} - k\ T_\infty$$

or $$c_1 = k\ R\left(T_R - T_\infty\right)$$

The temperature profile is

$$-k\,T = -\frac{k\,R\,(T_R - T_\infty)}{r} - k\,T_\infty$$

or $$\left(T_\infty - T\right) = -\frac{R\,(T_R - T_\infty)}{r}$$

The above expression gives the temperature T as a function of radial direction from the centre of the sphere.

The Heat Flux at the Surface

$$q' = \frac{c_1}{r^2}$$

$$c_1 = +\,k\,R\,(T_R - T_\infty)$$

$$q' = \frac{c_1}{r^2} = \frac{k\,R\,(T_R - T_\infty)}{r^2}$$

q' at the surface, i.e. at $r = R$

$$q' \text{ surface} = \frac{k\,(T_R - T_\infty)}{R}$$

Illustrations on Conduction of Heat in the Steady State Condition

1. The interior portion of an oven is maintained at the temperature of 1500°F by means of suitable control apparatus. The walls of the oven are 0.9 inch thick and constructed from a material having a thermal conductivity of 0.182 Btu/hr ft °F. Calculate the heat loss for each square foot of wall surface per hour.

Assume that the inside and outside wall temperatures are 1500°F and 400°F respectively.

Solution

The equation, $q = k\,A\,\dfrac{t_1 - t_2}{x_2 - x_1}$, holds good in this situation.

Substituting the necessary values and dimensions, particularly $\Delta x = 9/12$ ft, inch.the equation gives

$$q = 0.182(1)\,\frac{1500 - 400}{9/12} = 264 \text{ Btu/hr}$$

or $$q/A = 264 \text{ But/hr ft}^2$$

2. The wall of a furnace is made up of 9 inch of fire brick ($k = 0.72$ Btu/hr ft °F) 5 inch. of insulating brick ($k = 0.082$) and 7.5 inch. of red brick ($k = 0.52$).The inner and outer surface temperatures t_1 and t_4 of the wall are 1500°F and 150°F respectively. Neglecting the resistances of the mortar joints, compute the temperatures t_2 and t_3 at the interfaces.

Solution

The rate of heat flow through 1 ft² of the wall is obtained from the equation

$$q = \frac{A\,(t_1 - t_4)}{\dfrac{(\Delta x)_{12}}{k_{12}} + \dfrac{(\Delta x)_{23}}{k_{23}} + \dfrac{(\Delta x)_{34}}{k_{34}}}$$

This equation may be generalised for any number of layers. Calling the total temperature difference $\Sigma\,(\Delta t)$ and using the summation sign Σ also in the denominator gives

$$q = \frac{A\,\Sigma\,\Delta t}{\Sigma\,\dfrac{\Delta x}{k}}$$

In this case, there are three layers

$$q = \frac{1\,(1500 - 150)}{\dfrac{\frac{9}{12}}{0.72} + \dfrac{\frac{5}{12}}{0.082} + \dfrac{\frac{7.5}{12}}{0.52}}$$

$= 180$ Btu/hr

Since this value of q is equal to the heat flow through any section. The following equation will be used to calculate the temperature t_2 by applying it to the fire brick section.

$$q = \frac{k\,A\,(t_1 - t_2)}{x_2 - x_1}$$

$$q = \frac{(0.72)\,(1)\,(1500 - t_2)}{9/12}$$

Which yields $t_2 = 1312$ °F. In a similar way, one obtains $t_3 = 375$ °F

3. Illustration on conduction through a homogenous cylinder wall

Calculate the heat loss from 30 ft of 2 in nominal pipe covered with 1 inch of an insulating material having an average thermal conductivity of 0.0375 Btu/hr ft °F .Assume that the inner and outer surface temperature of the insulation are 380°F and 80 °F respectively.

Solution

The outside diameter of a 2 inch. nominal pipe is 2.375 inch. Adding 2 inch. to the pipe diameter, the insulation diameter is equal to 4.375 inch.

Applying the equation

$$q' = \frac{q}{L} = \frac{2\pi k\,(t_1 - t_2)}{ln\,\dfrac{r_2}{r_1}}$$

the heat loss per linear foot is equal to

$$q' = \frac{2\pi\,(0.0375)\,(380 - 80)}{ln\,\dfrac{4.375/12}{2.375/12}}$$

$= 116$ Btu/hr ft

The total heat loss for 30 ft of pipe is equal (116) (30) = 3480 Btu/hr.

4. Illustration on conduction through a composite cylinder wall—radial flow

Calculate the heat loss per linear foot from a 10 inch. nominal pipe (outside diameter = 10.75 inch.) covered with a composite pipe insulation consisting of 1 ½ inch.of insulation I placed next to the pipe and 2 inch. of insulation II placed upon insulation I. Assume that the inner and outer surface temperatures of the composite insulation are 700°F and 110°F respectively and that the thermal conductivity of material I is 0.055 Btu/hr ft °F and material II is 0.039 Btu/hr ft °F.

Solution

Using the equation

$$\frac{q}{L} = q' = \frac{2\pi\,(t_1 - t_4)}{\left(\dfrac{ln\,(r_2/r_1)}{k_{12}} + \dfrac{ln\,(r_3/r_2)}{k_{23}}\right)}$$

t_1 = inner surface temperature t_4 = outer surface temperature; the heat loss is found to be

$$q' = \frac{2\pi\,(700 - 110)}{\dfrac{ln\,\dfrac{13.750/12}{10.750/12}}{0.055} + \dfrac{ln\,\dfrac{17.750/12}{13.750/12}}{0.039}}$$

$= 323$ Btu/hr ft

The heat loss is 323 Btu/hr ft

Illustration on Conduction under Varying Heat Transfer area Condition

Consider an insulated rod with $x_1 = 0$. $T_1 = 50$ °C $x_2 = L = 8$ cm., $T_2 = 0$ °C. The area variation is linear $A_x = A_o\,(1 + x/L)$

$A_o = 0.05$ m^2. The thermal conductivity variation is exponential i.e. $k = k_o\, e^{T/T_o}$, $k_o = 1\text{-}5$ w/m °C (T in °C) and $T_0 = 100$°C. Calculate the heat transfer rate.

Solution

Under varying thermal conductivity and heat transfer area conditions, we can write the following expression.

$$\int_{T_1}^{T_2} k\,(T)\,dT = -q_x \int_{x_1}^{x_2} \frac{dx}{A} \qquad \text{...(1)}$$

We can define the average conductivity to rewrite above expression in another way taking that,

$$k_m = \frac{1}{T_2 - T_1} \int_{T_1}^{T_2} k\,(T)\,dT \qquad \text{...(2)}$$

$$q = \frac{k_m (T_1 - T_2)}{\int_{x_1}^{x_2} dx/A_x} \qquad \text{...(3)}$$

Compute k_m from equation (2)

$$k_m = \frac{1}{(0 - 50)} \int_{50}^{0} k_o\, e^{T/100}\, dT$$

$$= 2\,k_o\,(e^{0\text{-}5} - 1) = 1.297\,k_o$$

We can then evaluate the inverse area integral

$$\int_o^L \frac{dx}{A_o\,(1 + x/L)} = \frac{L}{A_o}\, ln\,(1 + x/L)\Big|_o^L$$

$$= \frac{0.08}{0.05}\, ln\,(2) = 1.11\, m^{-1}$$

Now substituting into (3), we obtain

$$q_x = \frac{(1.297)\,(1.5)\,(50 - 0)}{1.11} = 88\,\text{w}$$

Illustration on Steady State Heat Transfer—slab Geometry

A furnace wall has an inner layer of silica brick (k_1 = 1.07 w/m °K) and an outer layer of masonary brick (k_2 = 0.066 w/m °K) (see the figure given below).Furnace gas conditions are T_i = 320 °C h_i = 45.5 w/m² °K and outside is air is T_o = 25 °C h_o = 25.2 w/m² °K. It is desired that the interface temperature T_2 is to be 150°C. Calculate Δx_1 and Δx_2, the thicknesses of the walls.

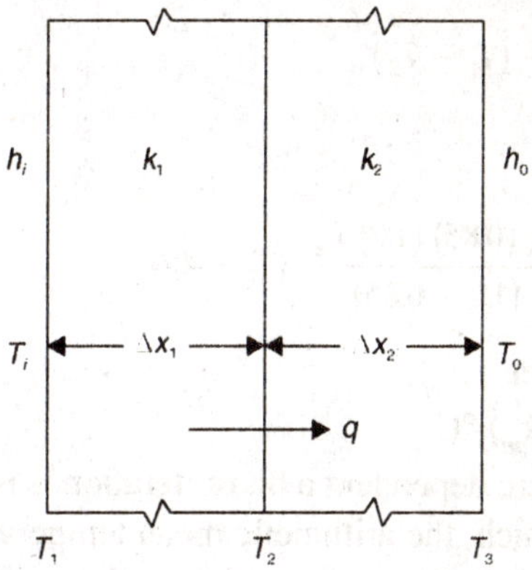

Fig. 2.5 *Steady state heat transfer through a furnace wall*

Solution

With T_2 known, we can apply the equation $q = \Delta T / \sum R$ in two stages. First from the furnace gas to the interface

$$q = \frac{T_i - T_2}{1/h_i + \Delta x_1/k_1} = \frac{320 - 150}{1/45.5 + \Delta x_1/1.07}$$

$$= 800 \text{ walts/m}^2$$

or $$\Delta x_1 = \frac{170 - 800/45}{800/1.07} = 0.204 \text{ m} = 20.4 \text{ cm}$$

Second, from the interface to the outside air

$$q = \frac{T_2 - T_0}{\Delta x_2/k_2 + 1/h_2} = \frac{150 - 25}{\Delta x_2/0.66 + 1/25.2} = 800 \text{ w/m}^2$$

$$\Delta x_2 = \frac{125 - 800/25}{800/0.66} = 0.077 \text{ m} = 7.7 \text{ cm}$$

Illustration on Conduction (steady state)—Radial Flow

An iron sphere has an outer diameter of 2m a wall thickness of 15 cm and an outer temperature of 40°C. The heat transfer is q = 250 kw from inside to outside (that is in the out ward direction). Accounting for variable $k(T)$ estimate the temperature of the inside surface. Take that a plot of k versus T for this material is available and can be used.

Solution

From the given data, $R_1 = 0.85$ m and $R_2 = 1.0$ m. we estimate the inner surface temperature as follows.

For a spherical geometry, we know

$$q = \frac{4\pi \; R_1 \; R_2 \; k_m}{R_2 - R_1} (T_1 - T_2)$$

In this situation

$$q = 250000 = \frac{4\pi \; (0.85) \; (1.0) \; k_m}{(1.0 - 0.85)} (T_1 - 40)$$

Solving for T_1, we get

$$T_1 = (40 + 3511/k_m) \; °C \qquad ...(1)$$

Since k_m is temperature dependent a bit of iteration is needed.Take an initial guess T_1 = 80°C from which, the arithmetic mean temperature T_m = (80 + 40)/ 2 = 60 °C. At 60°C the value of k can be read from the available plot. k is read as 74.4 w/m °K

Equation (i) given above, allows an improved estimate for T_1.

T_1 (new) = 40 +3511/74.4 = 87.2 °C. Repeating once T_m new = (87.2 + 40)/ 2 = 63.6 °C and k_m (new) from the plot, is 74.1 w/m °K which is hardly changed from the first iteration. This should be a sufficient convergence; T_1 (final) = (40 +3511/74.1) = 87.4 °C.

Examples on Conduction—Slab Geometry

For a steady—state heat conduction through a plane wall with dimensions and surface temperatures as shown in the following sketch, express

(i) The heat flow rate for a constant thermal conductivity k.

(ii) The heat transfer rate when the thermal conductivity of the wall material varies linearly with temperature according to the expression $k = k_o (1 + \beta T)$

Compare the result in (ii) with that in (i) if a single value of k evaluated at an arithmetic mean temperature, were used

Solution

For constant k, the x-directional form of the Fourier rate equation is

$$q_x = - \; k_A \; dt \, / \, dx$$

The above expression may be separated and solved easily as follows

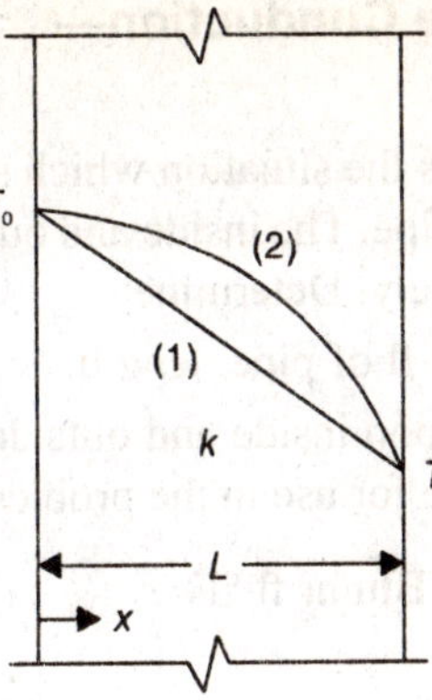

Fig. 2.6 *Steady state heat conduction through a plane wall*

$$q_x \int_o^L dx = -k A \int_{T_o}^{T_L} dT$$

$$q_x L = -k A (T_L - T_o)$$

$$= k A (T_o - T_L)$$

$$q_x = k A / L (T_o - T_L)$$

For the variable thermal conductivity case, the given k expression substituted into the Fourier rate equation is

$$q_x = -k_o (1 + \beta T) A \, dT / dx$$

Separating the variables and solving we get

$$q_x \int_o^L dx = -k_o A \int_{T_o}^{T_L} (1 + \beta T) \, dT$$

$$q_x L = -k_o A \left[T_L - T_o + \beta / 2 \left(T_L^2 - T_o^2 \right) \right]$$

$$q_x = \frac{k_o A}{L} \left[(T_o - T_L) + \beta / 2 \left(T_o^2 - T_L^2 \right) \right]$$

$$q_x = A / L \left\{ k_o \left[1 + \beta / 2 \, (T_o + T_L) \right] \right\} (T_o - T_L)$$

In the above equation, we observe that the term in braces is the thermal conductivity calculated at the arithmetic mean temperature. Thus an equivalent from of the above equation is

$$q_x = \frac{k_{av.} A}{L} (T_o - T_L)$$

Example on Steady State Conduction—Cylindrical Geometry

The following figure illustrates the situation which steam is transported through 1½ m. schedule 80 mild steel pipe. The inside and outside pipe wall temperatures are 205°F and 145°F respectively. Determine

(i) the heat loss from 10 ft of pipe, length.

(ii) the heat flux based upon inside and outside surface areas. The following data are available for use in the problem.

k of mild steel = 24.8 Btu/hr ft °F

For 1½ in.schedule pipe

Outer Diameter = 1.900 inch.

Inner Diameter = 1.500 inch.

Wall thickness = 0.200 inch.

Solution

The Fourier heat transfer rate equation for a cylindrical geometry is $q = -k\,A\,\dfrac{dT}{dr}$

q_r= heat transfer rate in the radial direction.

Noting that $A = 2\pi\, r\, L$, the above equation turns out to be

$$q_r = -k\,(2\pi\, r\, L)\; dT\,/\,dr$$

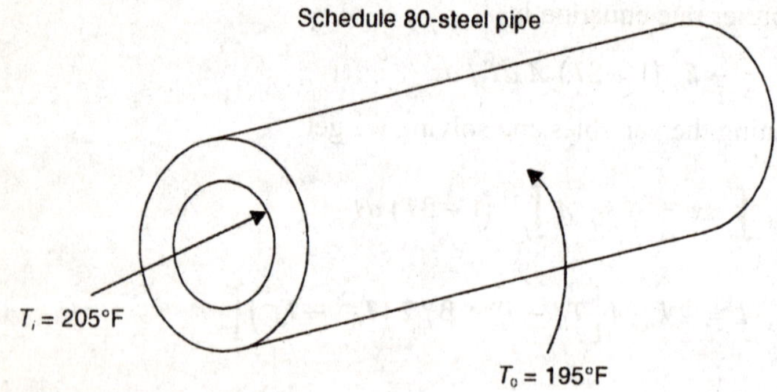

Fig. 2.7 *Heat loss from the mild steel pipe*

For the steady—state (situation), q_r is constant

Separating the variables and solving the above equation we get

$$q_r \int_{r_1}^{r_2} dr\,/\,r = -2\pi\; k\; L \int_{T_i}^{T_o} dT$$

$$q_r\; ln\; r_o\,/\,r_i = -2\pi\; k\; L\; (T_o - T_i)$$

$$= 2\pi\ L\ k\ (T_i - T_o)$$

$$q_r = \frac{2\pi\ k\ L}{ln\ r_o / r_i}(T_i - T_o)$$

Substituting the given values, and solving we get $q_r = \dfrac{2\pi\ (24.8)\ (10)\ (10)}{ln\ 1.9 / 1.5}$

= 65,000 Btu/hr

The inside and outside surface areas of the pipe are

$$A_i = \pi \left(\frac{1.500}{12}\right) 10 = 3.93\ \text{ft}^2$$

$$A_o = \pi \left(\frac{1.90}{12}\right) (10) = 4.98\ \text{ft}^2$$

The heat fluxes evaluated at both surface are thus

$$\frac{q_r}{A_i} = \frac{65{,}000}{3.93} = 16{,}700\ \text{Btu / hr ft}^2$$

$$\frac{q_r}{A_o} = \frac{65{,}000}{4.98} = 13{,}150\ \text{Btu / hr ft}^2$$

As can be seen from the above result, the two heat flux values differ by over 25% for the same heat flow rate. The importance of specifying the area upon which a given heat flux is based should be realized

Illustration of Heat Conduction—Radial Direction

1. A 12 inch nominal schedule 80 stainless steel (type 304) pipe carries water condensate from a condenser to a pump. The inside wall temperature is 40°C and the outside-wall temperature is 38°C. Determine the heat transfer through the pipe wall per unit length of the pipe. The *k* of steel is 14.4 w/m °K.Take the usual assumptions.

Solution

The specified wall temperatures will probably exist for only a short distance in the pipe. As fluid flows downsteam its temperature will decrease and the outside wall temperature will increase.

The assumptions made are (1) one- dimensional, (2) steady-state radial heat flow occurs.

Material properties are constant

we know $$q_r = \frac{2\pi\ k\ L}{ln\ R_2 / R_1}(T_1 - T_2)$$

For 12 nominal schedule 80 pipe.

$D_2 = 32.4$ cm

$D_1 = 29.53$ cm

Substituting into equation for heat flow, we get

$$\frac{q_r}{L} = \frac{2\pi\,(14.4)}{ln\,(32.4\,/\,29.53)}\,(40 - 38)$$

Solving $\frac{q_r}{L} = 1957.4$ w / m

$= 1.96$ k w/m

2. Refrigeration tubing is used in a domestic air conditioning system. In one such installation, 1 standard type 7 copper tubing is soldered in place from the compressor/Condenser unit outside the building to the expansion value/evaporation unit in the attic. The refrigerant keeps the inside wall of the tubing at 40°F. In order that little heat is gained through the tube wall, it is covered with a sponge type of insulation that is ½ inch thick and has a thermal conductivity of 0.02 Btu/ hr °R. The outside surface temperature of the insulation is 70 °F. Determine the heat gain per unit length of for this situation tube.

Solution

Copper refrigeration tubing has specification on size that are similar to those for copper water tubes. The difference is that refrigeration tubing is ductile enough that it can bent rigid by hand while the water tubing is quite rigid.

The usual assumptions are made for convenience.

k copper = 231 Btu/hr ft °R

$D_2 = 1.125$ inch $= 0.0938$ ft

$D_1 = 0.0879$ ft

$R_2 = 0.0469$ ft

$R_1 = 0.044$ ft

The wall thickness of the insulation is 0.5 inch

Thus $R_3 = 0.0469 + 0.5/12 = 0.0886$ ft

$$R_{k_1} = \frac{ln\,(R_2\,/\,R_1)}{2\pi\,k_1\,L} = ln\,\frac{(0.0469)\,/\,0.044}{2\pi\,(231)\,L}$$

or $\quad L\,R_{k_1} = 4.4 \times 10^{-5}$

$$\text{Similarly } L\,R_{k_2} = ln\,\frac{(0.0886\,/\,0.0469)}{2\cdot\pi\,(0.02)} = 5.06$$

As expected, the insulation offers the greater resistance to heat flow. The heat transferred by conduction then is

$$\frac{q_r}{L} = \frac{T_1 - T_3}{L \sum R_i} = \frac{40 - 70}{5.06}$$

$$q_r / L = -5.93 \text{ Btu / hr ft}$$

The negative sign here in indicative of the fact that the direction of heat flow is opposite to that assumed in the equation. Heat flow is thus from outside to inside.

Illustration on Heat Conduction through Composite Wall

A cold—storage warehouse is to be built of a number of individual rooms that are maintained at various temperatures. The walls of one room are constructed of 4 inch. brick masonry on the outside and ½ inch. thick plywood on the inside. Sandwiched between the brick and plywood is glass fibre insulation that is 3.5 inches thick. The inside-wall temperature is to be maintained at –10°F and outside—wall temperature is 70°F. As a first step in sizing the required cooling unit, determine the heat flow through the composite wall on a per square-foot basis. The k values are k_{brick} = 0.38 Btu/hr ft R

$k_{glass\ fibre}$ = 0.02 But/hr ft °R

$k_{plywood}$= 0.063 But/hr ft °R

Solution

The situation is represented schematically in the following sketch

The temperature with in the room will probably not be uniform, So the –10°F temperature will not be the inside-wall temperature at all locations. The heat flow near the corners is two-dimensional problem and will have an effect.

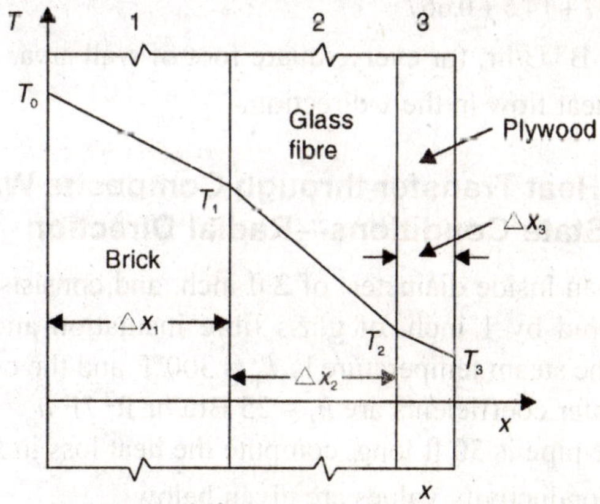

Fig. 2.8 *Steady state heat conduction through composite wall*

The assumption to be made are as follows

1. The inside and outside-wall temperatures are uniform and constant.

2. One-dimensional steady state heat conduction exists through the wall.
3. Material properties are constant

The heat flow is given by the equation

$$q_x = \frac{T_o - T_3}{\dfrac{\Delta x_1}{k_1\ A} + \dfrac{\Delta x_2}{k_2\ A} + \dfrac{\Delta x_3}{k_3\ A}}$$

The wall temperatures are uniform and constant. They are given as T_0 = 70°F T_3 = –10°F. The material thicknesses are $\Delta x_1 = 4/12 = 0.333$ ft,

$\Delta x_2 = \dfrac{3.5}{12} = 0.292$ ft, $\Delta x_3 = 0.5/12 = 0.042$ ft. Each resistance is

$$R_{k_1} = \frac{\Delta x_1}{k_1\ A} = \frac{0.333}{(0.38)\ (1)} = 0.877 \text{ hr °R / BTU}$$

$$R_{k_2} = \frac{0.292}{(0.02)\ (1)} = 14.6 \text{ hr °R / BTU}$$

$$R_{k_3} = \frac{0.042}{(0.063)\ (1)} = 0.667 \text{ hr °R / BTU}$$

Thus the glass fibre represents the greatest resistance to heat flow and so, compared to the other materials, we expect the greatest temperature difference to exist across the glass fibre. The heat flow becomes

$$q_x = \frac{70 - (-10)}{0.877 + 14.6 + 0.667}$$

q_x = 4.96 BTU/hr, for every square foot of wall area.

q_x is the heat flow in the *x*-direction.

Illustration on Heat Transfer through Composite Wall under Steady State Conditions—Radial Direction

A steam pipe has an inside diameter of 2.0 inch. and consists of 0.22 inch. of copper pipe covered by 1 inch. of glass fibre insulation and 0.008 inch. of aluminium foil. The steam temperature is T_1 = 300°F and the outside air is at T_0 = 80°F. Heat transfer coefficients are h_i = 25 Btu/hr ft² °F h_o = 12 Btu/hr ft² °F respectively. If the pipe is 30 ft long, compute the heat loss in Btu/hr

The thermal conductivity values are given below

$$\left.\begin{aligned} k_1\ (copper) &= 226 \text{ Btu / hr ft °F} \\ k_2\ (glass\ fiber) &= 0.022 \text{ Btu / hr ft °F} \\ k_3\ (aluminium) &= 136 \text{ Btu / hr ft °F} \end{aligned}\right\} \text{The average values in this temperature range}$$

Solution

For the given wall thickness, $r_1 = 1.00$ inch. $r_2 = 1.22$ inch., $r_3 = 2.22$ inch. $r_4 = 2.228$ inch.

The various thermal resistances are given by the following expressions

$$\text{Convective resistances}\begin{cases} R_{ci} = \dfrac{1}{h_i\ A_1} & A_1 = 2\pi\ r_1\ L \\ R_{co} = \dfrac{1}{h_o\ A_4} \cdot & A_4 = 2\pi\ r_4\ L \end{cases}$$

$$\text{Conductive resistances}\begin{cases} R_{k_1} = \dfrac{ln\,(r_2/r_1)}{2\pi\ L\ k_1} \\ R_{k_2} = \dfrac{ln\,(r_3/r_2)}{2\pi\ L\ k_2} \\ R_{k_3} = \dfrac{ln\,(r_4/r_3)}{2\pi\ L\ k_3} \end{cases}$$

$$q = \frac{\Delta T}{R_{ci} + R_{k_1} + R_{k_2} + R_{k_3} + R_{co}} = \frac{T_i - T_o}{\Sigma R}$$

Therefore

$$R_{ci} = \frac{1}{(25)\ 2\pi\ (1/12)\ 30} = 0.00255 \text{ hr } ^\circ\text{F/Btu}$$

$$R_{k_1} = \frac{ln\,(1.22/1.0)}{2\pi\ (30)\ (226)} = 0.000005 \text{ hr } ^\circ\text{F/Btu}$$

$$R_{k_2} = \frac{ln\,(2.22/1.22)}{2\pi\ (30)\ (0.022)} = 0.1444 \text{ hr } ^\circ\text{F/Btu}$$

$$R_{k_3} = \frac{ln\left(2.228/2.22\right)}{2\pi\ (30)\ (136)} = 0.0000001 \text{ hr } ^\circ\text{F/Btu}$$

$$R_{co} = \frac{1}{(12)\ 2\pi\left(\dfrac{2.228}{12}\right)(30)} = 0.00238 \text{ hr } ^\circ\text{F/Btu}$$

Thus the total resistance is $\Sigma R = 0.1493$ hr °F / Btu of which 97% is owing to the insulation

The heat transfer rate $\dfrac{T_i - T_o}{\Sigma R}$

$$= \frac{300 - 80}{0.1493} = 1470 \text{ Btu / hr}$$

Illustration on Compound Resistances in Series—Combined Convection Conduction—slab Geometry

A furnace wall is constructed with a 3″ of fire clay $k = 0.65$ Btu/hr °F next to the fire box and 1/4 inch of mild steel (k = 24 Btu/hr °F) on the outside. The inside surface of the brick is at 1200°F and the steel is surrounded by air at 80°F with an outside heat transfer coefficient of 12 Btu/hr ft^2 °F Determine

(a) the heat flux through each square foot of furnace wall

(b) the outside surface temperature of the steel

(c) the percent increase in heat flux if, in addition to the conditions specified, two 3/4 inch-diameter steel bolts extend through the composite wall per ft^2 of wall area.

Solution

For the composite wall, the analogous electrical circuit is

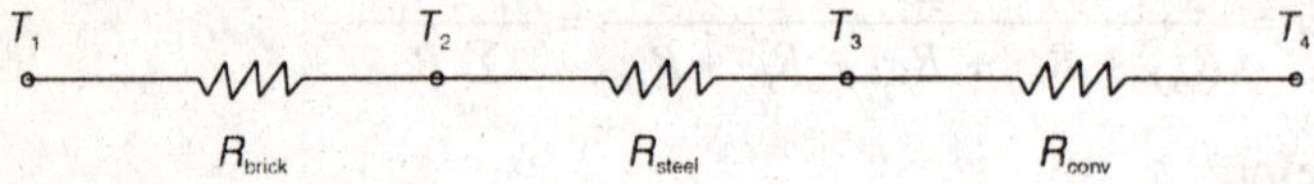

Fig. 2.9 *The analogous electrical circuit for the composite wall*

and the heat flux may be evaluated using the following equation

$$q = \frac{\Delta T}{R_{thermal}}$$

The thermal resistances are, in turn

$$R_{brick} = \left.\frac{L}{k\,A}\right|_{brick} = \frac{3/12}{(0.65)\;A} = \frac{0.385}{A} \text{ hr °F / Btu}$$

$$R_{steel} = \left.\frac{L}{k\,A}\right|_{steel} = \frac{0.25/12}{(24)\;(A)} = \frac{0.00087}{A} \text{ hr °F / Btu}$$

$$R_{conv} = \frac{1}{h\,A} = \frac{1}{(12)\;A} = \frac{0.0883}{A} \text{ hr °F / Btu}$$

The sum of the thermal resistances is

$$R = \frac{1}{A}\,(0.385 + 0.00087 + 0.0833)$$

$$= 0.469/A \text{ hr °F/Btu}$$

and the heat flux is

$$\frac{q}{A} = \frac{1200 - 80}{0.469} = \frac{1120}{0.469}$$
$$= 2390 \text{ But/hr ft}^2$$

with the heat flux now known, the steel surface temperature may be determined, employing either of the following equations.

$$1200 - T_3 = q.\ (R_{brick} + R_{steel})$$
$$T_3 - 80 = q.\ R_{conv}$$

The resulting value for T_3 is

$$T_3 = 1200 - (2390)\ (0.386)$$
$$= 80 + (2390)\ (0.0833) = 278°\text{F}$$

For the situation in which two additional bolts (steel) extend through the composite wall, there are now two paths whereby heat may flow from the inside of the furnace wall to the outside air. The equivalent electrical circuit in this case is presented below.

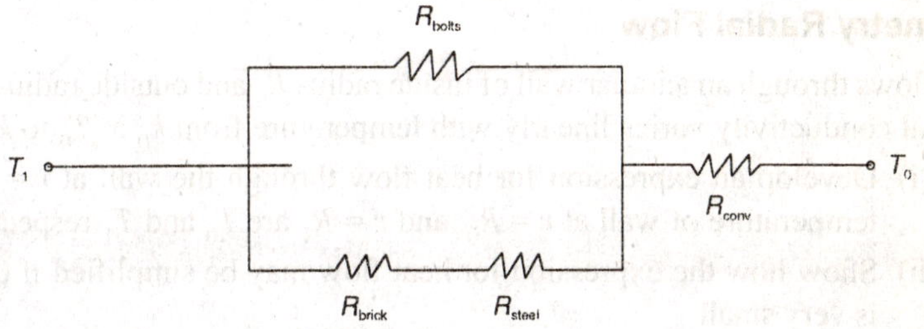

Fig. 2.10 *The electrical circuit equivalent to the situation of two bolts inclusion*

R_{brick}, R_{steel} and R_{conv} are all known. The thermal resistance for the steel bolts is calculated as

$$R_{bolts} = \frac{L}{kA} = \frac{\dfrac{\dfrac{3.25}{12}}{24}}{\left[2 \times \dfrac{\pi}{4}\left(\dfrac{0.75}{12}\right)^2\right] \cdot A}$$

$$= \frac{1.84}{A} \text{ hr °F/Btu}$$

The equivalent resistance of the parallel portion of the circuit is

$$R_{equiv} = \frac{1}{\dfrac{1}{R_{bolts}} + \dfrac{1}{R_{brick} + R_{steel}}}$$

$$= \frac{1}{(0.543 + 2.59)\,A} = \frac{0.319}{A}\ \text{hr °F/Btu}$$

The total thermal resistance for the wall and bolts is

$$R_{total} = R_{equiv} + R_{conv}$$

$$= \frac{0.319}{A} + \frac{0.0833}{A} = \frac{0.402}{A}$$

and the resulting heat flux is

$$\frac{q}{A} = \frac{\Delta T}{R} = \frac{1200 - 80}{0.402} = \frac{1120}{0.402}$$

$$= 2790 \text{ But/hr ft}^2$$

There is an increase of 400 Btu/hr ft² in the heat flux over the wall without bolts. The percentage increase is 16.7.

Examples on Conduction—Cylindrical Geometry Radial Flow

Heat flows through an annular wall of inside radius R_0 and outside radius R_1. The thermal conductivity varies linearly with temperature from k_0 at T_0 to k_1 at T_1

(i) Develop an expression for heat flow through the wall at $r = R_0$. The temperature of wall at $r = R_0$. and $r = R_1$ are T_0 and T_1 respectively

(ii) Show how the expression for heat flow may be simplified if $(R_1 - R_0)$ is very small

Solution

(i) we know $q = -\,k\,(2\pi\, r\, L)\,\dfrac{dt}{dr}$

or $$\frac{q}{r}\cdot dr = -(2\pi\, L)\, k\, dt$$

Since k varies linearly with temperature from k_0 at T_0, to k_1 and T_1, we can

write $k = k_0 + \left(\dfrac{k_1 - k_0}{T_1 - T_0}\right)(t - T_0)$

Therefore

$$q\,\frac{dr}{r} = -2\pi\, L\left[k_0 + \left(\frac{k_1 - k_0}{T_1 - T_0}\right)(t - T_0)\right] dt$$

Integrating between R_1 and R_0.

$$q\, ln\,\frac{R_1}{R_0} = -2\pi L\left[k_0\,(T_1 - T_0) + \left(\frac{k_1 - k_0}{T_1 - T_0}\right)\left\{\frac{T_1^2 - T_0^2}{2} - T_0\,(T_1 - T_0)\right\}\right]$$

$$= (2\pi L)(T_0 - T_1)\left[k_0 + \frac{k_1 - k_0}{2}\right]$$

$$= (2\pi L)\left(\frac{k_0 + k_1}{2}\right)(T_0 - T_1)$$

or $$q = (2\pi L)\left(\frac{k_0 + k_1}{2}\right)(T_0 - T_1) \times \left[ln\left(\frac{R_1}{R_0}\right)\right] \text{ Ans.}$$

(ii) If $(R_1 - R_0)$ is very small, we can write as follows

$$ln\frac{R_1}{R_0} = ln\left(1 + \frac{R_1 - R_0}{R_0}\right)$$

$$= \frac{R_1 - R_0}{R_0}$$

Therefore $$q_r = (2\pi L)(T_0 - T_1)\left(\frac{k_0 + k_1}{2}\right) \times \left(\frac{R_0}{R_1 - R_0}\right)$$

2. Obtain the dimensionless temperature distribution in a plain wall of uniform thickness '2a' meters generating heat at the rate of Q' k cal/hr m^3. The wall is exposed at both surfaces to a gas at uniform temperature t_g. The surface conductance is taken as h k cal/hr m^2 °C. (the surface conductance is the heat transfer coefficient)

Solution

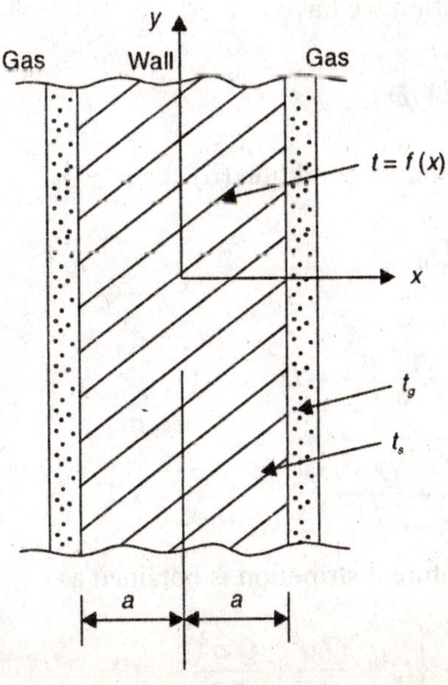

Fig. 2.11 *Schematic representation of the heat flow through a plain wall*

This situation can be represented schematically as above. Let the wall have infinite face area and the heat flow unidirectional along the thickness

The heat flow equation is

$\dfrac{\partial^2 t}{\partial x^2} + \dfrac{Q'}{k} = 0$ (can be deduced by the transport phenomena approach to be presented in the text later)

The boundary conditions are

(i) at $x = 0$, $dt/dx = 0$ because of the symmetrical nature of the geometry

(ii) The heat gained (k cal/m^2) by the gas is equal to that conducted across both the face surfaces, that is

$$2\,h\,(t_s - t_y) = Q' \quad (2a)$$

where t_s = surface temperature

or $$t_s = t_g + \frac{Q'a}{h} \text{ at } x = \pm a$$

Integrating Eq. (1), we obtain

$$dt/dx = -\frac{Q'}{k}\cdot x + c$$

Applying boundary condition (i) at $x = 0$, $dt/dx = 0$

we get $C = 0$: so $\dfrac{dt}{dx} = -\dfrac{Q'}{k}\cdot x$

On further integration we have

$$t = -\frac{Q'}{k}\cdot\frac{x^2}{2} + D$$

Now applying boundary condition (ii) at $x = \pm a$

$$t_s = t_g + \frac{Q'a}{h}$$

$$t_g + \frac{Q'\cdot a}{h} = -\frac{Q'}{k}\frac{a^2}{2} + D$$

or $$D = t_g + \frac{Q'a}{h} + \frac{Q'a^2}{2k}$$

Hence the temperature distribution is obtained as

$$t = -\frac{Q'}{k\cdot}\frac{x^2}{2} + \left\{t_g + \frac{Q'a}{h} + \frac{Q'a^2}{2k}\right\}$$

or $$t - t_g = -\left(\frac{Q' a^2}{2k} \cdot \frac{x^2}{a^2}\right) + \left(\frac{Q' a}{h} + \frac{Q' a^2}{2k}\right)$$

$$\frac{2k}{a^2 Q'}\left(t - t_y\right) = 1 - \left\{\left(\frac{x}{a}\right)^2 - \frac{(2k)}{ah}\right\}$$

$$-a \le x \le a$$

3. The two ends of a cylindrical metal wire of the length $2L$ which is perfectly insulated against radial heat flow are held at the same temperature t_0. The left end and the middle of the wire are connected with a source of electricity so that a current flows through the left half of the wire and produces heat at the uniform rate of Q . Assuming steady state conditions and k of wire to be independent of temperature.

(a) Express the temperature as a function of the distance from the left end. Two different equations will be found for the left half and right half of the wire.

(b) Will the temperature have a maximum value somewhere between the two ends? If so, at what distance from the left end?

(c) Express the maximum temperature, if any in the mathematical form.

Solution

The situation is represented pictorially represented as follows

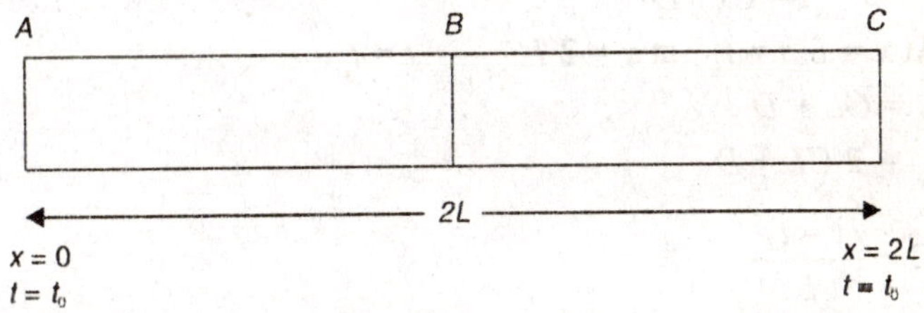

Fig. 2.12 *Schematic representation of heat flow through a cylindrical metal wire*

(a) For portion *AB*;

$$\frac{d^2t}{dx^2} = -\frac{Q'}{k}$$ (This equation can be arrived at by the transport phenomena approach)

or $$t = -\frac{Q'}{k} \cdot \frac{x^2}{2} + Ax + B$$

where *A* and *B* are constant of integration

Now at $x = 0, t = t_0$

$x = L, t = t_L$

so $B = t_0$

and $$t_L = -\frac{Q'}{k} \cdot \frac{L^2}{2} + AL + t_0$$

or $$AL = t_L - t_o + \frac{Q'}{k} \cdot \frac{L^2}{2}$$

$$A = \frac{t_L - t_o}{L} + \frac{Q' L}{2k}$$

or $$(t)_{AB} = -\frac{Q'}{k} \cdot \frac{x^2}{2} + \frac{(t_L - t_o)\,x}{L} + \frac{Q' Lx}{2k} + t_o$$

$$= \frac{Q'}{2k}(L - x)x + \frac{t_L - t_o}{L} x + t_o \qquad \text{...(a)}$$

Before eliminating t_2, let us write the differential equation for BC

$$\left(\frac{d^2 t}{dx^2}\right)_{BC} = 0$$

$$t = Cx + D$$

At $x = L$, $t = t_L$ at $x = 2L$, $t = t_o$

so $t_L = CL + D$

$t_o = 2\,CL + D$

or $$C = \frac{t_o - t_L}{L}$$

$$D = 2\,t_L - t_o$$

Thus $$(t_{BC}) = \frac{t_o - t_L}{L} x + (2t_L - t_o) \qquad \text{...(b)}$$

Now at B if the heat flow is considered, we get

$$-k\,A\left(\frac{dt}{dx}\right)_{\substack{AB\\x=L}} = -kA\left(\frac{dt}{dx}\right)_{\substack{BC\\x=L}}$$

From equations (*a*) and (*b*), we obtain

$$\frac{Q' L}{2k} - \left(\frac{Q' x}{k}\right)_{x=L} + \left(\frac{t_L - t_o}{L}\right) = \frac{t_o - t_L}{L}$$

or $$-\frac{Q' L}{2k} = \frac{t_o - t_L}{L} - \frac{t_L - t_o}{L}$$

$$= \frac{2(t_o - t_L)}{L}$$

$$(t_o - t_L) = -\frac{Q' L^2}{4k}$$

or $$(t_L - t_o) = \frac{Q' L^2}{4k}$$

$$t_L = t_o + \frac{Q' L^2}{4k}$$

From equation (*a*), the temperature distribution in *AB* is given by

$$t_{AB} = \frac{Q'}{4k}(L - x)x + \frac{Q' L}{4k}x + t_o$$

$$= \frac{Q'}{4k}\left(Lx + 2\,Lx - 2x^2\right) + t_o$$

$$= \frac{Q'}{4k}\left[x(3L - 2x)\right] + t_o$$

(ii) For portion *BC*

$$(t)_{BC} = -\frac{Q' L^2}{4kL}x + 2\left(t_o + \frac{Q' L^2}{4k}\right) - t_o$$

$$= \frac{Q' L}{4k}(2L - x) + t_o$$

Therefore $(t_{AB}) = \dfrac{Q'}{4k}x(3L - 2x) + t_o \ (x = 0, \text{ to } x = L)$

$(t)_{BC} = \dfrac{Q' L}{4k}(2L - x) + t_o \ (x = L, \text{ to } x = 2L)$

(b) For maximum or minimum

$$\frac{dt}{dx} = 0$$

Now $\left(\dfrac{dt}{dx}\right)_{AB} = \dfrac{Q'}{4k}[3L - 4x] = 0$

or $\quad 3L = 4x$

or $\quad x = 3/4\ L$

and $$\left(\frac{dt}{dx}\right)_{BC} = \frac{Q' L}{4k}(-1)$$

obviously BC has no maximum or minimum but curve AB has a maximum.

That, it is a maximum is seen easily

$$\frac{d^2 t}{dx^2} \text{ is } -ve$$

(c) $$(t)_{max} = \frac{Q'}{4k}\left[3L - 2 \times \frac{3}{4} \cdot L\right]\frac{3}{4}L + t_o = \frac{Q'}{4k}\left(\frac{3L}{2}\right)\left(\frac{3L}{4}\right) + t_o$$

or $$(t)_{max} = \frac{9}{32}\frac{Q'}{k}L^2 + t_o \text{ at } x = \frac{3L}{4}$$

Heat Transfer in Composite Wall in Series and Parallel Flows—slab Geometry

In numerous practical situations combination of series and parallel connected heat flow path are encountered.An example of such a case is illustrated by the composite wall shown in Fig. 2.13.

An approximate solution can be arrived at by assuming that the heat flow is essentially one dimensional. The composite wall can be divided into three sections. The thermal resistance of each section can be determined with the help of the thermal circuit shown in the figure. The intermediate layer consists of two separate thermal paths in parallel and its thermal conductance is the sum of the individual conductances. For a wall section of height $(b_1 + b_2)$ (refer to the figure) the conductance is

$$K_2 = \frac{k_2\, b_1}{L_1} + \frac{k_1\, b_2}{L_2} \text{ per unit length of the wall}$$

Using the equation

$$U\,A = \frac{1}{R_1 + R_2 + \ldots\ldots + R_n}$$ is the overall unit conductance (The concept of U to be elaborated later in the text; U is the overall heat transfer coefficient) from surface is

$$U = \frac{1}{(b_1 + b_2)(R_1 + R_2 + R_3)} = \frac{1}{\dfrac{L_1}{k_1} + \dfrac{b_1 + b_2}{\left(\dfrac{h_1\, b_1}{L_2}\right) + \left(\dfrac{k_2\, b_1}{L_2}\right)} + \dfrac{L_3}{k_3}}$$

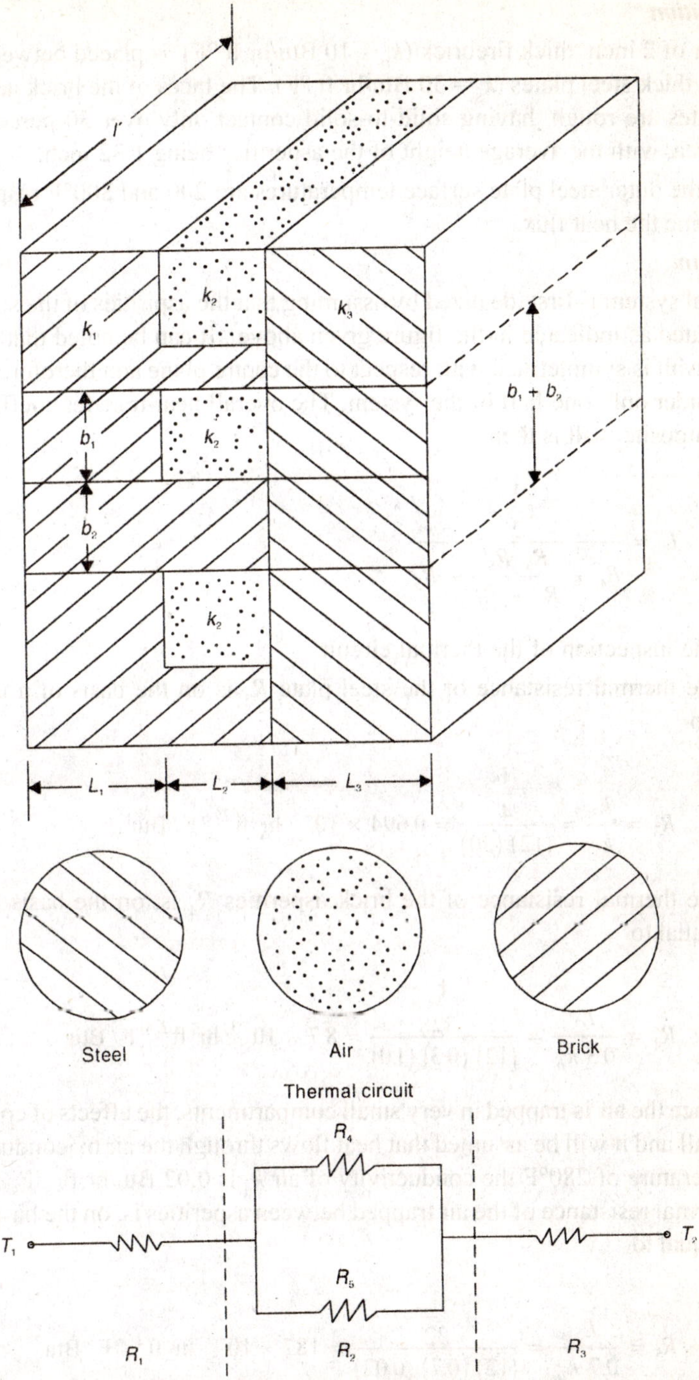

Fig. 2.13 Thermal circuit for a parallel-series composite wall L_1 = 1 inch, L_2 = 1/32 inch, L_3 = 1/4 inch for the following example

Illustration

A layer of 2 inch. thick firebrick (k_b = 10 Btu/hr ft °F) is placed between two 1/4 inch. thick steel plates (k_s = 30 Btu/hr ft °F). The faces of the brick adjacent to the plates are rough, having solid-to-solid contact only over 30 percent of the total area, with the average height of the asperities being 1/32 inch.

If the outer steel plate surface temperatures are 200 and 800°F respectively, determine the heat flux.

Solution

The real system is first idealized by assuming that the aspirities of the surface are distributed as indicated in the figure given above . It can be noted that the composite wall is symmetrical with respect to the centre plane and therefore we need to consider only one-half of the system. The overall heat-transfer coefficient for the composite wall is then

$$U = \frac{\frac{1}{2}}{R_1 + \frac{R_4 R_5}{R_4 + R_5} + R_3}$$

from the inspection of the thermal circuit

The thermal resistance of the steel plate R_3 is on the basis of a unit area, equal to

$$R_3 = \frac{L_3}{k_s} = \frac{\frac{1}{4}}{(12)(30)} = 0.694 \times 10^{-3} \text{ hr ft}^2 \text{ °F/Btu}$$

The thermal resistance of the brick asperities R_4 is, on the basis of a unit area, equal to

$$R_4 = \frac{L_2}{0.3\, k_b} = \frac{\frac{1}{32}}{(12)(0.3)(1.0)} = 8.7 \times 10^{-3} \text{ hr ft}^2 \text{ °F/Btu}$$

Since the air is trapped in very small compartments, the effects of convection are small and it will be assumed that heat flows through the air by conduction. At a temperature of 280°F the conductivity of air k_a is 0.02 Btu/hr ft^2 °F. Then R_5, the thermal resistance of the air trapped between asperities is, on the basis of unit area, equal to

$$R_5 = \frac{L_2}{0.7\, k_a} = \frac{\frac{1}{32}}{(12)(0.7)(0.02)} = 187 \times 10^{-3} \text{ hr ft}^2 \text{ °F/Btu}$$

The factors of 0.3 and 0.7 inch R_4 and R_5 respectively, represent the percentage of the total area for the two separate heat-flow paths.

The total thermal resistance for the two paths, R_4 and R in parallel, is

$$R_2 = \frac{R_4 \; R_5}{R_4 + R_5} = \frac{(8.7)(187) \times 10^{-6}}{(8.7 + 187) \times 10^{-3}}$$

$$= 8.3 \times 10^{-3} \text{ hr ft}^2 \text{ °F/Btu}$$

The thermal resistance of one half of the solid brick, R_1, is

$$R_1 = \frac{1}{2} \frac{L_1}{k_b} = \frac{1}{2} \frac{2}{(12)(1.0)}$$

$$= 83.5 \times 10^{-3} \text{ hr ft}^2 \text{ °F/Btu}$$

and U, the overall heat transfer coefficient is

$$U = \frac{\frac{1}{2} \times 10^3}{83.5 + 8.3 + 0.69}$$

$$= 5.4 \text{ Btu/hr ft}^2 \text{ °F}$$

A look at the values for the various thermal resistances show that the steel offers a negligible resistance, while the section although only 1/32 inch. thick, contributes 10 percent to the total resistance.

The rate of heat flow per unit area that is the heat flux is

$$\frac{q}{A} = U \; \Delta T = 5.4 \, (800 - 200)$$

$$= 3250 \text{ But/hr ft}^2$$

Examples in Steady State Heat Conduction

1. The conductivity of an 85% magnesia insulating material is taken to be a function of temperature. Determine B_k and k_o for a linear approximation [$k(T) = k_o(1 + B_k T)\, dT$] say between the temperature range 100 and 300°F.

(b) Estimate the rate of heat flow per unit area between these temperatures for a slab of 3 inch thickness.

Solution

(a) By means of the graphical method the slope of the straight line connecting the thermal conductivity curve between 100 and 300°F is found to be + 0.000031. The ordinate intercept of 0° degrees is 0.031

Thus we have

$$k(T) = 0.031\,(1 + 0.001\,T) \text{ for } 100\text{°F} < T < 300\text{°F}$$

The mean temperature is 200°F and the mean value of the thermal conductivity is

$$k_m = 0.031\,(1 + 0.001 \times 200)$$

$$= 0.0372 \text{ But/hr ft °F}$$

(b) The rate of heat flow per unit area is

$$q_k = \frac{\Delta T}{\dfrac{L}{A\,k_m}}$$

or

$$\frac{q_k}{A} = \frac{\Delta T}{\dfrac{L}{k_m}} = \frac{200}{\dfrac{3/12}{0.0372}}$$

$$= 29.8 \text{ But/hr ft}^2$$

2. Compute the heat loss from 10 ft of a 3 inch. nominal-diameter pipe covered with 1½ inch. of an insulating material having a thermal conductivity of 0.040 Btu/hr ft °F. Take that the inner and outer surface temperatures of insulation are 400 °F and 80 °F respectively.

Solution

The outside diameter of a nominal 3 inch. pipe is 3.50 inches. This is also the inside diameter of the insulation. The outside diameter is 6.5 inches.

The logarithmic mean area is

$$\bar{A} = \frac{A_o - A_i}{ln\,\dfrac{A_o}{A_i}} = \frac{10\pi\,\dfrac{(6.50 - 3.50)}{12}}{ln\left(\dfrac{6.50}{3.50}\right)}$$

$$= \frac{7.85}{0.62} = 12.70 \text{ ft}^2$$

Since $r_o/r_i < 2$, the arithmetic mean area would be an acceptable approximation

and $$\frac{A_o + A_i}{2} = \frac{10\pi\,(6.50 + 3.50)}{2\,(12)} = 13.10 \text{ ft}^2$$

Applying the equation

$$q_k = \frac{T_i - T_o}{\dfrac{r_o - r_i}{k\,\bar{A}}} \qquad q_k = \text{rate of heat transfer by conduction}$$

In this case

$$q_k = \frac{400 - 80}{\left(\frac{1.5}{12}\right)(0.04)\,(12.7)}$$

$$= 1300 \text{ Btu/hr}$$

3. Compute the heat loss per hour through one square meter of a furnace wall which is 400 mm thick. The outside and inside surface temperatures are 27°C and 1000°C respectively. Assume that the average thermal conductivity of the material of the wall is 0.96 k cal/hr metre °C

Solution

Inside surface temperature t_1 = 1000°C

outside surface temperature t_2 = 27°C

Thermal conductivity k = 0.96 k cal/hr m °C

The equation to be applied is

$$q = \frac{k\,(t_1 - t_2)}{x}$$

$$= \frac{0.96\,(1000 - 27)}{0.4}$$

$$= \frac{0.96 \times 973}{0.4}$$

$$= 2330 \text{ k cal/hr m}^2$$

i.e. 2330 kilo calories of heat will flow through each square metre of the furnace wall are per hour.

Illustration of Heat Conduction through Composite Plane Wall

Determine the rate of heat flow per unit area through a composite wall (of a room) made of stone and insulated with thermocole and hard board. The situation is represented by the following data

The outside surface temperature (t_1) = 40°C

The inside surface temperature (t_4) = 10°C

k of stone (k_1) = 2.8 w/m °C

Thickness of stone wall (x_1) = 500×100^{-3} m

k of thermocole (k_2) = 0.035 w/m °C

Thickness of thermocole (x_2) = 25×10^{-3} m

k of hard board k_3 = 0.175 w/m °C

Thickness of hard board x_3 = 5×10^{-3} m

Solution

The rate of heat transfer to the room per m². of the wall (q) is

$$q = \frac{t_1 - t_4}{\frac{x_1}{k_1} + \frac{x_2}{k_2} + \frac{x_3}{k_3}}$$

$$= \frac{(40 - 10)}{\left(\frac{500 \times 10^{-3}}{2.8}\right) + \left(\frac{25 \times 10^{-3}}{0.035}\right) + \left(\frac{5 \times 10^{-3}}{0.175}\right)} \text{ w/m}^2$$

$$= \frac{30}{(0.179) + (0.833) + (0.0294)}$$

$$q = 28.8 \times 3.6 = 103.75 \text{ kJ/hr m}^2$$

Example on Radial Conduction through Cylinders

A steam pipe having an outer diameter of 200 mm is covered with 25 mm thick layer of insulation material with an average k of 0.062 k cal/hr m. °C. The temperature of the pipe surface is 280°C and that of the outer surface of the insulation is 45°C. Determine the heat loss per hour if the length of the type is 10 metres.

Solution

Temperature of the inner surface t_1 = 280°C

Temperature of the outer surface t_2 = 45°C

k of insulation = 0.06 k cal/hr m °C

Inner radius of insulation, r_1 = 0.1 m

Outer radius of insulation, r_2 = 0.125 m

Axial length of insulation L = 10 m

(Pipe length)

$$q = \frac{2\pi\, k\, L\,(t_1 - t_2).}{ln \frac{r_2}{r_1}}$$

$$= \frac{2\pi \times 0.062 \times 10\,(280 - 45)}{ln\left\{\frac{0.125}{0.1}\right\}}$$

= 3980 k cal/hr

The heat loss per hour from the insulation of the steam pipe = 3980 k cal/hr

Example in Steady State Radial Heat Conduction

What is the quantity of heat that will be lost per metre length from a 6 inch. standard pipe covered with a 10 cm radial thickness of 85% magnesia when the outside temperature of the pipe is 450°C and the temperature of the outside surface of the insulation is 50°C? *k* of 85% magnesia is 0.0847 w/m °C at 250°C.

Solution

We known that $\dfrac{Q}{L} = \dfrac{2\pi\, k\,(T_i - T_o)}{ln\,\dfrac{D_o}{D_i}}$

$$k = 0.0847 \text{ w/m °C},\ T_i = 450 \text{ °C},\ T_o = 50 \text{ °C}$$

$$D_i = 6.625 \times 2.54 = 16.83 \text{ cm}$$

$$D_o = 16.83 + 2\ (10) = 36.83 \text{ cm}$$

$$D_o/D_i = 36.83/16.83 = 2.188$$

Substituting these values in the above expression

$$\frac{Q}{L} = \frac{2\pi\,(0.0847)\,(450 - 50)}{0.783} \text{ w/m}$$

$$= 270 \text{ w/m}$$

The above value is the quantity of heat that will be lost per metre length for this situation.

Example in Conduction of Heat —Plane Geometry

1. Determine the amount of heat that will be conducted per square metre of wall surface through a wall of a building with 22°C temperature drop between the inside and outside surfaces. The wall is built with a 20 cm thick concrete core covered on the inside with a 2 cm thick layer of gypsum plaster and an external layer of face brick (10 cm thick) attached to the concrete wall with a 1 cm thick layer of cement mortar. The *k* values are given below for the materials in question.

$$K_{\text{concrete}} = 1.36 \text{ w/m °C}$$

$$K_{\text{gypsum plaster}} = 0.48 \text{ w/m °C}$$

$$K_{\text{cement mortar}} = 1.16 \text{ w/m °C}$$

$$K_{\text{face brick}} = 1.32 \text{ w/m °C}$$

Solution

The situation is represented schematically below

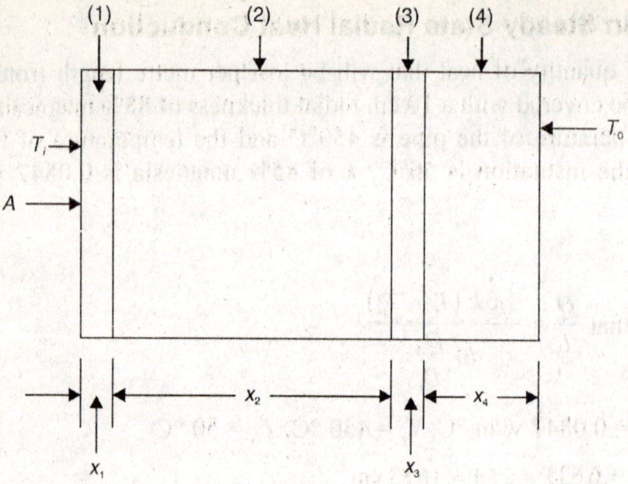

Fig. 2.14 *Heat conduction through the wall built with different materials*

1. $\text{Gypsum}_{\text{plaster}}$
2. Concrete
3. Cement mortar
4. Face brick

we know $Q = \dfrac{\Delta T}{R_{total}}$

$$R_{total} = R_1 + R_2 + R_3 + R_4 \ \text{°C/w}$$

$$\Delta T = 22 \text{ °C}$$

$$A = 1 \text{ m}^2 \qquad x_1 = 0.02 \text{ m (2 cm)}$$

$$R_1 = \frac{x_1}{A_1 \ k_1} \qquad k = 0.48 \text{ w/m °C}$$

$$R_1 = \frac{0.02}{1 \times 0.48} \text{ °C/w}$$

$$R_1 = 0.042 \text{ °C/w}$$

$$R_2 = x_2/A_2 \ k_2 \text{ °C/w}$$

$$x_2 = 20 \text{ cm} = 0.20 \text{ m}$$

$$k_2 = 1.36 \text{ w/m °C} \qquad A_2 = 1 \text{ m}^2$$

$$R_2 = \frac{0.20}{1.36 \times 1} = 0.147 \text{ °C/w}$$

$$R_3 = \frac{x_3}{A_3\, k_3}\ °\mathrm{C/w}$$

$x_3 = 1\ \mathrm{cm} = 0.01\ \mathrm{m}$

$k_2 = 1.15\ \mathrm{w/m\ °C}$ $\qquad A_3 = 1\ \mathrm{m}^2$

$$R_3 = \frac{0.01}{1 \times 1.15} = 0.009\ °\mathrm{C/w}$$

$$R_4 = \frac{x_4}{A_4\, k_4} \quad x_4 = 10\ \mathrm{cm} = 0.1\ \mathrm{m}, \quad k_4 = 1.32\ w/°C\ \mathrm{m}$$

$A_4 = 1\ \mathrm{m}^2$

$$R_4 = \frac{0.1}{1 \times 1.32} = 0.076\ °\mathrm{C/w}$$

$R_t = 0.042 + 0.147 + 0.009 + 0.076 = 0.274\ °\mathrm{C/w}$

Therefore, $\dfrac{Q}{A} = \dfrac{22}{1 \times 0.274}$

$= 80\ \mathrm{w/m^2}$

2. In the previous example (a) what is the section with the largest thermal resistance (b) What percentage of the total resistance is this section (c) What is the temperature drop across the concrete section of the wall?

Solution

(a) The concrete section (with $R = 0.147$) is the one with largest thermal resistance

(b) $R_2 = 0.147\ R_t = 0.274$

The percentage is $= \dfrac{0.147}{0.274} \times 100 = 54\ \%$

(c) $\Delta T = \dfrac{Q}{R_2}$ $\quad Q = 80$ w (from the previous example)

$R_2 = 0.147$

$\Delta T = 80 \times 0.147$

$= 11.8\ °\mathrm{C}$

Heat Conduction through a Spherical Wall

A furnace of spherical shape has 300 mm inner diameter is and 350 mm outer diamter. Its inner and outer surface temperatures are 450°C and 50°C respectively. The furnace is made of a refractory material whose mean thermal conductivity is

0.4 k cal/hr m °C. Determine the heat loss per hour. If the furnace is covered with 25 mm thick insulating material whose k is 0.05 k cal/hr m °C and the temperature of the insulating surface is 35°C, estimate the reduction in the heat loss per hour with the insulating material.

Solution

Inner radius r_1 = 300 mm (or 0.3 m)

Outer radius r_2 = 300 mm (or 0.35 m)

Mean thermal conductivity of the material k = 0.4 k cal/hr m °C.

Inner surface temperature of the furnace t_1 = 450 °C.

Outer surface temperature of the furnace t_2 = 50 °C.

The expression for the heat lost through the outer surface of the sphere /hr is

$$q = \frac{4\pi\ k\ r_1 r_2\ (t_1 - t_2)}{(r_2 - r_1)}$$

$$= \frac{4\pi \times 0.4 \times 0.3 \times 0.35 \times (450 - 50)}{0.05}$$

= 4200 k cal/hr

The thermal resistance of the spherical shell is given by the following expression

$$R_t = \frac{r_2 - r_1}{4\pi\ k \cdot r_1\ r_2}$$

In this situation

$$R_t = \frac{0.05}{4\pi \times 0.4 \times 0.3 \times 0.35} = 0.095 \text{ hr° C}/k \text{ cal}$$

Thermal resistance of insulation is given by the expression $R_{ins} = \dfrac{r_3 - r_2}{4\pi\ k \cdot r_3\, r_2}$

where r_3 represents the radial distance corresponding to the insulation layer.

For this case $R_{ins} = \dfrac{0.025}{4\pi \times 0.05 \times 0.375 \times 0.35} = 0.31$ hr °C/k cal

Total thermal resistance is

$R_t + R_{ins} = 0.095 + 0.31 = 0.405$ hr °C/k cal

Heat lost with insulation in place that is when the insulation is provided

$$= \frac{450 - 35}{0.405} = 1028\ k\text{ cal/hr}$$

Reduction in the heat loss due to the presence of insulation is

= 4200 – 1028 =3172 k cal/hr

Illustration on Steady State Conduction

Pressurised water is to be transported through pipe embedded in a 4-feet thick wall. It is desired to locate the pipe in the wall where the temperature is 250°F.One surface on the wall is held at 425°F and the other remains at 160°F as shown in the following sketch. If the wall material has a thermal conductivity which varies with temperature according to the expression

$$k = 0.5\ (1 + 0.06\ T)$$

where T is in °F and k is in But/hr ft °F how far from the hot wall should the pipe be located?

Solution

For this situation, the linear temperature profile is not valid due to the temperature dependence of k. with k written as a linear function of T in the form

$$k(T) = \alpha(1 + \beta T)$$

The rate equation becomes

$$q_x = -\alpha\,(1 + \beta T)\,A\,\frac{dT}{dx} \quad \text{(in the x-direction)}$$

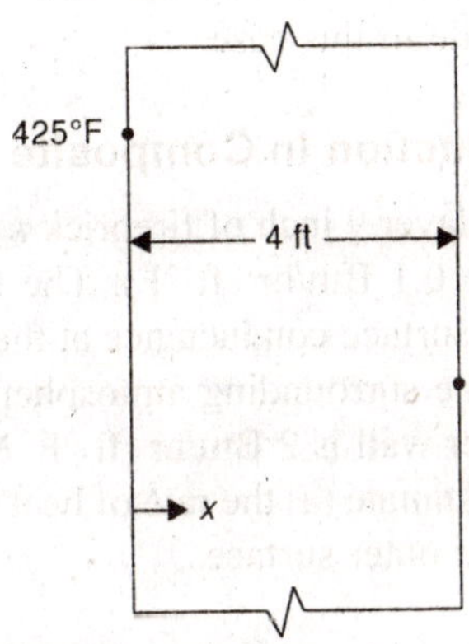

Fig. 2.15 *Steady state conduction-varying k*

Separating the variables and integrating.

We get $$\frac{q_x}{A}\int_0^x dx = -\alpha\int_{T_o}^{T_x}(1 + \beta T)\,dT$$

or $$\frac{q_x}{A}\cdot x = \alpha\left[T_o - T_x + \frac{\beta}{2}\left(T_o^2 - T_x^2\right)\right]$$

When integrating from $x = 0$ to a general position x. If we integrate from x to $x = L$ we get the relation

$$\frac{q_x}{A}(L-x) = \alpha\left[T_x - T_L + \frac{\beta}{2}\left(T_x^2 - T_L^2\right)\right]$$

For the situation under consideration, q_x/A constant, $L = 4$ ft, $T = 250°F$, $T_0 = 425°F$ $T_L = 160°F$ and α and β are the coefficients in the expression for k

Equating expression for q_x/A. we get

$$T_o - T_x + \frac{\beta}{2}\left(T_0^2 - T_x^2\right)$$

$$= \frac{x}{L-x}\left[T_x - T_L + \frac{\beta}{2}\left(T_x^2 - T_L^2\right)\right]$$

and inserting the known values of for T_x, T_0, T_L β, $x = 2.88$ ft.

It is interesting to note that had the temperature dependence of k been ignored and the linear temperature variation assumed the pipe would have being located at 2.64 ft from the hot wall. At this location the actual temperature is 294°F. The saturation pressure of water at 250°F is 29.82 psi, and at 294°F the value is 61.20 psi, a factor of over 2 to 1. Obviously the inclusion of the temperature dependence of k was worthwhile in this case.

Illustration on Heat Conduction in Composite Structures

A furnace wall consists of two layer 9 inch of firebrick $k = 0.8$ Btu/hr ft °F and 5 inch of insulating brick ($k = 0.1$ Btu/hr ft °F), The temperature inside the furnace is 3000°F and the unit-surface conductance at the inside wall is 12 Btu/hr ft °F. The temperature of the surrounding atmosphere is 80°F and the unit surface conductance at the outer wall is 2 Btu/hr ft °F. Neglecting the thermal resistance of the mortar joints estimate (a) the rate of heat loss per ft^2 of wall and the temperature at the inner and outer surfaces.

Solution

The rate of flow of heat has to be found out

We know q is given by

$$q = \frac{T_i - T_o}{\sum_{n=1}^{n=\text{anynumber}} R_n}$$

$$\frac{q}{A} = \frac{3000 - 80}{\frac{1}{12} + \frac{\frac{9}{12}}{0.8} + \frac{\frac{5}{12}}{0.1} + \frac{1}{2}}$$

$$= \frac{2920}{0.083 + 0.94 + 4.17 + 0.50}$$

Therefore, the heat flux = 2920/5.69 = 513 Btu/hr ft^2

It may be relevant here to mention that the insulating brick while representing only about 1/3 of the wall thickness, accounts for three quarters of the total thermal resistance.

Temperature at the interior wall

we know $$q = \frac{\Delta T_1}{R_1} = \frac{\Delta T_2}{R_2} \text{ etc.}$$

Applying the above expression, the temperature drop between the furnace gases and the interior surface $T_i - T_1 = q.\ R = (513)\ (0.083) = 43°F$.

This comparatively small temperature difference is in accordance with previous consideration indicating that the thermal resistance of the first section in the circuit is negligible. Thus heat can flow without a large potential and the temperature at the interior wall in nearly equal to that of the gases; in other words $T_i = T_1 - 43°F = 3000 - 43 = 2957°F$

The Temperature of the Outer Surface

The temperature of the outer surface can be computed in the aforementioned way and it turns out to be 336°F for this situation.

Illustration on Radial Heat Flow through Concentric Cylinders

Calculate the heat loss per linear foot from a 3 inch. steel scheduled 40 pipe (3.07 inch. inner diameter, 3.50 inch outer diameter, $k = 25$ Btu/hr ft °F) covered with a ½ inch thickness of asbestos insulation ($k = 0.11$ Btu/hr ft °F). The pipe transports a fluid at 300°F with an inner unit-surface conductance (heat transfer coefficient)of 40 Btu/hr ft^2 °F and is exposed to ambient air at 80°F with an average outer unit-surface conductance of 4.0 But/hr ft^2 °F. (heat transfer coefficient).

Solution

We know that

$$q = \frac{T_i - T_o}{\dfrac{1}{2\pi\, r_1\, L\, h_i} + \dfrac{\ln r_2 / r_1}{2\pi\, k_1\, L} + \dfrac{\ln r_3 / r_2}{2\pi\, k_2\, L} + \dfrac{1}{2\pi\, r_3\, L\, h_o}}$$

$$= \frac{T_i - T_o}{\sum_{n=1}^{n=4} R_n}$$

Using the above equation, the rate of heat transfer for $L = 1$ is

$$q = \frac{T_i - T_o}{R_1 + R_2 + R_3 + R_4}$$

$$= \frac{220}{\dfrac{1}{\pi\left(\dfrac{3.07}{12}\right)40} + \dfrac{ln\,(3.5/3.07)}{2\pi\,(25)} + \dfrac{ln(4.5/3.5)}{2\pi\,(0.11)} + \dfrac{1}{\pi\left(\dfrac{4.5}{12}\right)4}}$$

$$= \frac{220}{0.0312 + 0.00085 + 0.363 + 0.212}$$

$$= 362 \text{ Btu/hr ft}$$

It is to be noted that the thermal resistance is concentrated in the insulation and in the low surface conductance (the heat transfer coefficient) at the outer surface, while the resistance of the metal wall is negligible. If the pipe in question, were bare the heat loss would be 1040 Btu/hr ft. or nearly thrice as large as with insulation.

Summary of One—Dimensional Conduction Results

Many practical heat transfer problems are characterised by one-dimensional steady state conduction in plane (slab geometry), cylindrical or spherical walls without heat generation. Key results for these three geometries are summarized in the following table in which ΔT refers to the temperature difference. $T_{s,1} - T_{s,2}$ between the inner and outer surfaces of the respective geometry. In each case starting with the heat equation, one can derive the corresponding expression for the temperature distribution, heat flux, heat transfer rate and the thermal resistance.

Table 2.3 *One-dimensional, steady—state solutions to the heat transfer equation with no generation of heat*

	Plane wall (slab geometry)	*Cylindrical wall*	*Spherical wall*
Heat equation	$\dfrac{d^2T}{dx^2} = 0$	$\dfrac{1}{r}\dfrac{d}{dr}\left(r\dfrac{dT}{dr}\right) = 0$	$\dfrac{1}{r^2}\dfrac{d}{dr}\left(r^2\dfrac{dT}{dr}\right) = 0$
Temperature distribution	$T_{s,1} - \Delta T\dfrac{x}{L}$	$T_{s,2} + \Delta T\dfrac{ln\,(r/r_2)}{ln\,(r_1/r_2)}$	$T_{s,1} - \Delta T\left[\dfrac{1-(r_1/r)}{1-(r_1/r_2)}\right]$

Contd.

	Plane wall (slab geometry)	Cylindrical wall	Spherical wall
Heat flux (rate of heat transfer unit area)	$k\dfrac{\Delta T}{L}$	$\dfrac{k\,\Delta T}{r\,ln\left(\dfrac{r_2}{r_1}\right)}$	$\dfrac{k\,\Delta T}{r^2\left[\left(\dfrac{1}{r_1}\right)-\left(\dfrac{1}{r_2}\right)\right]}$
Rate of heat flow	$k\,A\dfrac{\Delta T}{L}$	$\dfrac{2\pi\,L\,k\,\Delta T}{ln\left(\dfrac{r_2}{r_1}\right)}$	$\dfrac{4\pi\,k\,\Delta T}{\left(\dfrac{1}{r_1}\right)-\left(\dfrac{1}{r_2}\right)}$
Thermal (R_t cond)	$\dfrac{L}{k\,A}$	$\dfrac{ln\left(\dfrac{r_2}{r_1}\right)}{2\pi\,L\,k}$	$\dfrac{\left(\dfrac{1}{r_1}\right)-\left(\dfrac{1}{r_2}\right)}{4\pi\,k}$

The critical radius of insulation is $r_{cr} = k/h$ for the cylinder and $r_{cr} = 2\,k/h$ for the sphere.

Illustration on Radical Heat Transfer in Spherical Geometry

A spherical, thin-walled metallic container is employed to store liquid nitrogen at 77°*K*. The diameter of the container is 0.5 m and the container is covered with an evacuated reflective insulation composed of silica powder. The insulation is 25 mm thick and its outer surface is exposed to ambient air at 300°*K*. The heat transfer coefficient is 20 w/m^2 °*K*. The latent heat of vaporization and the density of liquid nitrogen are respectively 2×10^5 joules and 804 kg/m^3. What is the rate of heat transfer to the liquids nitrogen. Calculate also the rate of the liquid boil-off *k* of evacuated silica powder at 300°*K* is 0.0017 w/m °*K*.

Solution

The usual assumptions made are

1. Steady state condition prevails.
2. Only radial direction heat transfer is occurring.
3. Negligible resistance to heat transfer through the container wall and from the container to the nitrogen holds good.
4. Constant properties.
5. No radiation heat transfer between outer surface of insulation and surroundings.

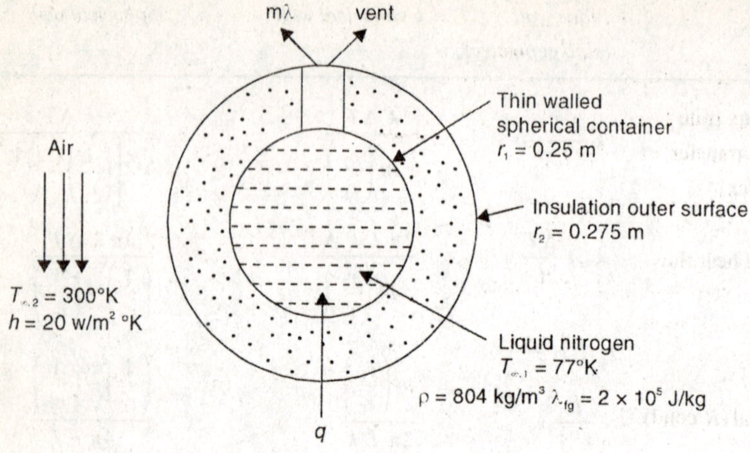

Fig. 2.16 *Radial heat transfer is spherical geometry*

The thermal resistance for conduction is

$$R_t, cond = \frac{1}{4\pi\, k}\left(\frac{1}{r_1} - \frac{1}{r_2}\right)$$

The thermal resistance for the convective heat transfer is

$$R_t, conv = \frac{1}{h\, 4\pi\, r^2}$$

The rate of heat transfer to the liquid nitrogen is then

$$q = \frac{T_{\infty,2} - T_{\infty,1}}{\left(\frac{1}{4\pi\, k}\right)\left[\left(\frac{1}{r_1}\right) - \left(\frac{1}{r_2}\right)\right] + \frac{1}{h\, 4\pi\, r^2}}$$

$$\text{Therefore } q = \frac{300 - 70}{\left[\frac{1}{4\pi\,(0.0017)}\right]\left[\frac{1}{0.25} - \frac{1}{0.275}\right] + \frac{1}{(20)\, 4\pi\,(0.275)^2}}$$

$$= \frac{223}{17.02 + 0.05} = 13.06 \text{ w}$$

2. Performing an energy balance for a control surface about the nitrogen, it follows that

Energy in – Energy out = 0

Energy in = q

Energy out = $m \lambda_{fg}$ associated with the latent energy due to boiling.

Therefore $q - m \lambda_{fg} = 0$

And the boil off rate m is

$$m = \frac{q}{\lambda_{fg}}$$

$$= \frac{13.06}{2 \times 10^5} = 6.53 \times 10^{-5} \text{ kg/sec}$$

The loss per day is

$$m = 6.53 \times 10^{-5} \times 3600 \times 24$$

$$= 5.64 \text{ kg/day}$$

or an volumetric basis

$$V = \frac{m}{\rho} = \frac{5.64}{804} = 0.007 \text{ m}^3/\text{day}$$

$$= 7 \text{ litres per day}$$

If the inner and outer surface temperatures are uniform, three practically useful body shapes satisfy the one-dimensional approximation

1. A thin plane slab 2. a long hollow cylinder and 3. a hollow sphere. These shapes with the corresponding thermal resistance are given below

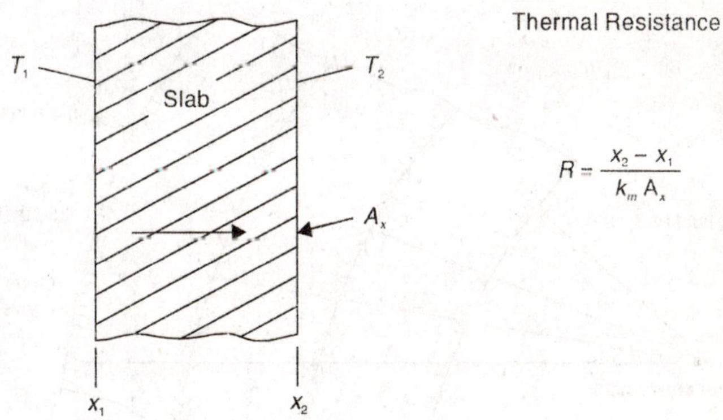

Fig. 2.17 *Three practically useful body shapes in heat conduction*

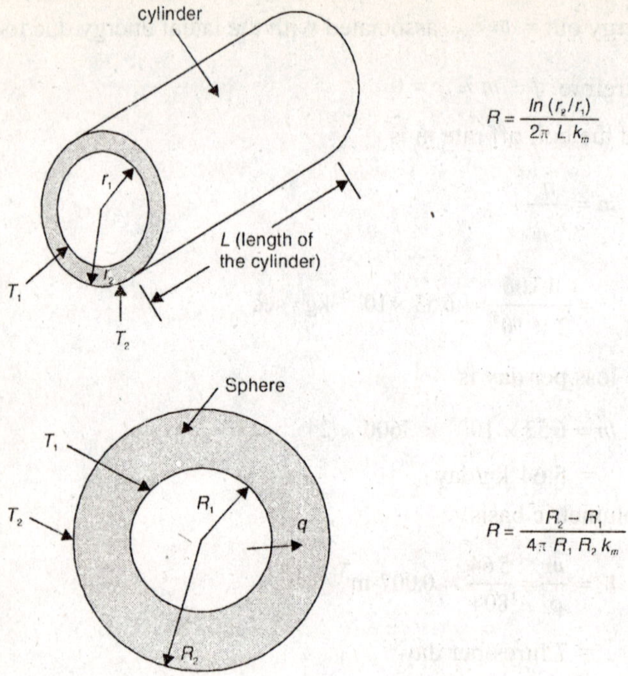

Fig. 2.17 (contd.) *Three practically useful body shapes in heat conduction*

The following sketch shows a three-layer cylinder with convection at both boundaries. The solution to such steady state heat transfer problems can be found by using the electric analogy.

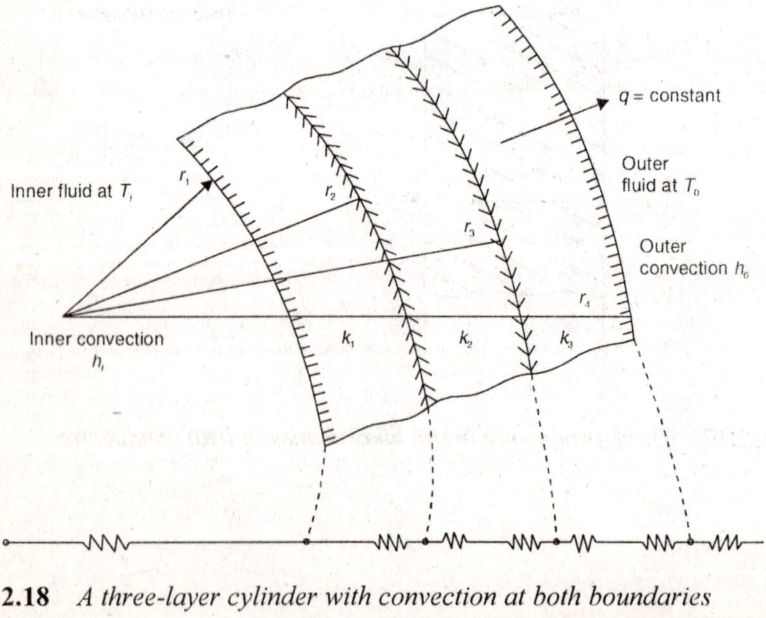

Fig. 2.18 *A three-layer cylinder with convection at both boundaries*

Relevance of Heat Transfer

Heat transfer phenomena play a vital role in many industrial and environmental problems.

As an example let us consider the important area of energy production and conversion there are numerous applications in this field that involve heat transfer effects in some manner. In the generation of electrical power, whether it be through nuclear fission or fusion, the combustion of fossil fuels, magneto-hydrodynamic processes or the use of geothermal energy sources there are numerous heat transfer problems that must be addressed and solved. These problems involve conduction, convection and radiation processes and relate to the design of system such as boilers, condensers and turbines. One is often confronted with the necessity to maximise heat transfer rates and to maintain the integrity of materials in high-temperature environments. One is also confronted with the thermal pollution problem associated with the discharge of large amounts of waste heat from a power generation plant to the environments. Numerous heat transfer considerations relate to the design of cooling powers to mitigate the environmental problems associated with this discharge.

In a smaller scale there are many heat transfer problems related to the development of solar energy conversion systems for space heating and air conditioning as well as for electric power generation. Heat transfer process also affects the performance of propulsion system such as the internal combustion and rocket engines. Heat transfer problems arise in the design of conventional space and water heating systems. In the design of incinerators and cryogenic storage equipment in the cooling of electronic equipment, in the design of refrigeration and air-conditioning system and in many other problems of industrial relevance. Heat transfer processes are also relevant to air and water pollution and strongly influence local and global climate.

The three modes of heat transfer are shown in the following figures

Conduction through a solid or a stationary fluid

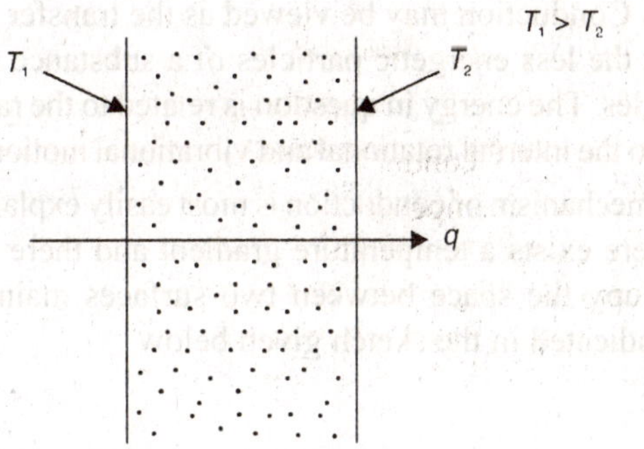

Fig. 2.19 *The three modes of heat transfer*

Convection from a surface to a moving fluid

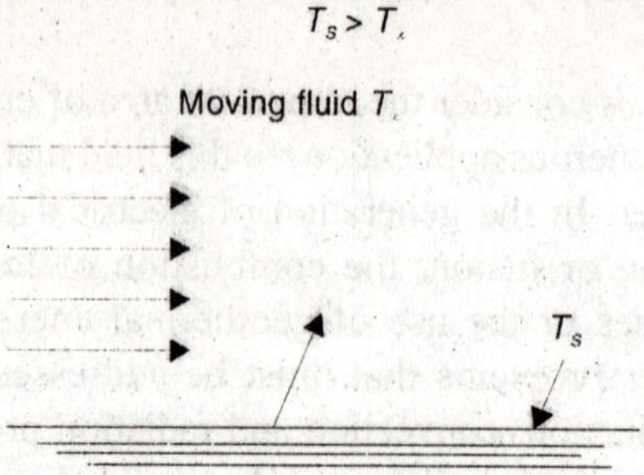

Net radiation heat exchange between two surfaces

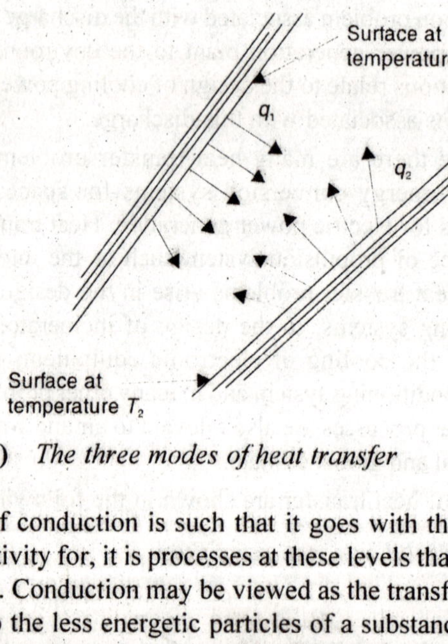

Fig. 2.19 (contd.) *The three modes of heat transfer*

The nature of conduction is such that it goes with the concepts of atomic and molecular activity for, it is processes at these levels that sustain this mode of heat transmission. Conduction may be viewed as the transfer of energy from the more energetic to the less energetic particles of a substance due to interactions between the particles. The energy in question is related to the random translational motion as well as to the internal rotational and vibrational motions of the molecules.

The physical mechanism of conduction is most easily explained by considering a gas in which there exists a temperature gradient and there is no bulk motion. The gas may occupy the space between two surfaces maintained at different temperatures as indicated in the sketch given below

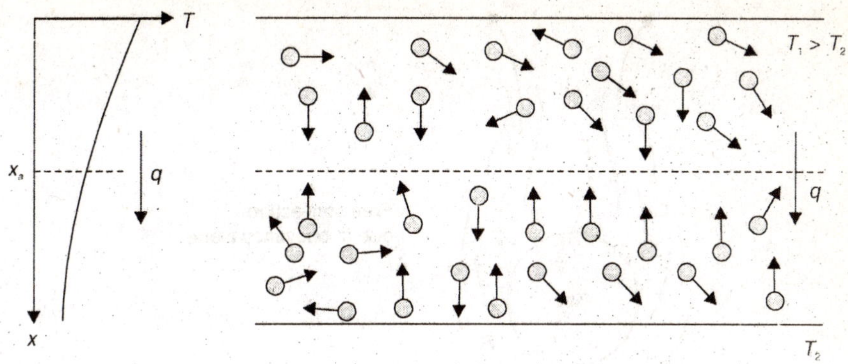

Fig. 2.20 *Association of conduction heat transfer with diffusion of energy owing to molecular activity*

The following sketch shows the convection regimes for flow near a hot sphere (a) forced convection, strong streaming motion.

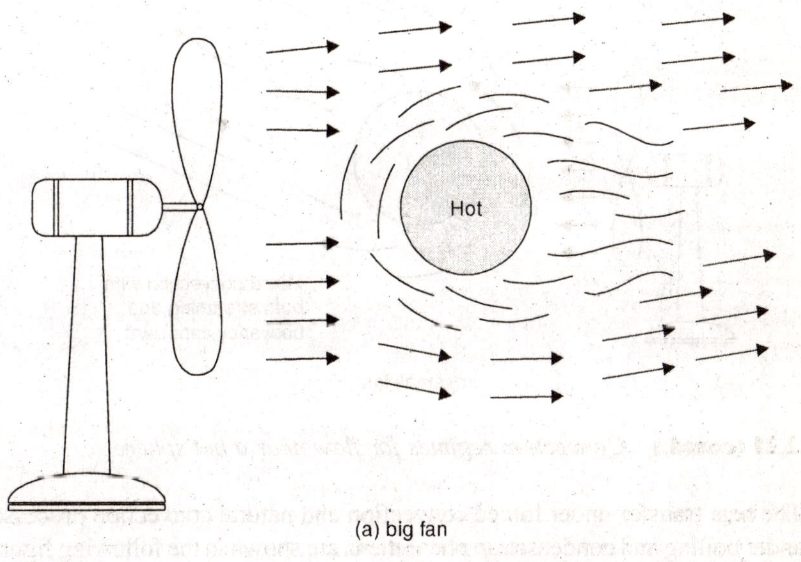

Fig. 2.21 *Convection regimes for flow near a hot sphere*

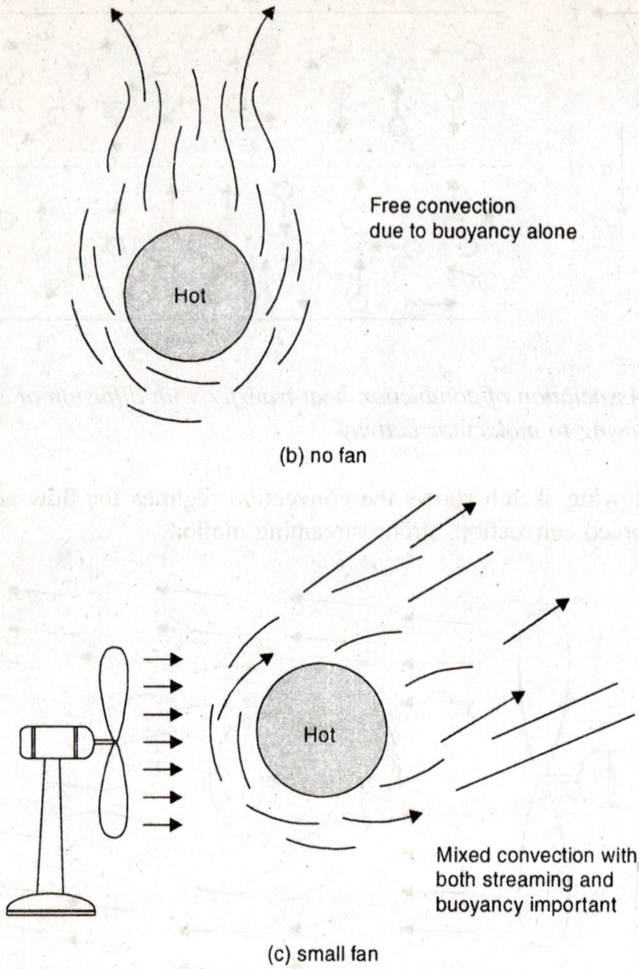

Fig. 2.21 (contd.) *Convection regimes for flow near a hot sphere*

The heat transfer under forced convection and natural convection processes and under boiling and condensation phenomena are shown in the following figure:

Forced convection

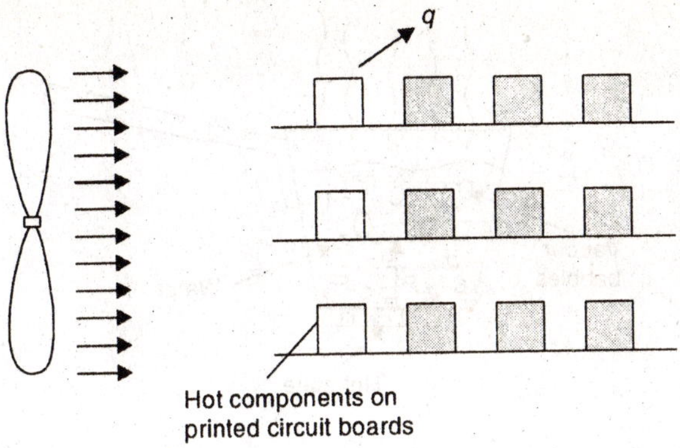

Free or natural convection

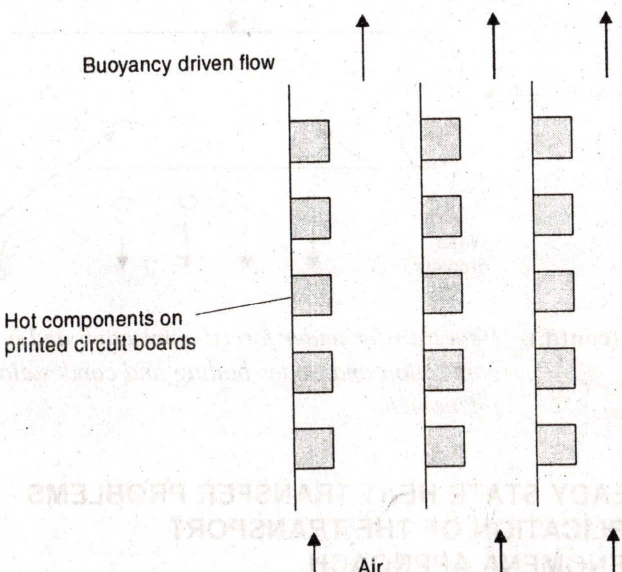

Fig. 2.22 *Heat transfer under forced convection and natural convection and under boiling and condesation phenomena*

Boiling phenomenon

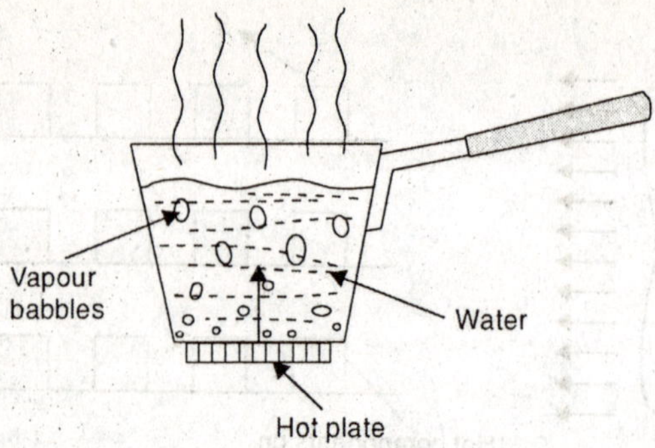

Condensation phenomenon

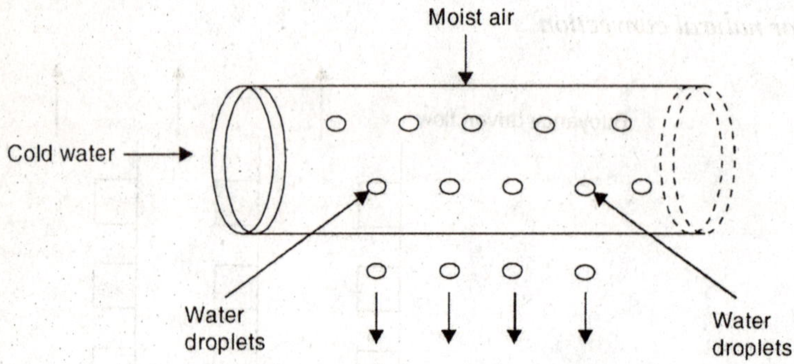

Fig. 2.22 (contd.) *Heat transfer under forced convection and natural convection and under boiling and condesation phenomena*

2.9 STEADY STATE HEAT TRANSFER PROBLEMS—APPLICATION OF THE TRANSPORT PHENOMENA APPROACH

The heat transport equation may be employed in the analytical treatment of many heat transfer problems. The actual steps involved in the analysis are

1. A volume is chosen and a balance on the transferent property is made with respect to the volume specified. The balance equation states that the rate of output of the property from the specified volume must equal the rate of influx or input plus the rate of production of the transferent property with in the volume. The transferent property is the property that gets transferred from one region to another.
2. The appropriate differential equation (transport) is substituted into the balance equation—equation for heat transfer/transport.

3. The differential equation is then integrated the constants of integration being evaluated from known boundary conditions.

It is likely that the physical property or the transport area varies with the distance or concentration of the transferent property in which case a relation must be established between the properties or distance or concentration, e.g. k varies with temperature

The mass diffusivity D, varies with distance in the case of diffusion

In some cases, the differential equation will be a little complex with the result these can not be integrated by simple analytical procedures. In such cases, it may be necessary to use more complex method like numerical or series approximations or graphical methods.

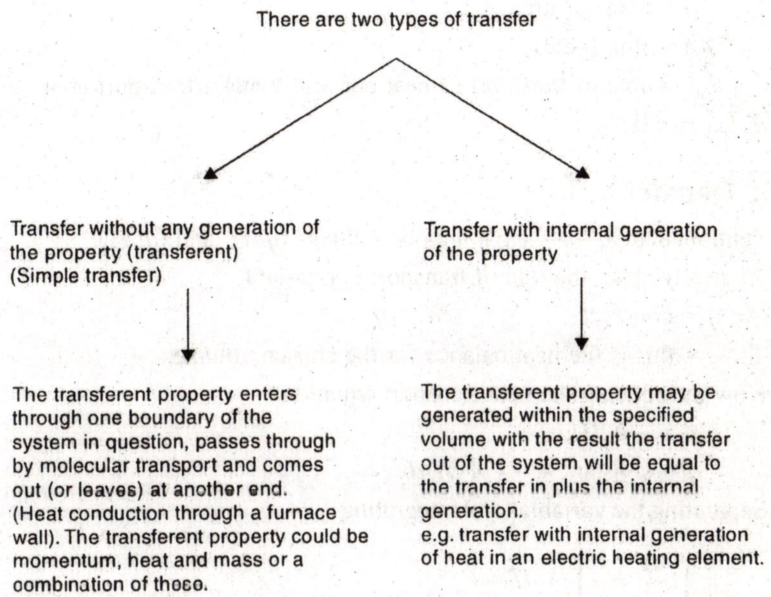

Fig. 2.23 *Types of transfer*

The heat that is generated by the element must be transferred form the element to the surroundings.

Generation should not be confused with creation—mass, heat or momentum generated with in the volume should have been present in a different form before it is transformed into the form in which it is generated.

Electrically heated element—electrical energy is transformed into thermal energy—change of the form of the energy.

There can be transfer of the property in multidirection—two and three dimensional transfers will involve complex mathematics.

Simple Transfer—Steady State Condition

(No generation of the transferent property)

The transferent property enters through one boundary of the system, passes through by molecular transport and leaves at another boundary. The rate of input of the transferent property into a volume must be the same as the rate of output: simple one-dimensional case.

Rate of input as the entrance boundary = rate of output at the exit boundary

No accumulation and generation of the property.

rate of transfer at any point in the transfer medium is constant with distance in the direction of transfer. In the case of heat transfer,

q = rate of heat transfer

A = transfer area

q/A = flux (heat)

= rate of transport of heat per unit transfer/transport area

$d/dx\ (q) = 0$

Heat Transfer

A_1— entrance area, A_2— exist area heat fluxes $(q/A)_1$ and $(q/A)_2$

At steady state, the rate of transport is constant

q = constant $\qquad q_1 = q_2$

= this is the heat balance for the chosen volume.

Now introducing the heat transport equation

$$q/A = -k\ dT/dx$$

i.e. $\quad q = \text{constant} = -k\ A\ dT/dx$

Separating the variables and integrating

$$q \int \frac{dx}{A} = -\int k\ dT + C_1$$

The above equation may be used to calculate the rate of heat transferred.

Transfer with Internal Generation

The transferent property being continuously generated must be continuously transferred to a boundary to maintain the steady state condition. The rate of transfer is not constant with distance but increases with distance as the boundary is approached.

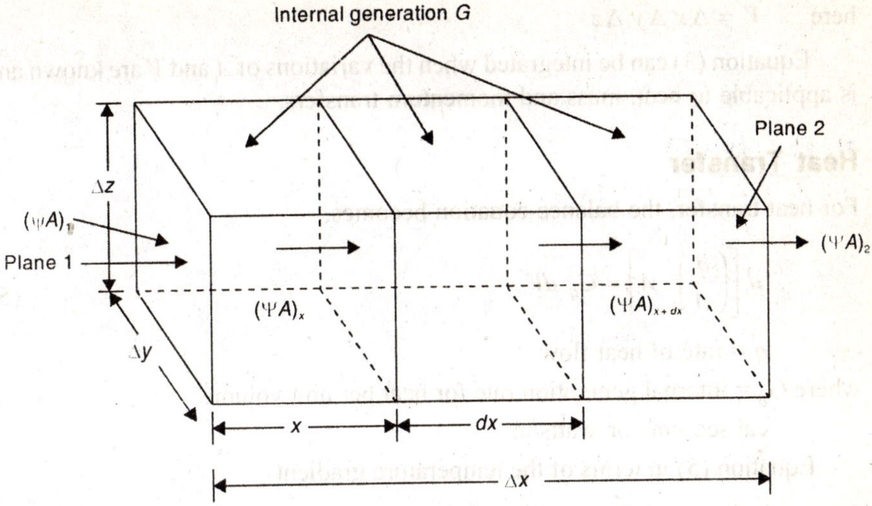

Fig. 2.24 *Transfer with internal generation of the transferent property*

ψ = flux of the transferent property

A = transfer area

G = rate of generation of property (per unit time) per unit volume.

Striking the property balance around the element of thickness dx

$$(\psi A)_x \quad + \quad G\,dV \quad = (\psi A)_{x+dx} \qquad ...(1)$$

↑ rate of influx into the element — ↑ rate of generation — ↑ rate of efflux of the transferent property

vol $dV = \Delta y\, \Delta z\, dx$

$$A = \Delta y\, \Delta z$$

A = transport area

$$(\psi A)_{x+dx} = (\psi A)_x = d(\psi A) \qquad ...(2)$$

Equation (1) becomes

$$d(\psi A) = G\,dV \qquad ...(3)$$

This equation states that the increase in rate of transfer across a volume element is equal to the rate of production within the element.

If G is constant for the entire volume between planes 1 and 2 equation (3) may be integrated to

$$(\psi A)_2 - (\psi A)_1 = GV \qquad ...(4)$$

here $\quad V = \Delta x \, \Delta y \, \Delta z$

Equation (3) can be integrated when the variations of A and V are known and is applicable to heat, mass and momentum transfers.

Heat Transfer

For heat transfer, the balance equation becomes,

$$d\left[\left(\frac{q}{A}\right)\cdot A\right] = G_q \, dV \qquad (5)$$

q = rate of heat flow

where G_q = internal generation rate for heat per unit volume.
cal/sec cm^3 or watts/m^3

Equation (5) in terms of the temperature gradient

$$d\left[-k\,A\,\frac{dT}{dx}\right] = G_q \, dV \qquad ...(6)$$

Integrating twice, we obtain

$$-kA\frac{dT}{dx} = \int G_q \, dV + C_1 \text{ and } \int k\,dT = \int \frac{\int G_q \, dV}{A} dx + \int C_1 \frac{dx}{A} + C_2 \qquad ...(7)$$

To perform the integration, the variations of k, V, A and G_q must be known functions of T or x.

Heat Flow in a Cylinder

Heat that is generated, uniformly within the cylinder (or the cylindrical element) is transferred in a radial direction to the surface. Equation (6) can be integrated for constant k and G_q.

So, $\quad V = \pi r^2 L \qquad A = 2\pi r L$

$dV = 2\pi r' L \, dr \quad dT/dx = dT/dr$

L = length of the cylinder

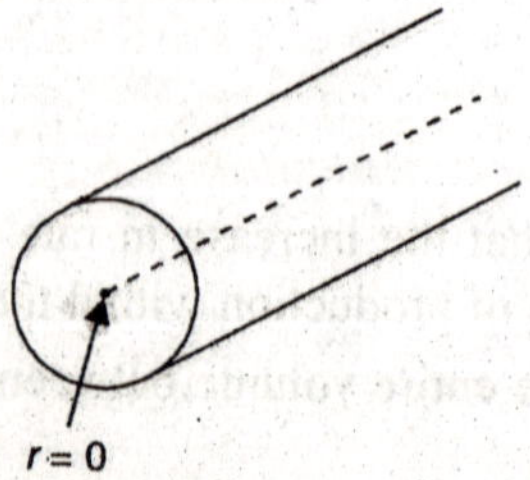

Fig. 2.25 *Heat flow in a cylinder*

Equation (6) becomes

$$d\left[-2\pi\, r\, L\, k\, \frac{dT}{dr}\right] = 2\pi\, L\, G_q \cdot r\, dr \qquad \text{...(8)}$$

Integrating once, we obtain

$$-2\,\pi\, r\, L\, k\,\frac{dT}{dr} = \pi\, L\, G_q\, r^2 + C_1$$

The temperature is maximum at the centre of the cylinder, if heat is uniformly generated.

At $r = 0$, $T = T_0$ or $dT/dr = 0$

This *b.c.* can be used to evaluate C_1

$$-2\pi\, L\, k\, r\,(0) = \pi\, L\, G_q\,(0) + C_1$$

or $C_1 = 0$

Equation (8) becomes

$$-k\left(\frac{dT}{dr}\right) = \frac{G_q\, r}{2} \qquad \text{(9)}$$

If dT/dr varies linearly with distance in the radial direction.

Integrating eqn. (9), we obtain

$$-k\, T = \frac{G_q \cdot r^2}{4} + C_2 \qquad \text{...(10)}$$

To evaluate C_2, the boundary condition (b.c) at the centre may be employed ie at $r = 0$, $T = T_0$

i.e. $-k\, T_0 = C_2$

$$-k\, T = \frac{G_q \cdot r^2}{4} - k\, T_0$$

Equation (10) becomes

$$k\,(T_0 - T) = \frac{G_q.\, r^2}{4} \qquad \text{.(11)}$$

If the temperature at the outside surface i.e. at $r = r_1$ is known, say $T = T_1$, then C_2 can be evaluated using this *b.c* also

$$-k\, T_1 = \frac{G_q.\, r_1^2}{4} + C_2$$

$$C_2 = -k\, T_1 - \frac{G_q\, r_1^2}{4}$$

Therefore the temperature profile becomes,

$$-kT = \frac{G_q r^2}{4} - kT_1 - \frac{G_q r_1^2}{4}$$

$$-k(T - T_1) = \frac{G_q}{4}\left(r^2 - r_1^2\right)$$

i.e. $$k(T_1 - T) = \frac{G_q}{4}\left(r^2 - r_1^2\right)$$

Any of the above two equations can be used to calculate the temperate at any point in the cylinder. The radial temperature distribution will be parabolic in a cylinder in which the heat is generated at a uniform rate.

Illustration

A long hollow cylinder has its inner and outer surfaces maintained at temperatures T_b and T_a respectively. The inner and outer radii are b and a respectively. Calculate the temperature profile in the solid section of the cylinder and determine the flux at both the surfaces. Are the fluxes the same ? Assume steady state condition.

This is a case of steady state transfer without heat generation.

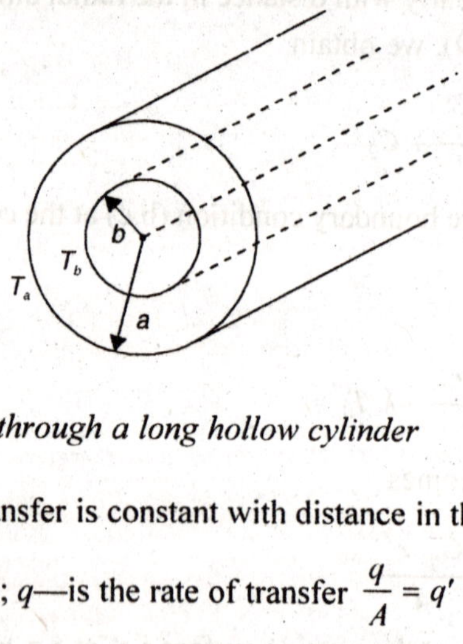

Fig. 2.26 *Heat flow through a long hollow cylinder*

Rate of heat transfer is constant with distance in the direction of transfer

i.e. the radial direction; q—is the rate of transfer $\frac{q}{A} = q' =$ flux (heat)

$$\frac{d}{dr}\left(\frac{q}{A} \cdot A\right) = 0 \qquad A = 2\pi r L$$

$$\frac{d}{dr}\left(\frac{q}{A} 2\pi r L\right) = 0$$

$$\frac{d}{dr}(q' r) = 0$$

Integrating $q\ r = C_1$

$$q' = C_1/r$$

Substituting Fourier's law of heat conduction

$$-k\frac{dT}{dr} = \frac{C_1}{r}$$

or
$$-k\ dT = \frac{C_1}{r}\cdot dr$$

Intg. $-k\ T = C_1\ ln\ r + C_2$

The boundary conditions are

At $r = b$; $\quad T = T_b$

$r = a \quad T = T_a$

$$\therefore -k\ T_a = C_1\ ln\ a + C_2 \qquad ...(1)$$

$$\therefore -k\ T_b = C_1\ ln\ b + C_2 \qquad ...(2)$$

Eliminating C_2 from the above two equations

$$k\,(T_a - T_b) = C_1\,(ln\,b - ln\,a)$$

$$= C_1\ ln\left(\frac{b}{a}\right)$$

$$C_1 = \frac{k\,(T_a - T_b)}{ln\left(\frac{b}{a}\right)}$$

$$C_2 = -k\ T_b - \frac{k\,(T_a - T_b)}{ln\left(\frac{b}{a}\right)}\ ln\ b$$

$$-k\,T = \frac{k\,(T_a - T_b)}{ln\left(\frac{b}{a}\right)}\ ln\ r - k\ T_b - \frac{k\,(T_a - T_b)}{ln\left(\frac{b}{a}\right)}\ ln\ b$$

$$T = -\frac{T_a - T_b}{ln\left(\frac{b}{a}\right)}\ ln\ r + T_b + \frac{T_a - T_b}{ln\left(\frac{b}{a}\right)}\ ln\ b$$

or
$$T = T_b + \frac{T_a - T_b}{ln\left(\frac{b}{a}\right)}\ ln\left(\frac{b}{r}\right)$$

The hear flux is given by, $q' = C_1/r$

$$C_1 = \frac{k\,(T_a - T_b)}{ln\left(\frac{b}{a}\right)}$$

$$\therefore q' = \frac{k\,(T_a - T_b)}{ln\left(\frac{b}{a}\right)}\left(\frac{1}{r}\right)$$

Therefore

$$\frac{q'_r}{q'_{r/r=a}} = \frac{k\,(T_a - T_b)}{ln\left(\frac{b}{a}\right)}\cdot\left(\frac{1}{a}\right)$$

at $$r = b,\; q'_r = \frac{k\,(T_a - T_b)}{ln\left(\frac{b}{a}\right)}\cdot\left(\frac{1}{b}\right)$$

The fluxes are not equal; but we can show that the rate of energy transfer, q at both the surfaces are equal

$$A = 2\pi\, r\, L$$

at $$r = a,\quad A = 2\pi\, L\, a$$

$$\therefore q = \frac{k\,(T_a - T_b)}{ln\,\frac{b}{a}}\cdot\frac{1}{a}\,2\pi\, a\, L$$

$$= \frac{k\,(T_a - T_b)}{ln\,\frac{b}{a}}\,2\pi\, L$$

Similarly, at $r = b$

$$q = \frac{k\,(T_a - T_b)}{ln\,\frac{b}{a}}\cdot\left(\frac{1}{b}\right)2\pi\, b\, L$$

For the given situation both are same; That is q is constant.

Problems

A hollow thick sphere has its inner ($r = b$) and outer ($r = a$) surfaces maintained at temperatures T_b and T_a respectively. Obtain an expression for the temperature profile in the solid section, at steady state conditions. Determine also the heat flux at both surfaces.

By the heat balance equation.

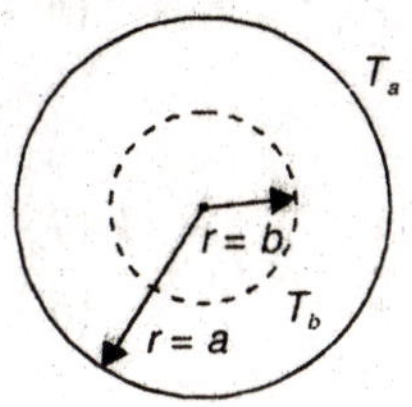

Fig. 2.27 *Heat transfer in a hollow thick sphere*

$$\frac{d}{dr}(q' \cdot A) = 0$$

$A = 4\pi r^2$ (surface area of the sphere; also known as transport area)

$$d/dr\,(q'\,r^2) = 0$$

$$q'\,r^2 = C_1$$

$$q' = C_1/r^2$$

$$-k\,dT/dr = C_1/r^2$$

or

$$-k\,dT = C_1/r^2\,dr$$

Integrating $-k\,T = -C_1/r + C_2$

at $r = b$, $T = T_b$

$r = a$ $T = T_a$

$$-k\,T_a = C_1/a + C_2 \quad \text{...(1)}$$

$$-k\,T_b = C_1/b + C_2 \quad \text{...(2)}$$

(1) – (2) — i.e. subtracting (2) from (1)

$$k\,(T_a - T_b) = C_1\,[1/a - 1/b]$$

$$C_1 = \frac{k\,(T_a - T_b)}{\left(\dfrac{1}{a} - \dfrac{1}{b}\right)}$$

$$C_2 = -k\,T_a + \frac{k\,(T_a - T_b)}{\left(\dfrac{1}{a} - \dfrac{1}{b}\right)} \cdot \frac{1}{a}$$

$$= -k\,T_a + \frac{k\,(T_a - T_b)}{(b-a)} \cdot \frac{\not{a}b}{\not{a}}$$

Temperature profile

$$-k\,T = -\frac{k\,(T_a - T_b)}{\left(\frac{1}{a} - \frac{1}{b}\right)} \cdot \frac{1}{r} - k\,T_a + \frac{k\,(T_a - T_b)}{(b-a)} \cdot (b)$$

$$-k\,T = -\frac{k\,(T_a - T_b)}{\left(\frac{1}{a} - \frac{1}{b}\right)} \cdot \frac{1}{r} - k\,T_a + \frac{k\,(T_a - T_b)}{(b-a)}$$

at $r = b$

$$-k\,T = -\frac{k\,(T_a - T_b)}{(b-a)} \cdot a - k\,T_a + \frac{k\,(T_a - T_b)}{(b-a)}$$

$$= -k\,T_a + \frac{k\,(T_a - T_b)}{(b-a)}\,(b-a)$$

$$= -k\,T_a + k\,T_a - k\,T_b$$

$$T = T_b$$

$$q'_{r=a} = -\frac{k\,(T_a - T_b)}{(a-b)} \cdot \left(\frac{b}{a}\right)$$

$$q'_{r=b} = -\frac{k\,(T_a - T_b)}{(a-b)} \cdot \left(\frac{a}{b}\right)$$

Problem

A constant rate of energy per unit volume G_q in the form of heat is uniformly generated in a long solid cylinder. The temperature at the outer surface of the cylinder ($r = a$) is maintained at T_a. Calculate the temperature profile and the heat flux at the wall under steady state conditions.

Problems on Heat Conduction
Heat Flow through a Cylinder

1. A pipe 5 cm outer diameter is lagged with a 4 cm layer of asbestos ($k = 0.18$) which is followed by a 3 cm layer of cork $k = 0.05$ *k*.cal/m hr °C. If the temperature of the outer surface of the pipe is 145°C and the temperature of the outer surface of the cork of 45°C. Calculate the heat loss in k cal/hr. m of the pipe.

Solution

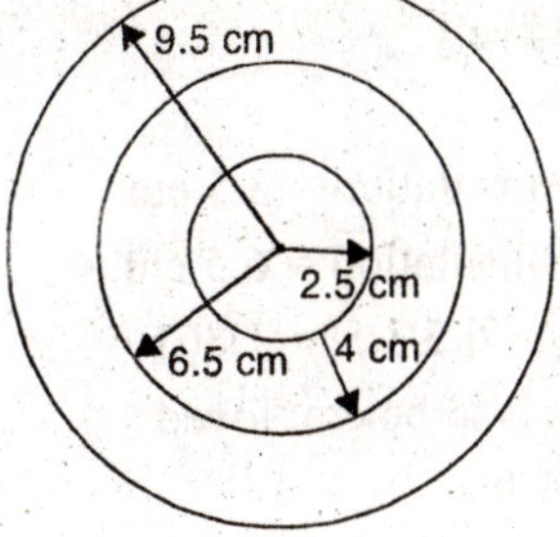

Fig. 2.28 *Heat flow through a cylinder*

We are using the log mean radius as the layers are very thick.

Asbestos

$$r_m = \frac{6.5 - 2.5}{2.303 \log \dfrac{6.5}{2.5}} = 4.18 \text{ cm}$$

For cork

$$r_m = \frac{9.5 - 6.5}{2.303 \log \dfrac{9.5}{6.5}} = 8 \text{ cm}$$

Asbestos layer

$$R = \frac{x_A}{k_A\ A} = \frac{\dfrac{4}{100}}{0.18 \times 2\pi \dfrac{4.18}{100} \times 1}$$

$$= 0.844$$

$$R_{cork} = \frac{x_B}{k_{B\cdot}\ A} = \frac{\dfrac{3}{100}}{(0.05)\ 2\pi \dfrac{8}{100} \times 1}$$

$$= 1.19$$

$$R_A + R_B = 2.03$$

$$\therefore q = \frac{145 - 45}{2.03} = 49.2\ k \text{ cals/hr}$$

2. Calculate the heat loss from 30 m of a 6 cm inner diameter pipe (outer diameter 7 cm) covered with 3 cm of an insulating material having a k value of 0.06 k cal/hr m °C. Take that the inner and outer surface temperature of the insulation are 200°C and 40°C respectively.

Solution

inner radius of the insulation = 3.5 cm

outer radius of the insulation = 6.5 cm

$$r_o/r_i = 6.5/3.5 = 1.861$$

∴ log mean radius is to be employed

$$r_m = \frac{6.5 - 3.5}{2.303 \log\left(\frac{6.5}{3.5}\right)} = 4.844 \text{ cm}$$

$$A_m = 2 \times \pi \times \frac{4.844}{100} \times 30 = 9.135 \text{ m}^2$$

$$\therefore q = \frac{k\, A_m\, (t_i - t_o)}{(r_o - r_i)}$$

$$= \frac{0.06 \times 9.135 \times (200 - 40)}{\frac{3}{100}}$$

$$= 2922 \text{ k cal/hr}$$

Steady State Conduction through a Spherical Wall

Let us consider the flow of heat through the wall of a hollow sphere whose inner radius is r_1 and outer radius is r_2. The k of the material with which this sphere is made is taken to be constant. The inner and outer surfaces are kept at constant temperatures t_1 and t_2 respectively ($t_1 > t_2$).

The temperature varies only in the radial direction.

To deduce the heat transfer rate expression for this situation. Let us take a thin layer between the outer and inner walls with a radius of r and thickness dr

Applying Fourier's law for the heat flow through the layer

$$q = -k\, A\, dt/dr \qquad ...(1)$$

Transport area in the case of a sphere is $4\pi\, r^2$ (surface area)

$$= -k\, 4\pi\, r^2 \frac{dt}{dr}$$

Separating the variables

$$-dt = \frac{q}{4\pi\, k} \frac{dr}{r^2} \qquad ...(2)$$

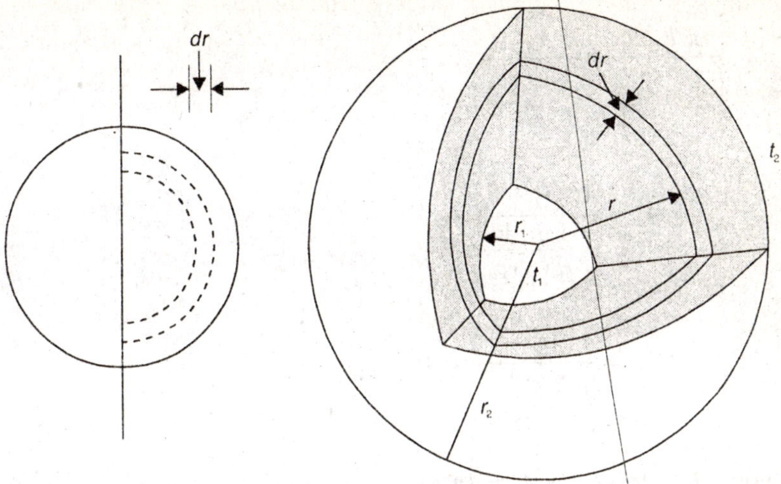

Fig. 2.29 *Heat conduction through a spherical wall*

Integrating between appropriate limits

$$-\int_{t_1}^{t_2} dt = \frac{q}{4\pi\, k}\int_{r_1}^{r_2}\frac{dr}{r^2} = \frac{q}{4\pi\, k}\left[-\frac{1}{r}\right]_{r_1}^{r_2}$$

$$= \frac{q}{4\pi\, k}\left(-\frac{1}{r_2}+\frac{1}{r_1}\right)$$

$$(t_1 - t_2) = \frac{q}{4\pi\, k}\left(\frac{1}{r_1}-\frac{1}{r_2}\right) \qquad ...(3)$$

or $$q = \frac{4\pi\, k\,(t_1 - t_2)}{\left(\frac{1}{r_1}-\frac{1}{r_2}\right)} = \frac{2\pi\, k\,\Delta t}{\frac{1}{d_1}-\frac{1}{d_2}} \qquad ...(4)$$

$$= \frac{\pi\, k\,\Delta t\, d_1\, d_2}{\left(\frac{d_2 - d_1}{2}\right)}$$

Expressing q in terms of A_m

$$q = \frac{k\, A_m\, \Delta t}{(r_2 - r_1)} \qquad ...(5)$$

By equating eq^{ns} (4) and (5), we get

$$\frac{4\pi\, k\, \Delta t}{\left(\frac{1}{r_1} - \frac{1}{r_2}\right)} = \frac{k\, A_m\, \Delta t}{r_2 - r_1}$$

or

$$A_m = \frac{4\pi\,(r_2 - r_1)}{\frac{1}{r_1} - \frac{1}{r_2}}$$

$$= \frac{4\pi\,(r_2 - r_1)}{\frac{r_2 - r_1}{r_1\, r_2}} = 4\pi\, r_1\, r_2$$

we know $A_1 = 4\pi\, r_1^{\,2}$ $\quad A_2 = 4\pi\, r_2^{\,2}$

$$\therefore A_m = \sqrt{A_1\, A_2}$$

i.e.

$$q = \frac{k\,\sqrt{A_1\, A_2}\,\Delta t}{(r_2 - r_1)}$$

In the case of a sphere, the average area to be employed in calculating the heat flow is the geometric mean of the internal and external areas.

Problem 1

Determine the heat loss through the wall of a spherical container with an inner diameter of 1.2 m and total wall and insulation thickness of 100 mm. Temperature at the inner surface is 140°C and that at the outer surface is 40°C. The equivalent k is 0.1 k cal/m. hr °C

$$q = \frac{4\pi\, k(t_1 - t_2)}{\left(\frac{1}{r_1} - \frac{1}{r_2}\right)}$$

$$= \frac{4 \times 3.14 \times 0.1 \times (140 - 40)}{\left(\frac{1}{0.6} - \frac{1}{0.7}\right)}$$

$= 530$ k cal/hr

Problem 2

Determine the rate of evaporation of liquid o_2 from a spherical container of 1.8 m outer diameter covered with 0.3 m of asbestos insulation. The following data may be used.

Temperature at the inner surface of insulation = – 183°C

Temperature at the outer surface of insulation = 0°C
Latent heat of vapourization of O_2 = 1640 k cal/kg mole
k of the insulation material at O°C = 0.135 k cal/hr. m °C
k of the insulation material at – 183°C = 0.108 k cal/hr. m °C

Solution

$r_1 = 0.9$ m $\qquad r_2 = 1.2$ m

In this case the heat flow is from outside to inside

$$\text{We know } q = \frac{4\pi\, k\,(t_1 - t_2)}{\left(\frac{1}{r_1} - \frac{1}{r_2}\right)} = \frac{4\pi\, k\, \Delta t}{\left(\frac{1}{r_1} - \frac{1}{r_2}\right)}$$

$$k_{av} = 0.1215$$

$$q = \frac{4 \times 3.14 \times 0.1215 \times 183}{\left(\frac{1}{0.9} - \frac{1}{1.2}\right)}$$

$$= 1007 \text{ k cal/hr}$$

$$\text{Evaporation loss} = \frac{1007}{1636}$$

that is,

(rate of evaporation) = 0.615 kg moles/hr
= 19.7 kg/hr

Problem 3

A steam pipe 160 mm inner diameter and 170 mm outer diameter is covered with two layers of insulation. The thickness of the 1st layer is 30 mm and that of the 2nd is 50 mm. The k values are 50, 0.15, and 0.08 k cal/m. hr °C for the pipe, 1st and 2nd insulation respectively. The temperature of the inner surface of the steam pipe is 300°C and that of the outer surface of the insulation layer is 50°C. Determine the quantity of heat lost per meter length of steam pipe and the layer contact temperatures (interface temperatures).

Solution

$d_1 = 0.16$ m $\quad d_2 = 0.17$ m $\quad d_3 = 0.23$ m and $d_4 = 0.33$ m

$$ln\frac{d_1}{d_2} = \frac{1}{0.06} \quad \text{or} \quad ln\frac{d_2}{d_1} = 0.06$$

$$ln\frac{d_3}{d_2} = 0.303 \qquad ln\frac{d_4}{d_3} = 0.362$$

$$q = \frac{2 \times \frac{22}{7} \times (300 - 50)}{\frac{0.06}{50} + \frac{0.302}{0.15} + \frac{0.362}{0.08}}$$

$$= 240 \text{ k cal/hr. m}$$

$$t_2 = 300 - \frac{240}{2\pi} \times 0.0012 \approx 300 - 0.46 \approx 300 \text{ °C}$$

$$t_3 = 300 - \frac{240}{2\pi} \times 2.02$$

$$= 223 \text{ °C}$$

Problem 4

Brickwork of a furnace is built of layer of fire clay and redbrick and the space between them is filled with common brick. Fire clay layer is 120 mm thick, C_1 common brick is 50 mm thick; *k* for each material is 0.93, 0.13 and 0.7 k cal/hr. m °C respectively.

Find the thickness of the red brick layer if common brick layer is to be absent and the heat transfer rate remains the same. Take that the fire clay thickness remains the same.

We know that

$$\frac{q}{A} = \frac{\Delta T}{\frac{x_A}{k_A} + \frac{x_B}{k_B} + \frac{x_C}{k_C}}$$

$$= \frac{\Delta T}{\frac{0.12}{0.93} + \frac{0.05}{0.13} + \frac{0.25}{0.7}}$$

When the common brick layer is absent we have

$$\frac{\Delta T}{\frac{0.12}{0.93} + \frac{x}{0.7}} = \frac{q}{A} = \frac{\Delta T}{\frac{0.12}{0.93} + \frac{0.05}{0.13} + \frac{0.25}{0.7}}$$

in which *x* is the thickness of the red brick layer (in metres)

Solving for *x*, we get

$$x = 0.5192 \text{ m}$$

$$= 519.2 \text{ mm}$$

Problem 5

The flat wall of a steel tank with area 5 m^2 is covered with two layers of insulating materials. Steel wall is 8 mm thick with k =46.5 k cal/hr. m °C. First insulation layer has a thickness of 50 mm and k for this given by $k = 0.144 + 0.00014\ t$. Where k is k cal/hr m °C and t in °C; second layer of insulation is 10 mm thick and its k is 0.698 k cal.hr. m °C. Temperatures of the inside wall is 250°C and that of the outer is 50°C. Find (a) the amount of heat transfer through the wall and (b) the boundary temperature for insulation layer.

Solution

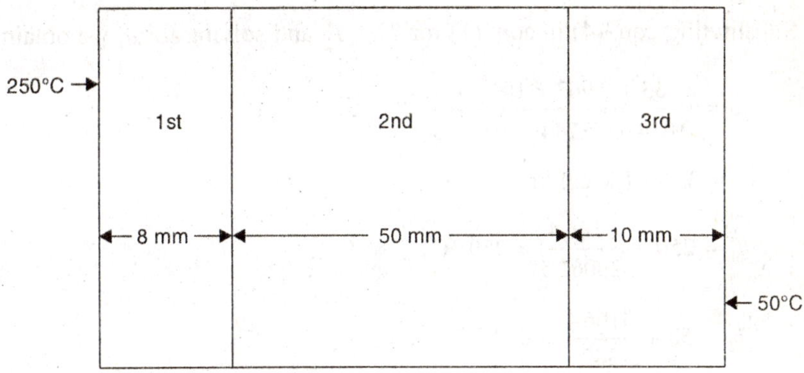

Fig. 2.30 *Heat transfer through a flat wall*

$$q = \frac{250 - 50}{\dfrac{0.008}{46.5 \times 5} + \dfrac{0.05}{k_{av} \times 5} + \dfrac{0.01}{0.698 \times 5}}$$

in which $k_{av} = 0.144 + 0.00014 \left(\dfrac{T_1 + T_2}{2}\right)$

$$= 0.144 + 0.7 \times 10^{-5}\ (T_1 + T_2)$$

$$q = \frac{200}{2.8997 \times 10^{-3} + \dfrac{0.01}{0.144 + 7 \times 10^{-5}\ (T_1 + T_2)}}$$

We have to solve for q; T_1 and T_2. To do this we formulate two more equations. For the first layer (T_1 is the interface temperature between layers (1 and 2)

$$q = \frac{(250 - T_1) \times 46.5 \times 5}{0.008} = 29060.5\ (250 - T_1) \qquad ...(2)$$

Similarly for the third layer

$$q = \frac{(T_2 - 50) \times 0.698 \times 5}{0.01} = 34.9\ (T_2 - 50) \qquad ...(3)$$

From equation (2) $250 - \dfrac{q}{29060.5} = T_1$

From eqn. (3) $50 + \dfrac{q}{349} = T_2$

Adding the above two, we obtain

$$T_1 + T_2 = 300 + 2.831 \times 10^{-3} q \qquad ...(4)$$

Substituting eqn. (4) in eqn. (1) for $T_1 + T_2$ and solving for q, we obtain

$$q = \frac{33 + 3.963 \times 10^{-5}\, q}{0.0141 + 5.744 \times 10^{-10}\, q}$$

$$= 3166.4 \text{ k cal/hr}$$

$$T_1 = 250 - \frac{3166.4}{29062.5} = 250\ °C$$

$$T_2 = 50 + \frac{3166.4}{349} = 59.1\ °C$$

Problem 6

The two faces of a slab at $x = 0$ and $x = L$ are kept at t_1 and t_2 °C. The k of the material is given as a temperature dependent value by $k = k_0\,(t^2 - t_0^2)$. Where t_0 and k_0 are constants. Deduce the expression for (i) heat flow per unit area and (b) k_{av} for this material.

We know $q/A = -k\, dt/dx$

$$\frac{q}{A}\,dx = -k\,dT = k_0\left(t^2 - t_0^2\right) dT$$

$$= k_0\,(t_0^2 - t^2)\,dt$$

Integrating with appropriate limits we get

$$\frac{q}{A}\int_o^L dx = k_0 \int_{t_1}^{t_2} \left(t_0^2 - t^2\right) dT$$

$$\frac{q\,L}{A} = k_0\left[t_0^2\,(t_2 - t_1) - \frac{1}{3}\left(t_2^3 - t_1^3\right)\right]$$

$$= k_0\,(t_2 - t_1)\left[\left(t_0^2 - \frac{1}{3}\left(t_2^2 + t_2\,t_1 + t_1^2\right)\right)\right]$$

$$\frac{q}{A} = \frac{k_{av}\,(t_2 - t_1)}{L}$$

where $k_{av} = k_0\left[t_0^2 - \frac{1}{3}\left(t_2^2 + t_2\, t_1 + t_1^2\right)\right]$

2.10 HEAT FLOW THROUGH A COMPOSITE CYLINDER WALL; CRITICAL THICKNESS OF INSULATION

We know that heat flow through concentric cylinder of different k's is met in practical situations in industry. An insulated pipe with a hot fluid flowing inside and exposed to colder medium on the outside is a typical example of this situation. If the pipe is relatively long, it can be safely assumed that the heat transfer occurs only radially. In the steady state , the rate of heat flow through each section is identical and is represented by

$$q = 2\pi_1\, l\, h_i\, (T_i - T_1)$$

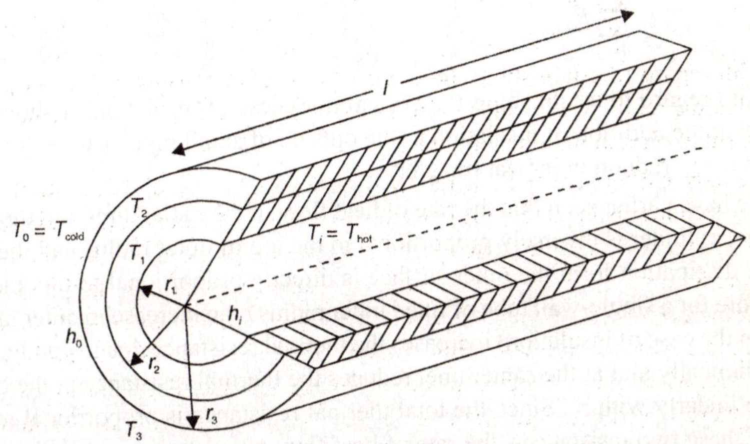

Fig. 2.31 *Heat flow through a composite cylinder wall: critical thickness of insulation*

$$= \frac{T_{hot} - T_1}{R_1} \qquad \text{...(1) (for the inner surface)}$$

$$q = \frac{2\pi\, k_1\, l}{ln\, \frac{r_2}{r_1}}(T_1 - T_2) = \frac{T_1 - T_2}{R_2} \qquad \text{...(2) for the inner cylinder}$$

$$q = \frac{2\pi\, k_2\, l}{ln\, \frac{r_3}{r_2}}(T_2 - T_3) = \frac{T_2 - T_3}{R_3} \qquad \text{...(3) for the outer cylinder}$$

$$q = 2\pi\, r_3\, h_0\, l\,(T_3 - T_0) = \frac{T_3 - T_{cold}}{R_4} \qquad \text{...(4) for the outer surface}$$

Since $q = \dfrac{\Delta t}{R_1 + R_2 + R_3 + R_4}$ ...(5) it turns out

$$q = \frac{T_i - T_o}{\dfrac{1}{2\pi\, r_1\, l\, h_i} + \dfrac{ln\,\dfrac{r_2}{r_1}}{2\pi\, k_1\, l} + \dfrac{ln\,\dfrac{r_3}{r_2}}{2\pi\, k_2\, l} + \dfrac{1}{2\pi\, r_3 l\, h_0}} \qquad \text{...(6)}$$

$$q = \frac{T_{hot} - T_{cold}}{\sum_{n=1}^{n=4} R_n} \qquad \text{...(7)}$$

With regard to establishing the critical thickness of insulation, it should be noted that the addition of insulation to the outside of small pipe or wire does not always entail reduction in heat transfer.

We have earlier seen that the rate of heat flow in the radial direction through a hollow cylinder is inversely proportional to the log of outer radius and the rate of heat dissipation form the outer surface is directly proportional to this radius. Therefore for a single-wall tube of fixed inner radius r_i an increase in outer radius r_0 (as in the case of insulation) increases the thermal resistance due to conduction logarithmically and at the same time, reduces the thermal resistance at the outer surface linearly with r_0. Since the total thermal resistance is proportional to the sum of these two resistances, the rate of heat flow may increase as insulation is added to a bare pipe or wire. If the insulation thickness is then increased further the heat loss gradually drops below the loss for a bare surface.

The relation between heat transfer and insulation thickness can be analysed quantitatively with the help of eqn. (6). In many practical situations the thermal resistance is concentrated in the insulation and at outer surface.

∴ Equation (6) is simplified—by taking that T_1 is the temperature at the inner surface of the insulation

Then $$q = \frac{2\pi\, k\, l\,(T_i - T_o)}{ln\left(\dfrac{r_0}{r_i}\right) + \dfrac{k}{h_0\, r_0}}$$

The first two terms in the denominator of eqn. (6) are ignored and the other two terms are divided by $2\pi\, k\, l$

where r_o= outer radius

r_i = inner radius

k = thermal conductivity of the insulation material.

For a fixed value of r_i, (the inner radius) the rate of flow of heat turns out to be a function of r_o i.e. $q = q\,(r_0)$ and will be a maximum at the value of r_0 for which $dq/dr_o = 0$

i.e. $$\frac{dq}{dr_o} = \frac{-2\pi\, k\, l\,(T_i - T_o)\left[\dfrac{1}{r_o} - \dfrac{k}{h_o\, r_o^{2}}\right]}{\left(ln\,\dfrac{r_o}{r_i} + \dfrac{k}{h_o\, r_o}\right)^2}$$

$$= 0 \qquad \ldots(9)$$

From equation (9) the radius for maximum heat transfer is termed the critical. radius i.e. $r_{oc} = k/h_o$.

Problem on the Effect of Temperature on '*k*'— in Spherical Coordinates

Liquified gases are sometimes stored in well-insulated spherical containers vented to the atmosphere. Develop an expression for the steady-state heat transfer rate through the walls of such a container; call the radii of the inner and outer wall r_0 and r_l. Let it be assumed that the temperature T_0 and T_l (at $r = r_0$ and $r = r_l$)are known. Assume that the k of the insulation material varies linearly with the temperature according to the relation

$$k = k_0 + (k_1 - k_0)\,\frac{T - T_0}{T_1 - T_0}$$

Solution

$$d/dr\,(q'\,r^2) = 0$$

$$q'\,r^2 = C_1$$

$$q' = C_1/r^2$$

$$-k\,dT/dr = C_1/r^2$$

$$-\int k\,dT = C_1 \int \frac{dr}{r^2}$$

$$-\int \left[k_0 + (k_1 - k_0)\,\frac{T - T_0}{T_1 - T_0}\right] dT = \frac{C_1}{r} + C_2$$

$$-\int \left[k_0 + \frac{k_1 - k_0}{T_1 - T_0}\,(T - T_0)\right] dT = \frac{C_1}{r} + C_2$$

$$-\int\left[k_0\,T+\left(\frac{k_1-k_0}{T_1-T_0}\right)\left(\frac{T^2}{2}-T\cdot T_0\right)\right]=\frac{C_1}{r}+C_2$$

at $r = r_0$, $T = T_0$

$$-\left[k_0\,T_0+\left(\frac{k_1-k_0}{T_1-T_0}\right)\left(\frac{T_0^{\,2}}{2}-T_0^{\,2}\right)\right]=\frac{C_1}{r}+C_2 \qquad \text{...(1)}$$

at $r = r_1$, $T = T_1$

$$-\left[k_0\,T_1+\left(\frac{k_1-k_0}{T_1-T_0}\right)\left(\frac{T_1^2}{2}-T_0\,T_1\right)\right]=\frac{C_1}{r}+C_2 \qquad \text{...(2)}$$

Subtracting (2) form (1)

$$-\left[k_0\,T_0+\left(\frac{k_1-k_0}{T_1-T_0}\right)\left(\frac{T_0^{\,2}}{2}-T_0^{\,2}\right)\right]+\left\{k_0\,T_1+\left[\frac{k_1-k_0}{T_1-T_0}\right]\left(\frac{T_1^2}{2}-T_0\,T_1\right)\right\}=C_1\left(\frac{1}{r_0}-\frac{1}{r_1}\right)$$

$$k_1\,T_1-k_0\,T_0+\left(\frac{k_1-k_0}{T_1-T_0}\right)\left[\frac{T_1^2}{2}-T_0\,T_1-\frac{T_0^2}{2}+T_0^{\,2}\right]=C_1\left(\frac{1}{r_0}-\frac{1}{r_1}\right)$$

$$k_1\,T_1-k_0\,T_0+\left(\frac{k_1-k_0}{T_1-T_0}\right)\left[\frac{T_1^2}{2}-T_0\,T_1+\frac{1}{2}\,T_0^{\,2}\right]=C_1\left(\frac{1}{r_0}-\frac{1}{r_1}\right)$$

$$\therefore C_1=\frac{k_0\,(T_1-T_0)+\left(\dfrac{k_1-k_0}{T_1-T_0}\right)\left(\dfrac{T_1^2}{2}-T_0\,T_1+\dfrac{T_0^{\,2}}{2}\right)}{\left(\dfrac{1}{r_0}-\dfrac{1}{r_1}\right)}$$

$$=\frac{k_0\,(T_1-T_0)+\left(\dfrac{k_1-k_0}{T_1-T_0}\right)\left(\dfrac{T_1-T_0}{2}\right)^2}{\dfrac{1}{r_0}-\dfrac{1}{r_1}}$$

$$=k_0\,(T_1-T_0)+\frac{\dfrac{(k_1-k_0)\,(T_1-T_0)}{2}}{\left(\dfrac{1}{r_0}-\dfrac{1}{r_1}\right)}$$

$$= (T_1 - T_0)\frac{\left[k_0 + \frac{k_1}{2} - \frac{k_0}{2}\right]^2}{\left(\frac{1}{r_0} - \frac{1}{r_1}\right)}$$

$$= \frac{(T_1 - T_0)\left(\frac{k_0 + k_1}{2}\right)}{\frac{1}{r_0} - \frac{1}{r_1}}$$

$$q' = c_1/r^2$$

$$q'_{outside} = c_1/r_1^2$$

$$C_1 = \frac{(T_1 - T_0)\left(\frac{k_0 + k_1}{2}\right)(r_0\, r_1)}{(r_1 - r_0)}$$

$$q = \frac{4\pi\, r_1^2\,(T_1 - T_0)\left(\frac{k_0 + k_1}{2}\right)(r_0\, r_1)}{(r_1 - r_0)\, r_1^2}$$

$$= 4\pi\, r_1\, r_0\left(\frac{k_0 + k_1}{2}\right)\left(\frac{T_1 - T_0}{r_1 - r_0}\right)$$

$$q_{outside} = 4\pi\, r_1\, r_0\left(\frac{k_0 + k_1}{2}\right)\left(\frac{T_1 - T_0}{r_1 - r_0}\right)$$

Heat conduction with internal generation of heat in the case of a rectangular slit

The *b.c*'s are (boundary conditions)

At $x = 0$, $T = T_1$

$x = 2L$, $T = T_2$

(a) Derive the expression for the temperature profile in this case of a rectangular slit where in heat is generated at uniform volumetric rate of G_q watts/m^3

(b) For k =200 w/m °k

G_q = 40 Mw/m3 (uniform heat generation)

T_1 = 160°C at $x = 0$

T_2 – 100°C at $x = 2L$

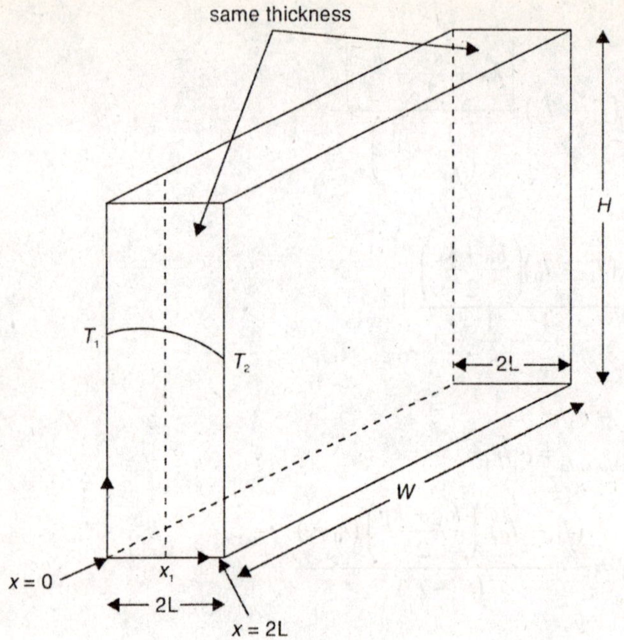

Fig. 2.32 *Heat conduction through a rectangular slit*

Slit thickness is to be taken as 2 cm. Determine

(a) T as a function of x i.e. $T(x)$

(b) q/A at the left face (flux)

(c) flux at the right face

(d) flux at the slit centre

Solution

$$\frac{d}{dx}(q) = G_q dV$$

$$V = 2\,L\,H\,W$$

x is a variable

$$dV = H\,W\,dx$$

$$A = W\,H$$

$$q = -\,k\,A\,dt/dx$$

$$= -\,k\,W\,H\,dT/dx$$

$$\frac{d}{dx}\left(-\,k\,W\,H\,\frac{dT}{dx}\right) = G_q\ dV = G_q\ HW\ dx$$

$$-k\,W\,H\,\frac{dT}{dx} = G_q\ HW\ x + C_1$$

$$-dT = \frac{G_q\ HW}{k\ W\ H}\ x\ dx + C_1'\ dx$$

$$-T = \frac{G_q}{k} \cdot \frac{x^2}{2} + C_1'\ x + C_2$$

$$= \frac{G_q\ x^2}{k \cdot 2} + C_1'\ x + C_2$$

B.c's

at $x = 0 \quad T = T_1; -T_1 = C_2$

at $x = 2L \quad T = T_2$

$$-T_2 = \frac{G_q}{2k} 4L^2 + C_1\ 2L - T_1$$

$$C_1 = \left[-T_2 + T_1 - \frac{2\ G_q\ L}{k}\right] \cdot \frac{1}{2L}$$

$$C_1 = \frac{T_1 - T_2}{2L} - \frac{G_q\ L}{k}$$

$$-T = \frac{G_q\ x^2}{2k} + \frac{(T_1 - T_2)}{2L}\ x - \frac{G_q . L}{k}\ x - T_1$$

$$-T = x\left\{\frac{T_1 - T_2}{2L} + \frac{G_q}{2k}\left(\frac{x}{2} - 2L\right)\right\} - T_1$$

$$T = T_1 + x\left[\frac{T_2 - T_1}{2L} + \frac{G_q}{2k}(2L - x)\right]$$

$$T = \left[\frac{100 - 160}{0.02} + \frac{(4 \times 10^7)(0.02 - x)}{2 \times 200}\right] x + 160 \quad G_q = 40\ \text{Mw/m}^3$$

$$T(x) = 160 - 10^3 x - x^2\ 10^5$$

(b) dT/dx at $x = 0$, has to be found out and substituted into Fouriers,s law

$$\left(\frac{dT}{dx}\right)_{\text{at } x = 0} = -10^3 - 2 \times 10^5 \times x \ \text{°K/m}$$

$$= -10^3 \ \text{°K/m}$$

$$\left.\frac{q}{A}\right|_{x=0} = -k \left.\frac{dT}{dx}\right|_{x=0}$$

$$= -200 \times -10^3 = 200 \text{ kw/m}^2$$

(c) $\frac{dT}{dx}$ at $x = 2L$ $\quad = -5\left(10^3\right)$

$$\left.\frac{q}{A}\right|_{x=2L} = -200 \times -5\left(10^3\right)$$

$$= 1 \text{ MW/ m}^2$$

(d) $\frac{dT}{dx}$ at $x = L$ $\quad = -10^3 - 2\left(10^5\right)(0.01)$

$$= -3 \times 10^3 \ {}^\circ\text{K/m}$$

$$\left.\frac{q}{A}\right|_{x=L} = -200 \times -3 \times 10^3$$

(slit centre)

$$= 600 \text{ kw/ m}^2$$

Illustration

Heat is generated within a sphere at 2.07×10^8 watts/ m^3. The sphere is 8 cm in diameter. The surface temperature is 370°K

(a) Calculate the temperature at the centre of the sphere

(b) Calculate the temperature at r =2 cm

k of the material is 47.7 w/m °K and is taken to be invariant.

$$A = 4\pi\, r^3 \qquad V = \frac{4}{3}\pi\, r^3$$

$$dV = \frac{4}{3}\, 3\pi\, r^2\, dr = 4\pi\, r^2\, dr$$

$$\frac{d}{dr}(q) = G_q\, dV$$

$$d\left(-k\, 4\pi\, r^2\, \frac{dT}{dr}\right) = 4\pi\, r^2\, G_q \cdot dr$$

$$-k\, 4\pi\, r^2\, \frac{dT}{dr} = \frac{4}{3}\pi\, r^3\, G_q + C_1$$

at $r = 0, \quad \dfrac{dT}{dr} = 0$

$$\therefore C_1 = 0$$

$$-k\, 4\pi\, r^2 \frac{dT}{dr} = \frac{4}{3}\pi\, r^3\, G_q$$

or
$$-k \frac{dT}{dr} = G_q \cdot \frac{r}{3}$$

Integrating $-k\,T = \dfrac{G_q\, r^2}{6} + C_2;$

B.C.

at $r = r_1, \quad T = T_1$

$$-k\, T_1 = \frac{G_q\, r_1^2}{6} + C_2$$

$$C_2 = -\frac{G_q\, r_1^2}{6} - k\, T_1$$

$$-k\, T = \frac{G_q\, r^2}{6} - \frac{G_q\, r_1^2}{6} - k\, T_1$$

$$-k\,(T_1 - T) = \frac{G_q}{6}\left(r^2 - r_1^2\right)$$

$$G_q = 2.07 \times 10^8 \text{ watts/m}^3$$

$$k = 47.7 \text{ w/m °k}; \; T_1 = 370\text{°k}$$

at $r = 0$,

$$47.7\,(370 - T) = \frac{2.07 \times 10^8}{6}\left(-r_1^2\right)$$

$$= -\frac{2.07 \times 10^8}{6}(0.04)^2$$

$$(370 - T) = -1157$$

$$T = 1527 \text{ °K}$$

Similarly at $r = 2$, the value of T can be determined.

Problem

An electrical resistance wire has a meeting point of 2500 °F. The electrical input to a wire 10 ft long and 1/4" diameter gives a uniform volumetric heat generation totalling 1400000 Btu/hr. The surface temperature is 1500 °F and k is 10 Btu/hr ft °F. Calculate the temperature at the centre of the wire. (Ans. 2612 °F)

A Summary of Results for One Dimensional Steady—State Heat Conduction

Configuration	*Heat flow Rate* q_x	*Thermal resistance* R_t	*Shape factor* S
Plane wall	$q_x = \frac{k\,A}{L}\Delta T$	$\frac{L}{k\,A}$	$\frac{A}{L}$
Hollow Cylinder	$q_r = \frac{2\pi\,k\,L}{ln\frac{r_o}{r_i}}\cdot\Delta T$	$\frac{ln\frac{r_o}{r_i}}{2\pi\,k\,L}$	$\frac{2\pi\,L}{ln\frac{r_o}{r_i}}$
Hollow Sphere	$q_r = \frac{4\pi\,k\cdot r_o r_i}{r_o - r_i}\Delta T$	$\frac{r_o - r_i}{4\pi\,k\,r_i\,r_o}$	$\frac{4\pi\,k\,r_i\,r_o}{r_o - r_i}$

Illustrations on conduction (Steady state)

1. Steam at 120 °C flows through a pipe which has an inner diameter of 10 cm and an outer diameter of 10.5 cm. The film coefficient between the steam and the inside pipe wall is 2800 w/m^2 °C. The outside of the 1% carbon steel pipe is exposed to atmospheric air at 26 °C with a surface coefficient of heat transfer of 9 w/m^2 °C. Steady state could be maintained and there is no internal generation of heat. Find the heat transfer rate per meter from the steam to the air across the pipe. k =43 w/m °C

Solution

We know

$$q = \frac{T_i - T_o}{\frac{1}{2\pi\,h_1\,r_1} + \frac{ln\left(\frac{r_2}{r_1}\right)}{2\pi k} + \frac{1}{2\pi\,h_0\,r_2}}$$

$$= \frac{2\pi\,(120 - 26)}{\frac{1}{2800\,(0.05)} + \frac{ln\frac{5.25}{5}}{43} + \frac{1}{9\,(0.0525)}}$$

$$= \frac{2\pi\,(94)}{0.0071 + 0.00113 + 2.12}$$

$$= 278 \text{ w/m}$$

Illustration

The wall of a guest house is constructed as indicated in the following figure, with two different materials forming the inside wall to achieve a wainscotting effect. The temperature of the air in the room is 70 °F and the heat transfer coefficient between the room air and the wall is 1.1. Btu/hr °F ft^2. The outside temperature is 20 °F with an outside surface coefficient of heat transfer of 3 Btu/hr ft^2 F,. The wall is 7 ft high and 13 feet deep. The lower material extends three feet from the floor. The thickness and thermal conductivities are Δx_a = 1/2″ k_a = 0.09 Δx_b = 1/2″ k_b = 0.12, Δx_c = 2″ k_c = 0.023 Δx_d = 1″ and k_d = 0.10″. Compute the heat transfer rate across the wall under the steady state conditions.

Solution

This is a case of combined conduction and convection. The situation can be represented by the following sketch

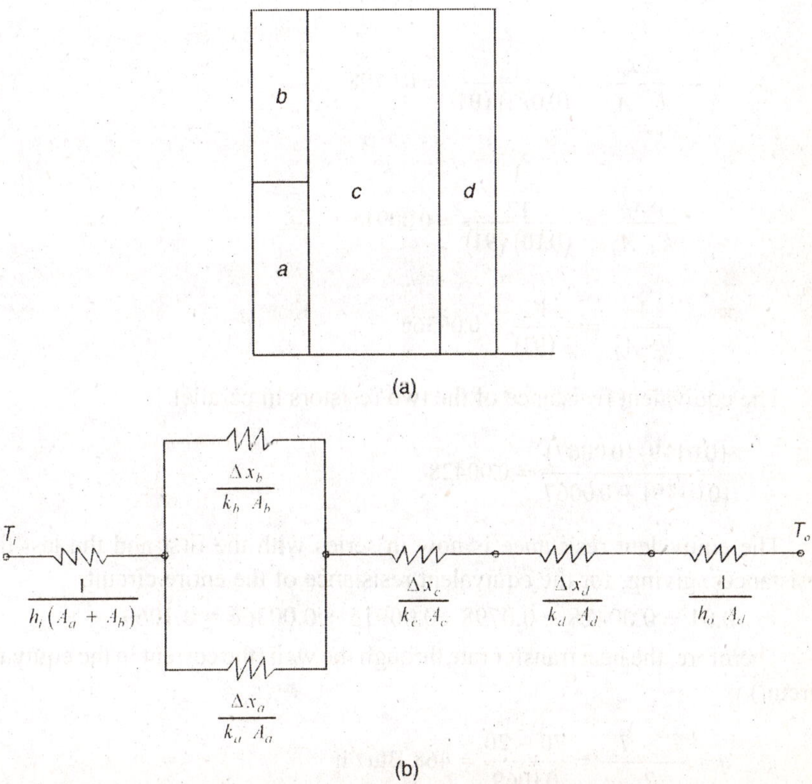

Fig. 2.33 *Heat transfer by combined conduction and convection*

Since

$$A_a = 3\ (13) = 39\ \text{ft}^2$$

$$A_b = 4\ (13) = 52\ \text{ft}^2$$

and $A_d = 39 + 52 = 91\ \text{ft}^2$, the resistance are

$$\frac{1}{h_i\ (A_a + A_b)} = \frac{1}{(1.1)\ (91)} = \frac{1}{100} = 0.01,$$

$$\frac{\Delta x_a}{k_a\ A_a} = \frac{\frac{1}{2}}{\frac{12}{(0.09)\ (39)}} = 0.0119$$

$$\frac{\Delta x_b}{k_b\ A_b} = \frac{\frac{1}{2}}{\frac{12}{(0.12)\ (52)}} = 0.0067$$

$$\frac{\Delta x_c}{k_c\ A_c} = \frac{\frac{2}{12}}{(0.023)\ (91)} = 0.0798$$

$$\frac{\Delta x_d}{k_d\ A_d} = \frac{\frac{1}{12}}{(0.10)\ (91)} = 0.00915$$

$$\frac{1}{h_0\ A_d} = \frac{1}{3\ (91)} = 0.00366$$

The equivalent resistance of the two resistors in parallel

$$\frac{(0.0119)\ (0.0067)}{(0.0119) + 0.0067} = 0.00428$$

The equivalent resistance is now in series with the first and the last three resistances, giving, for the equivalent resistance of the entire circuit.

$$0.01 + 0.00428 + 0.0798 + 0.00915 + 0.00366 = 0.1069$$

Therefore, the heat transfer rate through the wall (the current in the equivalent circuit) is

$$q = \frac{T_1 - T_\infty}{R_{eq}} = \frac{70 - 20}{0.1069} = 468\ \text{Btu/h}$$

Illustration

A steel tube $k = 25$ Btu/hr °F ft of 2″ *id* and 3″ outer diameter is covered with a 1″ layer asbestos insulation of $k = 0.12$ Btu/hr °F ft. The inside surface of the tube gets heat by convection from a hot gas at $t_a = 315.6$ °C that is 600 °F with a heat transfer coefficient of $h_a = 50$ Btu/hr ft² °F. While the outer surface of the insulation is exposed to the ambient air at $t_b = 100$ °F with a heat transfer coefficient $h_b = 3$ Btu/hr ft² °F. Determine for the situation (a) heat loss the ambient air per 10 ft length (b) the temperature drop across the tube material and the insulation layer. Radiation effects are negligible for this situation.

Solution

(a) The radial heat transfer through the tube is given by

$$q = \frac{(t_a - t_b)}{R_a + R_1 + R_2 + R_b}$$

$$R_a = \frac{1}{\pi \frac{2}{12} \times 10 \times 50} = 0.382 \times 10^{-3} \text{ hr °F/Btu}$$

$$R_1 = \frac{1}{2\pi \times 10 \times 25} \ln \frac{3}{2} = 0.258 \times 10^{-3} \text{ hr °F/Btu}$$

(using log mean radius or diameter)

$$R_2 = \frac{1}{2\pi \times 10 \times 0.12} \ln \frac{5}{3} = 0.678 \times 10^{-1} \text{ hr °F/Btu}$$

and
$$R_b = \frac{1}{\frac{\pi \times 5}{2} \times 10 \times 3} = 0.255 \times 10^{-1} \text{ hr °F/Btu}$$

The total resistance R is given by the sum of the individual resistances

$$R = 0.973 \times 10^{-1} \text{ hr °F/Btu}$$

∴ The heat loss from the tube becomes $q = 500/R = 5140$ Btu/hr.

(b) The temperature drop across the tube material and the asbestos layer are determined as

$$\Delta t_{\text{tube}} = R_1/R\,(t_a - t_b)$$

$$= \frac{R_1}{R} \times 500 = 1.33 \text{ °F}$$

$$\Delta s_{\text{asbestos}} = \frac{R_2}{R} \times 500 = 348 \text{ °F}$$

We observe that the temperature drop across the tube material << that across the insulation layer.

2. Determine the amount of heat loss/m^2 of a boiler brick-setting of 250 mm thickness if the gas temperature is 600 °C and room air temperature is 30°C. The heat transfer coefficient on the hot side is 20 and that on the cold side is 8 k cal /hr m^2 C. The k of the wall material is 0.7 kcal/hr m °C. Determine also the wall surface temperatures. Take the heat transfer areas to be the same.

Solution

$$q \text{ (per unit area)} = \frac{\Delta t}{\frac{1}{h_1} + \frac{L}{k} + \frac{1}{h_2}}$$

$$q = \frac{570}{\frac{1}{20} + \frac{0.25}{0.7} \times \frac{1}{8}}$$

$$= 1072 \text{ k cal/hr}$$

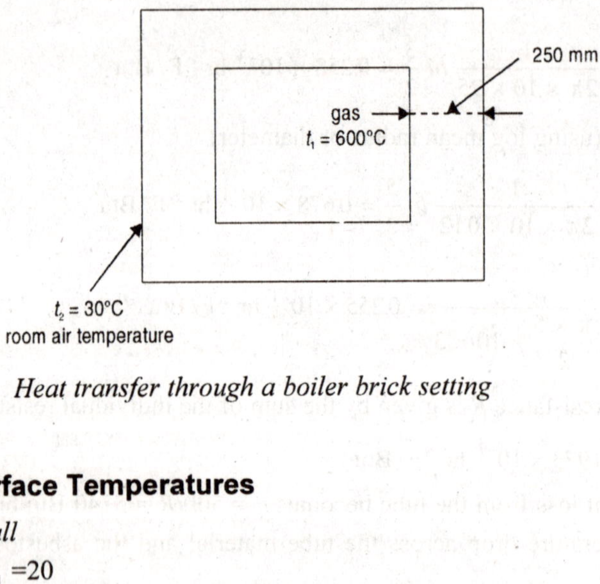

Fig. 2.34 *Heat transfer through a boiler brick setting*

Wall Surface Temperatures

Insider wall

$h_1 = 20$

$$q = \frac{\Delta t_1}{R_1} = \frac{\Delta t_1}{\frac{1}{h_1}} = \frac{\Delta t}{0.05}$$

$$\Rightarrow \Delta t_1 = 1072 \times 0.05 = 53.6 \text{ °C}$$

$$\Rightarrow T_{\text{inside wall}} = 546.4 \text{ °C}$$

Outside wall

$$\text{Similarly } 1072 = \frac{\Delta t_2}{\frac{L}{k}} = \frac{\Delta t_2}{0.357}$$

$$\Delta t_2 = 382.7$$

$$\text{or } T_{\text{outside wall}} = 163.7\ °\text{C}$$

Critical Radius of Insulation

There are many real situations in which the addition of insulation to the outside surface of a pipe does not always reduce the heat loss. To establish the conditions under which this occurs, we consider an insulation layer (i.e. the layer of insulation material) in the form of a hollow cylinder of length L. The inside surface of the cylinder at $r = r_i$ is kept at a uniform temperature T_i while the outside surface at $r = r_o$ dissipates heat by the phenomenon of convection into the environment at temperature $T\infty$ with a heat transfer coefficient h_0.

The rate of heat transfer 'q' through this insulation layer is

$$q = \frac{T_i - T_\infty}{R_{ins} + R_0} \quad \text{...(1)}$$

where the thermal resistances R_{ins} and R_0 of the insulation and of the outside surface are defined as $R_{ins} = \frac{1}{2\pi L k} ln \frac{r_o}{r_i}$ and $R = \frac{1}{2\pi r_o L h_0}$ ...(2)

in which k is the thermal conductivity of the insulating material.

If is assumed that T_i, T_∞, k, L, h_0 and r_i remain constant while r_0 varies, we observe that as r_o increases, the term R_o decreases but R_{ins} increases. Therefore it is expected that q may have a maximum magnitude for a certain value of $r_o = r_{oc}$. This critical value of the radius r_{oc} is found by differentiating equation (1) w.r.t r_o and setting the resulting expression equal to zero.

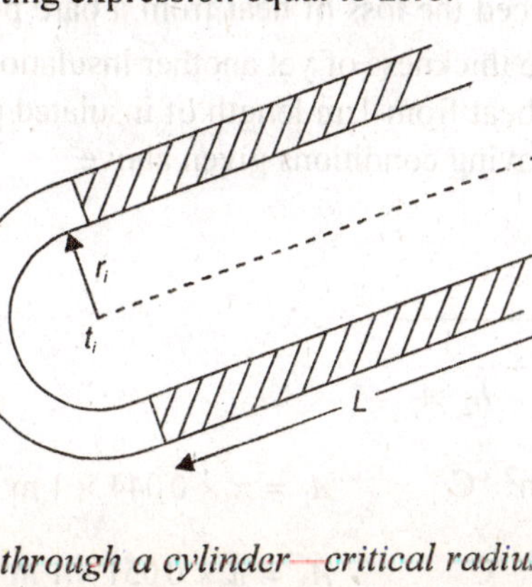

Fig. 2.35 *Heat flow through a cylinder—critical radius of insulation*

$$\frac{dq}{dr_o} = -\frac{2\pi k L (T_1 - T_\infty)}{\left(ln\left(\frac{r_o}{r_i}\right) + \frac{k}{h_0 r_0}\right)^2}\left(\frac{1}{r_0} - \frac{k}{h_0 r_o^2}\right) \quad ...(3)$$

$$= 0$$

The solution of eqn. (3) for r_0, gives the critical radius r_{oc} of insulation at which the heat transfer rate is a maximum.

we observe that $r_{oc} = \frac{k}{h_0}$ ...(4)

The physical significance of the above result in practice is as follows. If insulation is to be added on a pipe or wire whose outer radius is less than the critical radius of the insulation and the outside surface is kept at a uniform temperature, the heat loss from the pipe will increase as the insulation is added until the outside radius of the insulation becomes equal to the critical radius, r_{oc}, as the thickness of the insulation is increased beyond r_{oc}, the heat loss from the pipe will start decreasing.

Problem on the Determination of the Effect of Insulation Thickness on Heat Transfer, Critical Radius of Insulation

1. An oil pipe line with inner diameter of 44 mm and outer diameter of 51 mm is covered with a concrete layer of 80 mm thickness. k for the pipe and concrete are 50 and 1.28 w/m °C respectively. Mean temperature of oil in the pipe is 120 °C and the ambient air is at 20 °C. h of the oil is 100 w/m² °C and that of air is 10 w/m² °C.

(a) Find the rate of heat loss per metre length of pipe when it is (i) bare and (ii) insulated with concrete.

(b) What should be the magnitude of k of some insulation material to ensure that the loss of heat from the pipe line covered with this insulation of the same thickness (80 mm) does not exceed the loss of heat from a bare pipe'

(c) What should be the thickness of yet another insulation whose $k = 0.008$ w/m °C so that the loss of heat from 1 m length of insulated pipe, would be 1/3rd of that of the bare pipe having conditions given above.

Solution

(a) (i) $\dfrac{\Delta T}{\dfrac{1}{h_1 A_1} + \dfrac{x}{k A_m} + \dfrac{1}{h_2 A_2}}$

$h_1 = 100$ w/m² °C $\qquad A_1 = \pi \times 0.044 \times 1 \text{ m}^2$

$h_2 = 10$ w/m² °C $\qquad A_2 = \pi \times 0.051 \times 1 \text{ m}^2$

$$x = \frac{0.051 - 0.044}{2} = 0.0035 \text{ m}$$

k of the metal = 50 w/m °C

$$A_m = \pi\, d_m \times 1; \text{ in this case since } \frac{d_2}{d_1} \text{ is} < 1.5 \qquad \left(\frac{0.051}{0.044}\right)$$

arithmetic average diameter can be used

$$\Rightarrow d_m = 0.0475 \text{ m}$$

$$\therefore q = \frac{120 - 20}{\dfrac{1}{100 \times \pi \times 0.044 \times 1} + \dfrac{0.0035}{50 \times \pi \times 0.0475 \times 1} + \dfrac{1}{10 \times \pi \times 0.051 \times 1}}$$

$$= 143.5 \text{ w/m length}$$

(a) (ii)Insulated with a concrete layer

$$q = \frac{\Delta T}{\dfrac{1}{h_1\, A_1} + \left(\dfrac{L_m}{k_m\, A_m}\right)_{metal} + \left(\dfrac{L}{k\, A_m}\right)_{ins} + \dfrac{1}{h_2\, A_2}}$$

$\left(\dfrac{1}{h_1\, A_1}\right)\left(\dfrac{L_m}{k_m\, A_m}\right)$ both will be the same as in the previous section.

For the concrete layer

$$\left(\frac{d_2}{d_1}\right)_{concrete} = \frac{211}{51} > 1.5$$

$$d_{\substack{concrete\\insulation}} = \frac{0.0211 - 0.051}{ln\, \dfrac{211}{51}} = \frac{0.160}{ln\left(\dfrac{211}{51}\right)}$$

$$\Rightarrow A_{concrete} = \pi \times \frac{0.211 - 0.051}{ln\, \dfrac{211}{51}} \times 1$$

Now the outer diameter is 0.211 m

(i.e. for the last term in the denominator)

$$\Rightarrow q = \frac{100}{\dfrac{1}{100 \times \pi \times 0.044 \times 1} + \dfrac{0.0035}{50 \times \pi \times 0.0475 \times 1} + \dfrac{0.08\, ln\, \dfrac{211}{51}}{1.28 \times \pi \times 0.16 \times 1} + \dfrac{1}{10 \times \pi \times 0.211 \times 1}}$$

$$= 250 \text{ w/m length}$$

Replacing the concrete layer with another insulation material (whose k is to be determined) such that

$$(\text{heat loss})_{bare} = (\text{heat loss})_{insulation}$$

thickness of the insulation layer = 80 mm

$$143.5 = \frac{100}{\text{I term} + \text{II term} + \dfrac{0.08\, ln \dfrac{211}{51}}{k \times \pi \times 0.16 \times 1} + \text{iv term}}$$

↑ to be found

Solving for k, we get

$$k_{insu.} = 0.478 \text{ w/m °C}$$

(c) To determine the thickness of get another insulation whose $k = 0.08$ w/m °C

Let this thickness to 't'

Again the first two terms in the denominator will be the same. The other two can be expressed in terms of this 't'

q is now one-third (1/3) of the quantity corresponding to the bare condition

$$\frac{143.5}{3} = \frac{120 - 20}{\text{I} + \text{II} + \dfrac{t}{(0.08) \times \pi \dfrac{2t}{\log\left(\dfrac{0.051 + 2t}{0.051}\right)}} + \dfrac{1}{10 \times \pi \times (0.05 + 2t) \times 1}}$$

or

$$\frac{1}{10\pi\,(0.051 + 2t)} + \frac{ln\left(\dfrac{0.051 + 2t}{0.051}\right)}{0.16\pi} = 2.018$$

't' has to estimated by an iterative procedure

$t = 0.0363$ m or 36.3 mm in this set up and the outer diameter now becomes

$= 0.051 + 2 \times 0.0363$

$= 123.6$ mm

Problem

A pipe 20 mm inner diameter and 30 mm outer diameter carries a fluid at 448 °K. The ambient temperature is 293 °K. 'h' on the inner and outer sides are 22 and 5.0 w/m^2 °K respectively. Calculate the heat loss in the pipe in w/m neglecting the pipe resistance. What will be the heat loss if a layer of insulation of 50 mm thickness whose $k = 0.08$ w/m °K is applied to the pipe?

2.11 HEAT CONDUCTION WITH A VISCOUS HEAT SOURCE

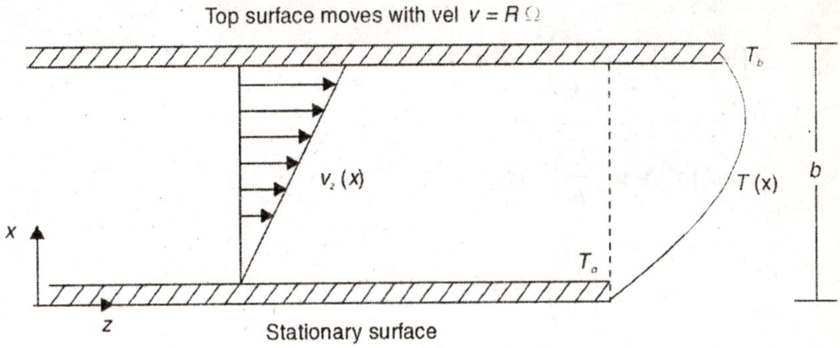

Fig. 2.36 *Conduction of heat with a viscous heat source*

b—width of the slit

L—length

The rubbing together of adjacent layers of fluid produces heat; that is mechanical energy is steadily degraded into thermal energy. The volumetric heat source resulting from this viscous dissipation is designated by S_v. Its magnitude will depend on the local velocity gradient, the more rapidly two adjacent layers move w.r.t. one another, the greater will be the viscous dissipation heating.

T is a function of *x*

Let us take the viscous heat source to be given by

$$S_v = -T_{xz}\left(\frac{dv_z}{dx}\right) = \mu\left(\frac{dv_z}{dx}\right)^2$$

For the steady laminar flow of a fluid with constant viscosity in a slit as shown in the sketch, the velocity profile is linear

$$V_z = \left(\frac{x}{b}\right) v$$

∴ The rate of viscous heat production per unit volume is

$$S_v = \mu\left(\frac{v}{b}\right)^2$$

This is what is to be substituted for our G_q—volumetric heat generation rate. We have now to make the thermal energy balance

$$d\left[\frac{q}{A}\cdot A\right] = G_q\; dv$$

Transfer area = W.L.

Vol. element for heat generation

$$dv = W\,L\,dx$$

$$d\,[W\,L\,q'] = \mu\left(\frac{V}{b}\right)^2 dx\,W\,L$$

$$d\,[q'] = \mu\left(\frac{V}{b}\right)^2 dx$$

$$q' = \mu\left(\frac{V}{b}\right)^2 x + C_1$$

We know nothing about the heat flux at any value of x and hence we can not evaluate C_1 at the stage

Inserting Fourier's law of heat conduction $-k\dfrac{dT}{dx} = \mu\left(\dfrac{V}{b}\right)^2 x + C_1$;

k is the thermal conductivity of the fluid.

Integrating the above equation (w.r.t x) for constant k, we get

$$T = -\left(\frac{\mu}{k}\right)\left(\frac{v}{b}\right)^2 \frac{x^2}{2} - \frac{C_1\,x}{k} + C_2$$

C_1 and C_2 are now to be evaluated from the boundary conditions.

at $\quad x = 0, \quad T = T_0$

$\quad\; x = b \quad T = T_b$

When the constants are thus determined and substituted back into the above equation for T.

We obtain

$$\frac{T - T_o}{T_b - T_o} = \left(\frac{x}{b}\right) + \frac{1}{2}\,Br\left(\frac{x}{b}\right)\left[1 - \left(\frac{x}{b}\right)\right]$$

where $\quad Br = \dfrac{\mu\,v^2}{k\,(T_b - T_o)}$

Br is the Brinkman number which is a measure of the extent to which the viscous heating is important relative to the heat flow resulting form the impressed temperature difference $(T_b - T_0)$

Examples of Viscous Heat Dissipation

The situation is important in engineering work where large velocity changes occur over short distance. Examples of such situations include

(a) flow of lubricant between fast moving parts
(b) flow of air in the boundary layer near an earth satellite or rocket

$$1 \quad \text{Btu} \Rightarrow \quad 1.05 \times 10^3 \ \text{kg}\frac{\text{m}^2}{\text{sec}^2}$$

$$S_v = T_{xz}\left(\frac{dv_z}{dx}\right)$$

T_{xz} = momentum flux
= Newtons/m^2
= kg/m. sec^2

$$\frac{dv_z}{dx} = \frac{\text{m}}{\text{sec}\cdot\text{m}} = \frac{1}{\text{sec}}$$

$$S_v = \frac{\text{kg}}{\text{m}\cdot\text{sec}^2}\cdot\frac{1}{\text{sec}} = \frac{\text{kg}}{\text{m}\cdot\text{sec}^3}$$

Volumetric generation rate of heat = Btu/hr. ft^3

Problems on steady state conduction of heat

1. A pipe standard schedule 40, two-inch steel pipe (inner diameter. 2.064″ and wall thickness 0.154″) carrying steam is lagged with 2″ of 85% magnesia covered in turn with 2″ of cork. Estimate the heat loss/hr./ft of pipe if the inner surface of the pipe at 250 °F and the outer surface of the cork is at 90 °F. The '*k*' of the substances concerned are

Steel	26.1 Btu/hr ft °F
85% magnesia	0.09 do.
cork	0.03 do.

Answer 24 Btu/hr ft

2. Heat is flowing through an annular cylinder wall of inside radius r_0 and outside radius r_1. The '*k*' of the metal varies linearly with temperature from k_0 at T_0 to k_1 at T_1. Develop an expression for the heat flow through the wall at $r = r_0$.

Answer $q_0 = 2\pi L\,(T_0 - T_1)\left(\dfrac{k_0 + k_1}{2}\right)\left(ln\,\dfrac{r_1}{r_0}\right)^{-1}$

2.12 THE TWO-DIMENSIONAL HEAT CONDUCTION PATH

The classic, two-dimensional heat conduction path element is that of the flow of heat through the wall of a thick walled pipe or a thick layer of insulation around the pipe.

In the section covered below the two-dimensional heat conduction path will be presented and discussed, followed by discussion on parallel and series two-dimensional path system.

The classic two-dimensional conduction path element is diagrammatically represented in the following figure

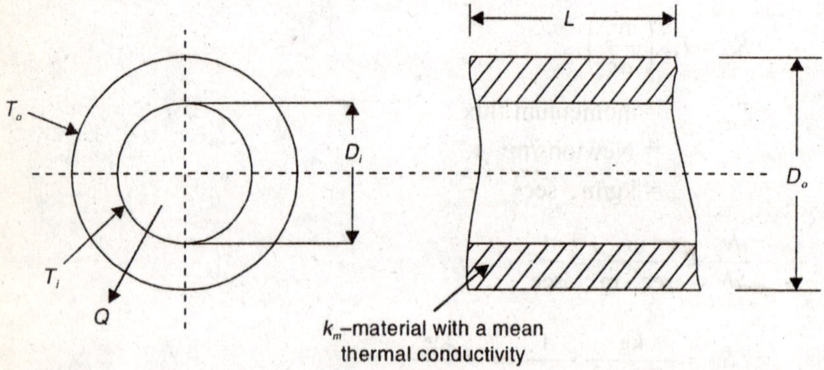

Fig 2.37 *Two-dimensional conduction path element*

The two equations that are commonly applied to the two-dimensional heat conduction path element are the (1) thick wall pipe equation and (2) thin wall pipe equation

The Thick Wall Pipe Equation

The thick wall pipe equation is an exact equation. It has been developed by integrating, in a classical manner, the thermal resistance of the conduction path between the inside and outside diameters of the pipe wall. The thick wall pipe equation is presented below. Because this equation is used most often with pipe, the heat flow parameter of greatest usefulness is the heat flow per unit length of pipe. Therefore the equation is written for Q/L.Note that Q/L is not influenced by the size of the pipe. Only the ratio of diameters (D_o/D_i) is involved

$$\frac{Q}{L} = \frac{2\pi \, k_m \, (T_i - T_o)}{ln\left(\frac{D_o}{D_i}\right)} \quad \text{w/m} \qquad ...(1)$$

Consistency must be adhered to as far as the usage of unit for the above quantities.

Q/L = heat flow/per unit length w/m

k_m = mean thermal conductivity w/m °C

T_i = temperature of inside wall, °C

T_o = temperature of outside wall, °C

D_o = outside diameter, m

D_i = inside diameter, m

$ln\,(D_o/D_i)$ = natural logarithm of the diameter ratio

When eqn. (1) is written in the form

$$\frac{Q}{L} = K_t\,\Delta T \qquad \text{...(2)}$$

where K_t is the thermal conductance per unit length of pipe

$$K_t = \frac{2\pi\,k_m}{ln\left(\frac{D_o}{D_i}\right)} \qquad \text{...(3)}$$

When eqn. (1) is written in the form

$$\frac{Q}{L} = \frac{1}{R_t}(\Delta T), \qquad \text{...(4)}$$

the thermal resistance per unit length of pipe is

$$R_t = \frac{ln\left(\frac{D_o}{D_i}\right)}{2\,\pi\,k_m} \qquad \text{...(5)}$$

Equation (3) and (5) indicate $\frac{m°C}{w}$ that the thermal conductance through the wall of a pipe is a function of the ratio of the outside to the inside diameters only and therefore is independent of the order of the magnitude of the pipe diameter.

The Thin Wall Pipe Equation

When natural logarithms are not conveniently available, the thin wall pipe equation can be used. This—equation is only an approximation which yields an erroneously higher thermal conductance. The thin wall pipe equation treats the pipe wall as a linear conduction path with constant heat flow path area through out the conduction path length. The heat flow path area is the average of the inside and outside circumference of the pipe. The heat flow path length is equal to the thickness of the pipe wall. Referring to the figure presented above, the thin wall pipe equation is $\frac{Q}{L} = \frac{\pi\,(D_o + D_i)}{D_o - D_i}\,k_m\,(T_i - T_o)$ w/m

The thermal conductance K being

$$K = \frac{\pi\,(D_o + D_i)}{D_o - D_i}\,k_m \text{ w/m °C}$$

Here again the units should be consistent.

When the thin wall pipe equation is employed the resulting heat flow rate Q/L or the thermal conductance K should be corrected by dividing the values obtained with the thin wall pipe equation by the error ratio appropriate for this situation.

However when insulation is applied to the pipe, the insulation diameter ratio usually becomes large enough with the result the error incurred by the use of the thin walled pipe equation is significant. If the thin walled pipe equation has been used for such configurations a suitable correction for the error should be incorporated.

Illustration

How much heat will be lost per hour from a 6 in standard pipe covered with 10—cm radial thickness of 85% magnesia when the outside temperate of the pipe is 450 °C and the temperature of the outside surface of the insulation is 50 °C' k_m = 0.0847 w/m °C at the average temperature of 250 °C.

Solution

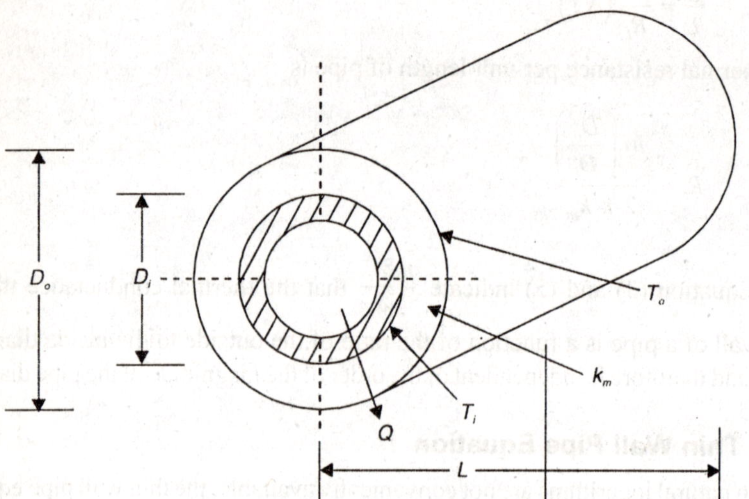

Fig. 2.38 *Heat loss from a pipe*

The thick wall pipe equation will be employed $\dfrac{Q}{L} = \dfrac{2\pi\, k_m\,(T_i - T_o)}{ln\,\dfrac{D_o}{D_i}}$ w/m

$k_m = 0.0847$ w/m °C

$T_i = 450$ °C

$T_o = 50$ °C

$D_i = 6.625 \times 2.54 = 16.83$ cm

$D_o = 16.83 + 2\,(10.0) = 36.83$ cm

$D_0/D_i = 36.83/16.83 = 2.188$

$\ln D_0/D_i = 0.783$

These quantities can be substituted in the above expression

$$\frac{Q}{L} = \frac{2\pi\,(0.0847)\,(450 - 50)}{0.7830}\ \text{w/m}$$

$$= 270\ \text{w/m}$$

Illustration

Solve the above illustration for the thick wall pipe equation using the thin wall pipe equation. The error factor is 1.048.

Solution

The thin wall pipe equation is

$$\frac{Q}{L} = \frac{\pi\,(D_o + D_i)\,k_m\,(T_i - T_o)}{(D_o - D_i)}$$

The quantities given are

$k_m = 0.0847$ w/m °C (at 250 °C average temperature)

$T_i = 450$ °C

$T_o = 50$ °C

$D_i = 6.625 \times 2.54 = 16.83$ cm

$D_o = 16.83 + 2\,(10.0) = 36.83$

$$\frac{Q}{L} = \frac{\pi\,(36.83 + 16.83)\,(0.0847)\,(450 - 50)}{(36.83 - 16.83)}$$

$$= \frac{\pi\,(53.66)\,(0.0847)\,(400)}{20.0}$$

$$= 285\ \text{w/m}$$

Thin wall equation error correction is done as follows

$$\left[\frac{D_o}{D_i} = \frac{36.83}{16.83} = 2.188\right]$$

The error factor, given in the problem statement is 1.048

Therefore the corrected Q/L = (285)/1.048 = 271 w/m

Two Dimensional Heat Flow

All heat conduction problems cannot be solved by analytical technique. The other method of arriving at the solution are

1. Graphical method (Flux plot)

2. Numerical solution (also called the relaxation method).

Two dimensional steady state heat flow problems can be solved by the aforementioned methods.

Graphical Method

This method is relatively inaccurate. Employing Fourier field equation the temperature map or the temperature field can be predicted. This map contains lines of constant temperature (isotherms) and a set of flow lines (adiabatics). Which are intersecting the isotherm at right angles. The net containing the two sets of lines have the following properties

(i) The isotherms will not intersect each other within the body

(ii) The flow lines will not intersect each other within the body

(iii) The isotherms and flow lines intersect at right angles.

(iv) The isotherms and flow lines meet the system boundaries at right angles to these boundaries

The figure given below illustrates the method of approach for flux plot. Along the boundaries, mark the points of equal temperature interval

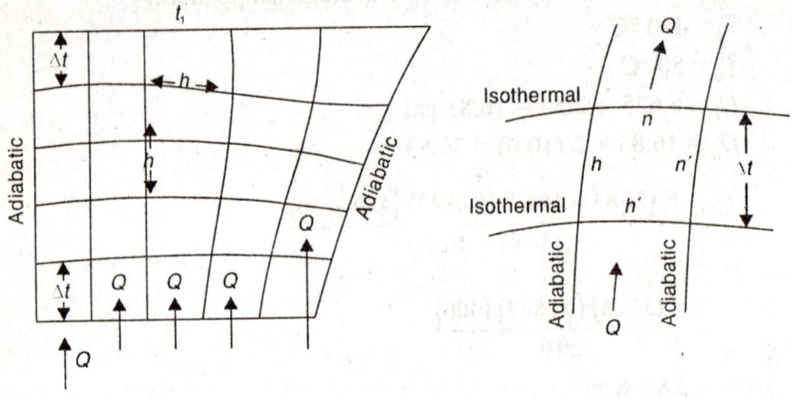

Fig. 2.39 *Two dimensional steady state heat conduction*

Connect these equal temperature points to get the isotherms. Construct the quadrilaterals formed by adjacent flow lines and isotherms, having equal length curved sides with right-angle corners.Consider one such little square drawn as above. If '*n*' is the side of the square, the temperature gradient across this square is $\frac{\Delta t}{n}$

The heat transfer rate equation is

$$Q = k \frac{1 - n}{n} \Delta t = k \Delta t$$

If there are M intervals between the hot and cold faces, then

$$Q = Q_1 + Q_2 + Q_3 + \ldots + Q_n$$

and each equal temperature interval = $\dfrac{t_1 - t_2}{M}$

where t_1 and t_2 are the surface temperatures of the hot and cold faces respectively If there are 'N' flow tubes

$$Q_{total} = N\,k\,\Delta t = \frac{N\,k\,(t_1 - t_2)}{M}$$

Therefore the shape factor $S = \dfrac{N}{M}$

This method of approach can be adopted for conduction through sections of complicated shapes.

Numerical Solution (Relaxation Method)

This method is best suited for two or three dimensional steady state heat flow situations. Consider a body and transform it to a square grid net-work and estimate the temperature at each node of the grid. In steady state no heat can be stored or released in any node assuming that the heat is conducted in the fictitious rods which form the grid. In the relaxation method, the temperature difference at different steps can be corrected and the above condition is obtained.

Consider the figure given below. Heat is assumed to be transmitted parallel to the X axis in the homogeneous body. Divide the body into a square grid net-work. Imagine heat sink at each node of the lattice. In the figure presented below, point 2 represents a heat sink

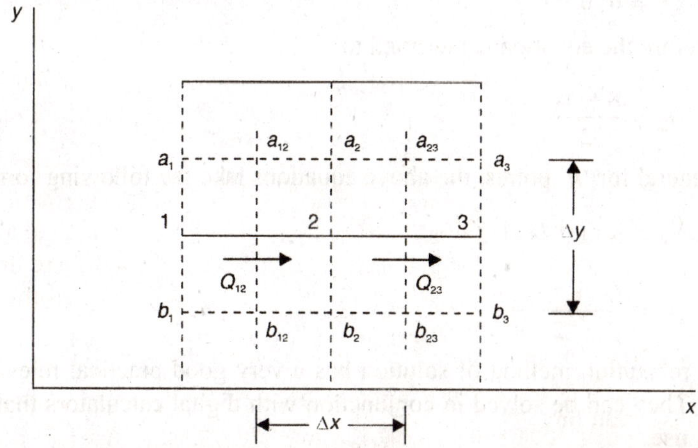

Fig. 2.40 *Numerical method*

The lines $a_1\, b_1$, $a_2\, b_2$, $a_3\, b_3$ etc are at temperature t_1, $t_2\, t_3$ respectively

$$Q_{12} = \frac{k\, z\, \Delta y}{\Delta x}(t_1 - t_2)$$

in which '*z*' represents the depth of the section

The heat flow from sink z is

$$Q_{23} = \frac{kz\, \Delta y}{\Delta x}.(t_2 - t_3)$$

Therefore the heat stored in the sink is

$$Q_2 = Q_{12} - Q_{23}$$

$$Q_2 = \left\{\frac{kz\, \Delta y}{\Delta x}(t_1 - t_2)\right\} - \left\{\frac{kz\, \Delta y}{\Delta x}(t_2 - t_3)\right\}$$

In case of a square grid, $\Delta x = \Delta y$

$$Q_2 = kz\,(t_1 + t_3 - 2t_2)$$

$$\frac{Q_2}{kz} = (t_1 + t_3 - 2t_2)$$

Putting $\frac{Q_2}{kz}$ as θ_2

$$\therefore \theta_2 = t_1 + t_3 - 2t_2$$

For steady state condition according to the earlier discussion, we have

$$Q_2 = 0; \theta = 0$$

Therefore the equation is modified to

$$t_2 = \frac{t_1 + t_3}{2}$$

In general for '*n*' points, the above equations take the following forms

$$\theta_n = t_{n-1} + t_{n+1} - 2t_n$$

and $$t_n = \frac{t_{n-1} + t_{n+1}}{2}$$

This relaxation method of solution has a very good practical relevance in industry. They can be solved in conjunction with digital calculators that are in common use.

In both the flux plot and the relaxation methods, we have the advantage of getting the complete map of the temperatures and the shape factor can be found out for any complex shape. The application of relaxation method to two dimensional heat flow is as follows. At the nodes (sink) , it is assumed that heat will be flowing in – 'x' and 'y' directions.

One such point t_{b_2} is indicated in the following figure

Let 'z' be thickness of the body. Considering only the two-dimensional heat flow, the heat flow into sink b_2 from the various directions can be expressed as follows

$$Q_i = \frac{k\,z\,\Delta y}{\Delta x}(t_{b1} - t_{b2})$$

$$Q_{ii} = \frac{k\,z\,\Delta y}{\Delta x}(t_{b3} - t_{b2})$$

$$Q_{iii} = \frac{k\,z\,\Delta x}{\Delta y}(t_{c2} - t_{b2})$$

$$Q_{iv} = \frac{k\,z\,\Delta x}{\Delta y}(t_{c2} - t_{b2})$$

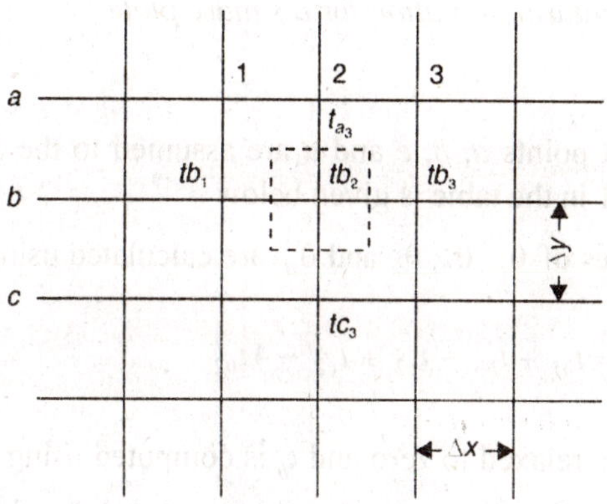

Fig. 2.41 *Application of the relaxation method*

The net heat transfer at the node 'b_2' is given by

$$k\,z\frac{\Delta y}{\Delta x}(t_{b1} - t_{b2}) + k\,z\,\frac{\Delta y}{\Delta x}(t_{b3} - t_{b2}) + k\,z\,\frac{\Delta x}{\Delta y}(t_{c2} - t_{b2}) + k\,z\,\frac{\Delta x}{\Delta y}(t_{c2} - t_{b2})$$

In the case of a square grid, $\Delta x = \Delta y$

and $\quad Q_0 = k\,z\,(t_{b1} + t_{b3} + t_{c2} + t_{c2} - 4\,t_{b2})$

Let $\theta_0 = \frac{Q_0}{k_z} = (t_{b1} + t_{b3} + t_{c2} + t_{c2} - 4\,t_{b2})$

For steady state conditions

$$Q_0 = 0,\ \theta_0 = 0 \text{ and hence } t_{b2} = \frac{t_{b1} + t_{b3} + t_{c2} + t_{c2}}{4}$$

Illustration

Estimate the temperature at the point shown in the figure given below for a square plate for which the boundary temperatures are given as 400 °C and 100 °C.

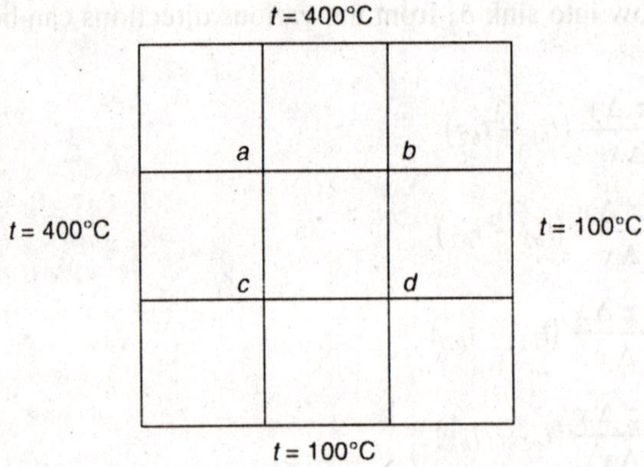

Fig. 2.42 *Temperature estimation for a square plate*

Solution

The temperature at points *a*, *b*, *c* and *d* are assumed to the 300, 200, and 160 respectively. Step 1 in the table is given below

In step 2, values of θ_a, θ_b, θ_c and θ_d are calculated using the equation

$$\theta_0 = \frac{Q_0}{kz} = t_{b1} + t_{b3} + t_{c2} + t_{c2} - 4t_{b2}$$

In step 3, θ_b is relaxed to zero and t_b is computed using the equation

$$t_{b_2} = \frac{t_{b1} + t_{b3} + t_{c2} + t_{c2}}{4}$$

Keeping t_a and t_d the same as got in step 2. Now the equation

$$\theta_0 = \frac{Q_0}{kz} = t_{b1} + t_{b3} + t_{c2} + t_{c2} - 4t_{b2}$$

will yield new values of θ_c and θ_d. Now in step 4, θ_d is relaxed to zero which gives a new values of t_d.

The procedure is continued until θ_a θ_b, θ_c and θ_d become zero. The set of values of temperature for these will give the actual temperatures of the four points *a*, *b*, *c* and *d*.

The complete procedure is indicated in the following Table 2.4.

Table 2.4

Step	Point a		Point b		Point c		Point d	
	θ_a	t_a	θ_b	t_b	θ_c	t_c	θ_d	t_d
1	–	300	–	200	–	200	–	160
2.	400 + 400 + 200 + 200 – 4 × 300 = 0	300	400 + 160 + 300 – 800 + 100 = 160	200	400 + 100 + 160 + 300 – 800 =160	200	200 + 100 + 100 + 200 – 4 × 160 = – 40	160
3.	400 + 400 + 240 + 240 – 4 × 300 = 80	300	0	[300 + 400 + 100 + (160)]/4 = 240	0	[300 + 400 + 100 + 160]/4 = 240	240 + 240 + 100 + 100 – 4 × 160 = 4	160

Contd.

Step	*Point a*		*Point b*		*Point c*		*Point d*	
	θ_a	t_a	θ_b	t_b	θ_c	t_c	θ_d	t_d
4	80	300	300 + 400 + 100 + 170 − 4 × 240 = 10	240	300 + 400 + 100 + 170 − 4 × 240 = 10	240	0	(240 + 240 + 100 + 100) 4 = 170
5.	0	320	30	240	30	240	0	*170
6.	14	320	0	247	0	243	14	*170
7.	0	323.5	9	247	9	247	0	*173.5
8.	4	323.5	0	249	0	249	4	*173.5
9.	0	324.5	3	249	3	249	0	*174.5
10.	2	324.5	0	250	0	250	2	*174.5
11.	0	325	0	250	0	250	0	*175

Heat Generation in Hollow Cylinder

For the case of a hollow cylinder the conduction equations in cylindrical co-ordinates is more appropriate. The equation, for the case of one-dimensional (radial) heat flow, as $\frac{\partial}{\partial r} r\left(\frac{\partial t}{\partial r}\right) = -\frac{Q'}{k}(r)$

Integrating this equation twice, we obtain

$$r\frac{\partial t}{\partial r} = -\frac{Q'}{k}\frac{r^2}{2} + C_1$$

or $$t = -\frac{Q'}{k}\frac{r^2}{4} + C_1 \ln r + C_2 \quad \text{...(1)}$$

where C_1 and C_2 are the constants of integration whose values can be determined by the boundary conditions. Let us assume that the boundary conditions are

(i) The outer surface is insulated that is

$$\left(\frac{\partial t}{\partial r}\right)_{r_2} = 0 \quad \text{...(2)}$$

(ii) The inner surface is cooled by a fluid having temperature t_g. This mean that the heat generated inside the cylinder, that is at the cylindrical wall, by the internal heat sources is completely transferred through this surface to the cooling fluid. Thus the heat generated per unit length

$$Q' = Q\,\pi\left(r_2^2 - r_1^2\right) = -k\left(\frac{\partial t}{\partial r}\right)_{r_1}(2\pi\, r_1)$$

$$= h\,(2\pi\, r_1)\left(t_{s_1} - t_g\right), \text{ at } r = r_1 \quad t = t_{s_1}$$

or $$t_{s_1} = \frac{Q'\left(r_2^2 - r_1^2\right)}{2\,h\,r_1} + t_g \quad \text{...(3)}$$

where r_1 and r_2 are inner and outer radii of the cylinder

From the first boundary condition we get

$$C_1 = \frac{Q'}{k}\frac{r_2^2}{2}$$

so, $$t = -\frac{Q'}{k}\frac{r^2}{4} + \frac{Q'}{k}\frac{r_2^2}{2}(\ln r) + C_2 \quad \text{...(4)}$$

at $r = r_1 \;;\; t = t_{s_1}$

or, $$t_{s_1} = -\frac{Q'}{k}\frac{r_1^2}{4} + \frac{Q'}{k}\frac{r_2^2}{2}(ln\, r_1) + C_2 \qquad ...(5)$$

Thus from Equation (3) and (5), we have

$$\frac{Q'}{2h\,r_1}\left(r_2^2 - r_1^2\right) + t_g = \frac{Q'}{k}\left\{\frac{r_2^2\, ln\, r_1}{2} - \frac{r_1^2}{4}\right\} + C_2$$

or $$C_2 = t_g + \frac{Q'}{2h\,r_1}\left(r_2^2 - r_1^2\right) - \frac{Q'}{2k}\left\{r_2^2\,(ln\, r_1) - \frac{r_1^2}{2}\right\}$$

Therefore, the temperature distribution in the cylinder is given by

$$t = -\left(\frac{Q'}{k}\frac{r^2}{4}\right) + \left(\frac{Q'}{k}\frac{r_2^2}{2}ln\, r\right) + \left[t_g + \frac{Q'}{2h\,r_1}\left(r_2^2 - r_1^2\right) - \frac{Q'}{2k}\left\{r_2^2\, ln\, r_1 - \frac{1}{2}r_1^2\right\}\right]$$

$$= t_g + \frac{Q'\, r_1^2}{4k}\left[\frac{2k}{h\,r_1}\left\{\left(\frac{r_2}{r_1}\right)^2 - 1\right\} + 1\left(\frac{r}{r_1}\right)^2 + 2\left(\frac{r_2}{r_1}\right)^2 ln\,\frac{r}{r_1}\right] \qquad ...(6)$$

The temperature at the outer surface can be obtained by putting $r = r_2$ in eqn. (6)

$$t_0 = t_g + \frac{Q'}{4}\frac{r_1^2}{k}\left[\left(\frac{2k}{h\,r_1} - 1\right)\left\{\left[\left(\frac{r_2}{r_1}\right)^2 - 1\right]\right\} + 2\left(\frac{r_2}{r_1}\right)^2 ln\,\frac{r_2}{r_1}\right] \qquad ...(7)$$

and that at the inner surface is obtained by putting $r = r_1$ in Eq. (6)

$$t_1 = t_g + \frac{Q'}{2}\frac{r_1}{h}\left\{\left(\frac{r_2}{r_1}\right)^2 - 1\right\} \qquad ...(8)$$

As a special case of the above is a heat generating solid cylinder cooled from the outside. Here $r_1 = 0$ and the fluid at temperature t_g flows at the outer surface with a film coefficient, h'.

The temperature distribution, will then be given by

$$t = t_g + \frac{Q'}{4}\frac{r_2^2}{k}\left\{\frac{2k}{h'\,r_2^2} + 1 - \left(\frac{r}{r_2}\right)^2\right\} \qquad ...(9)$$

and the temperature at the centre line given by

$$t_c = t_g + \frac{Q'}{4}\frac{r_2^2}{k}\left(\frac{2k}{h\,r_2} + 1\right) \qquad ...(10)$$

Illustration on Conduction with Heat Generation

1. A thin hollow tube with 3 mm outer diameter and 2 mm inner diameter is subjected to a current of 1000 amp while water at 30°C circulates inside the same. The resistivity of the material is 0.1 ohm-mm²/m. The heat transfer coefficient of the water film may be estimated as 30,000 k cal/m² hr °C. Calculate the surface temperature of the tube.

Owing to the flow of current in the tube there will be heat generation at a uniform rate given by

$$Q' = \frac{(0.864)\, I^2\, R_e}{A\, L}$$

where I = current, amp

R_e = electrical resistance, ohms

A = cross sectional area of conductor, m²

L = length, m

Solution

$$\text{Therefore } Q' = \frac{(0.864)\,1000^2}{\pi\left(3^2 - 2^2\right) \times 10^{-6}} \times \left\{0.1 \div \pi\left(3^2 - 2^2\right)\right\}$$

$$= 3.5 \times 10^8 \text{ k cal/m}^3 \text{ hr}$$

Now the outside and inside surface temperatures may be obtained by using the equations given by below

$$t_0 = t_g + \frac{Q'\, r_1^2}{4\ \ k}\left[\left(\frac{2k}{h\, r_1} - 1\right)\left\{\left(\frac{r_2}{r_1}\right)^2 - 1\right\} + 2\left(\frac{r_2}{r_1}\right)^2 ln\,\frac{r_2}{r_1}\right]$$

and $$t_1 = t_g + \frac{Q'\, r_1}{2h}\left\{\left(\frac{r_2}{r_1}\right)^2 - 1\right\}$$

and the data. r_2 = 3mm; r_1 = 2 mm ; t_g = 30 °C

h = 30000 k cal/m² hr °C ; $Q = 3.5 \times 10^8$

so

$$t_0 = 30 + \frac{3.5 \times 10^8 \times 2^2 \times 10^{-6}}{4 \times 15}\left[\left\{\left(\frac{3}{2}\right)^2 - 1\right\} \times \left(\frac{2 \times 15}{(30000) \times 2 \times 10^{-3}} - 1\right) + 2\left(\frac{3}{2}\right)^2 ln\frac{3}{2}\right]$$

$$= 57.8 \text{ °C}$$

and $$t_1 = 30 + \frac{3.5 \times 10^8 \times 2 \times 10^{-3}}{2 \times 30000}\left\{\left(\frac{3}{2}\right)^2 - 1\right\}$$

$$= 44.6\ °C$$

2. A copper wire 0.1 cm in diameter is insulated with plastic to an outer diameter of 0.3 cm and is exposed to an environment at t_e = 40 °C. The heat transfer coefficient from the outer surface of the plastic to the surroundings is h_e = 7.5 k cal/hr m² °C. What is the maximum steady current in amperes that this wire can carry without heating any part of the plastic above 95 °C. The thermal conductivities of plastic and copper are 0.3 and 330 k cal/hr m °C respectively. The electrical conductivity of copper is 5.1×10^5 ohm^{-1} cm^{-1}.

Solution

Let r_1 and r_2 denote the inside and outside radii of the plastic insulation. Then, the heat flow rate per unit length of the wire is given by

$$q = \frac{2\pi\,(t_1 - t_e)}{\left(\dfrac{ln\,\dfrac{r_2}{r_1}}{k}\right) + \left(\dfrac{1}{r_2\,h_e}\right)}$$

$$= \frac{2\pi\,(95 - 40)}{\dfrac{ln\left(\dfrac{0.15}{0.05}\right)}{0.3} + \dfrac{1}{\left(0.15 \times 10^{-2}\right)(7.5)}}$$

$$= \frac{2\pi \times 55}{3.67 + 89}$$

$$= 3.74\ \text{k cal/hr m}$$

$$= 4.47\ \text{k watt/m length}$$

In the steady state all the heat produced on account of current flow must be transferred to the surroundings. Here the rate of heat production is 4.47 watts per m. length of the wire. The heat production per unit volume is therefore

$$Q' = \frac{4.47}{\pi\left[(0.05) \times 10^{-2}\right]^2 \times 1}$$

$$= 5.7 \times 10^6\ \text{watts/m}^3$$

Since $Q' = I^2/k_e$

$$I^2 = Q'\ ke$$

$$= 5.7 \times 10^6 \times 5.1 \times 10^7$$

or $$I = 1.705 \times 10^7 \text{ amp/m}^2$$

$$\text{current} = \left(1.705 \times 10^7\right) \pi \left(0.05 \times 10^{-2}\right)^2$$

$$= 13.4 \text{ amp}$$

3. A chemical reaction is being carried out at constant pressure in a packed bed between two coaxial cylinder with radii 1.14 cm and 1.27 cm. The entire inner wall is at a uniform temperature of 500 °C and there is almost no heat transfer through this surface. The reaction releases heat at a uniform rate of 480, 000 k cal/hr m² through out the reactor The effective thermal conductivity of the packed bed may be taken as 0.45 k cal/hr m °C. What is the temperature of the outer wall'?

Solution

For this case, from the transport phenomena approach

$$\frac{1}{r}\left\{\frac{d}{dr}\left(r\frac{dt}{dr}\right)\right\} = -\frac{Q'}{k}$$

or $$\frac{dt}{dr} = -\frac{Q'r}{2k} + \frac{C_1}{r}$$

Since $dt/dr = 0$, at $r = R_0$ (inside radius)

$$C_1 = \frac{Q' R_0^2}{2k}$$

$$\frac{dt}{dr} = \frac{Q' R_0^2}{2\ k\ r} - \frac{Q'r}{2k}$$

$$\therefore t = \frac{Q'}{2k}\left(R_0^2 \ln r - \frac{r^2}{2}\right) + C_2$$

Since $t = T_0$ at $R = R_0$

$$C_2 = T_0 - \frac{Q' R_0^2}{2k}\left(\ln R_0 - \frac{1}{2}\right)$$

or $$t = T_0 - \frac{Q' R_0^2}{2k}\left(\ln\frac{r}{R_0}\right) - \frac{Q' R_0^2}{4k}\left(\frac{r^2}{R_0^2} - 1\right)$$

Substituting the numerical values

$$t = 500 + \frac{480000 \times (1.14 \times 10^{-2})^2}{2 \times 0.45}\left(ln\frac{1.27}{1.14}\right) - \frac{480000 \times (1.14 \times 10^{-2})^2}{4 \times 0.45}\left[\frac{1.24^2}{1.14^2} - 1\right]$$

$= 500 + 7.28 - 8.38$

$\therefore\ t = 498.9$ °C

PROBLEMS ON CONDUCTION

Problems on Steady State Heat Conduction

1. A pipe carries condensing steam at 9.35 kg/cm^2. The unit surface conductance of the steam side film is 4850 k cal/m^2 hr °C and that of the outside film at pipe is 7 k cal/m^2 °C. The thermal resistance of the pipe per unit area is 2×10^{-4} m^2 hr °C/k cal. If the pipe is laid in a room at temperature of 25 °C, calculate the temperature at the outer surface of the pipe.
2. A steam pipe with ID and OD as 100 and 170 mm is covered with two layers of insulation, 30 mm and 50 mm thick. Thermal conductivities of the insulating materials are 0.15 and 0.08 k cal/m hr °C respectively while that of steel is 50. The inner surface of the pipe is at 300 °C while the outer layer surface is at 50 °C. Determine the heat loss from the pipe and the layer contact temperatures.
3. A 15 cm OD pipe is covered with 5 cm of insulation of $k = 0.08$ k cal/hr m °C and 4 cm of insulation of $k = 0.06$ k cal/hr m °C. Calculate the heat loss from 100 m of pipe if the inner and outer surface temperatures of the insulation are 400 °C and 30 °C respectively. What is the heat loss per 100 m^2 of outer pipe surface?
4. The outside and inside surface temperatures of a 20 cm OD and 18 cm ID pipe ($k = 40$ k cal/hr m °C) are 400 and 406 °C respectively. Calculate the reduction in heat loss if a 5 cm thick layer of insulation ($k = 0.06$ k cal/hr m °C) is put on the pipe. Assume that the inner and outer surface temperatures of the insulation are 405 °C and 140°C. What is the inside surface temperature of this pipe in this case?
5. A standard 5 cm steel pipe carrying steam is insulated with a 7.5 cm layer of glass wool. The outer surface temperature of the pipe is 200 °C and that of the covering is 65 °C. Estimate the heat loss from the pipe. Assume the value of k of steel 39.0 k cal/m hr °C
6. A 30 cm outside diameter pipe diameter pipe is covered with two layers of insulation ($k_1 = 0.09$. $k_2 = 0.06$ k cal/m hr °C). The better insulating material is on the outside and it is 4 cm thick. The other insulating material is of 5 cm thickness. The inner and outer surface temperatures of the insulation are 350 °C and 50 °C. Determine

(a) Heat loss per m per hour

(b) Heat loss per m^2 of outer insulation surface

(c) Heat loss per m^2 of pipe surface

(d) Temperature of the surface between the two layers of insulation

7. A 5 cm outer diameter pipe is to be covered with two layers of insulation each 5 cm thick. If the thermal conductivity of one material is 4 times that of the other, and the inner and outer surface temperatures of the composite insulation are fixed what will be the ratio of the heat transfer when the better insulating material is next to the pipe to the heat transfer when the better insulating material forms the outer layer.
8. A 15 cm steam main has inside diameter of 12 cm and an outside diameter of 16 cm. It is insulated at the outside with asbestos. The steam temperature is 150°C and the air temperature is 20°C, h (steam side) =100 k cal/m^2 hr °C, h (air side) = 20 k cal/m^2 hr °C, k (asbestos) = 0.8′, k (steel) = 42 k cal/m hr °C. How thick should the asbestos be provided in order to limit the heat loss to 5 k cal/hr m^2.
9. Determine the heat transfer between two fluids separated by a copper pipe 3 mm thick, 20 mm outside diameter and 1.5 meters long. If the inner fluid (water) temperature is 15°C and the outer fluid (steam) temperature is 100°C. The water side coefficient is 1200 k cal/hr m^2 °C and the steam side coefficient is 10,000 k cal/hr m^2 °C and the conductivity for copper is 300 k cal/m^2 hr °C. Determine the log mean area and the arithmetic mean area of conducting surface and show if arithmetic mean can be used in place of log mean area.
10. Device the heat conduction equation for a hollow spherical shell, taking the flux at the inner surface and the temperature at the outer surface to have constant values. Solve the resulting equation for the temperature distribution in the shell.
11. Calculate the heat loss per m^2 of outside surface area of a heated sphere, 20 cm in diameter covered with 5 cm of insulation k = 0.06 k cal/m hr °C. The inside and outside surface temperatures of the insulation are 300°C and 50°C respectively.
12. An electrical cable of 1 cm OD is to be insulated with rubber (k = 0.05 k cal/m hr °C). The cable is to be located in air (h = 6 k cal/m hr °C) at 18°C Investigate the effect of insulation thickness on heat dissipation assuming the cable surface temperature of 85°C. Determine the percentage increase in heat flow by providing the critical insulation.
13. Calculate the heat loss from the surface of a furnace wall 25 cm thick. The inner and outer surface temperatures are 300°C and 50°C respectively. The conductivity of the wall material is given by $k = 0.002T - 10^{-6} T^2$ where T is in °C and k in k cal/m hr °C. Also determine the average value of the conductivity in this range.

14. Compute the heat loss per m^2 of surface area of a wall 25 cm thick. The inner and outer surface of the wall are at 300 °C and 40 °C respectively. The variation of k with temperature is given by

$$k = 0.03 + 5 \times 10^{-6}\, t^3$$

where t is in °C and k is in k cal/m hr °C

15. Compute the heat loss per m^2 of surface area of a wall 40 cm thick having surface temperature of 300 °C and 50 °C if the thermal conductivity of the wall material is given by

$$k = 0.005t - 5 \times 10^{-6}\, t^2$$

16. The thermal conductivity of a material is given by

$$k = A + Bt + Ct^2$$

Derive an expression giving the heat loss per m length of a hollow cylinder made of this material and having inner radius R_1 and outer radius R_2. Consider heat transfer in the radial direction only.

17. A steel tube is placed in quiet air. The tube is heated at one end at a given constant rate of heat flow and perfectly insulated at the other end. Develop expressions for the temperatures at either end of the tube, assuming that no appreciable heat flow occurs in the air inside the tube and that radial temperature difference in the steel are negligible. Assume, further the thermal conductivity of steel is independent of the temperature and that the coefficient of heat transfer is the same over all the outer tube surface.

The result should be given in the of a mathematical formulas and for the following special case.

Tube length = 2.54 m

Outer diameter of tube = 0.0305 m

Inner diameter of to be 0.0190 m

Thermal conductivity of steel = 43.7 k cal/hr m °C.

Temperature of air outside the tube = 23.9 °C

Coefficient of heal transfer on the outside = 6.85 k cal/hr m^2 °C

Rate of heat flow = 2.52 k cal/hr.

18. A 1.25 cm diameter iron rod (k = 37.4 k cal/hr m °C) 30 cm long is giving up heat to the environment at 20 °C. The ends of the cylinder are connected to two heat sources at 130 °C. Determine

(i) the temperature distribution in the rod

(ii) the temperature at the middle of the rod.

(iii) the heat flowing out of each heat source.

(iv) the heat flowing at the middle of the rod.

Assume that h = 20 k cal/hr m^2 °C

19. A 1.25 cm diameter iron rod (k = 37.4 k cal/hr m °C) 15 cm long protrudes out of a heat source at 130 °C into environment at 20 °C. The free end of the rod is insulated. If h = 20 k cal/hr m^2 °C find
    (a) the temperature distribution in the rod
    (b) the temperature at the insulated end
    (c) the heat flowing out of the heat source
20. One end of the long rod is inserted into a furnace while the other projects into ambient air. Under steady state the temperature of the rod is measured at two points 7.5 cm apart and found to be 125 °C and 88.5 °C when the ambient temperature is 20 °C. If the rod is 2.5 cm in diameter and h = 20 k cal/hr m^2 °C find the thermal conductivity of the rod material.
21. A thin rod of copper (k = 298 k cal/hr m °C); 1.22 cm diameter connect two parallel plates 15 cm apart, Air flows in the space between the plates (h = 49 k cal/hr m^2 °C), The plates are at 70 °C while the air temperature is 15 °C. Find the temperature at the centre of the rod.
22. What is the temperature distribution in a plane wall 7.5 cm thick in which heat is generated uniformly at the rate of 1050 k cal/hr m^3. One surface of the wall is maintained at – 18 °C and the other at 100°C. Assume k = 1.5 k cal/hr m °C. Also determine the magnified and direction of heat flow through each face per unit area.
23. Heat is generated at the interface between two slabs both 50 mm thick, one is of steel of which k = 30 k cal/hr m °C and the other is brass of k = 65 k cal/hr m°C, The temperature of the outside of the steal slab is 95 °C, while the of outside of the brass is 47 °C. When the rate of heat generation is 15,000 k cal/hr m^2 of contact area, calculate the heat flow rate through the outer surfaces and the interface temperatures. Also show how could such a condition be obtained.
24. A flat plate electric resistance heater 30 cm × 60 cm × 1.25 cm is used to keep the temperature of a bath at 100 °C. If the temperature at the mid-plane of the heater is maintained at 115 °C, calculated the rate of heat flow through both faces. Assume that k = 4.5 k cal/hr m °c and that heat is generated uniformly per unit volume. Neglect the temperature difference between heater surface and bath liquid and also heat flow through the edges.
25. Show that the steady state temperature distribution in a sphere in which heat is produced at a uniform rate per unit volume is given by Q', is given by

$$(t - t_o) = \frac{Q' R_0^2}{6k}\left[1 - \left(\frac{R}{R_o}\right)^2\right]$$

where R_o = outer radius and t_o = outer surface temperature.

26. The average heat production during ripening of oranges is estimated to be 320 k cal/m^3 hr. If the orange is assumed to be homogeneous with k = 0.15 k cal/hr m °C, compute the temperature at the centre of the orange and the heat flow from the outer surface. Assume that the diameter is 8.5 cm and the external surface temperature is 10 °C.

27. In a cylindrical fuel not of a nuclear reactor, the internal heat generation is

$$Q' = Q_0\left\{1-\left(\frac{r}{r_o}\right)^2\right\}$$

where Q' = local heat generation per unit volume at radius r

r_o = outside radius r

r_o = outside radius

Q_0 = rate of heat generation per unit volume at the centre line

Calculate the temperature drop from the centre line to the surface of a 2.5 cm OD rod having a thermal conductivity of 20 k cal/m hr °C if the rate of heat removal from its surface is 2,500,000 k cal/m^2 hr °C.

28. The inside dimension of an oven made of brick (k = 1.2 k cal/hr m °C) are 2m by 1.8 m by 0.6 m. The wall is 10 cm thick. The temperatures of the inner and outer surfaces are 60 °C and 100 °C. Calculate the loss of heat per hour.

29. The walls of a house are constructed with a 4 inch thick layer of brick. $\frac{1}{2}$ inch layer of cellotex, an air space $3\frac{5}{8}$ inch thick, another $\frac{1}{2}$ inch layer of cellotex and $\frac{1}{4}$ inch of wood paneling. The temperature at the outside surface of the brick is 30 °F and that at the inside of the wood paneling (exposed to room air) is 74 °F.

Find the heat flux through this wall if

(a) the air space transfers heat by conduction only

(b) there are convection currents in the air space making the equivalent conductance of this space 2.0 Btu/hr ft^2 °F.

(c) the space between inside and outside walls is filled with glass wool

k brick = 0.38 Btu/hr ft °F

k cellotex = 0.027 Btu/hr ft °F

k wool = 0.12 Btu/hr ft °F

k glass wool = 0.027 Btu/hr ft °F

30. Consider the case of steady state one-dimensional conduction through a medium having a cross sectional area decreasing linearly from the value of A_0 at $x = 0$, to A_L at $x = L$. Evaluate the heat flow rate. The temperatures at $x = 0$ and $x = L$ are T_0 and T_L respectively.

31. A composite wall is to be made of a $\frac{3}{8}$ inch stainless steel, 3 inch of corkboard and $\frac{3}{4}$ inch of plastic (k = 1.3 Btu/hr ft °F). The outside surface of the stainless steel is to be 250 °F and the temperature of the plastic surface will be 60 °F.
    (a) Determine the thermal resistance of each layer of material.
    (b) Determine the heat flux
    (c) Find the temperature at each surface of the corkboard.

32. In the previous problem the same composite wall is exposed to gases at 250 °F and 70 °F with convective heat transfer coefficients of 45 Btu/hr ft^2 °F and 6 Btu/hr ft^2 °F at the steel and plastic surfaces, respectively. Find
    (a) the heat flux
    (b) the surface and the interfacial temperatures for the wall.
    (c) which individual thermal resistance is controlling?

33. A $1\frac{1}{2}$ inch schedule-80 steel pipe with its outside surface at 400 °F is to be located in air at 90°F with the convective heat transfer coefficient between the steel surface and to surrounding air equal to 1.6 Btu/hr ft^2 °F. It is proposed to add 85% magnesia insulation to the pipe surface to reduce the heat loss to one-half of that for the bare pipe. What thickness of insulation should be used if the heat transfer coefficient, h_0 and outside steel surface temperature retain the some values given above.

34. A plane wall has thickness L and its two surfaces maintained at temperatures T_0 and T_L respectively. If the thermal conductivity of the wall material varies according to $k = k_0\left(1 + \alpha T + \beta T^2\right)$, develop an expression for the steady-state one-dimensional heat flux.

35. A composite wall is constructed of $\frac{1}{2}$ inch of aluminium, $\frac{1}{2}$ inch of corkboard and $\frac{1}{8}$ inch of plastic (k = 1.3 Btu/hr ft^2 F). The outside temperature of the aluminium is 250 °F and the plastic surface is held 80 °F. How much heat will be transferred per square foot of the wall

surface under these conditions? What will be the temperatures on either side of the corkboard?

36. The composite wall described in the above problem is exposed to 325°F air on the aluminium side ($h = 60$ Btu/hr ft^2 °F) and to 60 °F air on the plastic side $h = 6$ Btu/hr ft^2 °F. What will be the heat flux and the temperatures at the aluminium and plastic wall surfaces?

37. A pipe carrying 300 psi saturated steam passes through a room in which the air temperature is 80°F. If the values of 'h' on the inside and outside pipe walls are 1700 Btu/hr ft^2 °F and 7 Btu/hr ft^2 °F respectively what will be the heat loss per foot of pipe. The pipe is $2\text{-}\frac{1}{2}$ inch schedule-80 mild steel.

38. The steady-state rate of heat conduction through a plane wall is $q = \left(\frac{kA}{L}\right)\Delta T$. For steady-state conduction through a cylindrical wall the heat flow rate is given by the equation $q = \frac{2\pi k L}{ln \frac{r_0}{r_i}} \Delta T$

   (a) Show that the second equation given above can be put into the form of the first equation given above where the effective area $\bar{A}$ in the expression

   $$q = \frac{k \bar{A}}{r_0 - r_i} \Delta T$$

   is the logarithmic mean area given by $A_{lm} = \frac{2\pi L(r_0 - r_i)}{ln \frac{r_0}{r_i}}$

   (b) If the arithmetic mean area $\pi L(r_0 + r_i)$ rather than the logarithmic mean is used for a cylindrial wall what will be the percent error for values of $\frac{r_0}{r_1}$ 1.5, 3 and 5?

39. Find the appropriate value of $\bar{A}$ for a spherical wall when heat flow is expressed in the form.

   $$q = \frac{k \bar{A}}{r_0 - r_i} \Delta T$$

   Repeat (b) of the above problem for the above problem for the spherical case.

40. The composite wall of an oven consists of three material two of which are of known thermal conductivity $k_A = 20$ w/m °k and $k_c = 50$ w/m °k and known thicknesses, $x_A = 0.30$, $x_C = 0.15$m. The third material *B* which is sandwiched between materials *A* and *C* is of known thickness $x_B = 0.15$ m but its thermal conductivity is not known under steady-state operating conditions. Measurements reveal that an outer surface temperature of $T_{s,0} = 20$°C and inner surface temperature of $T_{s,i} =$ 600°C and an oven air temperature of $T_\infty = 800\ °C$. The inside heat transfer coefficient (convection) *h* is known to be 25 w/m²°K. Determine the thermal conductivity of material *B*.

3

Film Coefficients

3.1 CONDUCTION THROUGH LIQUIDS

This situation is rarely met in practice except in heat flows through thin films. In these cases, however, thickness of the film is not known and so the equations developed earlier can not be applied. This difficulty is obviated by the use of surface coefficient which will be elaborated later. Any body of fluid of appreciable size through which heat is flowing will develop convection currents of such a magnitude that heat is carried by both convection and conduction. It is rather difficult to have conduction alone and this may be one of the reasons for our not having accurate data *'k'* of fluids.

Convection

The character or nature of flow changes from viscous to turbulent condition when ${}^{N}R_e$ exceeds a certain value. Even in turbulent flow there is at the boundary a residual film that persists or remains in laminar flow. Turbulence may be caused by a stirrer or an agitator of any type or by pumping the fluid through a pipe (forced convection) or by the natural convection currents set up when a body of fluid otherwise stationary is heated. If heat is passing through the retaining wall to the fluid, the comparatively stagnant film is of great importance in determining the rate of heat transfer. This follows from the fact that all the heat reaching the bulk of the fluid must pass through this by conduction and the *k*'s of fluids are low, so that despite the film being thin the resistance offered by it to the heat flow is large. On the other hand beyond the film, the turbulence (or eddies) brings out a rapid equalization of temperature. In other words, it is the boundary film that offers a significant resistance to the heat flow from the wall to the bulk of the fluid.

Temperature Gradients in Forced Convection

The temperature distribution across a column of fluid in forced convection that is at the same time being cooled or heated is closely connected with the velocity distribution across the same column of the fluid. The following figure represents the temperature gradients for the situation in which heat is flowing from hot fluid through a metal wall into a cold fluid.

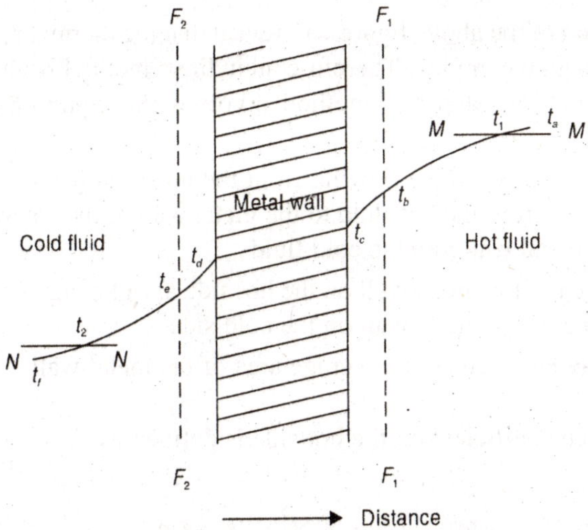

Fig. 3.1 *Temperature gradients in forced convection*

The dotted lines $F_1 F_1$ and $F_2 F_2$ on each side of the metal wall represent the boundaries of the films in viscous flow and all portions of the fluid to the right of $F_1 F_1$ and to the left of $F_2 F_2$ are in turbulent flow. The temperature gradient from the bulk or main body of the hot fluid to the metal wall is represented by the curved line $t_a\ t_b\ t_c$. The temperature t_a is the maximum temperature in the hot fluid is t_b the temperature at the boundary between the turbulent and laminar regimes

$t_c \rightarrow$ temperature at the actual interface between fluid and solid $t_d\ t_e\ t_f$'s significance is similar.

It is customary and convenient in heat transfer computations to use an average temperature of the fluid rather than t_a or t_b. This average temperature would be the one corresponding to completely mixed condition. This average temperature will be somewhat less than t_a and is shown as a straight line marked MM. The same analysis is applicable to the cold fluid side where the average temperature is t_2 represented by the line NN. If the fluid is not viscous and the pipe is not too large, these average temperatures are the ones obtained when a thermometer is inserted into the pipe so that its bulb is near the middle of the stream of fluid. To determine the actual course of the curve $t_a\, t_b\, t_c$ one has to carry out very careful measurements by fine thermocouples. The temperature gradient $t_c\, t_d$ is caused by the flow of heat in pure conduction and it turns out in most cases that it is a small fraction of the total temperature difference $t_a\, t_f$

Surface Coefficients
(Individual Heat Transfer Coefficients/Film Coefficients)

An examination of the above figure will reveal that the thermal resistances in the two fluids are quite complex. Therefore an indirect method is always employed for their calculations and this method involves the computation of surface coefficients.

If q is the heat (k cal/hr) flowing from the hot fluid to the cold one then q k cal/hr will pass from the hot fluid to the metal wall & the same q k cal/hr will pass from the metal wall into the cold fluid

If A_1—area of the metal wall on the hot side at right angles to the heat flow

and A_2—area of the metal wall on the cold side

A_m—can be taken as the average area of the metal wall

The surface coefficient on the hot side is defined by $$h_1 = \frac{q}{A_1(t_1 - t_c)} \quad ...(1)$$

comparing equation (1) with equation $q = \dfrac{k_m\ A\ \Delta t}{L}$, it will be seen that h_1 is analogous to k_m/L and therefore $\dfrac{1}{h_1\ A_1}$ turns out to be the thermal resistance.

Although the individual heat transfer coefficient is calculated and discussed as if it were an actual quantity by itself, reference to the figure will indicate that h_1 contains the effect of both the viscous film and the thermal resistance of the turbulent core that causes the temperature different $t_a - t_b$

Similarly h_2 is defined as $$h_2 = \frac{q}{A_2(t_d - t_2)} \quad ...(2)$$

So the situation represented by the above figure may be considered to consist of three resistances on series.

Ist : $\dfrac{1}{h_1\ A_1}$ on the hot fluid side

IInd : $\dfrac{L \text{ (or } x)}{k\ A_m}$ resistance of the metal wall

IIIrd : $\dfrac{1}{h_2\ A_2}$ on the cold fluid side

If these three thermal resistances are substituted for R_1, R_2, and R_3, in the

$$\text{exp}^n.\ q = \frac{At}{R_1 + R_2 + R_3}$$

this results,

$$q = \frac{\Delta t}{\frac{r}{h_1 A_1} + \frac{L\,(\text{or } x)}{k\,A_m} + \frac{1}{h_2 A_2}} \qquad ...(3)$$

$$A_m = \frac{2\pi L\,(r_0 - r_i)}{ln \frac{r_0}{r_i}}$$

(for tubes, pipes)

Concept of Overall Coefficients

Assume that the numerator of the denominator of the r.h.s. of equation (3) are multiplied by any one of the transport areas. For e.g. of A_1 is selected, equation (3) turns out to be

$$q = \frac{A_1\,\Delta t}{\frac{1}{h_1} + \frac{A_1}{A_m}\frac{L}{k} + \frac{A_1}{h_2 A_2}} \qquad ...(4)$$

The overall coefficient 'U_1' is defined by the exp^n

$$U_1 = \frac{1}{\frac{1}{h_1} + \frac{A_1 L}{A_m k} + \frac{A_1}{h_2 A_2}} \qquad ...(5)$$

Equations (5) and (4) on comparison show

$$q = U_1 A_1 \Delta t \qquad ...(6)$$

Equation (6) states that the rate of heat transfer is the product of three factors: overall heat transfer coefficient, temperature drop and area of the heating surface.

It should be pointed out in this context that if either of the other two areas viz. A_m or A_2 had been selected, there would have resulted coefficients based on these areas and designated by U_m and U_2. Before a 'U' is established, a definite area must be selected and the coefficient so determined is consequently based on the chosen area. In general choice is arbitrary.

Equation (5) for U_1 can be written in a convenient form for the case of heat transfer through tube walls. In this case L is thickness.

A_1 : proportional to the corresponding diameter D_1

Similarly $A_2 \propto D_2$

∴ Equation (5) becomes in this case

$$U_1 = \frac{1}{\frac{1}{h_1} + \frac{D_1}{D_m}\frac{L}{k} + \frac{D_1}{D_2 h_2}} \quad ...(7)$$

Similarly expressions can be written for U_m and U_2

In some situations, one particular area is more convenient than the other areas. Suppose that one surface coefficient say h_2 is large in magnitude when compared to the other coefficient h_1. In this case the term $\frac{D_1}{D_2 h_2}$ becomes small in comparison with the 2nd term in the denominator which represents the resistance of the tube wall. The resistance of the tube can be ignored or disregarded reducing equation (7) to

$$U_1 = \frac{1}{\frac{1}{h_1} + \frac{L}{k} + \frac{1}{h_2}} \quad ...(8)$$

In a case like the above it will be advantageous to base the U on that area that corresponds to the highest resistance or the lowest value of h, the individual heat transfer coefficient.

In situations involving thin-walled tubes of larger diameter, flat plates or any other case where a negligible error will be caused by employing a common area A for A_1, A_m and A_2, equation (8) may be used in lieu of equation (7). This implies that U_1, U_m and U_2 are identical.

In certain cases h_1 is very small in comparison with both h_2 and $\frac{L}{k}$ and the term $\frac{1}{h_1}$ in equation (7) is very large when compared to the other terms in the denominator. This will mean $U_1 \approx h_1$.

The problem of predicting the rate of flow of heat from one fluid to another through a retaining wall boils down essentially to the problem of predicting values of surface coefficients (h's) of the fluids concerned.

Accounting for Scale, Dirt etc.

When fouling or dirt factors (due to scale formation on the inside and outside of the tube walls or surfaces) have to be accounted for, then 'q' becomes:

$$q = \frac{\Delta t}{\dfrac{1}{A_1 h_{d_1}} + \dfrac{1}{h_1 A_1} + \dfrac{L}{k A_m} + \dfrac{1}{A_2 h_{d_2}} + \dfrac{1}{h_2 A_2}} \quad ...(9)$$

h_{d_1} and h_{d_2} are the fouling factors

A for the inside dirt factor will be take as A_1 and likewise for outside

$$U_1 = \frac{1}{\dfrac{1}{h_{d_1}} + \dfrac{1}{h_1} + \dfrac{A_1}{A_m} \cdot \dfrac{L}{k} + \dfrac{A_1}{A_2 h_{d_2}} + \dfrac{A_1}{h_2 A_2}} \quad ...(10)$$

Heat transfer surfaces may not be clean and may contain solid deposits of scale, dirt etc, the effect of such deposits being to add to the resistance for the flow of heat. The mechanism could be as one of simple dirt mechanically deposited or crystalline scale deposited because of the inverted solubility curve of dissolved constituents present.

Illustration

Methanol flowing in the inner pipe of a steel double-pipe heat exchanger is cooled by H_2O flowing in the outer pipe. The inner pipe is made from 2.5 cm nominal diameter pipe. The 'k' of steel is 40 k cal/hr m °C. The film coefficients and fouling factors are; alcohol film coefficient—$h_1 \to 900 \text{ k cal / m}^2 \text{ hr °C}$

H_2O film coefficient, $h_2 \to 1500 \text{ k cal / m}^2 \text{ hr °C}$

Inside fouling factors $hd_1 \to 5000 \text{ k cal / m}^2 \text{ hr °C}$

Outside fouling factors $hd_2 \to 2500 \text{ k cal / m}^2 \text{ hr °C}$

What is the overall coefficient based on outside area of the inner pipe?

Solution

The diameter and wall thickness of a 2.5 cm nominal diameter pipe are

$d_i = 2.40$ cm $\qquad d_o = 3.0$ cm

x_w (wall thickness) = 0.3 cm

$\therefore D_{av}$ (the one corresponding to the heat transfer through metal wall) = 2.7 cm

$$U_2 = \frac{1}{\dfrac{1}{h_1} \cdot \dfrac{D_2}{D_1} + \dfrac{1}{h_{d_1}} \cdot \dfrac{D_2}{D_1} + \dfrac{x_w}{k} \dfrac{D_2}{D_w} + \dfrac{1}{h_{d_2}} + \dfrac{1}{h_2}}$$

$$= \frac{1}{\dfrac{1}{900} \cdot \dfrac{3}{2.40} + \dfrac{1}{5000} \cdot \left(\dfrac{3}{2.4}\right) + \dfrac{\frac{0.3}{100}}{40} \dfrac{3}{2.7} + \dfrac{1}{2500} + \dfrac{1}{1500}}$$

$= 349.1 \text{ k cal / hr m}^2 \text{ °C}$

The following table presents the range of the magnitude of heat transfer coefficients in typical situations.

Table 3.1

Type of flow or situation	*h* k cal/hr m² °C	*h* w/m² °C
Free or natural convection	5–25	6–28
Turbulent forced convection inside pipes involving air	5–500	6–570
Turbulent forced convection inside pipes involving water	250–15,000	280–17,000
heat transfer in boiling of H_2O	2,500–50,000	2900–57000
Condensation of vapours/steam	5000–1,00,000	5,600–115,000

Illustration

What would be the overall coefficient for surface condenser made from 18-gauge. Admiralty metal 1″ tubes if the steam side coefficient is 2000, the deposit factors are 2000 on each side and the water side coefficient is 1800? What would U_o be, if the tubes were cleaned on both sides and only on the inside.

Solution

From the given information

wall thickness of the tube = 0.049″ id = 0.902″

mean tube diameter = 0.951″ $k = 63$ (in the appropriate units)

$$U_0 = \frac{1}{\dfrac{D_0}{D_i\, h_{d_1}} + \dfrac{D_0}{D_i\, h_1} + \dfrac{x_w}{k_m} \cdot \dfrac{D_0}{\overline{D}_L} + \dfrac{1}{h_0} + \dfrac{1}{h_{d_0}}}$$

$$= \frac{1}{\dfrac{\frac{1}{12}}{\left(\frac{0.902}{12}\right)2000} + \dfrac{\frac{1}{12}}{\left(\frac{0.902}{12}\right)1800} + \dfrac{\frac{0.049}{12}\,\frac{1}{12}}{(63)\left(\frac{0.951}{12}\right)} + \dfrac{1}{2000} + \dfrac{1}{2000}}$$

$U_0 = 446$ Btu/hr sq. ft., of the outside surface °F

If the scale deposits were removed from both sides

$U_0 = 845$ Btu/hr ft² °F

If only one side dirt is removed

$U_0 = 595$ Btu/hr ft^2 °F

Special Cases of *U*

Although the choice of area to be used as the basis of an U is arbitrary, sometimes one particular area is more convenient than the rest. Suppose, for example, one individual coefficient h_i is large numerically in comparison with the other that is h_o, and that the fouling factors are negligible. Also the term representing the resistance of the metal wall is usually small in comparison with $\frac{1}{h_0}$, the ratio $\frac{D_0}{D_i}$ and $\frac{D_0}{\overline{D}_L}$ have so little significance that they can be ignored and the equation

$$U_0 = \frac{1}{\frac{D_0}{D_i h_i} + \frac{x_w}{k_m} \cdot \frac{D_0}{\overline{D}_L} + \frac{1}{h_0}}$$

can be written in a simpler form, as

$$U_0 = \frac{1}{\frac{1}{h_i} + \frac{x_w}{k_m} + \frac{1}{h_0}}$$

In such a case it is advantageous to base on that area that corresponds to the largest resistance or the lowest value of 'h'.

For thin-walled tubes of large diameter, of flat plates negligible error is caused by using a common area for A_i, $\overline{A}_L$ and A_0 and the above equation can be used for U and U_i and U_0 are identical.

Sometimes one coefficient e.g. h_o is so very small in comparison with both $\frac{x_w}{k}$ and the other coefficient h_i, that the term $\frac{1}{h_0}$ is very large compared with other terms in the resistance sum. When this is true, it is sufficiently accurate to equate the overall coefficient to the small individual coefficient or in this case $U_0 = h_0$

Classification of '*h*'

The problem of predicting the rate of heat flow from one fluid to another through a retaining wall reduces to the problem of predicting the numerical values of the individual coefficients of the fluids concerned in the overall process.

The physical situations are

1. Heat flow to and from fluids inside tubes, without phase change

2. Heat flow to and from fluids outside tubes, without phase change
3. Heat flow to boiling liquids
4. Heat flow from condensing fluids.

Dependence of *U* on Temperature/ Effect of Temperature on *U*

$$Q = wc\,(t_2 - t_1) = WC\,(T_2 - T_1) \quad ...(1)$$

Let $B = \dfrac{wc}{WC} = \dfrac{T_2 - T_1}{t_2 - t_1}\,\dfrac{(\text{hot})}{(\text{cold})}$ ...(2)

Also $B = \dfrac{T - T_1}{t - t_1}$ ...(2*a*)

By a heat balance over the differential area dA

$$dQ = U\,(T - t)\,dA = wc\,dt \quad ...(3)$$

Equation (3) can be rewritten as

$$\frac{dt}{U\,(T - t)} = \frac{dA}{wc} \quad ...(4)$$

or

$$\int_{t_1}^{t_2} \frac{dt}{U\,(T - t)} = \frac{A}{wc} \quad ...(5)$$

For the situation under consideration

$$U = a\,(1 + bt) \quad ...(6)$$

Substituting the equations (2) and (6) into 5, yields

$$\int_{t_1}^{t_2} \frac{dt}{a(1 + bt)\left[T_1 - Bt_1 + (B - 1)t\right]} = \frac{A}{wc} \quad ...(7)$$

Separating into parts

$$\frac{1}{a(B - 1 - bT_1 + bBt_1)} \int_{t_1}^{t_2} \left[\frac{(B - 1)\,dt}{T_1 - Bt_1 + (B - 1)\,t} - \frac{b\,dt}{1 + bt}\right] = \frac{A}{wc} \quad ...(8)$$

Integrating, we obtain

$$\frac{1}{a(B - 1 - bT_1 + b\,Bt_1)}\, ln\left(\frac{T_1 - Bt_1 + (B - 1)t_2}{T_1 - Bt_1 + (B - 1)t_1}\right) - ln\left(\frac{1 + bt_2}{1 + bt_1}\right) = \frac{A}{wc} \quad ...(9)$$

Simplifying by the use of equation (2) and rearranging, we get

$$\frac{t_2 - t_1}{a(1+bt_1)\cdot(T_2 - t_2) - a(1+bt_2)(T_1 - t_1)} \; ln \frac{T_2 - t_2}{T_1 - t_1} - ln \frac{a(1+bt_2)}{a(1+bt_1)}$$

$$= \frac{A}{wc} \qquad \text{...(10)}$$

Substituting $U_1 = a(1 + bt_1)$ and

$$U_2 = a(1 + bt_2)$$

$$\Delta t_1 = T_1 - t_1, \Delta t_2 = T_2 - t_2$$

$$\frac{t_2 - t_1}{U_1 \Delta t_2 - U_2 \Delta t_1} \; ln \frac{U_1 \Delta t_2}{U_2 \Delta t_1} = \frac{A}{wc} \qquad \text{...(11)}$$

combining equations (11) and (1), we obtain

$$\frac{Q}{A} = \frac{U_1 \Delta t_2 - U_2 \Delta t_1}{2.303 \log \frac{U_1 \Delta t_2}{U_2 \Delta t_1}}$$

Illustration

5.0 kg/sec of an oil of specific heat 2.20 k J/kg °K is to be heated from 10 °C to 76 °C in a countercurrent heat exchanger with an U of 290 w/m^2 °K (inside). Water (sp. heat transfer 4.187 k J/kg °K) is available at 90 °C and at a flow rate of 3.0 kg/sec. Calculate the heat transfer area required.

Solution

An over all heat balance gives $5(76 - 10) \times 2.20 = 3 \times 4.187 \times \left(90 - T_1{}''\right) = Q$ where $T_1{}^{w}$ is the water outlet temperature and Q, is the total heat transfer

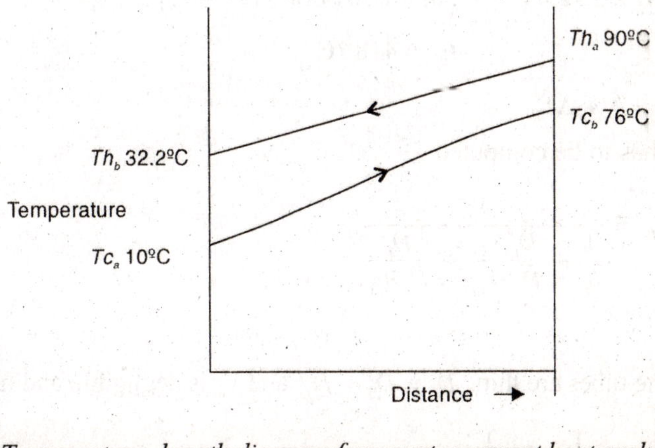

Fig. 3.2 *Temperature–length diagram for countercurrent heat exchanger*

$T_2'' = 32.2\ °C$

$Q = 726$ kw

The temperature difference at the ends of the heat exchanger are

$90 - 76 = 14\ °C$ and $32.2 - 10 = 22.2\ °C$

Substituting in the equation

$$Q = U_1\ A_i\ ln(\Delta T)$$

$$726{,}000 = 290\ A\left(\frac{22.2 - 14}{ln\dfrac{22.2}{14}}\right)$$

$A = 142\ m^2$ (heat transfer area)

Illustration

0.2 kg/sec of an alcohol flowing inside the pipes is to be cooled from 75°C to 35°C in a countercurrent flow h.e. Cooling water enters the exchanger at 12°C and at the rate of 0.16 kg/sec. The convection coefficient between the alcohol and the tube wall is 0.34 kw/m² °K and between the wall and the water is 0.225 kw/m² °K. The tubes may be assumed to be thin.

C_p data

C_p alcohol = 2.52 kJ/kg °K

C_p water = 4.187 kJ/kg °K

Compute the area of the heat exchanger surface.

The exit temperature of water may be estimated from the enthalpy balance equation. Heat lost by alcohol = heat gained by water

$$(0.2) \times 2.52 \times (75 - 35) = 0.16\ (4.187)\left(t_{c_2} - t_2\right)$$

$$t_{c_2} = 41.8\ °C$$

we know $q = UA\ \Delta t_m$

The U has to be computed

$$U_{alc} = \frac{1}{\dfrac{1}{h_i} + \dfrac{D_i\ x_w}{\overline{D}_L\ k_m} + \dfrac{D_i}{D_0\ h_0}}$$

(say inside)

Since the tubes are thin, $D_i = D_0 = \overline{D}_L$ and x_w is negligible and hence

$$\frac{1}{U_{alc}} = \frac{1}{h_{alc}} + \frac{1}{h_{water}}$$

$$\frac{1}{U_i} = \frac{1}{0.34} + \frac{1}{0.225}$$

$$U_i = 0.1355 \text{ kw/m}^2 \text{ °K}$$

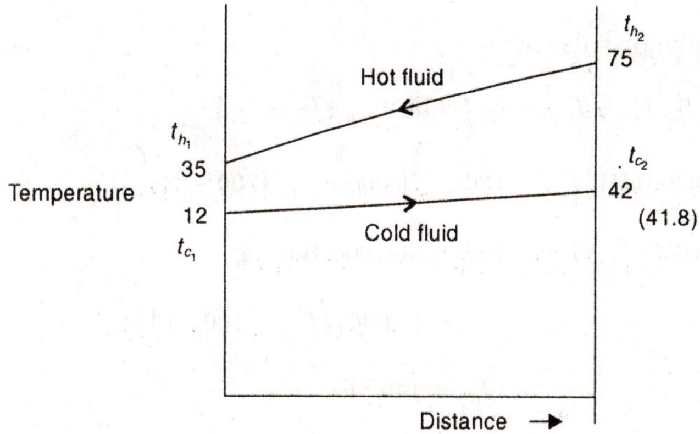

Fig. 3.3 *Temperature–length diagram for counter current exchanger*

$$\Delta t_2 = 75 - 41.8 = 33.2$$

$$\Delta t_1 = 35 - 12 = 23$$

$$\Delta t_{l_n} = \frac{\Delta t_2 - \Delta t_1}{2.303 \log \dfrac{\Delta t_2}{\Delta t_1}}$$

$$= \frac{33.2 - 23}{2.303 \log \dfrac{33.2}{23}} = 28 \text{ °K}$$

$$[\Delta t_2 = 348.2 - 315.0 = 33.2 \text{ °K}$$

$$\Delta t_1 = 308.2 - 285.2 = 23]$$

$$A = \frac{q}{U \, \Delta t_{lm}}$$

$$q = 0.2 \times 2.52 \times 40$$

$$= 20.2 \text{ k watts}$$

$$A = \frac{20.16}{(28)(0.1355} = 5.31 \text{ m}^2$$

Illustration

Calculate the required heat exchange area for the two types of flow arrangements viz. parallel flow and counterflow if 10,000 lbs/hr of water (C_p = 1 Btu/lb °F) is cooled from T_{h_1} = 200°F to T_{h_2} = 160°F by means of an equal rate of flow of water at 100°F if U = 400 Btu/hr ft^2 °F.

Solution

From the enthalpy balance

$$W_c\ C_{pc}\left(T_{C_2} - T_{C_1}\right) = W_h\ C_{ph}\left(T_{h_1} - T_{h_2}\right)$$

$$(10{,}000)\ (1)\left(T_{C_2} - 100\right) = 10{,}000 \times 1 \times (200 - 160)$$

or $$(10{,}000)\ (1)\ (200 - 160) = 400{,}000 \text{ Btu / hr}$$

$$= (10{,}000)\left(T_{C_2} - 100\right)\ (1)$$

$$T_{C_2} = 140\ °\text{F}$$

$$A = \frac{q}{U\ \Delta T_m} = \frac{400{,}000}{(400)\ (\Delta T_m)} = \frac{1000}{\Delta T_m}$$

(a) Parallel flow

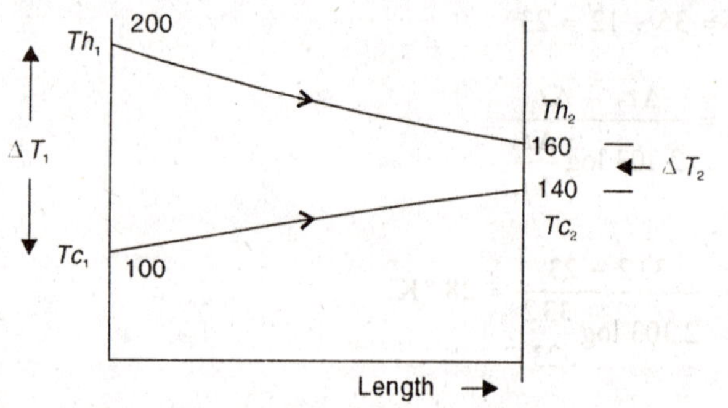

Fig. 3.4 *Temperature–length diagram for parallel flow situation*

$$\Delta T_m = \frac{\Delta T_1 - \Delta T_2}{2.303\ \log\left(\dfrac{\Delta T_1}{\Delta T_2}\right)}$$

$$\Delta T_1 = 100$$

$$\Delta T_2 = 20$$

$$\Delta T_m = \frac{100 - 20}{2.303 \log \frac{100}{20}}$$

$$= 49.6\ °F$$

$$A = \frac{1000}{49.6} = 20.2\ ft^2$$

(b) Counter flow

With equal flow rates ΔT along the heat exchanger is uniform at

$$160 - 100 = 60°F$$

$$\text{so, Area} = \frac{1000}{60} = 16.67\ ft^2$$

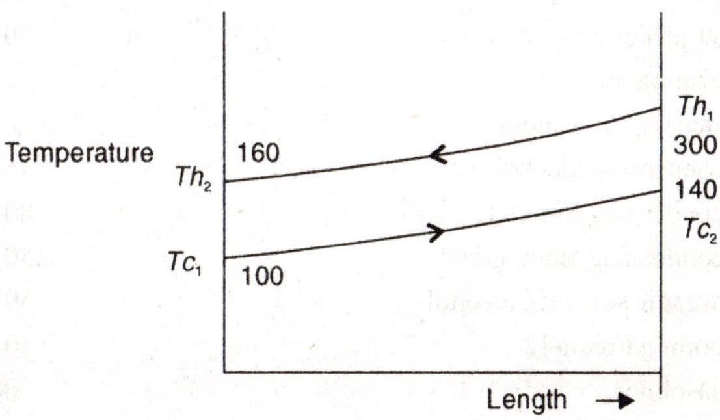

Fig. 3.5 *Temperature–length diagram for counter flow condition*

Problem

Determine the rate of heat transfer between two fluids separated by a copper tube 2 mm thick, 2 cm outer diameter and 1.5 m long if the inner fluid (H_2O) temperature is 30°C and the outer fluid (steam) temperature is 100°C. The H_2O side heat transfer coefficient is 2000 w/m^2 °K and the steam side heat transfer coefficient is 15,000 w/m^2 °K k of copper is 400 w/m °K.

Typical values of over-all heat transfer coefficients recommended for preliminary estimates are given in the following table.

Table 3.2 *Approximate over-all coefficients for preliminary estimates*

Duty	*Overall coefficients (Btu/hr ft² F)*
Steam to water	
Instantaneous heater	400–600
Storage-tank-heater	175–300
Steam to oil	
Heavy fuel	10–30
Light fuel	30–60
Light Petroleum distillater	50–200
Steam to aqueous solutions	100–600
Steam to gases	5–50
water to compressed air	10–30
Water to water jacket water coolers	150–275
Water to lubricating oil	20–60
Water to condensing oil vapours	40–100
Water to condensing alcohol	45–120
Water to condensing Freon-12	80–150
Water to condensing ammontia	150–250
Water to organic solvents, alcohol	50–150
Water to boiling Freon-12	50–150
Water to gasoline	60–90
Water to gas oil or distillate	35–60
Water to brine	100–200
Light organics to light organics	40–75
Medium organics to medium organics	20–60
Heavy organics to heavy organics	10–40
Heavy organics to light organics	10–60
Crude oil to gas oil	30–55

$\bar{h}_0$ = average unit surface heat transfer coefficient of the fluid on the outside of tubing in Btu/hr ft^2 °F

$\bar{h}_i$ = average unit surface heat transfer coefficient of the fluid inside tubing in Btu/hr ft^2 °F

R_0 = unit fouling resistance on outside of tubing in hr ft^2 °F/Btu

R_i = unit fouling resistance on inside of tubing in hr ft^2 °F/Btu

R_k = unit resistance of tubing in hr sq/ft (outside tube surface) °F/Btu

$\frac{A_0}{A_i}$ = ratio of outside tube surface to inside tube surface

For preliminary estimates of heat exchanger sizes and performance parameters it is often sufficient to know the order magnitude of the over all transmittance under average service conductions.

Fouling Factors in Heat Exchangers

The performance of heat exchangers under actual service conditions especially in the chemical and processing industry, cannot often be predicted from a thermal analysis alone. During operation with most liquids and some gases, a dirt film (a scale) gradually builds up on the heat transfer surface. This deposit or scale formation may be rust, boiler scale, silt coke or any number of things. It's effect which is often referred to as fouling, is to increase the thermal resistance. In other words due to the scale formation the performance of the heat exchanger will go down. The manufacturer can not usually predict the nature of the dirt deposit nor the rate of fouling. Therefore only the performance of clean exchangers can be guaranteed. The thermal resistance of the deposit can generally be obtained only from actual tests or from experience. If performance tests are made on a clean exchanger and repeated later after the unit has been in service for some time, the thermal resistance of the deposit can be determined from the relation.

$$R_d = \frac{1}{U_o} - \frac{1}{U}$$

in which U = overall heat transfer coefficient of the clean exchanger

U_o = overall heat transfer coefficient after the fouling has taken place that is under scale formation conditions

R_d = Unit thermal resistance of the scale

Fouling factors for various application have been complied by the Tubular Exchanger Manufacturers Association. And are available in their publication (TEMA 3/6 New York)

A few sample values are presented in the following table.

Table 3.3 *Table of Normal Fouling Factors*

Types of Fluid	*Fouling resistance hr. °F ft² /Btu*
Sea water below 125°F	0.0005
Sea water above 125°F	0.001
Treated boiler feed water above 125°F	0.001
Fuel oil	0.005
Quenching oil	0.004
Alcohol vapours	0.0005
Steam, non-oil-bearing	0.0005
Industrial air	0.002
Refrigerating oil liquid	0.001

The fouling factor should be applied as indicated in the following equation for the overall design transfer coefficient of unfinned tubes.

$$U_d = \frac{1}{\frac{1}{h_o} + R_o + R_k + \frac{R_i A_o}{A_i} + \frac{A_o}{h_i A_i}}$$

in which U_d = design overall coefficient of heat transfer in Btu/hr ft^2 °F based on a unit area of the outside tube surface.

Examples on Combined Mechanisms of Heat Transfer

An one inch nominal diameter steel pipe having its outside surface at 400°F is placed in still air at 90°F, with the convective heat transfer coefficient between the pipe surface and air equal to 1.5 Btu/hr ft^2 F. It is proposed to add 85% magnesia insulation to the outside surface of the pipe to to reduce the heat loss. What thickness of insulation will be required to reduce the heat loss by one-half if the pipe surface temperature and the heat transfer coefficient remain unchanged.

For 1 inch steel pipe OD = 1.315 in

$$= 0.1095 \text{ ft}$$

For bare pipe

$$q_o = h\, A\, \Delta T = (1.5)\,(\pi)\,(0.1095)\,(400 - 90)$$

$$= 160 \text{ Btu/hr per ft length of the pipe}$$

For insulated pipe, letting the outside diameter of the insulation be D_2

$$q = \frac{q_o}{2} = \frac{T_{atm} - T_{air}}{\frac{ln \frac{D_2}{D_1}}{2\pi k} + \frac{1}{\pi D_2 h}}$$

Putting all terms involving D_2 on one side of the equation and simplifying as much as possible, we have

$$\frac{h}{2k} ln \frac{D_2}{D_1} + \frac{1}{D_2} = \frac{2}{q_o} \Delta T \pi h$$

$$18.3\, ln \frac{D_2}{0.1095} + \frac{1}{D_2} = 18.25$$

The value of D_2 which satisfies this equality is 0.235 ft. Thus, the outside diameter of the insulated pipe is 2.82 in and the thickness of insulation required is slightly over the 0.75 inch.

Illustration on the Computation of the Overall Heat Transfer Coefficient-Plane Geometry

The wall of a building comprises the following: external wood (white pipe) sheathing 2.0 cm thick a vertical air space 10.0 cm wide and metal frame work and plaster 1 cm thick inside the surface finish. When the wind is blowing at 24 km/hour, what is the value of overall heat transfer coefficient (U) between the inside and outside air when the average temperature between inside and outside air is 16°C. The individual heat transfer coefficients h_{Ci} (air velocity and plaster) and h_{Co} [air velocity (= 24 km/hour) and smooth pine] are respectively 10.8 and 32.4 w/ m² °C. $k_{plaster\text{-}gypsum - metal\ frame\ work} = 0.47$ w/m °C, $k_{white\ pine} = 0112$ w/m °C.

Solution

This situation can be represented schematically as shown below.

$k_{white\ pine} = 0.112$ w/m °C (given)

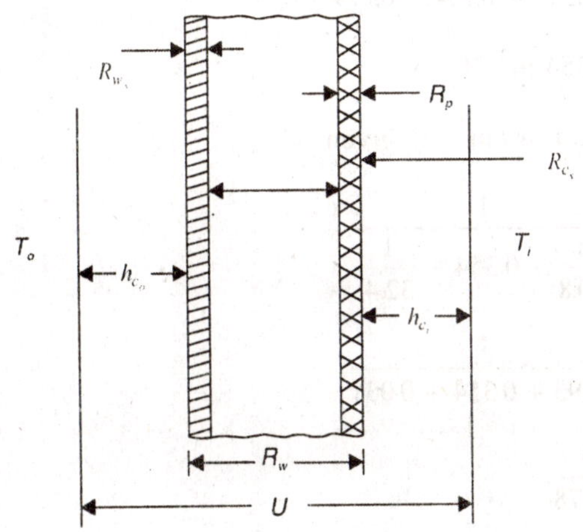

Fig. 3.6 *Heat transfer–plane goemetry*

We know that

$$U = \frac{1}{\frac{1}{h_{Ci}} + R_w + \frac{1}{h_{Co}}}$$

$$R_w = R_p + R_{as} + R_{ws}$$

$$h_{Ci} = 10.8 \text{ w / m}^2 \text{ °C}$$

$$h_{Co} = 32.4 \text{ w / m}^2 \text{ °C}$$

$$R_p = \frac{x_p}{k_p} \quad x_p = 1 \text{ cm } = 0.01 \text{ m}$$

$$k_p = 0.47$$

$$R_p = \frac{0.01}{0.47} = 0.0213 \text{ m °C / w}$$

$$R_{as} = \frac{1}{h_{as}} = \frac{1}{6.5} = 0.154 \text{ m}^2 \text{ °C / w}$$

$$R_{ws} = \frac{x_{ws}}{k_{ws}} \quad x_{ws} = 0.020 \text{ m} \quad k_{ws} = 0.112 \text{ w / m °C}$$

$$R_{ws} = \frac{0.020}{0.112} = 0.179 \text{ m}^2 \text{ °C / w}$$

$$R_w = 0.0213 + 0.154 + 0.179$$

$$= 0.354 \text{ m}^2 \text{ °C / w}$$

$$h_{Co} = 32.4 \text{ w / m}^2 \text{ °C given}$$

$$U = \frac{1}{\frac{1}{10.8} + 0.354 + \frac{1}{32.4}}$$

$$= \frac{1}{0.093 + 0.354 + 0.031}$$

$$= \frac{1}{0.478}$$

$$U = 2.09 \text{ w / m}^2 \text{ °C}$$

Illustration on Combined Conduction and Convection-Plane Geometry Systems

An ice chest contains a mixture of ice and water at 32°F, a temperature which is constant during the melting of ice. The walls of the chest are made of three materials sandwiched together as shown in the sketch given below. The outside shell is made of stamped sheet material (metal): 0.040-in-thick, low-carbon (IC) steel. The insulating material in the wall is $\frac{3}{4}$ inch thick styrofoam ($k = 0.02$ Btu/hr ft °R). The inside meterial is a fiber glass liner ($k = 0.09$ Btu/hr ft °R), $\frac{1}{4}$ inch thick.

Outside the ice chest is air at 90°F and the convection coefficient between the air and the vertical steei wall is 0.79 Btu/hr ft^2 °R. Inside, the convection coefficient between the ice water and the fiber glass is 150 Btu/hr ft^2 °R. Determine the heat-transfer rate through the wall of the ice chest based on a per-square-foot area. Determine also the overall heat transfer coefficient U for the wall. The k of IC steel is 24.8 Btu/hr ft °R.

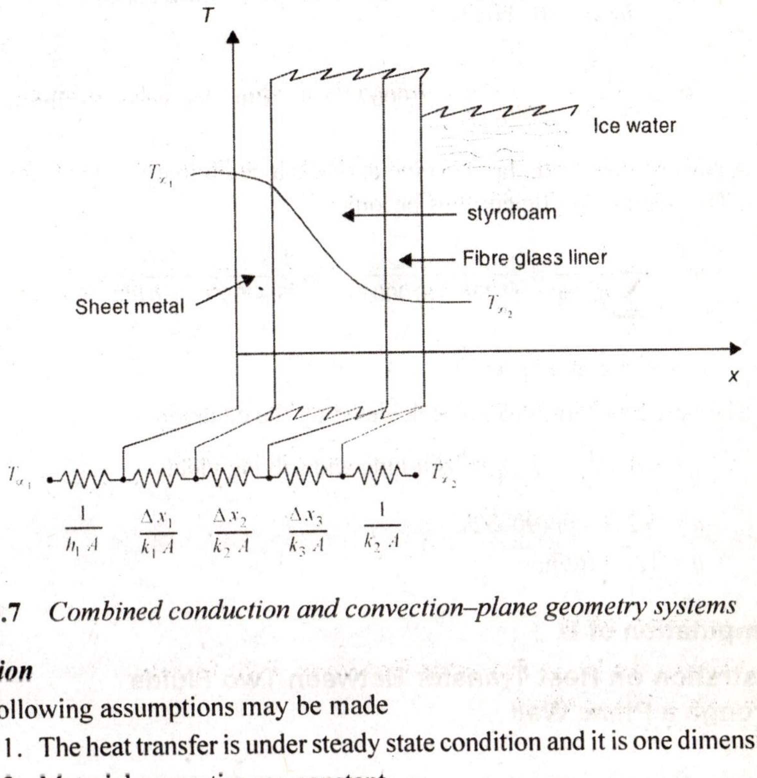

Fig. 3.7 *Combined conduction and convection–plane geometry systems*

Solution

The following assumptions may be made

1. The heat transfer is under steady state condition and it is one dimensional
2. Material properties are constant
3. Convection coefficients are constant and apply to the appropriate wall.

We can first determine U and then calculate the heat transferred

$$UA = \frac{1}{\sum_i R_i} = \frac{1}{\left(\underbrace{R_1}_{\text{Convection}} + \underbrace{R_2 + R_3 + R_4}_{\text{Conduction}} + \underbrace{R_5}_{\text{Convection}}\right)}$$

$$R_1 = \frac{1}{h_1\ A} = \frac{1}{0.79\ (1)} = 1.266\ °\text{F hr / Btu}\ \ \text{(air to sheet metal)}$$

$$R_2 = \frac{\Delta\ x_1}{k_1\ A} = \frac{\frac{0.04}{12}}{24.8\ (1)} = 0.0001\ °\text{F hr / Btu}\ \ \text{(sheet metal)}$$

$$R_3 = \frac{\Delta x_2}{k_2\ A} = \frac{\frac{0.75}{12}}{0.02(1)} = 3.125\ °\text{F hr / Btu (styroform)}$$

$$R_4 = \frac{\Delta x_3}{h_3\ A} = \frac{\frac{0.25}{12}}{(0.09)\ (1)} = 0.231\ °\text{F hr / Btu (fibre glass)}$$

$$R_5 = \frac{1}{h_2\ A} = \frac{1}{(150)\ (1)} = 0.0067\ °\text{F hr / Btu (ice water to fibre glass)}$$

As can be observed, the sheet metal offers a negligible resistance to the heat flow. The overall coefficient thus becomes

$$U = \frac{1}{\sum_i R_i\ A} = \frac{1}{(1.266 + 0.0001 + 3.125 + 0.231 + 0.0067)(1)}$$

$U = 0.216$ Btu/hr ft^2 °F

The heat transferred can be calculated by the equation

$q = UA\left(T_{\infty_1} - T_{\infty_2}\right)$ substituting and solving yields

$q = 0.216\ (1)\ (90–32)$

$q = 12.5$ Btu/hr

Computation of *U*

Illustration on Heat Transfer Between Two Fluids Through a Plane Wall

A heat exchanger wall consists of a copper plate $\frac{3}{8}$ inch thick. Supposing that the two surface coefficients are 480 and 1250 Btu/hr ft^2 °F respectively, calculate the overall heat transfer coefficient. Take that k for the given problem is 220 Btu/hr ft °F.

Solution

$$U = \frac{1}{\frac{1}{h_1} + \frac{\Delta x}{k} + \frac{1}{h^2}}$$

$$= \frac{1}{\frac{1}{480} + \frac{\frac{0.375}{12}}{220} + \frac{1}{1250}} = \frac{1}{0.00302}$$

$$= 331 \text{ Btu/hr ft}^2\ {}^\circ\text{F}$$

Calculation of Heat Transfer Area in a Heat Exchanger

Example

Calculate the outside tube area for a single-pass steam condenser to handle 731,300 lbs/hr dry saturated steam. The inlet steam pressure is 1.09 in Hg and hot-wall temperature is 81.7°F. The cooling water enters the inside of the tubes at 61.4°F and leaves at 69.9°F. The tubes are of 1 inch outside diameter by 0.902 in inside diameter, and the tube material has a thermal conductivity of 63 Btu/hr ft °F. The average water velocity in each tube is 7 ft/sec. Assume that the steam side film coefficient is 1000 Btu/hr ft² °F. The average water side coefficient is 1270 Btu/hr ft² °F.

Solution

The overall heat transfer coefficient is given by

$$U = \frac{1}{\dfrac{r_3}{h_1 r_2} + \dfrac{r_3 \ln \dfrac{r_3}{r_2}}{k} + \dfrac{1}{h_2}}$$

$$= \frac{1}{\dfrac{1}{(1.270)(0.902)} + \dfrac{0.5 \ln \dfrac{1}{0.902}}{12\,(63)} + \dfrac{1}{1000}}$$

$$= 516 \text{ Btu/hr ft}^2\ {}^\circ\text{F}$$

In the further solution of the problem it was assumed that the satuaration temperature (81.7°F) corresponding to the pressure of 1.09 m Hg prevails in the condenser

The LMTD is

$$(\Delta t)_{lm} = \frac{(81.7 - 61.4) - (81.7 - 69.9)}{\ln \dfrac{20.3}{11.8}}$$

$$= 15.7^\circ\text{F}$$

The rate of heat energy required to condense dry saturated steam at 1.09 inch Hg and 81.7°F is equal to the product of the change in enthalpy and the mass rate of steam condensed.

Thus

$$q = 731{,}300\,(1097.4 - 49.7)$$

$$= 765{,}000{,}000 \text{ Btu/hr}$$

The heat transfer area required is

$$A = \frac{765{,}000{,}000}{(516)\,(15.7)} = 94500 \text{ ft}^2$$

3. The following data were reported on a test run of a shell and tube water-to-water heat exchanger with a total outside tube area of 48.1 ft^2. The measured temperatures are given below.
Tube fluid inlet temperature = 59.5°F
Tube fluid outside temperature = 97.2°F
Shell fluid inlet temperature = 139.7°F
Shell fluid exit temperature = 108.1°F

The heat exchanger consisted of 98 brass tubes having an outside diameter of $\frac{3}{8}$ inch and a 0.049 inch wall. The shell 6.06 inch inside diameter had eleven half-moon baffles located approximately 4.3 inch apart. The dimension and spacing of the baffles were such as to allow a minimum outside water passage area between the tubes of approximately 6.67 sq. inch. The rate of water flowing through the tubes and shell were 18210 and 21780 lbs/hour respectively.

Calculate the Overall Heat Transfer Coefficient.

(a) from the test data and (b) using the equations which have been established by correlation of data by dimensional analysis.

Solution

In the present case, counterflow is assumed to exist although this is not exactly true. We know

$$(\Delta t)_{ln} = \frac{(\Delta t)_{max} - (\Delta t)_{min}}{ln\,\dfrac{(\Delta t)_{max}}{(\Delta t)_{min}}} \quad \text{(Log Mean Temperature Difference—LMTD)}$$

where $(\Delta t)_{max}$ and $(\Delta t)_{min}$ represent the maximum and minimum temperature differences in the heat exchanger

$$(\Delta t)_{ln} = \frac{(108.1 - 59.5) - (139.5 - 97.2)}{ln\,\dfrac{(108.1 - 59.5)}{(139.7 - 97.2)}} = 46\ °\text{F}$$

we know,

$$q = UA\,(\Delta t)_{ln}$$

$$= U\,(48.1)\,(46)$$

we also know that

$$q = m\,C_p\,(\Delta t)$$

$$= (18210)\,(1)\,(97.2 - 59.5) = 686{,}000 \text{ Btu/hr}$$

From the above two equations, we can get the value of U

$$U = 310 \text{ Btu/hr ft}^2\ {}^\circ\text{F}$$

(b) The individual heat transfer coefficient for the outside of the tubes (heating fluid) can be calculated by means of the following equation

$$\frac{hD}{k} = 0.33\left(\frac{Du\rho}{\mu}\right)^{0.6}\left(\frac{C_p\,\mu}{k}\right)^{0.33} \quad \text{(by dimensional analysis)}$$

which has been recommended by Colburn for correlating the variables which influence the heat transfer coefficients in the situation of heating of fluids flowing normal to tube banks.

In this problem the situation is similar if we consider the flow in the shell as a flow normal to tube banks. If will be assumed that the average tube temperature is 113°F. This will be checked again later.

The physical properties are to be evaluated at the average film temperature, which is given by

$$t_f = t + \frac{t_s - t}{2} = \frac{t + t_s}{2}$$

where t and t_s are the mean fluid and surface temperatures respectively.

The average film temperature in this problem is,

$$t_f = 123.9 + \frac{113 - 123.9}{2} = 118.4\ {}^\circ\text{F}$$

At this temperature, the pertinent thermo-physical properties of water in consistent units are

$$\rho = 61.7 \text{ lbs/c ft}$$

$$\mu = 1.38 \text{ lb/ft hr}$$

$C_p = 1$ Btu/lb °F (can be obtained from standard handbook on physical/ thermo-physical properties)

$$k = 0.366 \text{ Btu/hr ft}\ {}^\circ\text{F}$$

Further, we require

$$D = \frac{0.375}{12} = 0.03125 \text{ ft}$$

$$u = \frac{\dfrac{21780}{61.7}}{\dfrac{6.67}{144}} = 7610 \text{ ft / hr}$$

Then $N_{Re} = \dfrac{Du\rho}{\mu} = \dfrac{7610\,(0.03125)\,1.917}{0.0429}$

$$N_{P_r} = \frac{C_p\,\mu}{k} = \frac{(0.0429)\,(32.2)}{0.366} = 3.77$$

Substituting these dimensionless group in the equation

$$\frac{h\,D}{k} = 0.33\left(\frac{Du\rho}{\mu}\right)^{0.6}\left(\frac{C_p\,\mu}{k}\right)^{0.33}$$

$$N_{N_u} = \frac{hD}{k} = 0.33\,(10620)^{0.6}\,(3.77)^{\frac{1}{3}}$$

$= 133.8$ and from this, we can obtain h.

$$h = \frac{133.8\,(0.366)}{0.03125} = 1567 \text{ Btu / hr ft}^2$$

The individual coefficient 'h' for the inside of any of the 98 tubes (heated fluid) will be determined by the following eqution.

$$\frac{h\,D}{k} = 0.023\left(\frac{Du\rho}{\mu}\right)^{0.8}\left(\frac{C_p\,\mu}{k}\right)^{0.4} \quad \text{(Dittus-Boelter equation)}$$

At the average temperature of 78.3°F, the thermo-physical properties of water are

$$\rho = 62.2 \text{ lb / ft}^3$$

$$\mu = 2.13 \text{ lb / ft hr}$$

$$C_p = 1 \text{ Btu / lb °F}$$

$$k = 0.348 \text{ Btu / hr ft °F}$$

Further

$$D = \frac{0.277}{12} = 0.0231 \text{ ft}$$

$$u = \frac{\dfrac{18210}{62.2}}{98\,\dfrac{\pi\,(0.0231)^2}{4}} = 7140 \text{ ft / hr}$$

$$N_{R_e} = \frac{Du\rho}{\mu} = \frac{(7140)\,(0.0231)\,1.93}{0.0662} = 4810$$

$$N_{P_r} = \frac{C_p\,\mu}{k} = \frac{(0.0662)\,(32.2)}{0.348} = 6.10$$

By substituting in the Dittus-Boelter equation given above

$$N_{N_u} = \frac{hD}{k} = (0.023)\,(4810)^{0.8}\,(6.10)^{0.4}$$

$$= 41.78$$

and from this $h = \dfrac{41.78\,(0.348)}{0.0231} = 630\text{ Btu / hr ft}^2\ {}^\circ\text{F}$

Substitution of these quantities in the equation

$$U = \frac{1}{\dfrac{A_2}{h_1 A_1} + \dfrac{A_2\,ln\,\dfrac{r_3}{r_2}}{k\,2\pi\,L} + \dfrac{1}{h_2}}$$

gives the overall heat transfer coefficient. Assume that the k of brass is 58 Btu/hr ft°F. Then from the equation

$$U = \frac{1}{\dfrac{r_3}{h_1\,r_2} + \dfrac{r_3\,ln\,\dfrac{r_3}{r_2}}{k} + \dfrac{1}{h_2}}$$

$$U = \frac{1}{\dfrac{0.03125}{0.0231}\cdot\dfrac{1}{630} + \dfrac{0.03125}{2}\cdot\dfrac{1}{58}\,ln\,\dfrac{0.03125}{0.0231} + \dfrac{1}{156.7}}$$

$$= \frac{1}{0.00215 + 0.0000815 + 0.00064}$$

$$= 348\text{ Btu / hr ft}^2\ {}^\circ F$$

This value is 12.6% larger than the actual value. The deviation is not very great considering the simplifying assumptions which had to be made since the three terms in the denominator of the last equation are proportional to the three resistances involved, it is observed that the tube wall presents only about 3% of the total resistance. Therefore in the average temperature drop between the two fluids of about 45°F, only about 1.3°F is due to the heat resistance of the wall.

If the average temperature of the outer surface of the tubes is taken as t_3, then that for the innner surface would be $t_3 - 1.3$.

Then equating the heat flow between the the wall and the heated water, we get.

$$h_2\ A_2\,(t_3 - 123.9) = h_1\ A_1\left[78.3 - (t_3 - 1.3)\right]$$

or $$1567\ (0.03125)\ \pi\ L\ (t_3 - 123.9)$$

$$= 630\ (0.0231)\ \pi\ L\left[78.3 - (t_3 - 1.3)\right]$$

Evaluating gives $t_3 = 113.8$°F, which is in excellent agreement with the value 113°F assumed at the beginning. If there had been a considerable discrepancy, the calculation would have to be repeated with another estimate of t_3.

Illustration on the Computation of the Overall Heat Transfer Coefficient

1. Determine the overall heat transfer coefficient between water and oil if the water flows through a copper pipe 1.8 cm *ID* and 2.1 cm *OD* while the oil flows though the annulus between this pipe and a steel pipe. The water and oil side film coefficients are respectively 4000 and 1100 k cal/hr m^2 °C. The fouling factors on the water side and oil sides may be taken to be 0.0004 and 0.001 respectively. The thermal conductivity of the tube wall is 300 k cal/hr m °C.

Solution

We know

$$U = \frac{1}{\left(\frac{r_o}{r_i}\right)\frac{1}{h_i} + \left(\frac{r_o}{r_i}\right)\left(\frac{1}{h_{d_i}}\right) + \left(\frac{r_0}{k_w}\right) ln\frac{r_o}{r_i} + \frac{1}{h_{d_o}} + \frac{1}{h_o}}$$

Now $$\frac{r_0}{r_i} = \frac{2.1}{1.8} = 1.167,\ r_0 = 1.05 \times 10^{-2}$$

$$h_i = 4000$$

$$h_{d_i} = \left(\frac{1}{0.0004}\right)$$

$$h_{d_o} = \left(\frac{1}{0.001}\right) \text{ and } h_o = 1100$$

Therefore

$$U = \frac{1}{\left[\frac{1.167}{4000} + (1.667)\,(0.0004) + \frac{(1.05)\,10^{-2}}{300}\,ln\,(1.167) + 0.01 + \frac{1}{1100}\right]}$$

$$= \frac{1}{10^{-4}\left[2.92 + .4.668 + 0.056 + 10 + 9.09\right]}$$

$$= \frac{10^4}{26.734} = 374 \text{ k cal / hr m}^2 \text{ °C}$$

2. A single-pass steam condenser has to be designed to condense 35000 hg/hr of dry saturated steam at 2.77 cm Hg pressure (saturation temprature = 27.0°C). The cooling water enters the tubes at 16°C and leaves at 21°C. The tubes are 2.25 cm *ID* and 2.5 cm *OD*. The thermal conductivity of tube material is 90 k cal/hr m °C. The water and steam side film coefficients are 6,000 and 5,000 k cal/hr m^2 °C respectively. Determine the surface area of tubes required. Neglect the fouling factors.

Solution

In the absence of scale resistance, the following expression can be used for computing '*U*'

$$U = \frac{1}{\left(\frac{r_o}{r_i}\right)\frac{1}{h_i} + \left(\frac{r_o}{k}\right) ln\left(\frac{r_o}{r_i}\right) + \frac{1}{h_o}}$$

$$= \frac{1}{\left(\frac{2.5}{2.25}\right)\left(\frac{1}{6000}\right) + \left(\frac{1.25}{40 \times 100}\right) ln\left(\frac{2.5}{2.25}\right) + \frac{1}{5000}}$$

$$= \frac{1}{10^{-4}\left[1.85 + 0.145 + 2\right]}$$

or $$U = \frac{10^4}{3.995} = 2500 \text{ k cal / hr m}^2 \text{ °C}$$

The log mean temperature difference is

$$\Delta t_{lm} = \frac{(27.6 - 21) - (27.6 - 16)}{ln\left(\frac{6.6}{11.6}\right)}$$

$$= \frac{5}{ln\,(1.76)} = \frac{5}{0.5653} = 8.85 \text{ °C}$$

$$q = 35000 \times 582 \text{ k cal / hr}$$

we know $$A = \frac{q}{U\,\Delta t_m} = \frac{(35000)\,(582)}{(250) \times (8.85)} = 816 \text{ m}^2$$

Illustration in Heat Transfer by Combined Mechanisms—Compound Resistances

Calculate the rate of heat loss per unit area of the wall of rectangular combustion chamber. The walls of the the chamber are made up of 225 mm of red brick and 225 mm thick layer of magnesite brick. The gases inside the combustion chamber are at 1227°C. The convection coefficient on the inner surface of the chamber wall is 31 k cal/hr m^2 °C. The surrounding air temperature is 37°C. The convection coefficient at the outer surface of the wall is 16 k cal/hr m^2 °C. Determine also the outside surface temperature of the wall. Take the thermal conductivities of red brick and magnestic brick to be 0.57 and 3.3 k cal/hr m °C respectively. It may be mentioned that the radiation effects are included in the values of the convection coefficients given above.

Solution

The overall heat transfer coefficient U is given by (for a plane wall)

$$U = \frac{1}{\frac{1}{h_1} + \frac{x_1}{k_1} + \frac{x_2}{k_2} + \frac{1}{h_2}}$$

$$= \frac{1}{\frac{1}{31} + \frac{0.225}{3.3} + \frac{0.225}{0.57} + \frac{1}{16}}$$

$$= \frac{1}{(0.0322) + (0.0681) + (0.3950) + (0.0625)}$$

$$= 1.79 \text{ k cal / hr m}^2 \text{ °C}$$

$$\Delta t_{total} = 1227 - 37 = 1190 \text{ °C}$$

∴ The rate of heat loss per unit area of the wall surface is

$$q' = \frac{Q}{A} = U \, \Delta t_{total}$$

$$= 1.79 \times 1190 = 2130 \text{ k cal / hr m}^2$$

The ouside surface temperature say t_{w3} can be found as follows

$$t_{w3} - t_{f2} = \frac{q}{h_2} = \frac{2130}{16} = 133.1 \text{ °C}$$

t_{f2} is the surrounding air temperature.

Since t_{f2} is 37°C

$$t_{w3} = 37 + 133.1 = 170.1\text{°C}$$

Example in Heat Transfer From One Fluid to Another Fluid Through a Cylindrical Wall

A 150 mm diameter steam pipe with 165 mm outside diameter is covered with a layer of insulation 100 mm thick ($k = 0.08$ k cal/hr m °C). The temperature of the steam (t_{f1}) is 250°C and the surrounding air temperature t_{f2} is 35°C. the 'k' of the material of the pipe is 39 k cal/hr m °C. Take that the heat transfer coefficients of fluid on the hot and cold sides are 80 and 8 k cal/ hr m^2 °C respectively. Calculate the overall heat transfer coefficient and the rate of heat loss per unit length of the pipe. Determine also the temperature of the wall surfaces.

Solution

The data given are

$r_1 = 0.075$ m; $r_2 = 0.0825$ m; $r_3 = 0.1825$ m

$k_1 = 39.5$ k cal/hr m^2 °C; $k_2 = 0.084$ k cal/hr m^2 °C

$t_{f1} = 250$°C; $t_{f2} = 35$°C

$h_1 = 80$ k cal/hr m^2 °C; $h_2 = 8$ k cal/hr m^2 °C

The overall heat transfer coefficient per unit length of the pipe

$$U = \frac{2\pi}{\dfrac{1}{h_1 r_1} + \dfrac{ln\left(\dfrac{r_2}{r_1}\right)}{k_1} + \dfrac{ln\left(\dfrac{r_3}{r_2}\right)}{k_2} + \dfrac{1}{h_2 r_3}}$$

$$= \frac{2\pi}{(0.167) + (0.00244) + (9.9) + 0.685}$$

$$= 0.584 \text{ k cal/hr m °C}$$

The rate of heat flow per unit length of the cylinder/pipe.

$$q = U\left(t_{f1} - t_{f2}\right)$$

$$= 0.584\,(250 - 35) = 125.5 \text{ k cal / hr m}$$

If t_{w_f} is the temperature of the inner surface

$$q = h_1\, 2\pi\, r_1 \left(t_{f1} - t_{w1}\right)$$

i.e. $$125.5 = 80 \times 2\pi \times 0.075 \times \left(250 - t_{w_1}\right)$$

$$\therefore \left(250 - t_{w1}\right) = \frac{125.5}{(80)\,(2\pi)\,(0.075)} = 3.32 \text{ °C}$$

$$\therefore t_{w1} = 250 - 3.32 = 246.68 \text{ °C}$$

Also $q = h_2 \, 2\pi \, r_3 \left(t_{w3} - t_{f2}\right)$

$$125.5 = 8 \times 2\pi \times 0.1825 \left(t_{w_3} - 35\right)$$

$$\left(t_{w_3} - 35\right) = 125.5 \,/ (8 \times 2\pi \times 0.1825)$$

$$= 13.65\ °C$$

$$\therefore t_{w3} = 35 + 13.65 = 48.65\ °C$$

Illustration on Critical Radius of Insulation Calculation

A pipe with an outer diameter 34 mm which is at a temperature of 200°C is to be covered with an insulating material having a thermal conductivity of 0.2 k cal/m hr °C. If the surrounding air temperatu're is 30°C determine the heat loss for the pipe if the thickness of insulation is (i) 10 mm (ii) 20 mm (iii) 30 mm (iv) 55 mm and (v) 100 mm. What is the critical diameter of insulation and the corresponding maximum amount of heat loss? Take the heat transfer coefficient between the insulation and the fluid to be 4 k cal/hr m² °C.

Solution

(i) The heat loss per unit length of the pipe is given by the expression

$$q = \frac{t_{w2} - t_{f2}}{\left(\frac{1}{h_2} 2\pi r_3\right) + ln\left(\frac{\frac{r_3}{r_2}}{2\pi k_2}\right)}$$

t_{w2} = outer wall temperature of the pipe

t_{f2} = surrounding air temperature

In this case $t_{w2} = 200\ °C\ ;\ t_{f2} = 30\ °C$

$$h = 4 \text{ k cal / hr m}^2\ °C$$

and $r_2 = 17$ mm ; $r_3 = 17 + 10 = 27$ mm

$$q = \frac{(200 - 30)}{\left(\frac{1}{4} \times 2\pi \times 0.027\right) + \frac{ln\left(\frac{27}{17}\right)}{2\pi \times 0.2}}$$

$$= \frac{170}{1.47 + 0.369} = 92.5 \text{ k cal / hr m}$$

(ii) When thickness is 20 mm

$r_3 = 17 + 20 = 37$ mm

$$q = \frac{(200 - 30)}{\left(\frac{1}{4} \times 2 \times 0.037\right) + \frac{ln\left(\frac{37}{17}\right)}{2\pi \times 0.12}}$$

$$= \frac{170}{1.075 + 0.618} = 100.2 \text{ k cal / hr m}$$

(iii) For the situation in which, the thickness of the insulation is 30 mm

$r_3 = 17 + 30 = 47$ mm

$$\text{and } q = \frac{170}{(0.847) + (0.808)}$$

$$= 102.7 \text{ k cal / hr m}$$

(iv) When the thickness of the insulation is increased to 50 mm

$\therefore r_3 = 17 + 50 = 67$ mm

$$\text{and } q = \frac{170}{(0.594) + (1.09)} = 101 \text{ k cal / hr m}$$

(v) When the thickness of the insulation is 100 mm

$r_3 = 17 + 100 = 117$ mm

$$\text{and } q = \frac{170}{(0.34 + 1.535)} = 90.7 \text{ k cal / hr m}$$

The critical diameter of the insulation is

$$d_{c_r} = \frac{2k_2}{h_2} = \frac{(2 \times 0.2)}{4} = 0.1 \text{ meter or 100 mm}$$

Therefore the thickness of the insulation for the maximum heat loss from the pipe is (50 – 17) or 33 mm. And the corresponding maximum heat loss per unit length is

$$q_{max} = \frac{170}{(0.795 + 0.857)} = 100 \text{ k cal / hr m}$$

It may be noted from the above example that the heat loss from the pipe first increases and the reaches a maximum then again decreases.

Illustration on the Computation of *U*

A 6 inch nominal diameter steel (IC) pipe in the drying part of a car wash carries warmed air at a temperature of 140°F. The pipe is insulated with a 1 inch thick layer of glass wool. The convection coefficient between the air and the pipe wall is 12 Btu/hr ft^2 R. Outside the insulation the air temperature is 68°F and the convection coefficient between the glass wool and the ambient air is 1.5 Btu/hr ft^2 R. Determine the '*U*' for the system based on the inside diameter of the pipe.

The k for IC steel. $k = 24.8$ Btu/hr ft^2 R

k for glass wool, $k = 0.023$ Btu/hr ft^2 R

$D_2 = 6.625$ inch

$D_1 = 0.5054$ inch

Solution

Assumption

1. One-diamesional steady, radial heat flow occurs
2. The properties of the materials are constant
3. Convection coefficient are constant

The equation for the computation of the '*U*' modified slightly to account for the insulation is

$$U_1 = \frac{1}{\dfrac{1}{h_1} + \dfrac{R_1 \, ln \dfrac{R_2}{R_1}}{k_{12}} + \dfrac{R_1 \, ln\left(\dfrac{R_3}{R_2}\right)}{k_{23}} + \dfrac{R_1}{h_2 \, R_3}}$$

Note that the ratio of radii appearing with $\dfrac{1}{h_2}$ is $\dfrac{R_1}{R_3}$ because A_1 and A_3 are the areas through with convection takes place

Because the insulation is 1 inch thick

$D_3 = D_2 + 2$ inch $= 8.625$ inch $= 0.71875$ ft

Substituting the values concerned, in the equation for U_l, gives

$$U_1 = \frac{1}{\dfrac{1}{12} + \dfrac{0.5054 \, ln\left(\dfrac{0.552}{0.5054}\right)}{24.8} + \dfrac{0.5054 \, ln \dfrac{0.7185}{0.552}}{0.023} + \dfrac{1\,(0.5054)}{1.5\,(0.7187)}}$$

Evaluating each term in the denominator we get

$$U_1 = \frac{1}{0.0833 + 0.0018 + 5.80 + 0.469}$$

The resistance of the insulation, 5.80 is the greatest of the individual components, so we expect the gresatest temperature drop to exist across the insulation.

Solving $U_1 = 0.157$ Btu/hr ft^2 °R.

Problems on Film Coefficients

1. In a steel double-pipe heat exchanger, an organic liquid is flowing through the inner pipe.

   It is cooled by water flowing through the outer pipe. The inner pipe is made from a 3-cm nominal diameter pipe. The thermal conductivity of steel can be taken as 40 *k*. cal/hr m deg C. Data on the individual heat transfer coefficients and dirt factors are as under;

organic liquid film coefficient, h_1	= 950 k cal/m^2 hr deg C
water film coefficient, h_1	= 1425 k cal/m^2 deg C
inside dirt factor, hd_1	= 4500 k cal/m^2 deg C
outside dirt factor, hd_2	= 3000 k cal/m^2 deg C

   Determine the value of the overall heat transfer coefficient based inner pipe.

2. Determine the overall heat transfer coefficient for a condenser tube [whose '*k*' is 110 *k* cal/hr m °C having an inner diameter of 2.15 cm and an outer diameter of 2.40 cm. The water and steam side film coefficients are 4000 and 5250 *k* cal/hr m^2 °C respectively.

3. Steam is to be condensed at atmospheric pressure in a shell and tube exchanger consisting of 18-gauge Admiralty tubing [0.905 inch ID and 0.050 wall thickness]. Its thermal conductivity is 62 Btu/hr.ft °F. The individual heat transfer coefficients an *d* scale factors are as follows;

water side	= 850 Btu/hr ft^2 °F
steam side	= 2000 Btu/hr ft^2 °F
dirt factor on steam side	= 1800 Btu/hr ft^2 °F
dirt factor on water side	= 2200 Btu/hr ft^2

   Calculate the overall heat transfer coefficient based on the outside area of the tubes.

4. A shell and tube counterflow heat exchanger is to be designed for heating an oil from 90°F to 190°F by steam condensing at 215°F on the outside of the pipes. The steam side film coefficient is 1900 Btu/hr ft^2 °F and the outside heat transfer coefficient is 16 Btu/hr ft^2 °F. The '*k*' of the metal of the tubes is 17 Btu/hr ft °F. Find the overall heat transfer coefficient, *U*, based on the outer surface area of the tubes.

5. Saturated non-oil bearing steam at 1 atm. pressure enters the shell pass of the two-tube pass shell condenser. The tubes are made of schedule 160, ¾ inch steel pipe nominal diameter.

   The '*k*' of the steel may be taken as 20 Btu/hr ft °F. The condensing heat transfer coefficient is 2000 Btu/hr ft^2 °F and that of water in the tubes is 1400 Btu/hr ft^2 °F.

   Calculate the over all heat transfer coefficient based on the inside area of the tubes.

4

Unsteady State Heat Conduction

In the case of the steady state heat production the temperature at a given point is independent of time. The heat conduction said to be in the unsteady state when the temperature at a given point varies with time.

Situations in industry where the steady state heat conduction is involved include,

- quenching of billets
- annealing of solids
- manufacturing process of glass
- burning of bricks
- the steaming of wood
- vulcanisation of rubber

In the vulcanisation of rubber where chemical reactions occur during the heating period, heat transfer plays an important role because the rate of the reaction increases rapidly with small increases in temperature.

4.1 APPLICATION OF MOLECULAR TRANSPORT THEORY TO THE UNSTEADY STATE HEAT CONDUCTION

The concentration of the property (energy) at a point within the system varies with time and the change in concentration is accompanied by a variation in the flux and the rate of transport. Therefore the variation of the concentration of the transferent property with time will be expressed in the form of an equation.

For unsteady state transfer, the balance equation is

Input = out put + accumulation

Accumulation may have a positive or negative value depending on whether the property is generated or consumed (disappearing).

If the unsteady state problem is one which involves the transfer with internal generation, then the balance equation is input + generation = output + accumulation

The general unsteady state transport equation is

Rate of input – Rate of output = Rate of accumulation

δ = generalised diffusivity

T = concentration of the transferent property

ψ = flux of the transferent property

A = transfer area

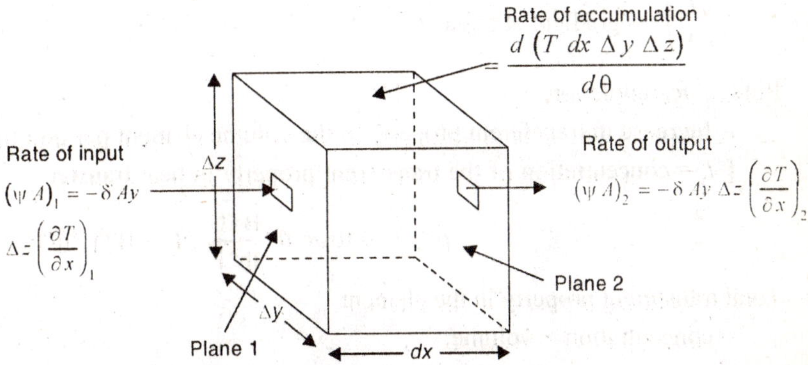

Fig. 4.1 *Unsteady state heat conduction—slab geometry*

Consider the volume element shown in the figure.

We consider the transport in the x-direction only

Rate of input into the volume element

$$= \psi A$$

$$(\psi A) = -\delta A \frac{dT}{dx}$$

The above expression is valid both for steady and unsteady states

For steady state $\rightarrow \frac{dT}{dx}$ and ψA (simple transfer) are constant and do not vary with distance

But at unsteady state conditions, they vary with time and position

The rate of input (at plane 1)

$$(\psi A)_1 = -\delta\,\Delta y\,\delta z\left(\frac{\partial T}{\partial x}\right)_1 \qquad ...(1)$$

$$A = \Delta y\, \Delta z$$

$$\left(\frac{\partial T}{\partial x}\right)_1 = \text{gradient of the transferent property at plane 1}$$

Rate of output (at plane 2)

$$(\psi A)_2 = -\delta\, \Delta y\, \Delta z \left(\frac{\partial T}{\partial x}\right)_2 \qquad ...(2)$$

$$\left(\frac{\partial T}{\partial x}\right)_2 = \text{gradient at plane 2}$$

Rate of accumulation

= Increase in transferent property in the volume element per unit time

[T = concentration of the transferent property in heat transfer

$$\rho c_p T = \text{lbs/c ft}\ \frac{\text{BTU}}{\text{lb° F}}\ \text{°F} = \text{BTU/ft}^3]$$

Total transferent property in the element

= concentration × volume

$= T\, dx\, \Delta y\, \Delta z$

$$\text{Rate of accumulation} = \frac{d\,(T\, dx\, \Delta y\, \Delta z)}{d\theta} \qquad ...(3)$$

The above three equations can be substituted in the overall balance equation

$$(\psi A)_1 - (\psi A)_2 = \frac{d\,(T\, \Delta y\, \Delta z\, dx)}{d\theta} \qquad ...(4)$$

or
$$-\delta\, \Delta y\, \Delta z \left(\frac{\partial T}{\partial x}\right)_1 + \delta\, \Delta y\, \Delta z \left(\frac{\partial T}{\partial x}\right)_2 = \frac{\partial T}{\partial \theta}\, \Delta y\, \Delta z\, dx \qquad ...(5)$$

or
$$-\left[\delta \left(\frac{\partial T}{\partial x}\right)_1\right] + \left[\delta \left(\frac{\partial T}{\partial x}\right)_2\right] = \left(\frac{\partial T}{\partial \theta}\right) dx \qquad ...(6)$$

Now the fluxes at the two planes may be related by

$$\left[\delta \left(\frac{\partial T}{\partial x}\right)_2\right] = \left[\delta \left(\frac{\partial T}{\partial x}\right)_1\right] + \frac{\partial}{\partial x}\left[\delta \left(\frac{\partial T}{\partial x}\right)\right] dx \qquad ...(7)$$

Equations (6) and (7) on combination yield

$$\frac{\partial}{\partial x}\left[\delta\left(\frac{\partial T}{\partial x}\right)\right]dx = \frac{\partial T}{\partial \theta}\,dx \qquad \text{...(8)}$$

If δ is taken to be constant, equation (8) becomes

$$\delta\left(\frac{\partial^2 T}{\partial x^2}\right) = \left(\frac{\partial T}{\partial \theta}\right) \qquad \text{...(9)}$$

The above equation is the general unidirectional unsteady-state molecular transport equation and it states that the change of concentration with time is proportional to the change in concentration gradient with position.

The unsteady state equation for heat transfer

$$\frac{\partial(\rho\, c_p\, T)}{\partial \theta} = \alpha\,\frac{\partial^2(\rho\, c_p\, T)}{\partial x^2}$$

$$\frac{\partial T}{\partial \theta} = \alpha\,\frac{\partial^2 T}{\partial x^2}; \quad \alpha = \frac{k}{\rho_{c_p}}$$

α is called thermal diffusivity, ft^2/hr

The above equation is Fourier's law for unsteady state heat conduction.

The unsteady state equation will reduce to the familiar steady-state equation, if the accumulation is zero

In the equation

$$\delta\left(\frac{\partial^2 T}{\partial x^2}\right) = \frac{\partial T}{\partial \theta}$$

$$\frac{\partial T}{\partial \theta} = 0$$

In which case $\delta\,\dfrac{\partial^2 T}{\partial x^2} = 0$

Integrating $\delta\left(\dfrac{\partial T}{\partial x}\right) = \text{constant}$

Which shows that the gradient is constant and is of the form corresponding to steady state transfer.

For heat transfer in all the three directions (x, y, z), the following equation can be developed.

$$\frac{\partial T}{\partial \theta} = \alpha \left[\frac{\partial^2 T}{\partial x^2} + \frac{\partial^2 T}{\partial y^2} + \frac{\partial^2 T}{\partial z^2}\right]$$

$$= \alpha \, \dot{\nabla}^2 \, T$$

where $\nabla = \left[\frac{\partial^2}{\partial x^2} + \frac{\partial^2}{\partial y^2} + \frac{\partial^2}{\partial z^2}\right]$

Similar equations can be derived in term of cylindrical and spherical coordinates.

4.2 SOLUTIONS OF THE UNSTEADY STATE EQUATIONS

Solution in many cases is difficult and of course is dependent on the geometry and boundary conditions. These equations have been solved for a few simple geometries and boundary conditions.

Fourier developed an analytical solution for heat transfer equation making use of what is known is Fourier series (infinite trigonometric series).

For a slab of thickness $2x_1$

T_0—initial temperature

The slab is suddenly exposed to the surroundings (at time zero) such that its surface temperature is T_1. Under these conditions the solution for the unsteady state heat transfer differential equation is

For symmetrical bodies

$$\frac{T_1 - T}{T_1 - T_0} = \frac{4}{\pi}\left[e^{-\alpha\pi^2\theta/4x_1^2} \sin\frac{\pi x}{2x_1} + \frac{1}{3}e^{-9\alpha\pi^2\theta/4x_1^2} \sin\frac{3\pi x}{2x_1} + \frac{1}{5}e^{-25\alpha\pi^2\theta/4x_1^2} \sin\frac{5\pi x}{2x_1} + \ldots\right]$$

The above expression can be evaluated for any value of x and θ.

T = Temperature at point x, at time θ

T_0 = initial temperature of the slab at all values of x

T_1 = few surface temperature of the slab at all values of θ

x = distance from centre plane to any point in the slab

x_1 = distance from centre plane to surface.

Use of Charts for Solving the Differential Equation

To facilitate calculation, solutions to the unsteady state equations have been prepared in graphical form for a few simple geometries

Large flat slab

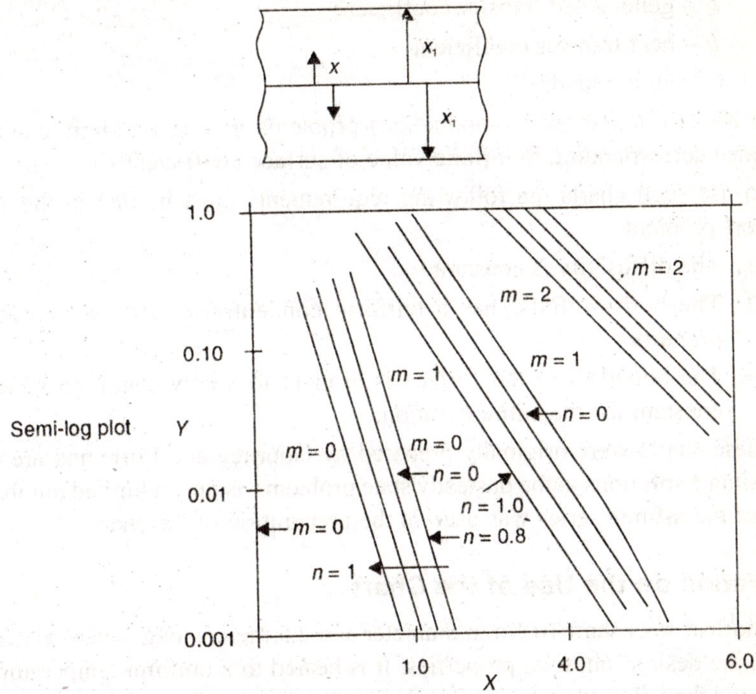

Fig. 4.2 *Plot for solutions for unsteady state heat transfer problems*

For the case of long cylinder and sphere similar curves for unsteady state transport are available. These charts can be used for solution of any unsteady state transport problems (heat, mass and momentum)

The following table presents the parameters involved in the calculation.

Notations	*Parameter symbol*	*General notation*	*Notation for heat transfer*
Accomplished change	Y	$\frac{T_1 - T}{T_1 - T_0}$	$\frac{T_1 - T}{T_1 - T_0}$
Relative time	X	$\frac{\delta\theta}{x_1^2}$	$\frac{\alpha\theta}{x_1^2}$
Relative position	n	$\frac{x}{x_1}$	$\frac{x}{x_1}$
Relative resistance	m	$\frac{\delta}{\in x_1}$	$\frac{k}{h\, x_1}$

x – distance from centre to any point. X, Y, m, and n – dimensionless

$\in$ – generalised transfer coefficient

h – heat transfer coefficient

x_1 – upto boundary

for all molecular transport (conduction) problems $m = 0$, (negligible surface resistance corresponding to infinite value of surface coefficient h).

To use such charts the following requirements must be met in the given transport problem.

1. The diffusivity is constant
2. The body initially has a uniform concentration of the transferent property.
3. The boundary to the surface is brought to a new condition which is constant for the entire duration.

These charts were originally prepared by Gourney and Lurie and are only approximate solutions to the unsteady state problems as only a limited number of terms of the infinite series was used in the preparation of the chart.

Illustration on the Use of the Chart

A cylindrical steel shaft 10 cm in diameter and 2.5 meters long is heat treated to give it, the desired physical properties. It is heated to a uniform temperature of 600°C and then plunged into an oil bath which maintains the surface temperature at 150°C. Calculate the radial temperature profile at 2 minutes and 3 minutes after the immersion. The values of c_p, ρ and k are available at these temperatures.

Solution

The temperature of the rod varies between 600°C and 150°C and hence it is desirable to check the value of α, the thermal diffusivity at these temperatures. Form the given values of c_p, ρ and k at these temperature, α can be calculated

$$\alpha = k / \rho\, c_p$$

$$\alpha \text{ at } 150°\text{C} = 0.110 \text{ cm}^2/\text{sec}$$

$$\alpha \text{ at } 600°\text{C} = 0.0625 \text{ cm}^2/\text{sec}$$

The difference is quite appreciable. Since no other, simple method is available we will use the average of these values assuming it to be constant.

$$\alpha = 0.0862 \text{ cm}^2/\text{sec}$$

The rod is initially at a uniform temperature of 600°C and the boundary maintained at the new temperature of T_1 = 150°C that is 423°k.

The chart given for the cylinder has to be used

For θ = 2 min $\qquad x_1 = 0.5$ cm

$$Y = \frac{T_1 - T}{T_1 - T_0} = \frac{423 - T}{423 - 873}$$

$$X = \frac{\alpha\,\theta}{x_1^2} = \frac{(0.0862)\,(120)}{25}$$

$$\simeq 0.42$$

$$n = \frac{x}{x_1},\; n = \frac{x}{5.0} \quad \text{where } x \text{ is in cm}$$

$$m = 0$$

At various radial positions

At $x = 2.5$ cm, $n = \frac{x}{5} = \frac{1}{2}$ and from the chart $Y = 0.10$

or $$0.10 = \frac{423 - T}{423 - 873}$$

or $$T = 200°\text{C or } 472°k.$$

Similarly additional points at min and 3 min may be calculated and the results shown on temperature profile presented below.

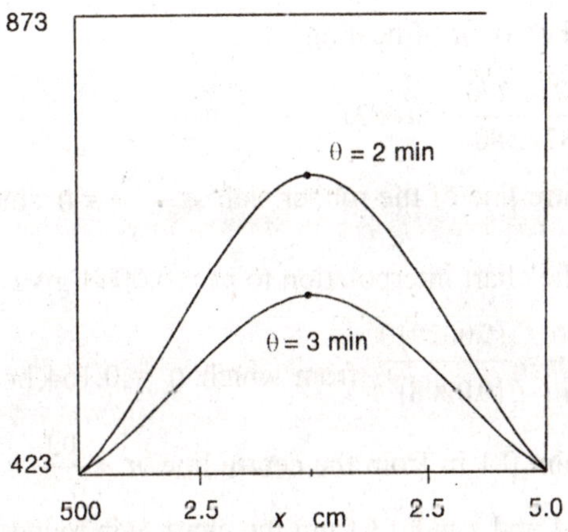

Fig. 4.3 *Temperature profile in cylindrical shaft*

Illustration 2

A flat slab of rubber 0.5 in thick initially at 80°F is to be placed between two electrically heated steel plates maintained at 287°F. The heating is to be discontinued when the temperature at the centre line of the rubber slab reaches 270°F.

(a) Calculate the length of the heating period.

(b) At the end of the run what would be the temperature of the rubber in plane 0.1 inch from the centre line?

(c) How long from the start of the heating, is required for the temperature to reach 270°F at the plane specified in part *b*.

(d) Repeat part (a) on the assumption that the rubber is heated from one face only the other being perfectly insulated.

Data for the rubber

$$k = 0.092 \frac{\text{Btu}}{\text{hr ft° F}}$$

$$\alpha = \frac{k}{\rho\, c_p} = 0.0029 \frac{\text{ft}^2}{\text{hr}}$$

Assume a constant coefficient h from metal to rubber of 1000 BTU/hr ft² F

Solution

(a) The mid-plane distance $= \dfrac{0.25}{12}$

$$x_1 = 0.0208 \text{ ft}$$

$$m = \frac{k}{h\, x_1} = \frac{0.092}{(1000)\,(0.0208)}$$

$$= 0.00442$$

At the end of θ hr of heating

$$Y = \frac{287 - 270}{287 - 80} = 0.0821$$

At the centre line of the rubber slab $n = \dfrac{x}{x_1} = 0$ since $Y = 0.0821$ and $n = 0$ on the chart interpolation to $m = 0.0044$ gives $X = 1.13$.

$$X = \frac{\alpha\,\theta}{x_1^2} = \frac{(0.0029)\theta}{(0.0208)^2} \text{ from which } \theta = 0.164 \text{ hr}$$

(b) For the point 0.1 in from the centre line $n = \dfrac{0.1}{0.25} = 0.4$ and as before $m = 0.0044$ and $x = 1.13$ from the chart, y is found to be 0.065

By definition, $Y = \dfrac{287 - t}{287 - 80}$ whence $t = 273.5$ F

(c) $Y = 0.0821$

$m = 0.0044$

$n = 0.4$

chart shows that $X = 1.03$

since by definition $X = \dfrac{\alpha\,\theta}{x_1^2}$

$$\theta = \frac{(1.03)\,(0.0208)^2}{0.0029}$$

$= 0.153$ hr

(d) When heated from one side only

$$x_1 = \frac{0.5}{12} = 0.0417 \text{ ft}$$

$$m = \frac{k}{h\,x_1} = \frac{0.092}{(1000)\,(0.0417)}$$

$$= 0.00221$$

At the centre of the slab, $n = 0.5$ and $Y = \dfrac{287 - 270}{287 - 80} = 0.0821$

From the chart

$$X = 0.955 = \frac{(0.0029)\,(\theta)}{(0.0417)^2} \text{ and therefore } \theta = 0.573 \text{ hr}$$

$\theta = 0.573$ hr when heated from one side only; as compared with 0.169 hour when heated from both sides.

4.3 UNSTEADY STATE TRANSFER WITH INTERNAL GENERATION OF THE TRANFERENT PROPERTY

If the unsteady state transfer involves also the internal generation of the transferent property, the balance equation is,

Input + generation = output + accumulation

The above equation can be expressed in terms of rate of each term

G = volumetric rate of generation of the transferent property

$$(\psi\, A)_1 + G\,(A\,dx) = (\psi\, A)_2 + \frac{d\,T}{d\theta} A\,dx$$

$$(\psi\, A)_1 = -\delta\,\Delta y\,\Delta z = \left(\frac{d\,T}{dx}\right)$$

$$(\psi\, A)_2 = -\delta\,\Delta y\,\Delta z \left(\frac{d\,T}{dx}\right)_2$$

by a similar development as adopted in the of unsteady state transfer without generation, we can show

$$\frac{d\,T}{d\theta} = \delta\left(\frac{\partial^2 T}{\partial x^2}\right) + G$$

This is the general equation for one dimensional transfer with internal generation of the transferent property.

For transfer/transport in x, y, z directions, the following equation results

$$\frac{dT}{d\theta} = \delta \left[\frac{\partial^2 T}{\partial x^2} + \frac{\partial^2 T}{\partial y^2} + \frac{\partial^2 T}{\partial z^2} \right] + G = \delta \nabla^2 T + G$$

For heat transfer

$$\frac{\partial T}{\partial \theta} = \alpha \nabla^2 T + \frac{G}{\rho c_p}$$

$$\frac{\partial (\rho c_p T)}{\partial \theta} = \alpha \nabla^2 (\rho c_p T) + G_q$$

The solutions of the unsteady state transport problems with generation is difficult even for simple geometrics.

4.4 NUMERICAL METHOD OF SOLVING ONE-DIMENSIONAL TRANSIENT CONDUCTION PROBLEM

This has been suggested by Dusinberre—based on his generalization of the increment method.

The sketch shows a cross section of a large slab of thickness *x*, having a uniform cross sectional area *A*.

The solid is divided into a number equal slices of thickness Δx, by temperature reference planes.

Let us consider the hatched zone *abcd* and write the heat balance on that. We have to know the temperature grandient.

The slope *–dt/dx at* plane *ad* approximated by the chord slope $\frac{t_0 - t_1}{\Delta x}$

For slab geometry

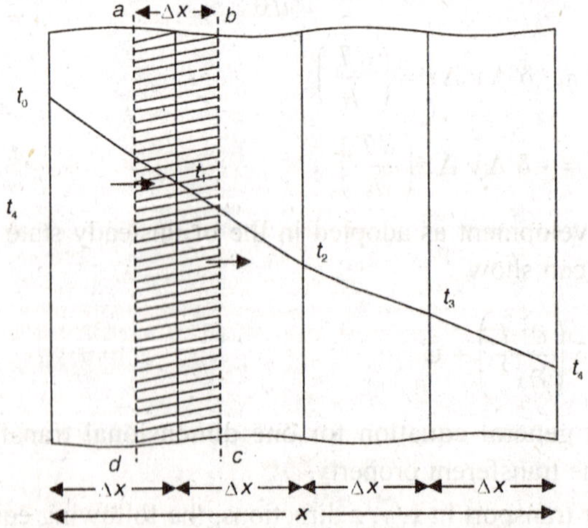

Fig. 4.4 *Plot for numerical method of solving transient conduction problem*

Similarly slope at plane $bc = \dfrac{t_1 - t_2}{\Delta x}$

The temperature value at plane 1 $\simeq$ the average temperature of the cross hatched area/zone.

The heat balance is $\dfrac{k\,A\,(t_0 - t_1)}{\Delta x} - \dfrac{k\,A\,(t_1 - t_2)}{\Delta x}$

$$= \frac{(A\,\Delta x)(\rho\,c_p)(t'_1 - t)}{\Delta\theta}$$

where t'_1 = the new temperature at plane 1, after the elapse of a finite time increment $\Delta\theta$.

We know $\alpha = \dfrac{k}{\rho\,c_p}$

Rewriting

$$k\,A\,(t_0 - t_1) - k\,A\,(t_1 - t_2) = \frac{A\,(\Delta x)^2\,\rho\,c_p\,(t'_1 - t_1)}{\Delta\theta}$$

or $$t_0 - t_1 - t_1 + t_2 = \frac{(\Delta x)^2\,(t'_1 - t_1)}{\alpha\,\Delta\theta}$$

$$\frac{\rho\,c_p}{k} = \frac{1}{\alpha}$$

Putting the dimension less ratio $\dfrac{(\Delta x)^2}{\alpha\,\Delta\,\theta}$ as M, the modulus

$$-t_0 - t_1 - t_1 + t_2 = M\left(t'_1 - t_1\right)$$

or $$M\,t'_1 = t_0 - 2t_1 + M\,t_1 + t_2$$

$$= t_0 + t_1\,(M - 2) + t_2$$

$$t'_1 = \frac{t_0 + t_1\,(M - 2) + t_2}{M}$$

Schmidt numerical method based on a simplification is as follows

Simplified approach by making $M = 2$, $t'_1 = \dfrac{t_0 + t_2}{2}$

that is after the elapse of one $\Delta\theta$, the new temperature at a given section is the arithmetic mean of the temperature previously prevailing at adjacent sections. This method can be applied in tabular form or graphically.

$$t'_2 = \frac{t_1 + t_3}{2} \text{ and likewise } t'_{n-1} = \frac{t_{n-2} + t_n}{2}$$

Illustration on Transient Conduction

A large slab of steel 1 ft thick initially has a uniform temperature of 700°F. Suddenly both faces are reduced to and maintained at 100 F. The steel a density of 440 lbs/c ft, a specific heat of 0.13 Btu/lb F and a k 25 Btu/ (hr) (ft) (F). It is desired to predict the temperature of the mid plane when 14.85 minutes have elapsed, using

(a) Schmidt tabular method based a M of 2, taking the initial surface temperature as $t_a = 100$°F.

(b) The same as in (a) but employing an initial temperature of 400 and.

(c) Using the analytical solution

$$\frac{T_1 - T}{T_1 - T_0} = \frac{4}{\pi}\left[e^{-\alpha\pi^2\theta/4x_1^2} \sin\frac{\pi x}{2x_1} + \frac{1}{3}e^{-9\alpha\pi^2\theta/4x_1^2} \sin\frac{3\pi x}{2x_1} + \frac{1}{5}e^{-25\alpha\pi^2\theta/4x_1^2} \sin\frac{5\pi x}{2x_1} + \ldots\right]$$

where T_0 = initial temperature of the slab at all values of x.

T = Temperature at any point x and time θ.

T_1 = new surface or surrounding temperature T

For the special case of a slab having a thickness $2x_1$ and a negligible surface resistance corresponding to an infinite value of 'h', the surface temperature changes to the temperature of the surroundings immediately at zero time.

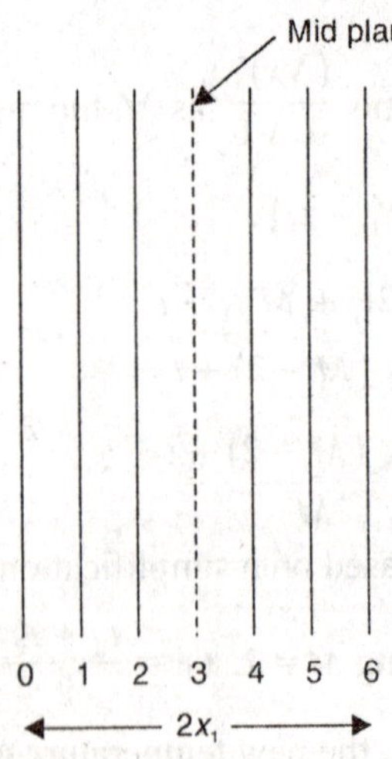

Fig. 4.5 *Solution to transient heat conduction*

Putting $a_1 = \left(\frac{\pi}{2}\right)^2$; $X = \frac{\alpha\theta}{x_1^2}$

$$\frac{T_1 - T}{T_1 - T_0} = \frac{4}{\pi}\left[e^{-a_1 X} \sin\frac{\pi x}{2x_1} + \frac{1}{3}e^{-9a_1 X} \sin\frac{3\pi x}{2x_1} + \frac{1}{5}e^{-25a_1 X} \sin\frac{5\pi x}{2x_1} + \ldots\right]$$

$$X = \frac{\alpha\theta}{x_1^2} = \frac{(0.392)\left(\frac{14.85}{60}\right)}{(0.5)^2} = 0.388$$

$$\alpha = \frac{k}{\rho c_p} = \frac{25}{(490)(0.13)} = 0.392$$

$$a_1 X = 0.957$$

$$\frac{T \text{ at mid plane} - T_1}{T_0 - T_1} \text{ is } \frac{T \text{ at mid plane} - 100}{700 - 100}$$

$$= \frac{4}{\pi}\left[e^{0.957} \sin\frac{\pi}{2} + \frac{1}{3}e^{-8.604} \sin\frac{3\pi}{2} + \frac{1}{5}e^{-25.9} \sin\frac{5\pi}{2} + \ldots\right]$$

$$= \frac{4}{\pi}(0.384)$$

T at mid plane = 393°F

Schmilt's Tabular Method

$$M = 2$$

$$_{\theta+\Delta\theta}T_x = \frac{_\theta T_{x+\Delta x} + {_\theta T_{x-\Delta x}}}{2}$$

$$t_1' = \frac{t_0 + t_2}{2}$$

$$\Delta\theta = \frac{(\Delta x)^2}{\alpha M}$$

Assume that there are six 2″ slices; that is we are dividing them into 6 (six) equal slices

$$\Delta x = 2'' = \frac{1}{6} \text{ foot}$$

$$\alpha = 0.392 \frac{\text{ft}^2}{\text{hr}}$$

$$\Delta\theta = \frac{\left(\frac{1}{6}\right)^2}{(0.392)(2)} = 0.0354 \text{ hr}$$

$$\text{Number of time steps } \frac{14.85/60}{0.0354} = 7$$
$$= N_{\Delta\theta}$$

If t_1, t_2, t_3 are temperatures at various positions in the slab and at time θ and $t'_1, t'_2, t'_3 \ldots$ at time $\theta + \Delta\theta$.

Let t_0 = temperature at sections 0 and 1

$t_0 = t_6$ that is at the two sufaces

t_a = temperature of the surroundings

t_s = temperature of surfaces

Since $\Delta x = \frac{1}{6}$ foot, there will be six 2″ slices.

That is the temperature values will be t_0, t_1, t_2, t_3, t_4, t_5, t_6. Since the system is symmetrical, both the surfaces are reduced to 100°F.

$t_0 = t_6$, $t_1 = t_5$, $t_2 = t_4$ and t_3 will be the temperature at the mid plane.

It is assumed that

$$t_0 = t_6 = 100°\text{F}$$

which may be incorrect

$$t'_1 = \frac{t_0 + t_2}{2} \qquad t'_2 = \frac{t_1 + t_3}{2} \qquad t'_3 = \frac{t_2 + t_4}{2}$$

$\Delta\theta$	t_0	t_1	t_2	t_3	t_4	t_5	t_6
0	100	700	700	700	700	700	100
1	100	400	700	700	700	400	100
2	100	400	550	700	550	400	100
3	100	325	550	550	550	325	100
4	100	325	438	550	438	325	100
5	100	269	438	438	438	269	100
6	100	269	354	438	354	269	100
7	100	227	354	354	354	227	100

$\Delta\theta = 7$, Corresponds to 14.85 minutes

The final midplane temperature is, 354 (39°F below the value obtained by analytical solution)

This discrepancy is caused by the ill chosen initial value of t_s (surface).

If the initial value is

$$t_0 = t_6 = \frac{t_0 + t_s}{2} = \frac{100 + 700}{2}$$

$= 400°F$

$t_3 = 395°F$

which is shown in the following table

$\Delta\theta$	t_0	t_1	t_2	t_3	t_4	t_5	t_6
0	400	700	700	700	700	700	400
1	100	550	700	700	700	550	100
2	100	400	625	700	625	400	100
3	100	362	550	625	550	362	100
4	100	325	493	550	493	325	100
5	100	296	437	493	437	296	100
6	100	268	395	437	395	268	100
7	100	248	352	395	352	248	100

$t_3 = 395°F$

Problems in Unsteady State Heat Conduction

1. A 8 cm long vertical glass tube is filled with water that is uniformly at a temperature of $T = 105°C$. The top is suddenly opened to the air at 1 atm. pressure. Plot the decrease of the height of water in the tube due to the evaporation process as a function of time until the bottom of the tube has cooled down to by 0.05°C.
2. A large flat electrical resistance strip heater is fastened to a fire brick wall uniformly at 17°C. When it is suddenly turned on, it releases heat at the uniform rate of 4500 w/m². Plot the temperature of the brick immediately under the heater as a function of time if the other side of the heater is insulated. What is the heat flux at a depth of 1 cm when the surface attains a temperature of 200°C?
3. A steel roller bearing of 2mm diameter and 2 cm length is hardened by heating to 1400°K uniformly and then quenching in a large oil tank at 350°K. Compute (a) the time for centres of cylinders to cool to 440°K and (b) the temperature at the surface at that time. Assume that $k = 34$ w/m°K

   $$\rho = 7200 \frac{kg}{m^3} \quad c_p = 620 \frac{J}{kg°\,K}$$

   and $h = 3450$ w/m²°K
4. A steel sphere of 10 cm in diameter is suddenly immersed in a tank containing an oil at 12°C. The initial temperature of the sphere is 250°C $h = 290$ w/m²°K. How long will it take the centre of the sphere to cool to 155°C.

   The required properties of steel may be taken from table.

5. A copper block 2 by 3 by 4 cm is at 250°C when immersed in a fluid with $h_\infty = 80 \dfrac{\text{W}}{\text{m}^2\text{K}}$ and $t_\infty = 15°\text{C}$. How long should the block take to cool to 35°C.
6. A long 4 inch diameter piece of an aluminum bar stock at 800°F undergoes quenching in a water stream at $h_\infty = 75 \dfrac{\text{Btu}}{\text{hr ft}}$ and $t_\infty = 90°\text{F}$. What is the bar temperature when it comes out of the bath after 3 minutes?
7. A steel cylinder which is 6 in long and whose diameter is 4 m. is to be quenched in a water bath from a temperature of 1500°F. Estimate the time required for the temperature at the centre to drop to 250°F. Take c_p as 0.17 and $\rho = 440$ at the appropriate units.
8. A plastic material in the form of a 2.5 in slab is to be cured in a flat-plane press having surfaces at a temperature of 350°F. If the material is initially at the temperature of 75°F, how long will it take to raise the temperature at 0.8 in below the surface to 275°F? Take the properties (average values) for the material as given below

   $k = 0.12$ $\rho = 110$ and $c_p = 0.40$ in the appropriate units.
9. In the vulcanization process of tyres the material is placed into a jig and steam at 300°F is admitted suddenly to both sides. If the tyre thickness is 1 inch and the initial temperature 75°F, estimate the time needed for the central layer to reach 275°F.

5

Convection

5.1 APPLICATION OF DIMENSIONAL ANALYSIS FOR CONVECTION EQUATIONS FOR FORCED CONVECTION AND NATURAL CONVECTION

Dimensional analysis is a method of deducing logical groupings of the variables involved in a process.

Dimensions

In engineering, the physical concepts are represented by symbols or dimensions e.g. length by L

Velocity by V

Through experience it has been learnt that a certain number of dimensions can be selected as fundamental and all the other dimensions may be expressed in terms of products of powers of these fundamental dimensions.

In momentum transfer, three fundamental dimensions would suffice to express all other quantities, the customary choice being either mass M, length L and time θ or force F, length L and time θ.

Either force or mass can be selected as the fundamental as it is really immaterial since they are directly related by Newton's second law

Force α product of mass and acceleration

$$F = \frac{mg}{g_c} = \frac{ML}{\theta^2 g_c}$$

g_c dimensional constant or conversion factor

It will be dimensions less if F is replaced by $\frac{ML}{\theta^2}$ or the dimension M by $\frac{F\theta^2}{L}$ g_c will be made dimensionless in SI system of units and its magnitude will be unity.

In heat transfer additional fundamental dimensions such as temperature T, and sometimes heat H are usually specified.

The dimension of heat can be eliminated since it can be expressed from the I law of thermodynamics as the dimensions of work-FL

First law $\Delta U = Q - w$

The word fundamental used here is quite arbitrary. The choice of a particular set of dimensions as fundamental in a given class of problems is dictated solely by convenience and practice.

The dimensions of commonly used quantities in heat transfer analysis are listed in the following table with reference to two alternative sets of fundamental dimensions, $M\,L\theta T$ and $M\,L\theta T\,H$.

It can be noted in this context that $F = \dfrac{ML}{\theta^2}$ and $M = \dfrac{F\theta^2}{L}$

Table 5.1

Quantity	*Symbol*	*Dimensions*	
Mechanical quantities		$M\,L\theta$	$F\,L\theta$
Diameter length	d, L	L	L
Area	A	L^2	L^2
Velocity	V	L/θ	L/θ
Acceleration	a, g	$\dfrac{L}{\theta^2}$	$\dfrac{L}{\theta^2}$
Mass	M	M	$\dfrac{F\theta^2}{L}$
Density	ρ	$\dfrac{M}{L^3}$	$\dfrac{F\theta^2}{L^4}$
Viscosity	μ	$\dfrac{M}{L\theta}$	$\dfrac{F\theta}{L}$
Kinenatic viscosity	ν	$\dfrac{L^2}{\theta}$	$\dfrac{L^2}{\theta}$
Mechanical quantities		$M\,L\theta$	$F\,L\theta$
Mass flow rate	W	$\dfrac{M}{\theta}$	$\dfrac{F\theta}{L}$
Force	F	$\dfrac{ML}{\theta^2}$	F

Contd.

Quantity	*Symbol*	*Dimensions*	
Pressure shear	p, T	$\frac{M}{L\theta^2}$	$\frac{F}{L^2}$
Surface tension	σ	$\frac{M}{\theta^2}$	$\frac{F}{L}$
Conversion factor ma/F	g_c	None	None
In $F\ M\ L\theta$ system $\rightarrow g_c = \frac{ML}{F\theta^2}$			
Heat transfer quantities		$M\ L\theta T$	$M\ L\theta T\ H$
Temperature	T	T	T
Heat transfer coefficient	h, U	$\frac{M}{\theta^2 T}$	$\frac{H}{\theta L^2 T}$
Thermal conductivity	k	$\frac{ML}{\theta^3 T}$	$\frac{H}{\theta LT}$
Mechanical equivalent of heat	J	none	$\frac{FL}{H}$
Thermal diffusivity	α	$\frac{L^2}{\theta}$	$\frac{L^2}{\theta}$
Specific heat (on mass basis)	c_p	$\frac{L^2}{\theta^2 T}$	$\frac{H}{MT}$
Coefficient of expansion	β	$\frac{I}{T}$	$\frac{I}{T}$
Heat transfer rate	q	$\frac{ML^2}{\theta^3}$	$\frac{H}{\theta}$
Quantity of heat	Q	$\frac{ML^2}{\theta^2}$	H
Enthalpy change (phase transformation)	λ	$\frac{L^2}{\theta^2}\left(H = \frac{ML^2}{\theta^2}\right)$	$\frac{H}{M}$
Kinetic energy equivalant of heat	K_H	none	$\frac{ML^2}{\theta^2 H}$

$$H = FL$$

$$F = \frac{ML}{\theta^2} \qquad k = \frac{H}{\theta LT}$$

$$H = \frac{ML^2}{\theta^2} \qquad = \frac{ML^2}{\theta^2}\,\frac{1}{\theta LT} = \frac{ML}{\theta^3 T}$$

It is allowable to add force F as a dimension, by including the conversion factor g_c in the dimensional analysis similarly the dimension of heat H may also be added by including either the mechanical equivalent of heat J (having dimensions $\frac{FL}{H}$) or the product Jg_c or K_H (the kinetic energy equivalent of heat which has the dimension $\frac{ML^2}{\theta^2 H}$.

Use of Dimensional Analysis

1. When the equations governing a process are known and solvable, dimensional analysis suggests logical grouping of quantities for presenting the results.
2. When the mathematical equations governing certain processes are unknown or too complex, dimensional analysis lays the foundation of an efficient experimental programme for obtaining the results, by reducing the number of vairables requiring investigation and by indicating a possible form of the semi-empirical correlations that may be formulated. It should be borne in mind, however, that dimensional analysis by itself cannot provide quanititative answers and thus can not be a substitute for the exact or the approximate mathematical solutions.

Methods of Finding the Dimensionless Groups

1. Algebraic
   (a) The classical method of Rayleigh
   (b) The improved π (P_i) theorem
2. Use of differential equations
3. Geometric, kinematic and dynamical similarities which indicate the significance of dimensionless groups.

Rayleigh's Algebraic Method

Illustration of the procedure

Example

For heating or cooling—a fluid flowing without any phase change in turbulent motion through heated or cooled tubes.

It is desired to determine the logical grouping of the factors affecting the coefficient of heat transfer, h.

The factors affecting h are

$k \rightarrow k$ of liquid because of the presence of laminar sub layer

Mass velocity, G – film thickness depends on mass velocity G

$$\rho u \ \frac{\text{gms}}{\text{cm}^3} \ \frac{\text{cm}}{\text{sec}}$$

D – tube diameter

μ – viscosity of the fluid

c_p – since for a given q, the specific heat affects the bulk temperature of the stream, therefore c_p should enter

These factors appear in the basic energy and hydromechanical equations.

Let us now take that we are using $M L \theta T$ system.

Letting ϕ represent any function, one can write

$$h = \phi\left(G, D, c_p, \mu, k\right)$$

For convenience, the foregoing is replaced by an infinite series

$$h = \alpha G^a D^b c_p^{\ e} \mu^f k^i + \alpha_1 G^{a1} D^{b1} c_p^{\ e1} \mu^{f1} k^{i'} + \ldots$$

Where the dimensionless factors α and α_1 and the dimensionless exponents may have any values required.

The constants α and α_1 are dimensionless therefore to be dimensionally consistent each term in the series must have the same dimensions as the term on the left side of the equation. Since the dimensions of each term in the series are identical (all terms of the series would be alike in form) only the dimensions of the first term of the above equation need be considered for dimensional honogenity

$$h = \alpha \ G^a \ D^b \ c_p^e \ \mu^f \ k^i$$

Let us take the $M L \theta T$ system

Substitution of the dimension gives

$$\frac{M}{\theta^3 T} = \left(\frac{M}{\theta L^2}\right)^a \left(L^b\right) \left(\frac{L^2}{\theta^2 T}\right)^e \left(\frac{M}{L\theta}\right)^f \left(\frac{ML}{\theta^3 T}\right)^i$$

Summation of the exponents of like dimensions gives the condition equations

$$\Sigma M = 1 = a + f + i \qquad \ldots(1)$$

$$\Sigma \theta = -3 = -a - 2e - f - 3i \qquad \ldots(2)$$

$$\Sigma T = -1 = -e - i \qquad \ldots(3)$$

$$\Sigma L = 0 = -2a + b + 2e - f + i \qquad \ldots(4)$$

From equation (3) $i = 1 - e$

From equation (1)

$$f = 1 - a - i$$
$$= 1 - a - 1 + e = e - a$$

From equation (4)

$$0 = -2a + b + 2e - f + i$$
$$= -2a + b + 2e - e + a + 1 - e$$
$$= -a + b + 1$$

or $a = b + 1$ or $b = a - 1$

Verification of equations 2,

$$-3 = a - 2e - f - 3i$$
$$-\cancel{3} = -\cancel{a} - 2\cancel{e} - \cancel{e} + \cancel{a} - \cancel{3} + 3\cancel{e}$$
$$= 0$$

substitution in the equation,

$$h = \alpha\, G^a\, D^b\, c_p^e\, \mu^f\, k^i$$

$$h = \alpha\, G^a\, D^{a-1}\, c_p^e\, \mu^{e-a}\, k^{1-e}$$

$$= \alpha \left(\frac{GD}{\mu}\right)^a \left(\frac{c_p\,\mu}{k}\right)^e \left(\frac{k}{D}\right)^1$$

$$\frac{hD}{k} = \alpha \left(\frac{GD}{\mu}\right)^a \left(\frac{c_p\,\mu}{k}\right)^e$$

or
$$\frac{hD}{k} = \phi\left[\left(\frac{DG}{\mu}\right)\left(\frac{c_p\,\mu}{k}\right)\right]$$

The unknown function has to be determined from the experimental data. Had we chosen $ML\theta TH$ system, as H and M appear in the factors, the dimensional constant $K_H = \dfrac{ML^2}{\theta^2 H}$ would have to be included.

Similarly

$$h = \phi\left(G, D, c_p, \mu, k, K_H\right)$$

or
$$h = \alpha\, G^a\, D^b\, c_p^e\, \mu^f\, k^i\, K_H^m + \alpha_1\, G^{a_1}\, D^{b_1}\, c_p^{e_1}\, \mu^{f_1}\, k^{i_1}\, K_H^{m_1} + \ldots$$

or
$$h = \alpha\, G^a\, D^b\, c_p^e\, \mu^f\, k^i\, K_H^m$$

Substitution of the dimensions gives.

$$\frac{H}{\theta L^2 T} = \left(\frac{M}{\theta L^2}\right)^a (L)^b \left(\frac{H}{MT}\right)^e \left(\frac{M}{L\theta}\right)^f \left(\frac{H}{\theta L T}\right)^i \left(\frac{ML^2}{H\theta^2}\right)^m$$

summation of the exponents of like dimensions would yield the following equations.

$$\Sigma H \qquad 1 = e + i - m$$

$$\Sigma M \qquad 0 = a - e + f + m$$

$$\Sigma L \qquad -2 = -2a + b - f - i + 2m$$

$$\Sigma \theta \qquad -1 = -a - f - i - 2m$$

$$\Sigma T \qquad -1 = -e - i$$

from which we can get,

$b = a - 1, f = e - a, i = 1 - e$ and $m = 0$, or $a = b + 1$

substitution of the above values gives

$$\frac{hD}{k} = \alpha \left(\frac{DG}{\mu}\right)^a \left(\frac{c_p \mu}{k}\right)^e$$

Exponents a and e, which were arbitrarily retained, were those appearing on G, and c_p.

Pi theorem: A more systematic way of determining the dimensionless groups.

Statement: If there are n_1 physical quantities of importance and n_2 fundamental dimensions then there exists a maximum number (n_{max}) of these quantities which by themselves cannot form a dimensionless group, where $n_{max} \le n_2$. Then by combining one each of the remaining quantities with the n_{max} quantities there can be formed n_π dimensionless groups where $n_\pi = n_1 - n_{max}$. The dimensionless groups thus formed are called "π terms" and identified by the symbols, $\pi, \pi_2 \ldots \pi_{n\pi}$

Rules that are Helpful in the Application of *Pi* Theorem

(a) Compile a tentative list of all the significant variables. Suppose in a particular problem, $n_1 = 7$. The physical equation can be expressed symbolically as

$$f(Q_1, Q_2, Q_3 \ldots Q_7) = 0$$

where the Q's represent the physical quantities (such as velocity, diameter, heat transfer coefficient).

If all these quantities can be expressed in terms of four fundamental dimensions say M, L, θ, T, then $n_2 = 4$.

(b) Next we select n_{max} physical quantities called primary Q's which by themselves cannot form a dimensionless group. Normally $n_{max} = n_2$, so a useful rule is to select the primary Q's so that each contains one of the fundamental dimensions at least once. Obviously, there is no unique set of primary Q's and experience will dictate the best choice.

(c) Form each π term by expressing the ratio of the remaining Q's to the product of powers of the primary Q's.

In, the case discussed here, the procedure yields three π terms and the original symbolic equation relating seven variables may now be replaced by

$$\phi\,(\pi, \pi_2, \pi_3) = 0$$

Heat Transfer in Forced Convection

Four fundamental dimensions are say M, L, θ and T

In forced convection there are seven physical variables involving the above four dimensional quantities.

Consequently three π terms will be obtained.

Four variables which together involve all the four dimensions and which do not themselves form dimensionless groups are V, d, μ and k.

Then h, ρ and c_p will each appear in a separate independent π term

The π terms are

$$|\pi_1| = \left|\frac{h}{V^a d^b \mu^c k^d}\right|$$

$$|\pi_2| = \left|\frac{\rho}{V^{a'} d^{b'} \mu^{c'} k^{d'}}\right|$$

$$|\pi_3| = \left|\frac{c_p}{V^{a''} d^{b''} \mu^{c''} k^{d''}}\right|$$

$$\pi_1 = \frac{M\theta^{-3}T^{-1}}{\left(\frac{L}{\theta}\right)^a (L)^b \left(\frac{M}{L\theta}\right)^c \left(\frac{ML}{\theta^3 T}\right)^d}$$

Equating the exponents of each dimension in the numerator and the denominator

$$M \rightarrow 1 = c + d \qquad \text{...(1)}$$

$$\theta \rightarrow -3 = a - 3d - c \quad \ldots(2)$$

$$L \rightarrow 0 = a + b - c + d \quad \ldots(3)$$

$$T \rightarrow -1 = -d \quad \ldots(4)$$

From the above four equations we obtain the following:

$$a = 0,\ b = -1,\ d = 1,\ c = 0$$

$$\text{Thus } \pi_1 = \frac{1}{d^{-1}\, k^{+1}} \text{ or } \frac{h}{d^{-1}\, k}$$

or $$\frac{hd}{k} = N_{Nu}$$

$$\text{Similarly } \pi_2 = \frac{\rho}{V^{a'}\, d^{b'}\, \mu^{c'}\, k^{d'}}$$

$$= \frac{M\,L^{-3}}{\left(\frac{L}{\theta}\right)^{a'} L^{b'} \left(\frac{M}{L\theta}\right)^{c'} \left(\frac{M\,L}{\theta^3\, T}\right)^{d'}}$$

$$M \rightarrow 1 = c' + d' \quad \ldots(1)$$

$$L \rightarrow -3 = a' + b' + c' + d' \quad \ldots(2)$$

$$\theta \rightarrow 0 = -a' - c' - 3d' \quad \ldots(3)$$

$$T \rightarrow 0 = -d'$$

or $$d' = 0,\ b' = -1\ a' = 1\ c' = 1$$

$$\frac{\rho}{d^{-1}\, V^{-1}\, \mu^{1}} \text{ or } \left(\frac{\rho\, V\, D}{\mu}\right)$$

$$\pi_2 = N_{R_e} = \frac{\rho\, V\, D}{\mu}$$

$$\text{Similarly } \pi_3 = \frac{c_p\, \mu}{k} = N_{Pr}$$

It is thus seen that heat transfer data in forced convection can be correlated in terms of these three dimensionless groups

$$N_{Nu} = \phi\,(N_{Re}\ N_{Pr})$$

Heat Transfer Coefficients Between Fluids and Tubes

Fluid Inside Tubes

Heat transfer by the combined mechanism of conduction and convection between clean smooth metal tubes and fluids flowing in turbulent flow inside the tubes:

Heat transfer coefficients as a function of flow characteristics and the physical properties of the stream.

We may write the expression, obtained by dimensional analysis, in the form

$$\frac{hD}{k} = z\left(\frac{DV\rho}{\mu}\right)^a \left(\frac{c_p \mu}{k}\right)^b$$

This form may be useful in demonstrating the correlation of experimental data.

Two series of experiments may be made-one-determining the relation of Nusselt number of (stanton number) with N_{Re} -stanton number is $\dfrac{N_{Nu}}{N_{Re}\ N_{Pr}}$ while the other group is maintained constant.

Another—determining the relation between N_{Nu} and N_{Pr}, maintaining N_{Re} constant.

The above form is useful in studying the influence of changes in velocity on h in a given system since each of these variables appears only once.

In determining the constants z, a and b from experimental data the normal procedure involves equating the heat transferred to the gain in enthalpy of the fluid.

With the proper mean temperature difference $(+\Delta T_m)$ substituted for $T_1 - T_2$ in the equation $q = hA(T_1 - T_2)$

where h is the heat transfer coefficient between the solid and the fluid.

$$q = hA(+\Delta T_m)$$

$$= G\ A_F\ c_p\ (T_{\text{out}} - T_{\text{in}})$$

$$(m\ c_p\ \Delta T)$$

where $G = \rho v =$ the mass velocity

A_F = cross-sectional area for the flowing fluid (flow area)

Rearranging,

$$\frac{h}{Gc_p} = \frac{A_F\ (T_{\text{out}} - T_{\text{in}})}{A(\Delta T)_m}$$

Thus the stanton number h/Gc_p evaluated directly from experimental data as indicated in the above equation

$$\text{Stanton number} = \frac{N_{Nu}}{N_{Re}\ N_{Pr}} = \frac{h}{c_p G}$$

$$\frac{hD}{k} = N_{Nu} = z\left(N_{Re}\right)^a \left(N_{pr}\right)^b$$

$$\frac{h}{c_p G} = z\left(\frac{DG}{\mu}\right)^{a-1}\left(\frac{c_p \mu}{k}\right)^{b-1}$$

$$\left(N_{Nu}\right)\left(N_{Re}\right)^{-1}\left(N_{Pr}\right)^{-1} = z\left(N_{Re}\right)^a\left(N_{Re}\right)^{-1}\left(N_{Pr}\right)^b\left(N_{Pr}\right)^{-1}$$

The values of the physical properties of the fluids vary with temperature and temperature is not constant either across or along the tube. The most convenient temperature for evaluating the physical properties is the arithmetic average temperature (value) of the inlet and outlet bulk temperatures of the fluid. The bulk temperature is the temperature that would be indicated if the flowing stream at any point were thoroughly mixed.

I Plot

A log – log plot is made.

Stanton number as a function of N_{Re} for selected values (constant) of N_{Pr} plotted on a log cross sectional paper

Heating and cooling, air, water, acetone, benzene, gas oil (N_{pr} = 29)

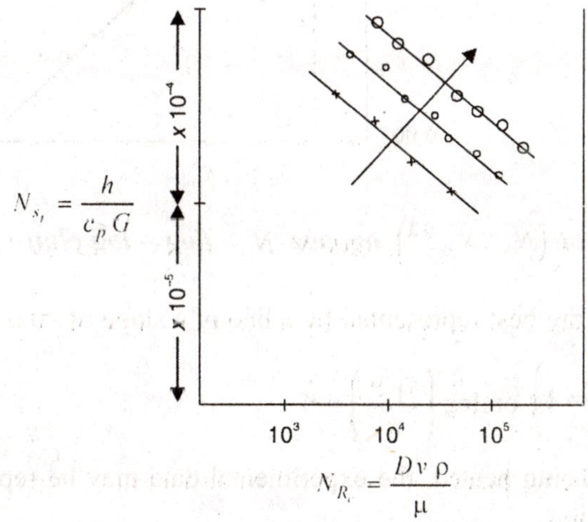

Fig. 5.1 *Plot of Stanton number versus Reynolds number*

Since the data would be represented by straight lines, all having the same slope (0.2), the use of the above equation

$$\frac{h}{Gc_p} = z\left(\frac{DV\rho}{\mu}\right)^{a-1}\left(\frac{c_p \mu}{k}\right)^{b-1}$$

$$\therefore a - 1 = -0.2$$

II Plot

The exponent $(b-1)$ of the $N_{Pr}\left(\frac{c_p\mu}{k}\right)$ is then determined in a similar manner by plotting a second graph.

$$\log\frac{h}{c_pG} = \log z + (a-1)\log\left(\frac{DG}{\mu}\right) + (b-1)\log\left(\frac{c_p\mu}{k}\right)$$

Plotting now $\left[\log\frac{h}{c_pG} + 0.2\log\frac{DG}{\mu}\right]$ as a function of $\log\frac{c_p\mu}{k}$, for the same data used in the above figure for different/various fluids listed in the above section.

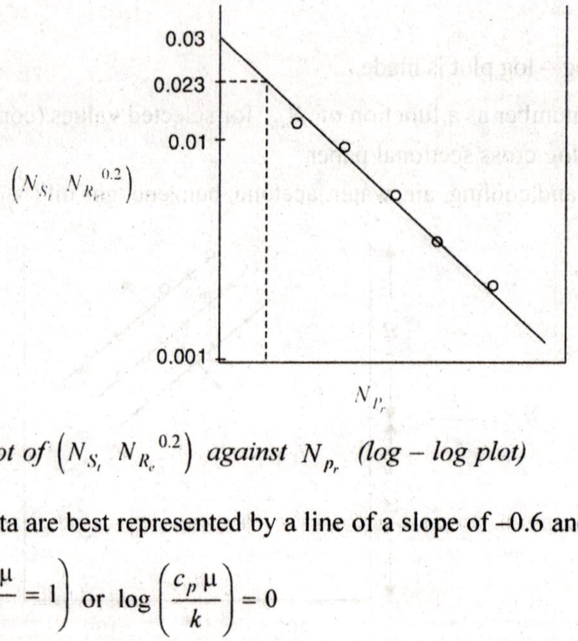

Fig. 5.2 *Plot of* $\left(N_{S_t}\ N_{R_e}^{\ 0.2}\right)$ *against* N_{P_r} *(log – log plot)*

These data are best represented by a line of a slope of –0.6 and intercept of 0.023 at $\left(\frac{c_p\mu}{k}=1\right)$ or $\log\left(\frac{c_p\mu}{k}\right)=0$

For fluids being heated, the experimental data may be represented by the following equation

$$\frac{h}{c_pG} = 0.023\left(\frac{DG}{\mu}\right)^{-0.2}\left(\frac{c_p\mu}{k}\right)^{-0.6}$$

or
$$\frac{h}{c_pG} = 0.023\left(\frac{\mu}{DG}\right)^{+0.2}\left(\frac{k}{c_p\mu}\right)^{+0.6}$$

If the fluid is being cooled instead of being heated, a similar procedure leads to the following correlation

$$\frac{h}{c_p G} = 0.023\left(\frac{\mu}{DG}\right)^{0.2}\left(\frac{k}{\mu c_p}\right)^{0.7}$$

Since the major part of the resistance to heat transfer is in the viscous layer, it would appear better to evaluate the physical properties at a temperature more representative of the viscous layer than the bulk temperature.

Colburn suggested that this temperature be chosen as the mean of the average bulk temperature and the average surface temperature.

$$\frac{h}{c_p G} = 0.023\left(\frac{\bar{\mu}}{DG}\right)^{0.2}\left(\frac{k}{c_p \bar{\mu}}\right)^{\frac{2}{3}}$$

where $\bar{\mu}$ = viscosity determined at the described mean temperature

The above correlation which holds good both for fluids being heated and cooled is better for fluids whose viscosities are temperature sensitive.

The other way to account for the variation in μ is by Sieder and Tate equation which is applicable for $N_{Re} > 8000$ situations

Effect of $\frac{L}{D}$ becomes negligible

$$\frac{h}{c_p G} = 0.027\left(\frac{\mu}{DG}\right)^{0.2}\left(\frac{k}{c_p \mu}\right)^{\frac{2}{3}}\left(\frac{\mu}{\mu_s}\right)^{0.14}$$

where μ = viscosity at average bulk temperature of the fluid

μ_s = viscosity at average temperature of the heating and cooling surface.

$\left(\frac{\mu}{\mu_s}\right)^{0.14}$ for water and glycerin when each fluid is heated by contact with surface at 225°F and at an average bulk temperature of 140°F; the term $\left(\frac{\mu}{\mu_s}\right)^{0.14}$ is 1.4 for glycerin and 1.12 for water.

Dittus-Boelter Equation

This is only a rearranged from of the aforementioned equation

$$N_{st} = \frac{N_{Nu}}{N_{Re}\,N_{Pr}} = \frac{h}{c_p G} = 0.023\left(\frac{DG}{\mu}\right)^{-0.2}\left(\frac{c_p \mu}{k}\right)^{-0.6}$$

$$\frac{hD/k}{N_{Re}\,N_{Pr}} = 0.023\left(\frac{DG}{\mu}\right)^{-0.2}\left(\frac{c_p\,\mu}{k}\right)^{-0.6}$$

(For fluids which are heated)

or $$\frac{hD}{k} = 0.023\left(\frac{DG}{\mu}\right)^{0.8}\left(\frac{c_p\,\mu}{k}\right)^{0.4}$$

Range for, N_{Re} = 10000 to 120,000

N_{P_r} = 0.7 to 120

For fluids which are cooled

N_{P_r} has the exponent, (often taken as) 0.3.

For a very large temperature difference $(T_w - T_b)$ with air as the fluid, Desmon and Sams found the following equation to correlate the experimental data.

$$\frac{hD}{k_b} = 0.020\left(\frac{VD\rho_f}{\mu_f}\right)^{0.8}\left(\frac{c_p\,\mu}{k}\right)_b^{0.4}$$

subscripts b and f represent the following.

Subscript $b \rightarrow$ bulk or mixed mean fluid temperature.

$f \rightarrow$ film temperature defined by $\frac{1}{2}(T_w + T_b)$

where T_w is the wall temperature.

The above correlation was found to be valid for $N_{Re} > 10^4$ and $\left(\frac{T_w}{T_b}\right)$ up to 3.55.

For Fluids of High Conductivity Such as Molten Metals

When molten metal such as mercury or sodium of high k and log N_{P_r} (< 0.1) are transferring heat, the thermal conductivity is more significant than the degree of turbulence in affecting heat transfer.

By making necessary assumptions and using velocity profile of the fluid, a relationship has been arrived between the dimensionless groups N_{Nu} N_{R_e} and N_{P_r}

For substances whose N_{Pr} < 0.1 the following equation is recommended

$$N_{N_u} = 7 + 0.025\,(N_{Pe})^{0.8}$$

(constant wall temperature) $N_{Pe} = (N_{Re})\,(N_{Pr})$

$$= \frac{Dv\rho c_p}{k}$$

with uniform heat flux $N_{Nu} = 0.625\left[(N_{Re})\,(N_{Pr})\right]^{0.4}$

Dimensional Analysis as Applied to Free or Natural Convection

In this case energy exchange is between a body and an essentially stagnant fluid. Fluid motion is due entirely to buoyancy force arising from density variations in the fluid which is found when any object is dissipating energy to its surroundings—may be intentional as in the cooling of some machine or electrical device or in the heating of a house or room by a convective heating system. It may also be unintentional as in the loss of energy from a steam pipe.

Fluid motion generated by natural convection may be laminar or turbulent.

The coefficient of thermal expansion β is a property of the fluid, defined as the fractional increase in volume, at constant pressure of the fluid /°F or

$$\beta = \frac{\left(\dfrac{\partial v}{\partial t}\right)_p}{v}$$

where v = specific volume of the fluid

ft³/lb (for liquids) or ft³/lb mole (for gases)

$\left(\dfrac{\partial v}{\partial t}\right)_p$ rate of change of specific volume with temperature at constant pressure

$$\frac{\text{ft}^3}{\text{lb}^\circ\text{F}} \text{ or } \frac{\text{ft}^3}{\text{lb mole}^\circ\text{F}}$$

The above expression for β may be written as

$$\beta = \frac{\dfrac{\Delta v}{\Delta t}}{\bar{v}};$$ β is constant over a definite temperature range.

where $\bar{v}$ = average specific volume $\dfrac{\text{ft}^3}{\text{lb}}$

In terms of density

$$\beta = \frac{\dfrac{1}{\rho_2} - \dfrac{1}{\rho_1}}{(t_2 - t_1)\left(\dfrac{1}{\rho_1} + \dfrac{1}{\rho_2}\right)\Big/2}$$

$$= \frac{\rho_1 - \rho_2}{\bar{\rho}_a\,(t_2 - t_1)}$$

$$\bar{\rho}_a = \frac{\rho_1 + \rho_2}{2}$$

in case of an ideal gas, since $v = \dfrac{RT}{p}$

$$\left(\frac{\partial v}{\partial t}\right)_p = \frac{R}{p}$$

therefore $\beta = \dfrac{R/p}{RT/p} = \dfrac{1}{T}$

β has the dimensions of reciprocal temperature.

Dimensional Analysis

We will now consider the nature of the body force.

ρ_s density of cold undisturbed air

ρ density of the warmer fluid

Δt the temperature difference between the undisturbed fluid and the heated surface.

The buoyancy force on unit volume

$$= (\rho_s - \rho)\,g$$

ρ_s is related to ρ, by

$$\rho_s = \rho\,(1 + \beta\,\Delta t) \qquad \boxed{\beta = \frac{\rho_s - \rho}{\Delta t \cdot \rho}}$$

where β is the coefficient of cubical expansion (thermal expansion) of the fluid.

Thus the buoyancy force is

$$= \left[\rho\,(1+\beta\,\Delta t) - \rho\right] g$$

$$= \left[\rho + \rho\,\beta\,\Delta t - \rho\right] g$$

$$= \rho\,\beta\,\Delta t\;g$$

The independent variables on which natural convection coefficient h depends may now be listed

– a buoyancy force term would appear in the differential equation of momentum, hence $\beta\,\Delta t$ and g appear in addition to,

L, ρ, μ, c_p and k (where L is the linear dimension that is the characteristic length.)

This is the dimension which would be used in the Reynalds number for a forced flow in the same direction as the natural convecting flow.

The heat transfer coefficient can be expressed as

$$h = \phi\left(\rho, \mu, c_p, k, x, g, \beta, \Delta t\right)$$

We will use the $M\,L\,\theta\,T$ system in the dimensional analysis.

Pi Theorem Method

The maximum number of dimensionally incompatible factors may be found, but the maximum number can not exceed the number of fundamental dimensions that is four in this case. All the necessary dimensions $(M\,L\,\theta\,T)$ must be included in the dimensionally incompatible factors.

One of the possible sets contains

x, ρ, β and μ

No combination whatever of these dimensionally incompatible factors would yield a dimensionless group.

Simpler Method to Find π's

(Suggested by Hunsaker and Rightmire)

When a large number of factors are involved (as in the present case) the time required to manipulate the equations is reduced considerably by solving for the dimensions in terms of the dimensionally incompatible factors

$$L = x;\; M = PL^3 = Px^3$$

$$T = \frac{1}{\beta};\; \theta = \frac{M}{L\mu} = \frac{\rho x^3}{x\mu} = \frac{\rho x^2}{\mu}$$

Each of the remaining factors $(h,\ k,\ c_p,\ \Delta t \text{ and } g)$ contains one or more of the dimensions and the incompatible factors contain all the dimensions. Hence each of the remaining factors must produce a dimensionless group when its dimensions are eliminated by use of one or more the above four equations.

$$\Delta t = T = \frac{1}{\beta};\ \beta\,\Delta t = \pi_1$$

$$g = \frac{L}{\theta^2} = \frac{x}{\dfrac{\rho^2 x^4}{\mu^2}} = \frac{\mu^2}{\rho^2 x^3} \text{ or } \frac{x^3 \rho^2 g}{\mu^2} = \pi_2$$

$$c_p = \frac{L^2}{\theta^2 T} = \frac{x^2}{\dfrac{\rho^2 x^4}{\mu^2}\,\dfrac{1}{\beta}} = \frac{\mu^2 \beta}{\rho^2 x^2}$$

or $$\frac{c_p \rho^2 x^2}{\mu^2 \beta} = \pi_3$$

$$h = \frac{M}{\theta^3 T} = \frac{\rho x^3}{\dfrac{\rho^3 x^6}{\mu^3}\,\dfrac{1}{\beta}} = \frac{\mu^3 \beta}{\rho^2 x^3}$$

or $$\frac{h\,\rho^2 x^3}{\mu^3 \beta} = \pi_4$$

$$k = \frac{M\,L}{\theta^3 T} = \frac{\rho\, x^3\, x}{\dfrac{\rho^3 x^6}{\mu^3}\,\dfrac{1}{\beta}} = \frac{\mu^3 \rho}{\rho^2 x^2}$$

or $$\frac{k\,\rho^2 x^2}{\mu^3 \beta} = \pi_5$$

It is thus seen that five dimensionless groups fully replace the nine physical factors.

$$\text{Now } \pi_2\,\pi_1 = \frac{x^3 \rho^2 g}{\mu^2}\,\beta\,\Delta t$$

= the Grashof (buoyancy) modulus or number

$$\frac{\pi_3}{\pi_5} = \frac{c_p \mu}{k} \rightarrow \text{Prandtl number}$$

$$\frac{\pi_4}{\pi_5} = \frac{hx}{k}$$ Nusselt number and therefore we see, that

$$N_{N_u} = \text{Nusselt number} = \phi\left(N_{G_r}, N_{P_r}\right)$$

The characteristic linear dimension is shown below

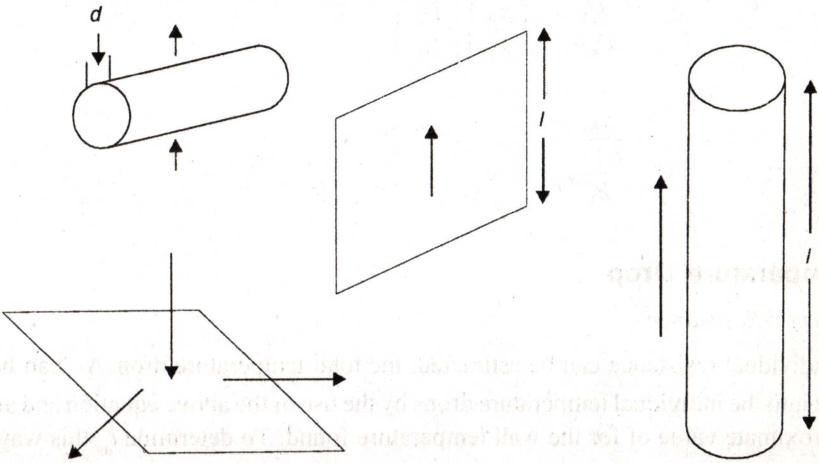

Fig. 5.3 *The characteristic linear dimensions*

l – characteristic linear dimension.
(length or height of the surface)

Cross Sections Other than Circular

The diameter D in both N_{Nu} and N_{Re} has to be replaced by equivalent diameter D_{eq} which is defined as four times the hydraulic radius, r_H

$$r_H = \frac{\text{Cross sectional area}}{\text{wetted perimeter of the section}}$$

for circular channel

$$r_H = \frac{\pi D^2}{4\pi D} = \frac{D}{4}$$

and D_{eq} = diameter of the channel

we know that $\frac{hD}{k} = 0.023\left(\frac{DG}{\mu}\right)^{0.8}\left(\frac{c_p\mu}{k}\right)^{\frac{1}{3}}\left(\frac{\mu}{\mu_w}\right)^{0.14}_{\text{(viscosity correction term)}}$

μ_w = viscosity value at the wall temperature

μ = viscosity of the fluid at the average fluid temperature

Estimation of t_w

$$\text{We know } \frac{\Delta t}{\frac{1}{U_0}} = \frac{\Delta t_i}{\frac{D_0}{D_i h_i}} = \frac{\Delta t_w}{\left(\frac{x_w}{k_w}\right)\left(\frac{D_0}{\overline{D}_L}\right)}$$

$$= \frac{\Delta t_0}{\frac{1}{h_0}}$$

Temperature Drop

Thermal Resistance

If individual resistance can be estimated, the total temperature drop Δt can be split into the individual temperature drops by the use of the above equation and an approximate value of for the wall temperature found. To determine t_w this way, the wall resistance $\frac{x_w}{k_m}\frac{D_0}{\overline{D}_L}$ could be neglected and the above equation is used as follows

From the first two terms of the above equation

$$\Delta t_i = \frac{\frac{D_0}{D_1 h_i}}{\frac{1}{U_0}}\Delta t$$

substituting for $\frac{1}{U_0}$ from the equation

$$\frac{1}{U_0} = \frac{D_0}{D_i h_i} + \underbrace{\frac{x_w}{k_m}\frac{D_0}{\overline{D}_L}}_{\text{ignored}} + \frac{1}{h_0}$$

yields $$\Delta t_i = \frac{\frac{1}{h_i}}{\frac{1}{h_i} + \frac{D_i}{D_0 h_0}} \cdot \Delta t$$

To use the above equation one has to know the preliminary value of the coefficients h_i and h_o which can be computed by employing Dittus-Boelter equation without the viscosity correction term.

Δt_i = temperature drop through inside fluid

Then t_w that is the wall temperature is got as follows

If the inside fluid is getting heated $t_w = t + \Delta t_i$

If the inside fluid is cooled $t_w = t - \Delta t_i$

where t is the average fluid temperature.

Illustration

Benzene is cooled from 130°F to 90°F in inner pipe of a double-pipe exchanger. Cooling water flows counter currently to benzene, entering the jacket at 60°F and leaving at 80°F. The exchanger consists of an inner pipe of $\frac{7}{8}''$ 16 BWG copper tubing jacketed with schedule 40 $1\frac{1}{2}''$ steel pipe. The inner velocity of benzene is 5 ft/sec and that of water is 4 ft/sec. Neglecting the resistance of the wall and the scale films compute the film coefficients of benzene and water and the overall coefficient based on the outside of the inner pipe. Apply the viscosity correction in the computation of the film coefficients.

Solution

The properties at the average temperatures are to be first noted

These are ρ, μ, k and c_p

The diameters of the inner tube are

$$D_{i_t} = \frac{0.745}{12} = 0.0621'$$

$$D_{o_t} = \frac{0.875}{12} = 0.0729'$$

The inside diameter of the jacket is $D_{i_j} = \frac{1.610}{12} = 0.1342$ ft

The equivalent diameter of the annular jacket space

$$\text{c.s. area} = \frac{\pi}{4}\left[0.1342^2 - 0.0729^2\right]$$

Wetted perimeter = $\pi\,[0.1342 + 0.0729]$

$$r_H = \frac{\frac{\pi}{4}\left[0.1342^2 - 0.0729^2\right]}{\pi\,(0.1342 + 0.0729)}$$

$$= \frac{1}{4} \times 0.0631$$

$$D_{Eq} = 0.0631 \text{ ft}$$

Benzene Section

$$N_{Re} = \frac{0.0621 \times 5 \times 3600 \times 53.1}{1.16}$$

$$= 5.12 \times 10^4$$

$$N_{P_r} = 5.67$$

Water Stream

$$N_{R_e} = \frac{D_{Eq}\, v\, \rho}{\mu}$$

$$= 2.35 \times 10^4$$

$$N_{P_r} = 6.76$$

$$h_{i\,(\text{benzene})} = 346$$

$$h_0 = 771$$

The temperature drop over the benzene resistance

$$\Delta t_1 = \frac{\frac{1}{346}}{\frac{1}{346} + \frac{0.0621}{0.0729 \times 771}}(110 - 70)$$

$$= 29°\text{F}$$

$$t_w = 110 - 29 = 81°\text{F}$$

viscosity at this temperature

$$\mu_{w\,\text{benzene}} = 1.45 \text{ lbs/hr ft}$$

$$\mu_{w\,\text{water}} = 2.06 \text{ lb/hr ft}$$

$$\left(\frac{\mu}{\mu_w}\right)^{0.14} = 0.969 \text{ for benzene and } 1.018 \text{ for water}$$

$h_i = 335$ Btu/hr ft^2 °F
$h_o = 785$ Btu/hr ft^2 °F
$U_0 = 209$ Btu/ft^2 hr °F

Alternate form of the Dittus-Boelter Equation and Colburn j_H Factor

$$\frac{hD}{k} = 0.023\left(\frac{DG}{\mu}\right)^{0.8}\left(\frac{c_p\mu}{k}\right)^{\frac{1}{3}}\left(\frac{\mu}{\mu_w}\right)^{0.14}$$

Multiplying both sides by $\dfrac{1}{N_{Re}}\dfrac{1}{N_{Pr}}$

$$\frac{hD}{k}\frac{k}{c_p\mu}\frac{\mu}{DG} = 0.023\left(\frac{DG}{\mu}\right)^{0.8}\left(\frac{c_p\mu}{k}\right)^{\frac{1}{3}}\left(\frac{\mu}{\mu_w}\right)^{0.14}\left(\frac{k}{c_p\mu}\right)\frac{\mu}{DG}$$

from the above, we get $\dfrac{h}{c_pG}\left(\dfrac{c_p\mu}{k}\right)^{\frac{2}{3}}\left(\dfrac{\mu_w}{\mu}\right)^{0.14} = 0.023\left(\dfrac{DG}{\mu}\right)^{-0.2}$

The left side of the above equation, is used to define j_H factor

i.e. $$j_H = \frac{h}{c_pG}\left(\frac{c_p\mu}{k}\right)^{\frac{2}{3}}\left(\frac{\mu_w}{\mu}\right)^{0.14}$$

We have j_H j factor for mass transfer also. The j_H factor is the counterpart of friction factor in friction correlations.

It could be seen from the above equation that j_H factor is a function of Reynalds number.

Problems

1. Determine the rate of heat transfer between the fluids separated by a copper rube 2mm thick 2 cm o.d and 1.5 m long, if the inner fluid (water) temperature in 30°C and the outer fluid (steam) temperature is 100°C. The water side film coefficient is 2000 w/m^2°k and the steam side film coefficient is 15000 w/m^2°k. The thermal conductivity of copper is 400w/m°k.
2. Water at a mean temperature of 37.8°C flow through a 2×10^{-2} m id smooth walled tube with a mean velocity of 1.52 m/sec. The tube wall is 17°C above the water temperature. Determine the heat transfer coefficient and the total heat transfer rate to the water stream for 3 m length of the tube assuming that the velocity and temperature profiles are fully developed. The given data are

$$v_{\text{at } 37.8°C} = 0.687 \times 10^{-6} \text{ m}^2/\text{sec}$$

$$k = 0.63 \frac{\text{W}}{\text{m}} °\text{C} \quad N_{P_r} = 4.52$$

3. 100,000 kg/hr of feed water enters a tabular economiser at 10°C. The inside diameter of the tube is 50mm. There are 20 tubes in each pass. Calculate the water film (heat transfer) coefficient using the expression

$$N_{Nu} = 0.023\,(N_{Re})^{0.8}\,(N_{Pr})^{0.4}$$

At the mean temperature of water, density = 1 gm/cc

c_p = 1 k cal/kg °C

viscosity = 2.44 kg/m hr

and k = 0.54 k cal/hr m°C

$$\text{mass flow rate} = \rho\, A_{\text{total flow area}}\, u$$

$$A_{\text{total}} = n \cdot \frac{\pi d^2}{4}$$

$$\therefore\ u = \frac{100{,}000}{1000 \times 0.0395} = \frac{100}{0.0395} \text{ m/hr}$$

n = no. of tubes in a pass

Simplified Expressions for Natural Convection in Air

In a majority of situations, the fluid involved in a natural convection process is at atmosphere pressure.

The general expressions have been modified to apply specifically to air and a simplified expression for the heat transfer coefficient written in the form

$$h = A\left(\frac{\Delta T}{L}\right)^b \quad \text{has been suggested.}$$

In the above expression A and b are constants and they depend on the geometry and flow L is the significant/appropriate length which is also a function of geometry and flow.

The following table lists the values suggested for various orientations, geometries and flow conditions as indicated by the magnitude of the product of $N_{Gr}\, N_{Pr}$. Values of h evaluated using the constants from the table in the above equation, have the dimensions of Btu/hr ft^2 °F. ΔT the temperature difference referred to is that between the surface and the bulk air with units of °F.

The example given below illustrates the savings to be realized when the above equation is used rather than one of the more involved correlations seen earlier.

Table 5.2 *Constants to be used in the equation given (above) for natural convection in air*

Geometry	*Applicable range*	*A*	*b*	*L*
Vertical surfaces	$10^4 < N_{G_{rL}} \ N_{P_r} < 10^9$	0.29	$\frac{1}{4}$	height
(planes and cylinders)	$10^9 < N_{G_{rL}} \ N_{P_r} < 10^{12}$	0.19	$\frac{1}{3}$	height
Horizontal	$10^3 < N_{G_{rD}} \ N_{P_r} \times 10^9$	0.27	$\frac{1}{4}$	diameter
Cylinders	$10^9 < N_{G_{rD}} \ N_{P_r} < 10^{12}$	0.18	$\frac{1}{3}$	diameter
Horizontal planes, hot plates facing up or	$10^5 < N_{G_{rL}} \ N_{P_r} < 2 \times 10^7$	0.27	$\frac{1}{4}$	length of side
cold plates facing down	$2 \times 10^7 < N_{G_{rL}} \ N_{P_r} < 3 \times 10^{10}$	0.22	$\frac{1}{3}$	length of side
Cold plates facing up or hot plates facing down	$3 \times 10^5 < N_{G_{rL}} \ N_{P_r} < 3 \times 10^{10}$	0.12	$\frac{1}{4}$	length of side

Problem on Natural Convection

A bare horizontal steam pipe with an outside diameter of 6 inches carries saturated steam at 230°F. The temperature of the air surrounding the pipe is 70°F. Calculate the quantity of condensate at the end of a 100 ft section of the pipe

$$\Delta H_v = 958.7 \text{ Btu/lb}$$

Solution

The systemmatic way of doing this problem is as follows

Given: A bare horizontal pipe 6 inches O.D carries saturated steam $v = 2.2 \times 10^{-4} \ \frac{\text{ft}^2}{\text{sec}}$ at 230°F. The temperature of the air surrounding the pipe is 70°F. At the average temperature (150°F) k = 0.0168 Btu/hr ft °F $N_{Pr} = 0.7$.

Objective: To determine the quantity of condensate at the end of a 100 ft section of the pipe.

Assumptions:

1. Steady state conditions prevail
2. The temperature of the outside surface of the pipe is the same as the steam temperature.
3. The heat loss due to radiation, from the pipe is negligible.

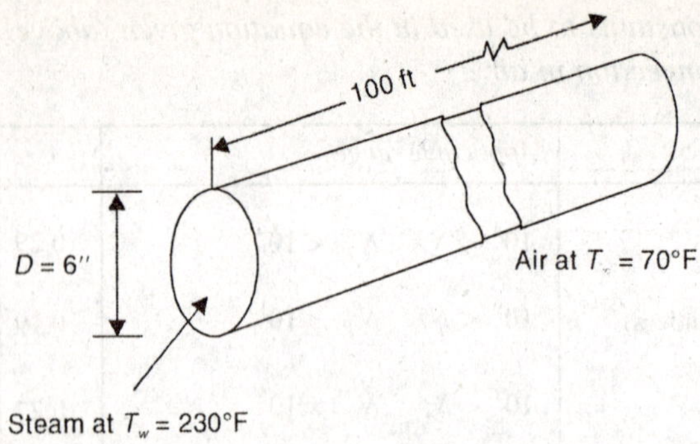

Fig. 5.4 *Free convection in horizontal pipe*

Principles/expressions: As a result of the heat transfer from the hot pipe surface to the surrounding air, some of the steam in the pipe will condense. Therefore in order to calculate the amount of condensate *m*, we have to calculate the rate of heat transfer due to natural convection from the pipe surface and then equate this value of heat transfer to the product of *m* and ΔH_v.

The rate of heat transfer from the pipe surface to the air is

$$q = h\,A\,(T_w - T_\infty)$$

where A = the surface area of the pipe = $\pi\,D L$

Also $q = m\,\Delta H_v$

where m = the rate at which the steam is condensed in L feet of the pipe.

ΔH_v = the latent heat of vapourization of steam at 230°F.

Properties:

D = 6 inches T_w = 230°F

L = 100 feet T_∞ = 70° F

Analysis:

$$\beta = \frac{1}{T_\infty} = \frac{1}{460 + 150} = 1.60 \times 10^{-3}\,\frac{1}{°R}$$

$$g = 32.2 \text{ ft/sec}^2$$

$$N_{Gr_D} = \frac{g\,\beta\,\Delta T\,D^3}{\nu^e}$$

$$= \frac{(32.2)\,(1.64)\,10^{-3}\,(230-70)\left(\frac{6}{12}\right)^3}{\left(2.2 \times 10^{-4}\right)^2}$$

$$= 2.18 \times 10^7$$

since $10^4 < N_{Re_D} < 10^9$

$$N_{Re_D} = N_{Pr}\, N_{Gr_D}$$

$$= 0.7 \times \left(2.18 \times 10^7\right)$$

$$= 1.52 \times 10^7$$

We use the equation $N_{Nu_{av}} = 0.53\left(N_{Re_D}\right)^{\frac{1}{4}}$ for $10^4 < N_{Re_D} < 10^9$ to compute $N_{Nu_{av}}$

$$N_{Nu_{av}} = 0.53\left(N_{G_r}\, N_{P_r}\right)^{\frac{1}{4}}$$

$$= (0.53)\left(1.52 \times 10^7\right)^{\frac{1}{4}} = 33.1$$

$$h_{av} = \frac{N_{N_{u_{av}}} \cdot k}{D} = \frac{(33.1)\,(0.0168)}{\frac{6}{12}}$$

Then the heat loss from a 100 ft length of the pipe is

$$q = \pi\, DL\, h_{av}\,(T_w - T)$$

$$= \pi\left(\frac{6}{12}\right)(100)\,(1.11)\,(230-70)$$

$$= 27900 \text{ Btu/hr.}$$

Since the latent heat of vapourization for steam at 230°F is 958.7 Btu/lb, the rate of condensate collection per hour equals

$$m = \frac{q}{\Delta H_v} = \frac{27900}{958.7}$$

$$= 29.1 \text{ lb/hr per 100 ft length}$$

Remarks: The rate of condensate collection is 29.1 lb/hr. If the radiative heat transfer from the surface of the pipe to the surroundings were accounted for, we would find that there is an additional heat loss of 3976 Btu/hr for a surface emissivity of 0.1. This would increase the condensate rate to 33.25 lbs/hr.

Illustration on Natural Convection

Five-hundred-pound saturated steam is transported through a 2 inch schedule 80 steel pipe with 1½ inch 85% magnesia insulation on the outside. It is desired to know the rate of heat loss per foot from the insulated pipe if the steam line is (a) horizontal and (b) vertical with a run of 20 ft. The pipe is surrounded by 40°F still air. The saturation temperature of 500 psi steam is 467°F. The data needed in this problem are given below

2 inch schedule 80 pipe

$ID = 1.939$ in

$OD = 2.375$ in

$1\frac{1}{2}$ inch insulation layer

$ID = 2.375$ in

$OD = 5.375$ in

Mild steel $k = 23$ Btu/hr ft °F

85% magnesia k = 0.041 Btu/hr ft °F

Solution

The thermal resistance per foot of length of each part of the heat transfer path is determined as follows

Inside wall (condensing steam)

$$R_i = \frac{1}{h_i\ A_i} \quad (h_i \text{ is large, thus this term can be neglected})$$

Steel pipe (conduction)

$$R_i = \frac{ln\left(\dfrac{D_o}{D_i}\right)}{2\pi\ k\ L}$$

$$R_i = \frac{ln\dfrac{(2.375)}{(1.939)}}{2\pi\ (23)\ (1)}$$

$$= 0.00141\ \frac{\text{hr° F}}{\text{Btu}}$$

Insulating layer (conduction)

$$R_t = \frac{ln\left(\frac{D_o}{D_i}\right)}{2\pi\, k\, L}$$

$$= \frac{ln\left[\frac{(5.375)}{(2.375)}\right]}{2\pi\,(0.041)\,(1)}$$

$$= 3.16\ \frac{\text{hr}^\circ\text{ F}}{\text{Btu}}$$

Outside surface (natural convection)

$$R_t = \frac{1}{h_0\ A_0} = \frac{1}{h_0\ (\pi)\left(\frac{5.375}{12}\right)(1)}$$

$$= \frac{0.71}{h_0}\ \frac{\text{hr}^\circ\text{ F}}{\text{Btu}}$$

Between steam and outside air, the heat loss per foot of pipe is

$$q = \frac{\Delta T}{R_t} = \frac{427}{3.161 + \frac{0.71}{h_0}} \qquad \text{...(1)}$$

The heat loss can also be expressed in terms of the conduction between inner and outer surfaces as

$$q = \frac{467 - T_0}{3.161} \qquad \text{...(2)}$$

or in terms of the convection at the outer surface, in the form

$$q = \frac{h_0}{0.71}\,(T_0 - 40) \qquad \text{...(3)}$$

In both of these latter expressions, T_o is the temperature at the outside surface of the insulation.

One should note that the heat transfer coefficient is a function of the unknown temperature T_0. The problem now becomes one of trial and error in which the surface temperature is assumed initially and then checked. The problem will be completed when assumed and calculated values of T_o agree.

We know that, in natural convection the heat transfer coefficient is relatively small. For a value of h_0 = 10 Btu/hr ft^2 °F corresponding value of q and T_0 are approximately 132 Btu/hr and 49°F respectively. An h of 1 Btu/hr ft^2°F will yield corresponding values of 110 Btu/hr and 118°F respectively. The range of values which might be considered in the assumption of T_0 is greatly reduced by this kind of preliminary calculation.

We shall choose $T_0 = 100$ °F for our initial assumption; corresponding values of N_{Gr_D} and N_{Pr} are

$$N_{Gr_D} = \frac{\beta g}{\nu^2} D^3 \Delta T$$

$$= \left[(2.29) \times 10^6\right]\left[\frac{5.375}{12}\right]^3 (60)$$

$$= 12.35 \times 10^6$$

$$N_{Pr} = 0.710$$

A film temperature of 70°F was employed in determining the needed properties of air.

According to the equation

$$N_{Nu_D} = 0.53\left[(N_{Gr_D})(N_{Pr})\right]^{0.25}$$

the value of N_{Nu_D}, applying to the case of a horizontal steam line, is

$$N_{Nu_D} = 0.53\left[(12.35 \times 10^6)(0.71)\right]^{0.25}$$

$$= 28.8$$

and the heat transfer coefficient is

$$h_0 = \frac{k}{D} N_{Nu} = \frac{(0.0149)(28.8)}{\frac{5.375}{12}}$$

$$= 0.959 \text{ Btu/hr ft}^2\text{°F}$$

The heat flow rate and outside surface temperature which correspond to this value of h are

$$q = 109.5 \text{ (Btu/hr)/ft}$$

and $\quad T_0 = 121$°F

with this value of T_0 used a second series of calculations, we obtain, for the horizontal situation

$$N_{Gr_D} = 15.05 \times 10^6$$

$$N_{Pr} = 0.708$$

$$N_{Nu_D} = 30.3$$

$h = 1.03$

$q = 111$ Btu/hr

$T_0 = 116.5$

It is quite clear that one more calculation will produce a refined value of T_0, however the heat loss value will be changed by a very small amount thus the answer to (a) of this problem is $q = 111$ Btu/hr per foot of pipe.

In the case of a vertical pipe the significant length in N_{Gr_L} and N_{Nu_L} becomes the height rather than the diameter. Equations (1), (2) and (3) still hold good.

If an initial value of 120°F is assumed for T_0, the important parameters are determined as follows

$$N_{Gr_L} = \frac{\beta g}{\nu^2} L^2 \Delta T$$

$$N_{Gr_L} = \frac{\left(2.09 \times 10^6\right)(20)^3 (80)}{1}$$

$$= 1.338 \times 10^{12}$$

$$N_{Pr} = 0.708$$

The Nusselt number may now be determined from the equation

$$N_{Nu_L} = 0.0210 \left(N_{G_r} N_{P_r}\right)^{\frac{2}{5}} \text{ for } N_{G_r} N_{P_r} > 10^9$$

$$= 0.0210 \left[\left(1.338 \times 10^{12}\right)(0.708)\right]^{\frac{2}{5}}$$

$$= 1300$$

and from this, the heat transfer coefficient is determined as

$$h = \frac{k}{L} N_{Nu_L}$$

$$= \frac{0.0152}{20}(1300)$$

$$= 0.988 \text{ Btu/hr ft}^2 \text{ F}$$

values of q and T_0 may now be determined from expressions (1) and (3) they are

$$q = \frac{427}{3.161 + \frac{0.71}{0.988}}$$

$$= 110 \text{ Btu/hr}$$

$$T_0 = 40 + \frac{0.71}{0.988}(q)$$

$= 119°F$

This agreement with the assumed value for T_0 of 120°F is quite satisfactory there is no need for additional calculation. The answer to (b) of this problem for a vertical steam line is q = 110 Btu/hr per foot of the pipe.

No significant difference in heat loss exists, for this case, between a vertical and horizontal configuration.

Example on Natural Convection

A $1\frac{1}{2}$ m OD pipe carrying slightly wet steam at 15 psig is installed in a location where it is covered by water after a heavy rain but is exposed to air under normal conditions. The k of water is 0.389 Btu/hr ft°F. The Nusselt number for water is 77. The temperature of the steam at 15 psig is 250°F. Compute the rate of heat loss by convection per foot length of the pipe.

Solution

The heat transfer coefficient for water is

$$\bar{h}_c = N_{N_u} \frac{k}{D}$$

$$= 77 \frac{0.389}{\frac{1.5}{12}}$$

$$= 236 \text{ Btu/hr ft}^2\text{°F}$$

The rate of heat loss by convection per foot length of the pipe

$$q_c = \bar{h}_c A (T_s - T_\infty)$$

$$= (236)\, \pi \left(\frac{1.5}{12}\right) (200)$$

$$= 18500 \text{ Btu/hr ft}$$

Empirical Equations for Free Convection Based on Experimental Results

The equation $N_{N_u} = \phi\left(N_{G_r}\right)\phi\left(N_{P_r}\right)$ represents generalised correlation for natural (free) convection. The form of the function of N_{G_r} and N_{P_r} may be determined for different situations with the help of experimental results. Some of the frequently used relationships are presented below.

Laminar Flow Over Vertical Planes and Cylinders

$$\left(N_{G_r} \cdot N_{P_r} = 10^4 \text{ to } 10^8\right)$$

For free convection from vertical plates, researchers have derived the following relations theoretically for the local Nusselt number $N_{N_{u_x}}$ and the average Nusselt number N_{N_u} for a plate of height L:

$$N_{N_{u_x}} = 0.508 \left(\frac{N_{P_r}}{0.952 + N_{P_r}} \right)^{\frac{1}{4}} \left(N_{G_{r_x}} \, N_{P_r} \right)^{\frac{1}{4}}$$

$$N_{N_u} = 0.677 \left(\frac{N_{P_r}}{0.952 + N_{P_r}} \right)^{\frac{1}{4}} \left(N_{G_r} \, N_{P_r} \right)^{\frac{1}{4}}$$

with these correlations, one can fit data for air and short vertical planes. In the above equation, the characteristic dimension is taken to be the height (L) of the plane.

However, Eckert and Jackson recommend

$$N_{N_{u_f}} = 0.555 \left\{ \left(N_{G_r} \right)_f \left(N_{P_r} \right)_f \right\}^{\frac{1}{4}}$$

$$\left(N_{G_r} \, N_{P_r} \right)_f = 10^5 \text{ to } 10^8$$

McAdams recommends the following equation for vertical planes and vertical cylinders (characteristic dimension height L)

$$\frac{h_L}{k_f} = 0.59 \left[\left(\frac{L^3 \rho_f^2 \, g \, \beta_f \, \Delta t}{\mu_f^2} \right) \left(\frac{\mu \, c_p}{k} \right)_f \right]^{\frac{1}{4}}$$

$$\left(N_{N_u} \right)_f = 0.59 \left\{ \left(N_{G_r} \right)_f \left(N_{P_r} \right)_f \right\}^{\frac{1}{4}}$$

for $N_{G_{rf}} \, N_{P_{rf}}$ from 10^4 to 10^9

For liquids and gases, Fischenden and Saunders suggest the following equation for the case of vertical planes and large diameter cylinders

$$\left(N_{N_{u_f}} \right) = 0.56 \left\{ \left(N_{G_r} \right)_f \left(N_{P_r} \right) \right\}^{\frac{1}{4}}$$

$$\left(N_{G_r} \, N_{P_r} \right)_f = 10^5 \text{ to } 10^8$$

Streamline Flow Over Horizontal Cylinders

$$\left(N_{G_r} \, N_{P_r} = 10^4 \text{ to } 10^8 \right)$$

The data on heat transfer from horizontal pipes to air and to water are represented by

$$\frac{hD}{k_f} = 0.53\left\{\left(\frac{D^3 \rho_f^2\ g\ \beta_f\ \Delta t}{\mu_f^2}\right)\left(\frac{c_p \mu}{k}\right)_f\right\}^{\frac{1}{4}}$$

or $$\left(N_{N_u}\right)_f = 0.53\left\{\left(N_{G_r}\right)_f \left(N_{P_r}\right)_f\right\}^{\frac{1}{4}}$$

$$\left(N_{G_r}\ N_{P_r}\right)_f = 10^3 \text{ to } 10^9$$

For laminar flow over cylinders larger than the wires in diameter, the data for all fluids metallic or non-metallic are well represented by

$$\left(N_{N_u}\right)_f = 0.53\left\{\left(\frac{N_{P_r}}{N_{P_r} + 0.952}\right)_f \left(N_{G_r}\ N_{P_r}\right)_f\right\}^{\frac{1}{4}}$$

while the data for liquid metals by

$$\left(N_{N_u}\right) = 0.53\left\{\left(N_{G_r}\right)_f \left(N_{P_r}\right)_f^2\right\}^{\frac{1}{4}}$$

Streamline Flow Over Horizontal Plates

Fischenden and Saunders have correlated data from horizontal square plates exposed to air; they recommend

(i) For horizontal heated plates facing up or cooled plates facing down

$$\left(N_{N_u}\right)_f = 0.54\left\{\left(N_{G_r}\right)_f \left(N_{P_r}\right)_f\right\}^{\frac{1}{4}}$$

$$\left(N_{G_r}\ N_{P_r}\right)_f = 10^5 \text{ to } 2\times 10^7$$

(ii) For horizontal heated plates facing down or cooled plates facing up

$$\left(N_{N_u}\right)_f = 0.27\left\{\left(N_{G_r}\right)_f \left(N_{P_r}\right)_f\right\}^{\frac{1}{4}}$$

$$\left(N_{G_r}\ N_{P_r}\right)_f = 3\times 10^5 \text{ to } 10^{10}$$

In the above two equations, the characteristic dimensions (to compute N_{G_r} and N_{N_u}) is taken to be the side of the plate.

Turbulent Flow Over Vertical Planes and Cylinders

$$\left(N_{G_r}\, N_{P_r}\right) = 10^9 \text{ to } 10^{12}$$

For this case the following correlation has been suggested by McAdams

$$\frac{hL}{k_f} = 0.13\left\{\left(\frac{L^3\,\rho_f^2\, g\,\beta_f\,\Delta t}{\mu_f^2}\right)\left(\frac{c_p\,\mu}{k}\right)_f\right\}^{\frac{1}{3}}$$

$$\text{or } \left(N_{N_u}\right)_f = 0.13\left\{\left(N_{G_r}\right)_f \left(N_{P_r}\right)_f\right\}^{\frac{1}{3}}$$

$$\left(N_{G_r}\, N_{P_r}\right) = 10^9 \text{ to } 10^{12}$$

This correlation may also be applied to liquids and gases being heated inside vertical tubes at low velocities.

Eckert and Jackson suggest the equation

$$\left(N_{N_u}\right) = 0.021\left\{\left(N_{G_r}\right)_f \left(N_{P_r}\right)_f\right\}^{0.4}$$

$$\left(N_{G_r}\, N_{P_r}\right)_f = 10^{10} \text{ to } 10^{12}$$

Turbulent Flow Over Horizontal Plates

For horizontal heated plates facing up on cooled plates facing down in air or gas, Fischenden and saunders suggest

$$\left(N_{N_u}\right) = 0.14\left\{\left(N_{G_r}\right)_f \left(N_{P_r}\right)_f\right\}^{\frac{1}{3}}$$

$$\left(N_{G_r}\, N_{P_r}\right)_f = 2 \times 10^7 \text{ to } 3 \times 10^{10}$$

Flow Over Spheres and Rectangular Blocks

In the case of spheres and rectangular blocks, the equation suggested for horizontal cylinders may be used, provided that the characteristic dimension L is computed by

$$\frac{1}{L} = \frac{1}{L_v} + \frac{1}{L_H}$$

where L_v = vertical dimension

L_h = horizontal dimension

For a sphere the characteristic dimension from the above equation turns out to be $D/2$.

Flow Inside Vertical or Horizontal Enclosed Air Spaces

In the case of enclosed air spaces, the Grashof and Nusselt numbers are computed by using the thickness δ of the air space (that is the clearance between the two enclosing surfaces) as the characteristic dimension.

Thus

$$\left(N_{N_u}\right)_f = \frac{h\delta}{k_f}$$

$$\left(N_{G_r}\right)_f = \left(\frac{\delta^3 \rho_f^2 \, g \, \beta_f \, \Delta t}{\mu_f^2}\right)$$

where δ = thickness of air space or clearance between the two surfaces

For vertical air spaces. Jakob suggests the following equations.

$$\left(N_{N_u}\right)_f = 1; \left(N_{G_r}\right)_f < 7 \times 10^7$$

$$\left(N_{N_u}\right)_f = 0.0384 \left(N_{G_r}\right)_f^{0.37} \left(N_{G_r}\right)_f = 10^4 \text{ to } 8 \times 10^4$$

$$\left(N_{N_u}\right)_f = 0.0317 \left(N_{G_r}\right)_f^{0.37}; \left(N_{G_r}\right)_f > 2 \times 10^5$$

For horizontal air spaces, Jacob suggests

$$\left(N_{N_u}\right)_f = 1; \left(N_{G_r}\right) < 10^3$$

$$\left(N_{N_u}\right)_f = 0.21 \left\{\left(N_{G_r}\right)_f \left(N_{P_r}\right)_f\right\}^{\frac{1}{4}} \left(N_{G_r}\right)_f = 10^4 \text{ to } 3.2 \times 10^5$$

$$\left(N_{N_u}\right)_f = 0.075 \left\{\left(N_{G_r}\right)_f \left(N_{P_r}\right)_f\right\}^{\frac{1}{3}} \left(N_{G_r}\right)_f = 3.2 \times 10^5 \text{ to } 10^7$$

For horizontal air spaces, Graaf and Held recommend

$$\left(N_{N_u}\right)_f = 1.0 \left(N_{G_r}\right)_f < 2 \times 10^3$$

$$\left(N_{N_u}\right)_f = 0.0507 \left(N_{G_r}\right)^{0.1}; \left(N_{G_r}\right)_f = 2 \times 10^3 \text{ to } 5 \times 10^4$$

$$\left(N_{N_u}\right)_f = 3.8; \left(N_{G_r}\right)_f = 5 \times 10^4 \text{ to } 2 \times 10^5$$

$$\left(N_{N_u}\right)_f = 0.0426 \left(N_{G_r}\right)_f^{0.37} \left(N_{G_r}\right)_f > 2 \times 10^5$$

Simplified Equations for Air

In the case of air $\left(N_{P_r} = 0.7\right)$, the calculation may be done by the following equations.

(i) Laminar flow $\left(N_{G_r} \; N_{P_r}\right) < 10^8$

$$h = c'\left(\frac{\theta}{L}\right)^{0.25} \text{ or } c'\left(\frac{\theta}{D}\right)^{0.25}$$

where $\theta = t_w - t_{air}$

L or D = characteristic length

c' = constant

(ii) Turbulent flow $\left(N_{G_r}\ N_{P_r}\right) > 10^8$

$$k = c''\,(\theta)^{0.25}$$

where c'' = a constant

Illustrations on Free Convection

1. A vertical plate is heated from one side and maintained at 96°C. In the other side is air at atmospheric pressure and 20°C. Calculate the local heat transfer coefficient at a distance of 20 cm from the lower edge and the average value over the 20 cm length.

Solution

$$t_w = 96°\text{C} \quad t_{air} = 20°\text{C}$$

$$\text{Hence } t_f = \frac{96 + 20}{2} = 58°\text{C}$$

At the temperature of 58°C the properties of air from tables are as given below.

$$\nu = 18.77 \times 10^{-6}\ \text{m}^2/\text{sec}$$

$$k = 2.48 \times 10^{-2}\ \text{kcal/m hr°C}$$

$$\beta = \frac{1}{273 + 58} = \frac{1}{331}\ °\text{K}^{-1}$$

$$N_{P_r} = 0.694$$

$$\text{Hence } N_{G_{rx}} = \frac{g\,\beta\,x^2\,\rho^2\,\Delta t}{\mu^2} = \frac{(9.806)\,(0.2)^3\,(96 - 20)}{331 \times \left(18.77 \times 10^{-6}\right)^2}$$

$$= 51 \times 10^6$$

using the equation

$$N_{N_{ux}} = 0.508\left(\frac{N_{P_r}}{0.952 + N_{P_r}}\right)^{\frac{1}{4}}\left(N_{G_{rx}} \cdot N_{P_r}\right)^{\frac{1}{4}}$$

we have

$$N_{N_{u_x}} = \frac{h_x\, x}{k} = 0.508\left(\frac{N_{P_r}}{0.952 + N_{P_r}}\right)^{\frac{1}{4}}\left(N_{G_{r_x}} \times N_{P_r}\right)^{\frac{1}{4}}$$

$$h_x = \frac{2.48 \times 10^{-2}}{0.2}\left\{0.508\sqrt{.694}\,(1.646)^{-\frac{1}{4}}\left(51 \times 10^6\right)^{\frac{1}{4}}\right\}$$

$$= 4.8 \text{ k cal/m}^2 \text{ hr °C}$$

The average value of heat transfer coefficient is

$$h = \frac{1}{L}\int_0^L h_x\, dx = \frac{4}{3} h_L$$

The average value of heat transfer coefficient is

$$= \frac{4}{3} \times 4.8 = 6.4 \text{ k cal/m}^2 \text{ hr°C}$$

Since $x = L = 0.2$ m in this illustration

It may be mentioned in this context that the same answer can be obtained by using the equation

$$N_{N_u} = 0.677\left(\frac{N_{P_r}}{0.952 + N_{P_r}}\right)^{\frac{1}{4}}\left(N_{G_r} \cdot N_{P_r}\right)^{\frac{1}{4}}$$

directly for h, the heat transfer coefficient.

2. Determine the rate of heat loss from a 100 mm diameter steam pipe placed horizontally in ambient air at 30°C. The length of the pipe is 4m and the wall temperature $t_w = 170$°C.

Solution

The properties of air at $t_f = 100$°C are given below

$k = 0.0276$ k cal/m hr°C

$\nu = 23.31 \times 10^{-6}$ m²/sec

$N_{P_r} = 0.69$

$$\beta = \frac{1}{373}\ °\text{k}^{-1}$$

$$\text{Therefore } N_{G_r} = \frac{(9.806)\,(0.1)^3\,(170 - 30)}{373 \times \left(23.31 \times 10^{-6}\right)^2}$$

$$= 6.94 \times 10^6$$

using the equation

$$\frac{hD}{k} = 0.53 \left\{ \left(\frac{D^3 \rho_f^2 \, g \, \beta_f \, \Delta t}{\mu_f^2} \right) \left(\frac{c_p \, \mu}{k} \right)_f \right\}^{\frac{1}{4}}$$

we obtain

$$N_{N_u} = 0.53 \left(6.94 \times 10^6 \times 0.69\right)^{\frac{1}{4}}$$

$$= 2.52$$

or $$h = \frac{N_{N_u} \, k}{D} = \frac{(25.2)\,(0.0276)}{0.1} = 6.96 \text{ k cal/m}^2 \text{ hr °C}$$

The heat loss $h \times A \times \Delta t = 6.96 \times \pi \times 0.1 \times 4 \times 140$

$$= 1220 \text{ k cal/hr}$$

3. A horizontal cylinder of 2.5 cm diameter and 0.6 cm long is suspended in water at 20°C. Calculate the rate of heat transfer if the cylinder surface is at 55°C. What would be the rate of heat transfer if the cylinder is suspended in light oil at 55°C. The properties of water and oil at 37.5°C are given below.

Table 5.3

	Water	*Oil*
$\rho \left(\frac{\text{kg}}{\text{m}^3}\right)$	992	905
$c_p \left(\frac{\text{k cal}}{\text{kg °C}}\right)$	1.0	0.45
$\mu \left(\frac{\text{kg}}{\text{hr m}}\right)$	2.47	82.0
$k \left(\frac{\text{k cal}}{\text{hr m °C}}\right)$	0.534	0.114
$\beta \left(°\text{K}^{-1}\right)$	3.96×10^{-4}	7.2×10^{-4}
N_{pr}	4.64	324

Solution

The Grashof number is given by

$$N_{Gr} = \frac{D^3 \rho^2 \, g \, \beta \, \Delta t}{\mu^2}$$

$$= \frac{\left(2.5 \times 10^{-2}\right)^3 (992)^2 (9.81 \times 3600 \times 3600) \left(3.96 \times 10^{-4}\right)(55-20)}{(2.47)^2}$$

$= 4.45 \times 10^6$

we may use the following equation

$$\left(N_{N_u}\right)_f = 0.53 \left\{\left(N_{G_r}\right)_f \left(N_{p_r}\right)_f\right\}^{\frac{1}{4}}$$

$$= 0.53 \left\{(4.45) \times 10^6 \times 4.64\right\}^{\frac{1}{4}}$$

$$= (0.53)(67.3) = 35.7$$

or $\quad h = \dfrac{35.7 \times 0.534}{0.025} = 762$ k cal/hr m² °C

or $\quad q = h A \Delta t$

$$= (762)(\pi \times 0.025 \times 0.6)(35)$$

$$= 1260 \text{ k cal/hr}$$

In the case of oil

$$N_{G_{r_f}} = \frac{\left(2.5 \times 10^{-2}\right)^3 (905)^2 (9.81 \times 3600 \times 3600) \left(7.2 \times 10^{-4}\right)(55-20)}{(82)^2}$$

$= 6100$

$$\therefore \left(N_{N_u}\right)_f = 0.53 \left\{(6100) \times (324)\right\}^{\frac{1}{4}}$$

$$= 0.53 \times 36.8 = 19.5$$

Therefore $h = \dfrac{19.5 \times 0.114}{0.025}$

$= 89$ k cal/hrm²°C

Therefore the heat transfer rate, $q = 89 (\pi \times 0.025 \times 0.6)\, 35$

$= 147$ k cal/hr

4. The parallel outer and inner walls of a building are 4m high and 5 m long. The walls are 10 cm apart. The inner surface of the inner wall is at 25°C and the inner surface of the outer wall is at 5°C.

(i) Calculate the total heat loss per hour.

(ii) If the air space is divided in half by a sheet of aluminium foil 0.025 mm thick parallel to the walls, what would be the heat loss per hour.

Solution

It would be in order to evaluate the properties of air at the mean temperature of the two surfaces is at $\frac{(5+25)}{2} = 15\,°C$

At the temperature of 15°C

$$\rho = 1.225 \ \text{kg/m}^3$$

$$\mu = 0.065 \ \text{kg/hr m}$$

$$c_p = 0.24 \ \text{k cal/kg °C}$$

$$k = 0.0216 \ \text{k cal/hr.m °C}$$

$$\beta = 1/288\ °\text{K}^{-1}$$

Therefore $N_{P_r} = \frac{(0.065)(0.24)}{0.0216}$

$= 0.72$

$$N_{G_{r\,f}} = \frac{\delta^3 \rho^2 g \beta \Delta t}{\mu^2}$$

$$= (0.1)^3 (1.225)^2 (9.81)(3600)^2 \times \left(\frac{1}{288}\right)(25-5)\frac{1}{(0.065)^2}$$

$= 3.15 \times 10^6$

(i) From the equation

$$(N_{N_u})_f = 0.07\left(\frac{L}{\delta}\right)^{\frac{-1}{9}} \left\{(N_{G_r})_f (N_{P_r})_f\right\}^{\frac{1}{3}}$$

we can calculate the heat transfer coefficient and the heat transfer rate

$$\frac{h\delta}{k_f} = 0.071 \times \left(\frac{L}{\delta}\right)^{\frac{-1}{9}} \left(N_{G_{r\,f}}\ N_{P_{r\,f}}\right)^{\frac{1}{3}}$$

$$= 0.071 \times \left(\frac{4}{0.1}\right)^{\frac{-1}{9}} \left\{(3.15) \times 10^6 \times 0.72\right\}^{\frac{1}{3}}$$

$$= 0.071 \times \frac{1}{1.505} \times 131.3$$

$= 6.18$

$$\therefore h = \frac{6.18 \times 0.0216}{0.1} = 1.34 \text{ k cal/hr m}^2 \text{ °C}$$

$$q = h\, A\, \Delta t$$
$$= 1.34\ (4 \times 5)\ 20$$
$$= 538 \text{ k cal/hr}$$

In case we employ the following equation

$$N_{N_{u f}} = 0.0317 \left(N_{G_{r f}} \right)^{0.37}$$

$$= (0.0317)\left(3.15 \times 10^6\right)^{0.37}$$

$$= 0.0317 \times 256$$
$$= 8.1$$

$$\therefore h = \frac{8.1 \times 0.0216}{0.1} = 1.75$$

and $\quad q = 1.75 \times 400 = 700$ k cal/hr

(ii) In the second situation there are 2 air spaces, each 5 cm thick. Since the temperature difference between the two surfaces is only 20°C, we may take that the properties of air in the two air spaces are the same and evaluate these properties at 15°C. Thus the resistance of each air space to heat flow will be the same and the temperature of the aluminimum foil will be midway between the two surfaces i.e. 15°C.

Then for each air space

$$\delta = 0.05 \text{ m}$$

$$\Delta t = 10 \text{ °C}$$

$$\left(N_{G_r}\right)_f = \frac{\delta^3\, \rho^2\, g\, \beta\, \Delta t}{\mu^2}$$

$$= \frac{(0.05)^3\ (1.225)^2\ (9.81 \times 3600 \times 3600)\ (10)}{(0.065)^2\ (288)}$$

$$= 1.97 \times 10^5$$

From the equation

$$\left(N_{N_u}\right)_f = 0.2 \left(\frac{L}{\delta}\right)^{\frac{-1}{9}} \left\{ \left(N_{G_r}\right)_f \left(N_{P_r}\right)_f \right\}^{\frac{1}{4}}$$

$$= 0.2\left(\frac{4}{0.05}\right)^{\frac{-1}{9}}\left\{1.97 \times 10^5 \times 0.72\right\}^{\frac{1}{4}}$$

$$= 0.2 \times \frac{1}{1.626} \times 19.4$$

$$= 2.39$$

$$\because h = \frac{2.39 \times 0.0216}{0.05} = 1.03 \text{ k cal/hr m}^2\ °\text{C}$$

$$\therefore q = h\ A\ \Delta t$$

$$= (1.03)\ (20)\ (10)$$

$$= 206 \text{ k cal/hr}$$

In this case, the use of the equation yields

$$\left(N_{N_u}\right)_f = 0.0317\left(1.97 \times 10^5\right)^{0.37}$$

$$= 2.89$$

$$h = 1.25 \text{ k cal/hr m}^2\ °\text{C}$$

$$q = 250 \text{ k cal/hr}$$

Examples in Free Convection Heat Transfer

1. A vertical pipe 75 mm in diameter and 2 metre high is kept at a constant temperature of 110°C (t_w). The pipe is surrounded by atmosphere air at a temperature of 30°C (t_∞). Determine the heat lost by natural convection from this cylinder. The cylinder may be taken as vertical plate for the purpose of calculation of heat loss. It may be noted that the equation for N_{N_u} for vertical surfaces (both plane and cylindrical) is given by

$$N_{N_u} = c\left[\left(N_{G_r}\right)\left(N_{P_r}\right)\right]^m$$

In which the values of c and m are to be selected depending on the value of the product of $\left(N_{P_r}\right)$ and $\left(N_{G_r}\right)$. If $\left(N_{P_r}\ N_{G_r}\right)$ lies between

10^4 and 10^9, c = 0.59 and m = 0.25

10^9 and 10^{12}, c = 0.129 and $\text{m} = \frac{1}{3}$

Solution

$$\text{The mean film termperature} = \frac{(t_w + t_\infty)}{2}$$

$$= \frac{(110 + 30)}{2} = 70\ °C$$

At 70°C, the properties of air from the appropriate table, are

$$k = 2.55 \times 10^{-2} \text{ k cal/hr m °C}$$

$$\nu = 20.02 \times 10^{-6} \text{ m}^2\text{/sec} \quad N_{p_r} = 0.694$$

And the value of β is computed at the temperature of the undisturbed fluid i.e. 110°C, from

$$\beta = \frac{1}{T} = \frac{1}{383} = 0.261 \times 10^{-2} \quad K^{0\ -1}$$

To choose the value of '*c*' and '*m*' for use in the equation given in the problem statement, N_{G_r} for the flow must be found out.

Using the equation

$$N_{G_r} = \frac{L^3\, g\, \beta\, \Delta t}{\nu^2}$$

$$= \frac{(2)^3 \times 9.81 \times 0.261 \times 10^{-2} \times 80}{\left(20.02 \times 10^{-6}\right)^2}$$

$$= 4.1 \times 10^{10}$$

The value of the product

$$\left(N_{G_r}\, N_{p_r}\right) = \left(4.1 \times 10^{10}\right)(0.694)$$

$$= 2.84 \times 10^{10}$$

For this range $c = 0.129$ and $m = \frac{1}{3}$

Therefore, the Nusselt number for this case is got by using the following equation

$$\left(N_{N_u}\right) = 0.129\left(2.84 \times 10^{10}\right)^{\frac{1}{3}}$$

$$= 0.393$$

Therefore

$$h = N_{N_u} \cdot \frac{k}{l}$$

$$= \frac{393 \times 2.55 \times 10^{-2}}{2}$$

$= 5$ k cal/hr m^2°C

Employing the well-known Newton's law of cooling expression

$$Q = h\,A\,(t_w - t_\infty)$$

$$= 5 \times \pi \times 75 \times 10^{-3} \times 2\,(110 - 30)$$

$$= 188.5 \text{ k cal/hr}$$

2. Calculate the heat transfer coefficient in free convection heat transfer between a horizontal wire (employed for heating the surrounding fluid by passing current) and air at 25°C. The surface of the wire may be taken to be at a temperature of 95°C. $D = 2.5$ mm. If may be mentioned in this context that plots of N_{N_u} as a function of the product of N_{G_r} and N_{P_r} are available for free convection in vertical plates and cylinders and free convection in horizontal cylinders which can be used to calculate/find out each other if one quantity is known. The properties of air at the mean fluid temperature which is equal to $\dfrac{(25+95)}{2} = 60\ °\text{C}$ are given below

$$k = 2.49 \times 10^{-2} \text{ k cal/hr m °C}$$

$$\nu = 18.97 \times 10^{-6} \text{ m}^2/\text{sec}$$

$$N_{P_r} = 0.696 \text{ and } \beta = \frac{1}{298}\ °\text{K}^{-1}$$

Solution

The Grashof number is

$$N_{G_r} = \frac{D^3\,g\,\beta\,\Delta t}{\nu^2}$$

$$= \frac{\left(2.5 \times 10^{-3}\right)^3 (9.81)\left(\dfrac{1}{298}\right) 70}{\left[(18.97)\left(10^{-6}\right)\right]^2}$$

$$= 100$$

and $\left(N_{G_r}\,N_{P_r}\right) = (100) \times (0.696) = 69.6$

The value of N_{N_u} can be obtained from the appropriate plot. It is read as 1.995 and

$$h = \left(N_{N_u}\right)\frac{k}{D}$$

$$= \frac{1.995 \times 2.49 \times 10^{-2}}{2.5 \times 10^{-3}}$$

$$= 19.90 \text{ k cal/m}^2\text{°C}$$

3. A horizontal pipe laid in the ground is 75 mm in diameter and 2 metre high. It is kept at a constant temperature of 110°C. The pipe is surrounded by atmosphere air at a temperature of 30°C. Calculate the heat lost by natural convection from this pipe. The fluid properties are

$k = 2.55 \times 10^{-2}$ k cal/hr m °C

$\nu = 20.02 \times 10^{-6}$ m²/sec, $N_{P_r} = 0.694$

It may be added here that N_{N_u} is given by

$$N_{N_u} = c\left[\left(N_{G_r}\right)\left(N_{P_r}\right)\right]^m$$

where $c = 0.525$ and $m = 0.25$ if $10^4 < \left(N_{G_r}\right)\left(N_{P_r}\right) < 10^9$

and $c = 0.129$ and $m = \frac{1}{3}$ if $10^9 < \left(N_{G_r}\right)\left(N_{P_r}\right) < 10^{12}$

Solution

In the computation of N_{G_r}, the characteristic length l is replaced by the diameter D. Therefore, Grashof number for this situation is

$$N_{G_r} = \frac{D^3 g\,\beta\,\Delta t}{\nu^2}$$

$$= \frac{\left(75 \times 10^{-3}\right)^3 (9.81)\left(0.261 \times 10^{-2}\right)(80)}{\left[(20.02)\left(10^{-6}\right)\right]^2}$$

$$= 2.16 \times 10^8$$

and $\left(N_{G_r}\right)\left(N_{P_r}\right) = \left(2.16 \times 10^8\right) \times (0.694)$

$$= 1.5 \times 10^8$$

Since this value lies between 10^4 and 10^9 the values of c and m as given in the problem statement are

$c = 0.525$ and $m = 0.25$

Therefore

$$N_{N_u} = c\left[\left(N_{G_r}\right)\left(N_{P_r}\right)\right]^{m}$$

$$= 0.525\left[1.5 \times 10^{8}\right]^{0.25}$$

$$= 58$$

Hence $h = \left(N_{N_u}\right) \cdot \dfrac{k}{D}$

$$= \frac{58 \times 2.55 \times 10^{-2}}{75 \times 10^{-3}}$$

$$= 19.7 \text{ k cal/hr m}^2 \text{ °C}$$

$$Q = h\ A\left(t_w - t_\infty\right)$$

$$= 19.7 \times 75 \times 10^{-2} \times 2 \times 80$$

$$= 744 \text{ k call/hr}$$

Thus the heat loss in the present case is roughly four times more than the case of vertical orientation.

Typical Values of *U*

The following table presents some typical values of *U*. The data have been assembled from literature.

Table 5.4 *Typical values or ranges of U*

Heat Exchange Configuration	*U* w/m^2 °C
Walls and roofs of dwelling with a 25 km/hr exterior wind velocity	
Frame walls	0.8–5
Finished masonary walls	0.5–6
Uninsolated roofs	1.2–4
Insulated roofs	0.3–2
Single-pane windows	~ 6 but main heat loss is by infiltration
Air to heavy tars and oils	As low as 45 As high as 600
Air to low-viscosity liquids	

Contd.

Heat Exchange Configuration	U w/m² °C
Air to various gases	60–550
Air condensers	350–780
Steam or water to oil	60–340
Liquids in coil immersed in liquids	110–2000
Steam-jacketed, agitated vessels	500–1900
Shell and tube ammonia condensers	800–1400
Steam condensers with 25°C water	1500–5000
Feed water heaters	110–8500
Condensing steam to high pressure boiling water	around 7000
Heat pipes	
Cryogenic	C1000
Water	300
Liquid metal	50,000

Some typical values of fouling resistance R_f are given in the following table

$$R_f = \frac{1}{U_{\text{old}}} - \frac{1}{U_{\text{new}}}$$

Table 5.5 *Some typical fouling resistances*

Fluid and situation	*Fouling Resistance* R_f *m² °C/w*
Distilled water	0.0001
Sea water	0.0001–0.0002
Clean river and lake water	0.0002–0.0006
About the worst waters used in heat exchanges	<0.0020
Treated boiled feed water	0.0001–0.0002
Transfer or lubricating oil	0.0002
Fuel oil	0.0010
Most industrial liquids	0.0002
Most refinery liquids	0.0002–0.0008
Non-oil-bearing steam	0.0001
Oil-bearing steam (e.g. turbine exhaust)	0.0002
Most stable gases	0.0005
Engine exhaust gases	0.002
Fuel gases	0.0020

Overall Heat Transfer Coefficients—Approximate Values

Table 5.6

Fluid combination	*U Btu/hr ft² °F*
Water to compressed air	10–30
Water to water, jacket water coolers	150–275
Water to brine	100–200
Water to gasoline	60–90
Water to gas oil or distillate	35–60
Water to organic solvents, alcohol	50–150
Water to condensing alcohol	45–120
Water to lubricating oil	20–60
Water to condensing oil vapours	40–100
Water to condensing or boiling Freon-12	50–150
Water to condensing ammonia	150–250
Steam to water,	
Instantaneous heater	400–600
Storage tank heater	175–300
Steam to oil,	
Heavy fuel	10–30
Light fuel	30–60
Light petroleum distillate	50–200
Steam to aqueous solutions	100–600
Steam to gases	5–50
Fluid combination	*U Btu/hr ft² °F*
Light organic to light organic	40–75
Medium organics to medium organics	20–60
Heavy organics to heavy organics	10–40
Heavy organics to light organics	10–60
Crude oil to gas oil	30–55

Approximate values of overall heat transfer coefficients

Physical situation	*Btu/ hr ft² °F*	*w/m²*
Brick exterior wall plaster interior un insulated	0.5	2.50
Frame exterior wall plaster interior un insulated	0.25	1.40
With rock-wool insulation	0.07	0.40
Plate glass window	1.10	6.20

Contd.

Physical situation	*Btu/ hr ft² °F*	*w/m²*
Double-plate glass window	0.40	2.30
Steam condenser	200–1000	1000–500
Feed water heater	200–1500	1000–800
Freon 12 condenser with water coolant	50–150	300–850
Water-to-water heat exchanger	150–300	850–
Finned-tube heat exchanger, water in tubes, air across tubes	5–10	30–35
Water-to oil heat exchanger	20`60	115–340

Approximate Relations for Film Coefficients in Natural Convection for Air

The approximate values for the heat transfer coefficients in natural convection for air can be obtained from the equations tabulated below. These equation have been cited in literature.

Table 5.7

Orientation of heated surface	'h' is k cal/hr m²°C, l and D are in meter Δt in °C	
	Laminar $10^9 > N_{G_r} \, N_{P_r} > 10^4$	*Turbulent* $N_{G_r} \, N_{P_r} > 10^9$
1. Horizontal cylinders d = diameter	$h = 2\left[\frac{\Delta t}{D}\right]^{\frac{1}{4}}$	$h = 1.07\,(\Delta t)^{\frac{1}{3}}$
2. Vertical plates and cylinders l = vertical dimension	$h = 2.15\left[\frac{\Delta t}{l}\right]^{\frac{1}{4}}$	$h = 1.13\,(\Delta t)^{\frac{1}{3}}$
3. Horizontal plates:	$h = 2\left[\frac{\Delta t}{l}\right]^{\frac{1}{4}}$	$h = 1.31\,(\Delta t)^{\frac{1}{3}}$
Heated pipes up, cooled plates down, heated plates down cooled plates up l = side dimension	$h = 0.89\left[\frac{\Delta t}{l}\right]^{\frac{1}{4}}$	

Physical Interpretation of Some of the Dimensionless Groups

Table 5.8

Name of the dimensionless group	*Group*	*Physical interpretation*
Biot (Bi)	$\frac{h L}{k}$	$\frac{\text{Internal resistance to heat conduction}}{\text{External resistance to heat conduction}}$
Euler (Eu)	$\frac{p}{\rho u_\infty^2}$	$\frac{\text{Pressure forces}}{\text{Inertia forces}}$
Fourier (Fo)	$\frac{k\,T}{\rho\, c_p\, L_c^2}$	dimensionless time for transient conduction
Froude	$\frac{u_\infty^2}{L_c\, g}$	$\frac{\text{inertia forces}}{\text{gravity forces}}$
Grashof (Gr)	$\frac{g\, L_c^2\, \beta\,(t_w - t_\infty)}{v^2}$	$\frac{(\text{buoyancy forces})\,(\text{inertia forces})}{(\text{viscous forces})^2}$
Nusselt (Nu)	$\frac{h\, D_c}{k}$	ratio of temperature gradients
Peclet (Pe)	$\frac{c_p\, \rho\mu_\infty\, L_c}{k}$	$\frac{\text{convective heat transfer}}{\text{conductive heat transfer}}$
Prandtl (Pr)	$\frac{c_p\, \mu}{k}$	$\frac{\text{momentum diffusivity}}{\text{thermal diffusivity}}$
Reynolds (Re)	$\frac{\rho\, u_\infty\, L_c}{\mu}$	$\frac{\text{inertia forces}}{\text{viscous forces}}$
Stanton (St)	$\frac{h}{u_\infty\, c_p\, \rho}$	$\frac{\text{heat transfer at wall}}{\text{convective heat transfer}}$

Illustration on the Application of Force Convection Correlations

A tube 4m long is held at 100°C by steam jacketing. A fluid flows through the tube at the rate 175 kg/hr. The tube diameter is 20mm. The mean temperature (or bulk temperature) of the fluid at 30°C. The properties of the fluid are given below.

Temperature	*30°C*	*65°C*	*100°C*
c_p (k cal/kg °C)	0.46	0.48	0.50
k (k cal/hr m °C)	0.13	0.12	0.11
ρ kg/m^3	875	850	825
ν (m^2/sec)	12×10^{-6}	5.1×10^{-6}	2.6×10^{-6}

Use the following correlations

$$N_{N_u} = \frac{hD}{k} = 3.65 + \frac{0.0668\left(\dfrac{D}{L}\right)\left(N_{R_e}\ N_{P_r}\right)}{1 + 0.04\left\{\dfrac{D}{L}\cdot N_{R_e}\ N_{P_r}\right\}^{\frac{2}{3}}} \qquad \text{...(1)}$$

(applicable for the situations of fully developed flow with uniform wall temperature to compute the average Nusselt number)

and $$N_{N_u} = 1.86\left[N_{R_e}\ N_{P_r}\ \frac{D}{L}\right]^{\frac{1}{3}}\left(\frac{\mu_b}{\mu_w}\right)^{0.14} \qquad \text{...(2)}$$

μ_b fluid viscosity at bulk temperature of fluid μ_w is that at tube wall temperature

Solution

In this case

$$t_b = 30\ °\text{C}$$

$$t_w = 100\ °\text{C}$$

$$t_f = \frac{t_b + t_w}{2} = \frac{30 + 100}{2} = 65\ °\text{C}$$

Let us first evaluate N_{P_r} and N_{R_e} by taking the property values at t_f

$$\left(N_{P_r}\right)_f = \left(\frac{c_p\ \mu}{k}\right)_f = \frac{(5.1)\times 10^{-6}\times 3600\times 850\times 0.48}{0.12}$$

$$= 62$$

$$N_{R_e} = \left(\frac{Du}{\nu}\right)_f$$

$$u = \frac{m}{\rho A} = \frac{175}{850\times\dfrac{\pi}{4}\times(0.02)^2}$$

$$= 655\ \text{m/hr}$$

$$\therefore \left(N_{R_e}\right)_f = \frac{655\times 0.02}{5.1\times 10^{-6}\times 3600} = 715$$

The flow is seen to be laminar

Employing equation (1)

$$N_{N_u} = 3.65 + \frac{0.0668\left(\frac{0.02}{4}\right)(175)(62)}{1 + 0.04\left\{\left(\frac{0.02}{4}\right)(715)(62)\right\}^{\frac{2}{3}}}$$

$$= 3.65 + 5.9 = 9.55$$

Hence $h = 9.55 \times \frac{0.12}{0.02} = 57.3$ k cal/hr m^2 °C

Using equation (2), we get

$$\frac{h\,D}{k_b} = 1.86\left(\frac{\mu_b}{\mu_w}\right)^{0.14}\left[\frac{D\,u\,\rho}{\mu_b}\left(\frac{c_p\,\mu}{k}\right)_b\frac{D}{L}\right]^{\frac{1}{3}}$$

$$\left(\frac{\mu_b}{\mu_w}\right)^{0.14} = \left(\frac{875 \times 12 \times 10^{-6}}{825 \times 2.6 \times 10^{-6}}\right) = (4.92)^{0.14} = 1.25$$

$$\frac{D\,u\,\rho}{\mu_b} = \frac{m}{\rho\,A}\,\frac{D}{\mu_b}\cdot\rho = \frac{175}{\left(\frac{\pi}{4}\right)0.02}\,\frac{1}{12 \times 10^{-6} \times 875 \times 3600}$$

$$\left(\frac{c_p\,\mu}{k}\right)_b = \frac{\left(875 \times 12 \times 10^{-6} \times 3600\right)(0.46)}{0.13} = 134$$

$$\frac{D}{L} = \frac{0.02}{4} = 0.005$$

so $$\frac{h_D}{k_b} = (1.86)(1.25)(295 \times 134 \times 0.005)^{\frac{1}{3}}$$

$$= (1.86)(1.25)(198)^{\frac{1}{3}}$$

$$= (1.86)(1.25)(5.84)$$

$$= 13.6$$

Hence $$h = \frac{13.6 \times 0.13}{0.02}$$

$$= 88.5 \text{ k cal/hr m}^2 \text{ C}$$

Illustration on Forced Convection/Turbulent Flow Over Plane Surfaces

The crankcase of an automobile is approximately 30 in long, 12 in wide and 4 in deep. Assuming that the surface temperature of the crankcase is 160°F, estimate the rate of heat flow from the crankcase to atmospheric air at 40°F at a road speed of 60 miles/hour. Assume that the vibration of the engine and the chassis induce the transition from laminar to turbulent flow so near the leading edge that, for practical purposes, the boundary layer is turbulent over the entire surface

The average convection heat transfer coefficient for the bottom and sides may be used for the front and rear surfaces. Radiation can be neglected

ρ_{air} = 0.072 lbs/ft³ (at 100 °F)

k_{air} = 0.0154 Btu/hr ft °F (at 100 °F)

Solution

The Reynolds number is

$$N_{R_{e_L}} = \frac{u_\infty \rho L}{\mu} = \frac{60 \times 0.072 \times \frac{30}{12}}{1.285 \times 10^{-5}}$$

$$= 1.21 \times 10^6$$

The equation to be employed is

$$N_{N_u} = 0.036 \quad N_{P_r}^{\frac{1}{3}} N_{R_e}^{0.8}$$

$$= (0.036)\,(896)\,(73{,}480)$$

$$= 2370$$

and the average convection heat transfer coefficient becomes

$$\overline{h_c} = \overline{N}_{N_{u_L}} \frac{k}{L}$$

$$= \frac{(2370)\,(0.0154)}{\frac{30}{12}}$$

$$= 14.55 \text{ Btu/hr ft}^2 \text{ F}$$

The overall area is 4.84 ft² and the rate of heat loss is hence

$$q = \bar{h}_c \; A\,(T_s - T_\infty)$$

$$= (14.55)\,(4.84)\,(160 - 40) = 8430 \text{ Btu/hr}$$

Example on the Heating of Fluids in Turbulent Flow Through Pipes

Calculate the average film coefficient of heat transfer at the water side of a single pass steam condenser. The tubes are 0.952 in inside diameter and the cooling water enters at 61.4°F and leaves at 69.9°F. The average water velocity is 7 ft/sec. The properties of water at a mean temperature of 65.7°F are

$\rho = 62.3$ lbs/c ft

$\mu = 2.51$ lb/ft hr

$c_p = 1.0$ Btu/lb °F

$k = 0.340$ Btu/hr ft °F

$$\text{Thus } N_{R_e} = \frac{D\,u\,\rho}{\mu} = \frac{25200\,(0.0752)\,1.94}{0.0780} = 47{,}100$$

$$N_{P_r} = \frac{c_p\,\mu}{k} = \frac{(0.0780)\,(32.2)}{0.340} = 7.39$$

$$N_{N_u} = \frac{h\,D}{k} = 0.023\,(47{,}100)^{0.8}\,(7.39)^{0.4}$$

$$= 280.3$$

and

$$h = \frac{(280.3)\,(0.340)}{0.0752} = 1267 \text{ Btu/hr ft}^2 \text{ °F}$$

In the cases of gases, the prandtl number changes very little. It is about 0.67 for monatomic gases, 0.70 for diatomic gases, 0.80 for triatomic gases.

Illustration on Convection

1. Water flows in a 50 mm diameter tube at a velocity of 80 cm/sec. The length of the tube is 3m. Determine the heat transfer coefficient if the tube wall temperature (t_w) is 70 °C and the bulk temperature of water is 30 °C. The properties of water at the bulk temperature and the film temperature are

	30 °C	*50 °C Film temperature* $t_f = \frac{30 + 70}{2}$
ρ kg/m^3	1000	988
c_p k cal/kg °C	1.0	1.0
k (k cal/hr m °C	0.54	0.557
ν (m^2/sec)	9.2×10^{-7}	5.56×10^{-7}
N_{P_r}	6.13	3.54

Solution

Let us first calculate N_{R_e} to ascertain whether the flow is laminar or turbulent

$$N_{R_e} = \frac{D\,u\,\rho}{\mu} = \frac{D\,u}{\nu}$$

$$= \frac{0.8 \times 0.05}{9.2 \times 10^{-7}} = 4.35 \times 10^4$$

The flow is turbulent. Using the Dittus-Boelter equation which is for forced convection

$$\frac{h\,D}{k} = (0.023)\left(\frac{D\,u\,\rho}{\mu}\right)^{0.8}\left(\frac{c_p\,\mu}{k}\right)^{0.4}$$

$$= (0.023)\left(4.35 \times 10^4\right)^{0.8}(6.13)^{0.4}$$

$$= (0.023)\,(5160)\,(2.07)$$

$$= 246$$

$$h = \frac{(246)\,(0.54)}{0.05} = 2650 \text{ k cal/hr m}^2\ {}^\circ\text{C}$$

We may also determine the value of h by employing the Colburn equation

$$\left(\frac{h}{c_p\,\rho u}\right)\left(\frac{c_p\,\mu}{k}\right)^{\frac{2}{3}} = \frac{0.023}{\left(\frac{D\,u\,\rho}{\mu}\right)^{0.2}}$$

$$\left(\frac{c_p\,\mu}{k}\right) = N_{pr} = 3.54$$

$$\left(\frac{D\,u\,\rho}{\mu}\right) = \left(\frac{Du}{\nu}\right) = \frac{0.8 \times 0.05}{5.56 \times 10^{-7}} = 7.2 \times 10^4$$

Therefore

$$\left(\frac{h}{1 \times 988.1 \times 0.8 \times 3600}\right)(3.54)^{\frac{2}{3}} = \frac{0.023}{\left(7.2 \times 10^4\right)^{0.2}}$$

$$h = \left(\frac{(0.023)\,(988.1)\,(0.8)\,(3600)}{(3.54)^{\frac{2}{3}}\left(7.2 \times 10^4\right)^{0.2}}\right)$$

$$= \frac{(18.25)\,(3600)}{(2.325)\,(9.4)}$$

$= 3500$ k cal/hr m^2 °C

3. In a heat exchanger water flows through a long copper tube of id 2.2 cm at a bulk velocity of 2m/sec. It is heated by steam condensing at 150 °C on the outside of the tube. The water enters the heat exchanger at 15 °C and leaves it at 60 °C. Find the heat transfer coefficient for water.

Solution

$$\text{Mean bulk temperature } (t_b) = \frac{15 + 60}{2} = 37.5°\text{C}$$

The physical properties of water at 37.5 °C are

$\rho = 990$ kg/m^3

$c_p = 0.0998$ k cal.kg °C

$\mu = 0.00069$ kg/m sec

$k = 0.547$ k cal/hr m °C

$$\text{Therefore } N_{R_e} = \frac{2 \times \left(\frac{2.2}{100}\right) \times 990}{0.00069}$$

$$= 63{,}200$$

$$N_{P_r} = \frac{(0.00069) \times (3600) \times 0.998}{0.547}$$

$$= 4.58$$

In view of the fact that the flow is turbulent and the tube is quite long we may calculate the heat transfer coefficient from Dittus-Boelter and Colburn equations

Using the former equation

$$h = \frac{(0.023)\,(0.54)}{\left(\frac{2.2}{100}\right)} (63200)^{0.8}\ (4.58)^{0.4}$$

$$= (0.566)\ (6950)\ (1.84)$$

$$= 7240 \text{ k C/hr m}^2\ °\text{C}$$

To employ the Colburn equation, we should know the film temperature. This, in turn, calls for a knowledge of the wall temperature on the water side. If we assume that the average coefficient for condensation is of the same magnitude as the heat transfer coefficient for water and neglect the temperature drop over the copper tube wall the wall temperature will be

$$\frac{(150 + 37.5)}{2} = 93.75$$

Hence, the water side film temperature will be

$$t_f = \frac{t_b + t_w}{2} = \frac{37.5 + 93.75}{2}$$

$$= 65.6°C$$

The properties of water at 65.6 °C are

$\rho = 980$ kg/m^3

$k = 0.576$ k cal.hr m °C

$c_p = 1.00$ k cal/kg °C

$\mu = 0.000438$ kg/m sec

Hence $N_{P_r} = \left(\frac{c_p \mu}{k}\right) = \frac{(0.000438)(3600)(1)}{0.576}$

$$= 2.74$$

$$N_{R_e} = \frac{D u \rho}{\mu} = \frac{2 \times 0.022 \times 980}{0.000438}$$

$$= 98{,}600$$

The equation to be employed is

$$\frac{h}{c_p \rho u} = \frac{0.023}{(N_{P_r})^{\frac{2}{3}} (N_{R_e})^{0.2}}$$

$$= \frac{0.023}{(2.74)^{\frac{2}{3}} (98.600)^{0.2}}$$

$$= 1.18 \times 10^{-3}$$

$$\because h = \left(1.18 \times 10^{-3}\right)(0.998)(980)(2 \times 3600)$$

$$= 8320 \text{ k cal/hr m}^2 \text{ °C}$$

In case we want to include the viscosity correction term in the computation of h, we can use the Seider-Tate equation

$$\frac{h}{c_p \rho u} = (0.023)\left(N_{P_r}\right)^{\frac{-2}{3}} \left(N_{R_e}\right)^{-0.2} \left(\frac{\mu_w}{\mu_f}\right)^{-0.14}$$

Here the N_{P_r} and N_{R_e} values are evaluated at the bulk temperature of the fluid as in the case of Dittus-Boelter equation. Hence

$$N_{R_e} = 63{,}200$$

$$N_{P_r} = 4.58$$

At the wall temperature of 93.75°C, the value of $\mu_w = 0.0003$ kg/m– sec

Therefore

$$h = (0.023)\,(990 \times 2 \times 3600 \times 0.998)\,(4.58)^{\frac{-2}{3}} \times (63{,}200)^{-0.2}\left(\frac{0.0003}{0.00069}\right)^{-0.14}$$

$$= 7340 \text{ k cal/hr m}^2\text{°C}$$

4. Calculate the heat transfer coefficient for mercury flowing through a 1.5 cm I.D. tube at the rate of 30000 kgs/hr. The bulk temperature of mercury may be taken as 40 °C. The following properties at 40 °C may be used.

For Mercury: $k = 7.5$ k cal/hr m °C

$c_p = 0.033$ k cal/kg °C

$\mu = 5.45$ kg/m– hr

Solution

$$N_{R_e} = \frac{D\,u\,\rho}{\mu} = \frac{G\,D}{\mu}$$

$$G = \frac{w}{A} = \frac{30000}{\frac{\pi}{4} \times \left(1.5 \times 10^{-2}\right)^2}$$

$$= 1.63 \times 10^8 \text{ kg/hr m}^2$$

$$N_{Re} = \frac{1.63 \times 10^8 \times 1.5 \times 10^{-2}}{5.4}$$

$$= 4.53 \times 10^5$$

$$N_{P_r} = \frac{c_p\mu}{k} = \frac{5.4 \times 0.033}{7.5} = 0.0238$$

$$N_{P_E} = N_{R_e} \times N_{P_r}$$

$$= 4.53 \times 10^5 \times 0.0238$$

$$= 10750$$

In view of the fact that $N_{R_e} > 10000$ and N_{P_r} is in the range 0.005 to 0.05, the following equation can be used. This equation agreed quite well with experimental results

$$N_{N_u} = 6.7 + 0.0041\left(N_{R_e}\,N_{P_r}\right)^{0.793}(e)^{41.8\,N_{P_r}}$$

$$= 6.7 + 0.0041\,(10750)^{0.793}\,e^{(41.8)\,(0.0238)}$$

$$= 6.7 + 0.0041\,(1570)\,(2.7)$$

$$= 24.1 \text{ or } \quad h = \frac{24.1 \times 7.5}{1.5 \times 10^{-2}} = 12050 \text{ k cal/hr m}^2\ \text{°C}$$

For design purposes any of the following equations may be used with reasonable accuracy

$$N_{N_u} = 7 + 0.025\, N_{Pe}^{0.8}$$

and $$N_{N_u} = 0.625\, N_{pe}^{0.4}$$

If we are employing the first one, we obtain

$$N_{N_u} = 7 + 0.025\, N_{Pe}^{0.8}$$

$$= 7 + 0.025\,(10750)^{0.8}$$

$$= 7 + 0.025\,(1680) = 49$$

and $$h = \frac{49 \times 7.5}{1.5 \times 10^{-2}} = 24500 \text{ k cal/hr m}^2\ \text{°C}$$

In case we employ the second one, the result turns out to be

$$N_{N_u} = 0.625(N_{Pe})^{0.4}$$

$$= (0.625)\,(41) = 25.6$$

and $$h = \frac{25.6 \times 7.5}{1.5 \times 10^{-2}}$$

$$= 12,\ 800 \text{ k cal/hr m}^2\ \text{°C}$$

5. 205 kg/hr of air are cooled from 100 °C to 30 °C by making the air flow through a 3.5 cm ID pipe coil bent into a helix of 0.6 m diameter. Calculate the value of the air-side heat transfer coefficient if the properties of air at 65 °C are

$k = 0.0255$ k cal/hr m °C

$\mu = 0.003$ kg/hr m

$N_{P_r} = 0.7$

Solution

$$G = \frac{205}{\frac{\pi}{4}(3.5)^2 \times 10^{-4}}$$

$$= 2.14 \times 10^4 \text{ kg/m}^2 \text{ hr}$$

From the Dittus-Boelter equation

$$\frac{hD}{k} = 0.023\left(\frac{DG}{\mu}\right)^{0.8}\left(\frac{c_p\,\mu}{k}\right)^{0.4}$$

$$= 0.023\left[\left(\frac{3.5 \times 10^{-2} \times 21.4 \times 10^{4}}{0.073}\right)\right]^{0.8}(0.7)^{0.4}$$

$$= 0.023\left(10^{4} \times 0.87\right) = 200$$

Therefore $h = \dfrac{200 \times 0.0255}{0.035}$

$$= 146 \text{ k cal/hr m}^2\ ^\circ\text{C}$$

For fully developed turbulent flow of air in a helical coil, the following relation may be used for the ratio of h_c (in the coil) to h in a straight pipe

$$\frac{h_c}{h} = 1 + 3.5\frac{D}{D_{He}}$$

In which D_{He} = helix diameter

In this case

$$\frac{h_c}{h} = 1 + 3.5\left(\frac{3.5}{60}\right)$$

$$= 1.204$$

$$\therefore h_{coil} = (1.204)(146)$$

$$= 176 \text{ k cal/hr m}^2\ ^\circ\text{C}$$

On the Magnitude of the Heat Transfer Coefficient for a Few Situations

It is better one has some idea of how big the heat transfer coefficient might be in a given situation. The following table provides some illustrative values of h that might have been observed or calculated.

Table 5.9 *Some illustrative values of convective heat transfer coefficients*

Situation	$\bar{h}$ *w/m² °C*
Natural convection, single phase	
• 0.3 m vertical wall in air $\Delta T = 30\ ^\circ$C	4.33
• 4 cm OD horizontal pipe in H_2o, $\Delta T = 30\ ^\circ$C	570
• 0.025 cm diameter wire in methanol $\Delta T = 50\ ^\circ$C	4,000

Contd.

Situation	$\bar{h}$ w/m² °C
Forced convection	
• Air at 30 m/sec over 1 m plate, $\Delta T = 70\,°C$	80
• Water at 2 m/sec over a 0.06 m plate, $\Delta T = 15\,°C$	590
• Aniline-alcohol mixture at 3 m/sec in a 2.5 cm ID tube $\Delta T = 80\,°C$	2,600
• Liquid sodium at 5 m/sec in a 1.3 cm ID tube at 370°C	75,000
Boiling water	
• During film boiling at 1 atm	300
• In a tea kettle	4,000
• A peak-pool boiling heat flux at 1 atm	40,000
• A peak-flow boiling heat flux at 1 atm	100,000
• Approximate maximum convective heat flux under optimum conditions	900,000
Condensing	
• Typical value in a horizontal cold water tube steam condenser	15,000
• Same but condensing benzene	1700
• Drop wise condensation of water	100,000

Illustration in Forced Convection Heat Transfer

1. Determine the average value of the heat transfer coefficient (convection film coefficient) on the inside surface of a tube of 500mm ID meant for heating water. The mass flow rate of the water is 25000 kgs/hr. The water enters at a temperature of 20°C and leaves at 60°C. Determine also the heat transferred per unit length of the tube, if the constant flux heat condition is maintained at the wall, which may be assumed to be 10°C above the average bulk temperature of the fluid. The physical properties of water at the arithmetic mean temperature of 40°C are given below

$\rho = 992.2$ kg/m3 $\quad k = 54.5 \times 10^{-2}$ k cal/hr m °C

$\nu = 0.659 \times 10^{-6}$ m²/sec $\quad N_{P_r} = 4.31$

Solution

Let us calculate the flow velocity and the Reynolds number to find out whether is the flow is laminar or turbulent

The mass flow rate $m = \rho\, A\, u$

Hence the average flow velocity u

$$= \frac{\left(\dfrac{25000}{3600}\right)}{\dfrac{\pi}{4} \times \left(50 \times 10^{-3}\right)^2 \times 992.2}$$

$= 3.57$ m/sec

$$N_{R_e} = \frac{u\,d}{\nu} = \frac{3.57 \times 50 \times 10^{-3}}{0.659 \times 10^{-6}}$$

$= 2.71 \times 10^5$

The flow is evidently turbulent

Therefore

$$N_{N_u} = 0.023\left(N_{R_e}\right)^{0.8}\left(N_{P_r}\right)^{0.4}$$

$$= (0.023) \times \left(2.77 \times 10^5\right)^{0.8} (4.31)^{0.4}$$

$$= 0.023 \times 2.22 \times 10^4 \times (1.795)$$

$= 915$

Hence the average heat transfer coefficient

$$\bar{h} = \frac{\left(N_{N_u}\right) k}{d} = \frac{(915) \times 54.5 \times 10^{-2}}{50 \times 10^{-3}}$$

$= 9960$ k cal/hr m^2 °C

The heat transferred per unit length is

$$q = \frac{Q}{L} = \frac{\bar{h}\,A\left(t_w - t_b\right)}{L}$$

$$= \bar{h}\,\pi d\left(t_w - t_b\right)$$

$$= 9960 \times \pi \times 50 \times 10^{-3} \times 10$$

$-$ 15650 k cal/hr m

2. Estimate the film coefficient for the water flowing through an annulus formed by two concentric tubes having 25 mm outside diameter (inner tube) and 40 mm inside diameter (outer tube) at a velocity of 3 m/sec. The temperature of the inner tube wall is maintained at a constant value of 40°C while the bulk temperature of the flowing fluid in the annulus is 80°C. Take the flow to be fully developed. The properties evaluated at the bulk temperature are given below

$k = 58 \times 10^{-2}$ k cal/m hr °C

$\nu = 0.365 \times 10^{-6}$ m^2/sec

$N_{P_r} = 2.21$

Solution

We can calculate the hydraulic diameter (equivalent diameter) using the expression

$$D_h = \frac{4\left(\frac{\pi}{4}\right)\left(d_2^2 - d_1^2\right)}{\pi\,(d_1 + d_2)}$$

$$\left(D_h = \frac{4\ (\text{cross sectional area})}{\text{wetted perimeter}}\right)$$

$$D_h = d_2 - d_1$$

D_h in this case is,

$$D_h = 40 - 25 = 15 \text{ mm}$$

$$= 15 \times 10^{-3} \text{ m}$$

$$N_{R_e} = \frac{u\,D_h}{\nu} = \frac{3 \times 15 \times 10^{-3}}{0.365 \times 10^{-6}}$$

$$= 1.23 \times 10^5$$

The flow is obviously turbulent. The expression to be employed is

$$N_{N_u} = 0.023\left(N_{R_e}\right)^{0.8}\left(N_{P_r}\right)^{0.3}$$

N_{P_u} has the power of 0.3 because the fluid in question is cooled as its flows past the wall.

$$N_{N_u} = (0.023)\left(1.23 \times 10^5\right)^{0.8}(2.21)^{0.3}$$

$$= 344$$

and
$$\bar{h} = \frac{N_{N_u} \cdot k}{d}$$

$$= \frac{344 \times 5.8 \times 10^{-2}}{15 \times 10^{-3}}$$

$$\bar{h} = 13{,}300 \text{ k cal/hr m}^2\ ^\circ\text{C}$$

3. Determine the heat transfer coefficient for ambient air at 20°C flowing normal to a 50 mm outside diameter pipe at a velocity of 5 m/sec. Take the average surface temperature of the pipe to be 120°C. The physical properties at the average film temperature are $k = 2.55 \times 10^{-2}$ k cal/hr m °C and $\nu = 20.05 \times 10^{-6}$ m² / sec. Take that in the case of heating fluids flowing normal to the single wires or tubes, the following form of the empirical relation can be employed $N_{N_u} = 0.3\left(N_{R_e}\right)^{0.57}$.

Solution

$$\left(N_{R_e}\right) = \frac{u\,D}{\nu} = \frac{5 \times 50 \times 10^{-3}}{20.02 \times 10^{-6}}$$

$$= 1.25 \times 10^4$$

using the equation/expression given in the problem statement.

$$N_{N_u} = 0.3\left(N_{R_e}\right)^{0.57}$$

$$= 0.3 \times \left(1.25 \times 10^4\right)^{0.57}$$

$$= 66$$

$$\therefore\ \bar{h} = \left(N_{N_u}\right)\left(\frac{k}{d}\right)$$

$$= \frac{66 \times 2.55 \times 10^{-2}}{50 \times 10^{-3}}$$

$$\bar{h} = 33.6 \text{ k cal/hr m}^2\ ^\circ\text{C}$$

4. Liquid sodium enters at 290°C and leaves at 310°C through a tube of 15 mm inside diameter in a certain section of an experimental fast reactor. The average velocity of flow through the tube is 5 m/sec and the flow throughout the tube is in the fully developed condition. If constant heat flux is maintained along the tube and the wall temperature is maintained 20°C higher than the bulk temperature of the liquid sodium

(i) Calculate the length of the tube required to effect the heat transfer.

(ii) Compare the values of heat transfer coefficients obtainable in a similar experimental setup having the same inlet and exit temperature conditions where the heat transfer fluid is

(a) Saturated water at 300°C

(b) Saturated steam at 300°C

If may be noted that in a fully developed flow regions of turbulent flow of liquid metals in smooth tubes with uniform wall heat flux conditions, the following equation recommended in the literature can be employed

$$\left(N_{N_u}\right) = \frac{h\,d}{k} = 0.625\left(N_{R_e}\,N_{P_r}\right)^{0.4} = 0.625\left(N_{P_e}\right)^{0.4}$$

The properties of sodium at the average bulk temperature (300°C) are given below

$k = 61$ k cal/hr m °C $\qquad \nu = 39.4 \times 10^{-8}$ m²/sec

$c_p = 0.306$ k cal/kg °C

$$N_{P_r} = 0.63 \times 10^{-2} \qquad \rho = 878 \text{ kg/m}^3$$

Solution

$$N_{R_e} = \frac{d\,u}{\nu} = \frac{5 \times 15 \times 10^{-3}}{39.4 \times 10^{-8}} = 1.902 \times 10^5$$

$$\therefore N_{P_e} = \left(1.902 \times 10^5\right)\left(0.63 \times 10^{-2}\right)$$

$$= 1.2 \times 10^3$$

using the equation given in the problem statement

$$N_{N_u} = \frac{h\,d}{k} = 0.625\left(N_{P_e}\right)^{0.4}$$

$$= 0.625\left(1.2 \times 10^3\right)^{0.4}$$

$$= 10.62$$

$$\therefore h = \frac{10.62 \times 61}{15 \times 10^{-3}} = 43{,}300 \text{ k cal/hr m}^2\ ^\circ\text{C}$$

we now calculate the total heat to be transferred so that we can evaluate the tube length

We know, mass flow rate $= \rho\, A\, u$

$$= 878 \times \frac{\pi}{4} \times \left(15 \times 10^{-3}\right)^2 \times 5$$

$$= 0.776 \text{ kg/sec}$$

$$Q = m\, c_p\, \Delta t$$

$$= (0.776 \times 3600)\ 0.306 \times 20$$

$$= 17{,}100 \text{ k cal/hr}$$

we know also that $Q = h\,A\,(t_w - t_f)$

$$= \frac{17100}{43300 \times 20} = 0.0198 \text{ m}^2$$

Also $A = \pi\, d\, L$ where L is the length through which the heat transfer is brought about. Computing this length we have

$$L = \frac{0.0918}{\pi \times 15 \times 10^{-3}} = 0.42 \text{ m}$$

(ii) Comparison of heat transfer coefficients

(a) For saturated water at 300°C the physical properties from the appropriate tables are .

$k = 46 \times 10^{-2}$ k cal/hr m °C $\nu = 0.128 \times 10^{-6}$ m²/sec $N_{Pr} = 0.97$

Calculating the N_{R_e} for this fluid, we have

$$\left(N_{R_e}\right) = \frac{5 \times 15 \times 10^{-3}}{0.128 \times 10^{-6}} = 5.86 \times 10^{5}$$

The heat transfer coefficient can be calculated from the correlation

$$N_{N_u} = 0.023 \left(N_{R_e}\right)^{0.8} \left(N_{P_r}\right)^{0.4}$$

$$= 0.023 \left(5.86 \times 10^{5}\right)^{0.8} (0.97)^{0.4}$$

$$= 937$$

$$\therefore\ h = \frac{\left(937 \times 46.4 \times 10^{-3}\right)}{\left(15 \times 10^{-3}\right)}$$

$$= 29000 \text{ k cal/hr m}^2\ °\text{C}$$

(b) For saturated Steam at 300°C, the physical properties from the appropriate table

$k = 5.39 \times 10^{-2}$ k cal/hr m °C

$\nu = 0.461 \times 10^{-6}$ m²/sec

$N_{P_r} = 2.13$

$$N_{R_e} = \frac{5 \times 15 \times 10^{-3}}{0.461 \times 10^{-6}} = 1.625 \times 10^{5}$$

Using the correlation given above

$$N_{N_u} = (0.023)\left(1.625 \times 10^{5}\right)^{0.8} (2.13)^{0.4}$$

$$= 460$$

∴ The heat transfer coefficient

$$h = \frac{460 \times 5.39 \times 10^{-2}}{15 \times 10^{-3}}$$

$$= 1654 \text{ k cal/hr m}^2 °\text{C}$$

Thus it is observed from the above example that the film coefficient in the case of liquid metals are approximately 1.5 times more then that of liquid water and 25 times larger than that of steam at the same temperature.

Illustration on Forced Convection

1. A solution of aniline and alcohol is flowing at a velocity of 10 fps through a long 1 in ID thin wall tube. On the outer surface of the tube, steam is condensing at atmosphere pressure and the temperature of the tube wall is 212°F. The tube is clean and there is no thermal resistance due to a scale deposit on the inner surface. Using the physical properties given below, determine the heat transfer coefficient between the fluid and the pipe by means of the equation

$$\frac{h_c}{c_p G} = 0.023\left(\frac{\mu_f}{DG}\right)^{0.2}\left(N_{P_r}\right)^{-0.66}$$

Take that the bulk temperature of the aniline solution is 68°F and ignore entrance effects.

Physical properties of the aniline solution are as follows.

Temp°F	*Viscosity c.p*	*k Btu/hr ft °F*	*sp gr*	c_p *Btu/lb °F*
68	5.1	0.100	1.03	0.50
140	1.4	0.098	0.98	0.53
212	0.6	0.095	–	0.56

Solution

Let us first determine the Reynolds number to find out whether the flow is turbulent or streamline. Evaluating the viscosity at the bulk temperature, we obtain

$$\left(N_{R_e}\right)_{\substack{\text{bulk}\\ \text{temp.}}} = \frac{D\,u\,\rho}{\mu} = \frac{(10)\left(\frac{1}{12}\right)(1.03)\,(62.4)}{(5.1)\,(0.000672)}$$

$$= 15600$$

The flow is obviously turbulent

The average film temperature according to the equation $T_f = 0.5\,(T_s + T_b)$

$$T_f = 0.5\,(T_s + T_b)$$

$$= 0.5\,(68 + 212) = 140\ °\text{F}$$

and the N_{R_e} is 56800 when the viscosity is taken at T_f

The Prandtl number, based on the property values at the bulk temperature is

$$N_{P_r} = \frac{c_p\,\mu}{k} = \frac{(0.5)\,(5.1)\,(0.000672)\,(3600)}{0.100}$$

$$= 61.5$$

(0.000672 is the conversion factor from centipoise to lb/sec when the properties are taken at T_f, the Prandtl number is 18.4 using the equation given in the problem statement, to evaluate the heat transfer coefficient, we have after some rearrangement

$$h_c = 0.023 \frac{k_f}{D} N_{R_e f}^{0.8} N_{P_r f}^{0.33}$$

$$= 0.023 \left(\frac{0.098}{1/12}\right) (6380)\ (2.64)$$

$$= 456 \text{ Btu/hr ft}^2 \text{ °F}$$

2. Determine the heat transfer coefficient for water flowing at a velocity of 10 ft/sec an annulus formed between a 1 in OD tube and a 1½ in ID tube. The water is at 180°F and is being cooled. The temperature of the inner wall is 100°F and the outer wall of the annulus is insulated. Neglect entrance effects and compare the results of the following equations

$$\frac{h_c}{c_p\, G} = 0.023 \left(\frac{\mu_f}{D\, G}\right)^{0.2} \left(N_{P_r}\right)^{-0.66} \quad \ldots(1)$$

$$N_{S_t} = \phi\left(N_{R_e}\right) \psi\left(N_{P_r}\right) \left(\frac{\mu_b}{\mu_s}\right)^n \quad \ldots(2)$$

For liquids having N_{P_r} larger that 1.0 the exponent n of the viscosity ratio

$\left(\frac{\mu_b}{\mu_s}\right)$ in the second equation given above is,

$n = 0.36$ for liquid heating

$n = 0.20$ for liquid cooling

the properties of water are given in table presented below

T°F	*μ lb/hr ft*	*kBtu/hr ft °F*	*ρ lb/ft³*	*c_p Btu/lb °F*
100	1.67	0.36	62.0	1.0
140	1.14	0.38	61.3	1.0
180	0.75	0.39	60.8	1.0

The product $\left(N_{S_t}\right)\left(\frac{\mu_s}{\mu_b}\right)^{0.2} = 16 \times 10^{-4}$

Solution

The hydraulic diameter D_H for this geometry is 0.5 in. The N_{R_e} based on the hydraulic diameter and the bulk temperature properties is

$$N_{R_e\, D_h} = \frac{D_H\, u\, \rho}{\mu}$$

$$= \frac{(10) \times \left(\frac{0.5}{12}\right)(60.8)(3600)}{0.75}$$

$$= 125{,}000$$

Based on the mean film temperature T_f, the Reynolds number is $N_{R_e\,D_f} = 82{,}000$. The Prandtl number at the bulk temperature is

$$N_{P_{rb}} = \frac{c_p\,\mu}{k} = \frac{(1.0)(0.75)}{0.39} = 1.92$$

and at T_f, we observe that $N_{P_{rf}} = 3.0$

According to equation (1) given in the problem statement, we have

$$N_{s_t} = \frac{h_c}{c_p\,u} = 0.023\,N_{R_e\,D_f}^{-0.2}\,N_{P_{r\,f}}^{-0.66}$$

$$= \frac{(0.023)}{(9.6)(2.08)} = 0.00115$$

so that $h_c = (1)(61.3)(10)(3600)(0.00115)$

$$= 2570 \text{ Btu/hr ft}^2 \text{ F}$$

The group $N_{s_t}\left(\frac{\mu_s}{\mu_b}\right)^{0.2}$ is given as 16×10^{-4}

Therefore the heat transfer coefficient is

$$h_c = \left(16 \times 10^{-4}\right)\frac{k_b}{D_H}\,N_{R_{eDb}}\,N_{P_{rb}}\left(\frac{\mu_b}{\mu_s}\right)^{0.2}$$

$$= \left(16 \times 10^{-4}\right)\left(\frac{0.39}{0.5/12}\right)(125{,}000)(1.92)(0.85)$$

$$= 3060 \text{ Btu/hr ft}^2\text{F}$$

We observe that the results obtained by two different methods of calculating h_c for turbulent flow in ducts agree within 16 percent.

5.2 HEAT TRANSFER FROM EXTENDED SURFACES OR FINS

As the name implies, it is the case of extension of heat transfer area with a solid of relatively small c.s area protruding from a large body into a fluid at a different temperature.

When the thermal resistance on the inside of metal tube is much lower than on the outside as when steam condensing in a pipe is being used to heat air, externally finned surfaces or extended heating surfaces are of immense value in increasing substantially the rate of heat transfer per unit length of the tube.

A typical case is heating of a particular gas, such as air, by condensing steam for air, h is = 10

While h steam = 1000–2000 in the same units

The $u = h$ of air and the capacity of a unit area of heating surface will be much lower and the length has to be increased very much to given a reasonable capacity.

To conserve space and to bring down the cost of the equipment in such cases, certain types of heat exchange surfaces called extended surfaces have been developed. In these units the outside area of the tube is multiplied or extended by fins. Thereby the outside area in contact with the fluid is made much larger than the inside area. The fluid stream having the lower heat transfer coefficient is brought into contact with the extended surface.

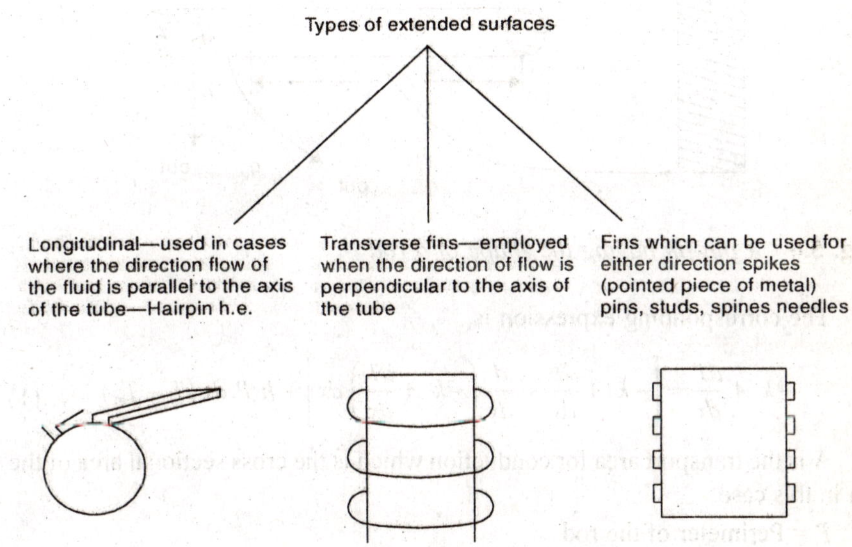

Fig. 5.5 *Types of extended surfaces*

Extended Surface of Uniform Cross Section

Let us consider as a simple illustration a pin-fin having the shape of a rod whose base is fixed to a wall at the surface temperature T_s as shown in the figure. The fin is cooled along its surface by a fluid at temperature T_∞. The fin has uniform *c.s.* area A, is made of a material whose k is uniform and the heat transfer coefficient between the surface of the fin and the fluid is 'h'. We will take that the transverse temperature gradients are too small to be considered and that the temperature at any cross section of the rod is uniform i.e. $T = T(x)$ only.

In order to derive an equation for the temperature profile (distribution) we strike a balance for a small element of the fin.

Heat flows by conduction into the left face of the element while it flows out of the element by conduction through the right face and by convection from the surface.

For steady state conditions

Rate of heat flow by cond^n into element at x

= (Rate of flow of heat by cond^n our of element at $x + dx$) + [Rate of heat flow by conv^n from surface bet. x and $(x + dx)$]

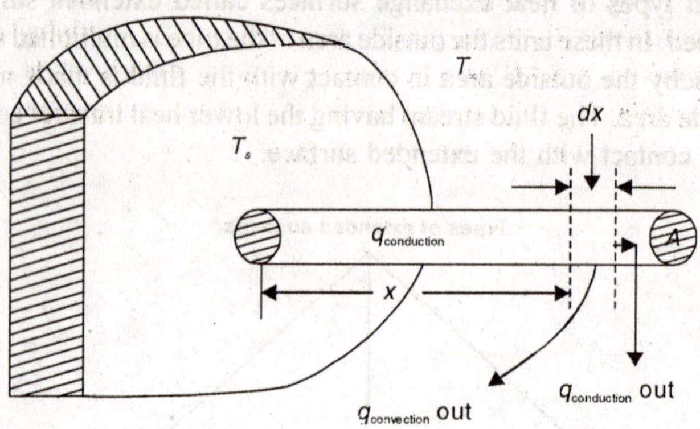

Fig. 5.6 *A pin-fin having the shape of a rod*

The corresponding expression is,

$$-k\,A\frac{dT}{dx} = \left[-k\,A\frac{dT}{dx} + \frac{d}{dx}\left(-k\,A\frac{dT}{dx}\right)dx\right] + h\,P\,dx\,(T - T_\infty) \quad \ldots(1)$$

A is the transport area for conduction which is the cross sectional area of the fin in this case.

P = Perimeter of the rod

$P\,dx \rightarrow$ represents the surface area ($h.t$ area) between sections x and $x + dx$ in contact with the surrounding fluid – heat flow by convection in the radial direction).

Taking k and h to be uniform, equation (1) can be reduced to

$$\frac{d^2T}{dx^2} = m^2\,(T - T_\infty) \quad \ldots(2)$$

$$= \frac{h\,P}{k\,A}$$

Equation (2) is a standard form of an ordinary 2nd order linear differential equation with general solution

$$T - T_\infty = c_1\, e^{mx} + c_2\, e^{-mx} \qquad ...(3)$$

In this c_1 and c_2 are constants of integration whose values must be determined from the *b.c*'s

B.C.I at $x = 0$, $T = T_s$ (the temperature at the base of the rod equals the temperature of the surface to which the rod is fixed)

This *b.c.* when substituted in equation (3) results in

$$T_s - T_\infty = c_1\, e^{m(o)} + c_2\, e^{-m(0)}$$

$$= c_1 + c_2 \qquad ...(4)$$

To solve for c_1 and c_2, we require another equation that is another *b.c.* and this *b.c.* is governed by the nature of the problem.

We shall consider 3 cases:

I Case

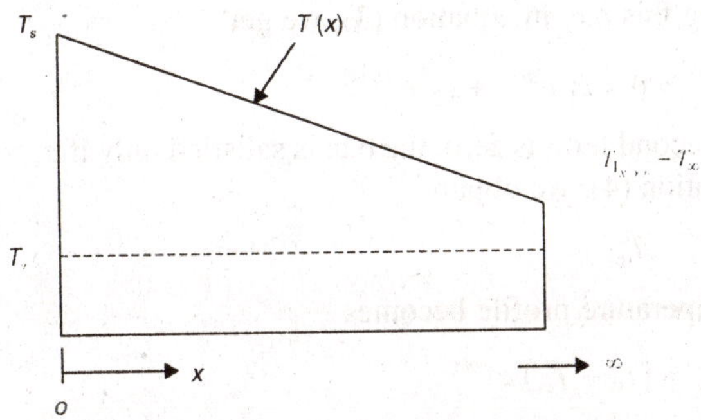

Fig. 5.7 *Temperature distribution in a rod with infinite length*

II Case

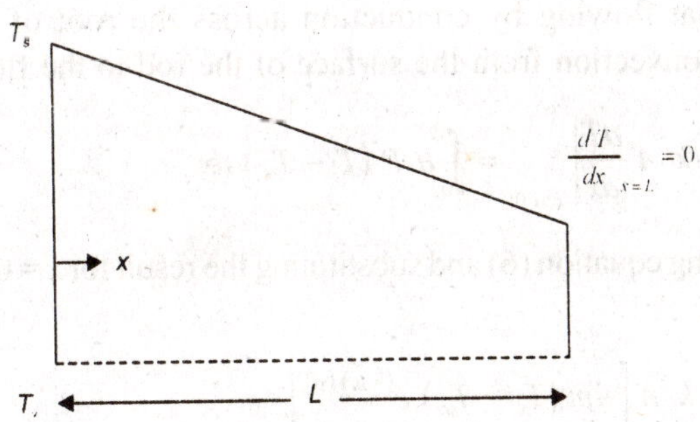

Fig. 5.8 *Temperature profile in a rod with finite length*

III Case

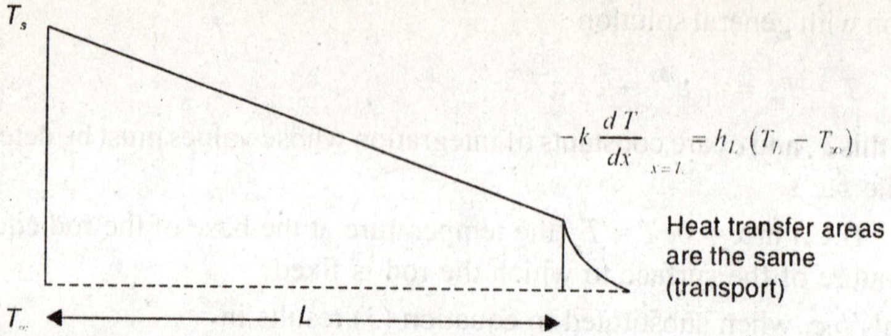

Fig. 5.9 *Temperature profile in the case of heat being lost by convection at the end of the rod with finite length*

I Case

If the rod is infinitely long the temperature will approach the temperature of the fluid, as $x \to \infty$, $T = T_\infty$

Substituting this *b.c.* in equation (3), we get

$$T - T_\infty = 0 = c_1\, e^{m\infty} + c_2\, e^{-m\infty} \qquad ...(5)$$

Since the second term is zero, the b.c. is satisfied only if $c_1 = 0$. Substituting 0 for c_1 in equation (4), we obtain

$$c_2 = T_s - T_\infty$$

∴ The temperature profile becomes

$$T - T_\infty = (T_s - T_\infty)\, e^{-mx} \qquad ...(6)$$

Determination of Heat Flow Rate

This can be done in two ways

(i) Since the heat flowing by conduction across the root of the fin must be transmitted by convection from the surface of the rod to the fluid

$$q_{\text{fin}} = -k\, A \left.\frac{dT}{dx}\right|_{x=0} = \int_0^\infty h\, P\, (T - T_\infty)\, dx \qquad ...(7)$$

Differentiating equation (6) and substituting the result for $x = 0$, in equation (7), gives

$$q_{\text{fin}} = -k\, A \left[-m\, (T_s - T_\infty)\, e^{(-m)(o)}\right]_{x=0} \qquad \boxed{m^2 = \frac{h\, p}{k\, A}}$$

$$= \sqrt{k\, P\, h\, A}\, (T_s - T_\infty) \qquad ...(8)$$

(ii) The same result is got by evaluating the convection heat flow from the surface of the rod.

$$q_{fin} = \int_0^{\infty} h\,P\,(T_s - T_\infty) e^{-mx}\, dx$$ by substituting the temperature profile exp^n. in the 3rd term of equation (1)

$$= -\frac{h\,P}{m}(T_s - T_\infty)\, e^{-mx}\Big|_0^{\infty}$$

$$= \sqrt{h\,P\,k\,A}\,(T_s - T_\infty) \qquad \ldots(9)$$

II Case

If the rod's length is finite, but the heat loss from the end of the rod is ignored or if the end of the rod is insulated, the second b.c requires that the temperature gradient at $x = L$ be zero i.e $dT/dx = 0$, at $x = L$.

Under these conditions

$$c_1 = \frac{T_s - T_\infty}{1 + e^{2mL}} \text{ and } c_2 = \frac{T_s - T_\infty}{1 + e^{-2mL}}$$

The temperature distribution is therefore

$$T - T_\infty = (T_s - T_\infty)\left[\frac{e^{mx}}{1 + e^{2mL}} + \frac{e^{-mx}}{1 + e^{-2mL}}\right] \qquad \ldots(10)$$

which in simplified dimensionless form, turn out to be

$$\frac{T - T_\infty}{T_s - T_\infty} = \frac{\cosh m(L - x)}{\cosh (mL)} \qquad \ldots(11)$$

The heat loss from the fin can be found from equation (7) by substituting the temperature gradient at the root

$$\left.\frac{dT}{dx}\right|_{x=0} = (T_s - T_\infty)\, m\left(\frac{1}{1 + e^{2mL}} - \frac{1}{1 + e^{-2mL}}\right) \qquad \ldots(12)$$

This equation can be rewritten in a more convenient form, by placing the final member (last term) of the r.h.s over a common denominator and noting that

$$\frac{e^{mL} - e^{-mL}}{e^{mL} + e^{-mL}} = \tanh (mL)$$

where tanh is the hyperbolic tangent

$$\therefore\ q_{rod} = -k\,A\left.\frac{dT}{dx}\right|_{x=0}$$

$$= \left(\sqrt{PhkA}\right)(T_s - T_\infty)\tanh(mL) \qquad ...(13)$$

III Case

If the heat lost by the end of the rod is by convection, the heat flowing by cond^n to the face at $x = L$ must be equal to the convective heat flow from the end section of the rod to the fluid or

Transport area is the same for both

$$-k\left.\frac{dT}{dx}\right|_{x=L} = h_L\left(T_{x=L} - T_\infty\right) \qquad ...(14)$$

It is to be noted at this juncture that h_L, the heat transfer coefficient, at the end face is likely to be different from h which is the coefficient over the circumferential surface of the rod.

$$\boxed{q_{x=L} = -k\left(c_1\, m\, e^{mL} - c_2\, e^{-mL}\, m\right)}$$

substituting for $\left.\frac{dT}{dx}\right|_{x=L}$ from equation (3), we obtain

$$q_{x=L} = -k\left(c_1\, m\, e^{mx} - c_2\, m\, e^{-mx}\right)_{x=L}$$

$$= h_L\left(T_{x=L} - T_\infty\right) \qquad ...(15)$$

For $T - T_\infty$, substituting $c_1\, e^{mL} + c_2\, e^{-mL}$ (equation 3)

or $$c_2\, e^{-mL} - c_1\, e^{mL} = \frac{hL}{k_m}\left[c_1\, e^{mL} + c_2\, e^{-mL}\right] \qquad ...(16)$$

Equation (4) and (16) can now be solved simultaneously to get c_1 and c_2 (the algebraic is tedious)

In substituting the values of c_1 and c_2 into equation for temperature distribution the dimensionless temperature distribution along the fin is

$$\frac{T - T_\infty}{T_s - T_\infty} = \frac{\cosh m(L - x) + \dfrac{hL}{mk}\sinh L\, m(L - x)}{\cosh mL + \dfrac{hL}{mk}\sinh mL} \qquad ...(17)$$

and $$q_{\text{fin}} = \sqrt{hPkA}\,(T_s - T_\infty)\,\frac{\sinh mL + \dfrac{hL}{mk}\cosh mL}{\cosh mL + \dfrac{hL}{mk}\sinh mL} \qquad ...(18)$$

$$= \sqrt{hPkA}\,(T_s - T_\infty)\frac{\left(1 - p\,e^{-2mL}\right)}{\left(1 + p\,e^{-2mL}\right)}$$

where $p = \dfrac{km - h}{km + h}$

Effectiveness of Fins

The heat transfer effectiveness of a fin is measured by a parameter or term called the fin efficiency η_{fin} which is defined as

$$\eta_{fin} = \frac{\text{actual heat transfer by the fin}}{\text{heat which would have been transferred if the entire fin were at the base temperature}}$$

for a circular pin fin of diameter D and length L with its end insulated is

$$\eta_{fin} = \frac{\tanh\sqrt{\dfrac{4L^2 h}{kD}}}{\sqrt{\dfrac{4L^2 h}{kD}}}$$

In the case of a fin of rectangular *c.s.*, of length L and thickness t, the fin effectiveness for an insulated end is

$$\eta_{fin} = \frac{\tanh\sqrt{\dfrac{hPL^2 h}{kA}}}{\sqrt{\dfrac{hPL^2}{kA}}}$$

i.e. $= \dfrac{\tanh h\,mL}{mL} \qquad \because m^2 = \dfrac{hP}{kA}$

if a rectangular fin is long, wide and thin $P/A = 2/t$.

Illustration

1. Estimate the increase in heat dissipation rate which could be got from a cylinder wall by using four pin-shaped fins per sq. in each having a diameter of $3/16''$ and ht. of $1''$. Take that the 'h' between the surface of the cylinder wall or a fin and the surrounding air is 25 Btu/hr ft^2 °F, the cylinder wall is at 600°F and the air is at 70°F. The wall and the fins are made of Aluminium whose k is 120 Btu/hr ft°F.

Solution

The fin is of finite length. It is assumed that $\frac{dT}{dx} = 0$ at $x = L$

The heat dissipation per sq. in of surface (*h.t.* area) without the fin is

$$q = \frac{25}{144}(600 - 70) = 92 \text{ Btu/hr}$$

The heat dissipation for a single fin can be determined using equation (13) that is

$$q_{\text{rod}} = \sqrt{PhkA}\,(T_s - T_\infty)\tanh mL$$

Here it should be noted that A the *c.s.* area of the pin shaped fin is its heat transfer area too

$$P = \frac{3}{16} \times \pi \times \frac{1}{2} = 0.0492 \text{ ft}$$

$$A = \frac{\pi}{4} \times \left(\frac{3}{16}\right)^2 \times \frac{1}{144} = 0.000192 \text{ ft}^2$$

$$k_{(\text{given})} = 120 \text{ Btu/hr ft °F}$$

$$\therefore \sqrt{hPkA} = \sqrt{25 \times 0.0492 \times 120 \times 0.000192}$$

$$= 0.167$$

$$mL = \frac{1}{12}\sqrt{\frac{hP}{kA}} = \frac{1}{12}\sqrt{\frac{25 \times 0.0492}{120 \times 0.000192}}$$

$$= 0.192 \qquad \tanh = \frac{e^{mL} - e^{-mL}}{e^{mL} + e^{-mL}}$$

$$q_{\text{fin}} = 0.167 \times 530 \times 0.188 \qquad = (2.7128)\ 0.192 \ldots$$

$$= 16.7 \text{ Btu/hr} \qquad = 0.188$$

since we have 4 fins/sq. in of the heating surface of the cylinder, the ht (heat) dissipation rate would be 66.8 Btu/hr.

The rate of *h.t.* from the remaining wall surface = area not occupied by fins × h × ΔT

$$= \frac{25}{144} \times (1 - 0.11)(600 - 70)$$

$$\uparrow 4 \times 0.000192 \times 144$$

$$= 92\ (1 - 0.11)$$

The total heat transfer rate per sq. in of the heat transfer surface area is

= 66.6 + 92 (1 – 0.11)

= 149 Btu/hr

∴ The increase in the heat dissipation rate is over 50%.

Illustration

2. It is desired to compare the efficiency of a plate fin of length L 1.524 cm and thickness 2.54 mm for the following cases

(i) Fin material is Al (k = 207.6 w/m °C) and the h is 283.9 w/m^2 °C.

(ii) Fin material is steel of h = 41.5 w/m °C and h is 510 w/m^2 °C.

Solution

The η_{fin} for a plate fin with negligible heat flow (loss) from the tip of the fin is given by

$$\eta_{fin} = \frac{\tanh mL}{mL}$$

(i) $mL = L\sqrt{\frac{2h}{kt}}$

i.e. $m = \sqrt{\frac{hP}{kA}}$ and $\frac{P}{A} = \frac{2}{t}$

$$m = \sqrt{\frac{2h}{kt}}$$

$$\therefore mL = 1.524 \times 10^{-2}\sqrt{\frac{2 \times 283.9}{207.6 \times 2.54 \times 10^{-3}}}$$

$$= 0.5$$

$$\because \eta_{fin} = \frac{\tanh 0.5}{0.5} = \frac{0.462}{0.5} = 92.4\%$$

(ii) $mL = 1.524 \times 10^{-2}\sqrt{\frac{2 \times 510.9}{41.5 \times 2.54 \times 10^{-3}}}$

$$= 1.5$$

$$\therefore \eta_{fin} = \frac{\tanh 1.5}{1.5} = \frac{0.905}{1.5}$$

$$= 60\%$$

Illustrations on Heat Transfer from Extended Surfaces

1. A long steel rod the material of which has a thermal conductivity of 45 w/m °C protrudes from a wall. Compute the heat transfer from the rod to a surrounding fluid under the following conditions

rod diameter = 2.5 cm

convective fluid temperature (T_c) = 40°C

wall temperature (T_0) = 200°C

convective heat transfer coefficient = 25 w/m²°C

Solution

This situation can be represented by the following sketch

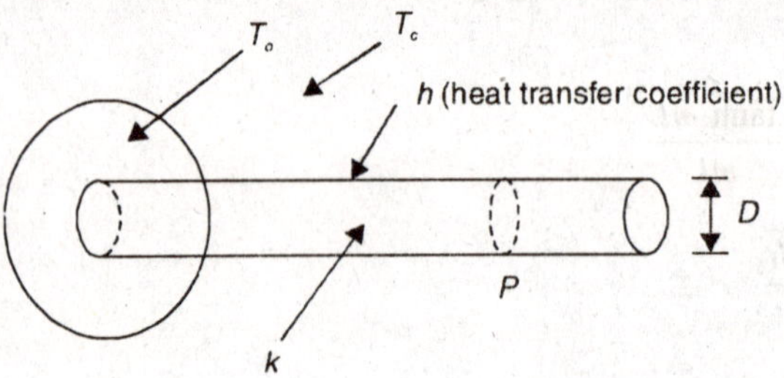

Fig. 5.10 *Heat transfer from a steel rod protruding from a wall*

The expression to be employed in the computation of heat transfer is

$$Q = \sqrt{h\,P\,k\,A}\,(T_o - T_c)$$

$h = 25$ w/m² °C $\qquad P = 2.5 \times 10^{-2} \times \pi = 0.079$ m

$k = 45$ w/m °C $A = (2.5)^2 \times \frac{\pi}{4} \times 1.0^{-4}$

$= 0.00049$ m²

$T_o = 200$ °C $\qquad T_c = 40$ °C

$$Q = \sqrt{25 \times 0.079 \times 45 \times 0.00049}\;(200 - 40)$$

$$= \sqrt{0.0435} \times 160$$

$$= 0.2086 \times 160 = 33.4 \text{ w}$$

2. In case the rod in the previous example were 15 cm long with the end insulated, how much heat would be transferred to the surroundings fluid?

Solution

This situation can be represented schematically as under

We know that for the configuration viz any length (of the rod) with the end insulated so that no heat is transferred from that end the heat transfer is given by the expression

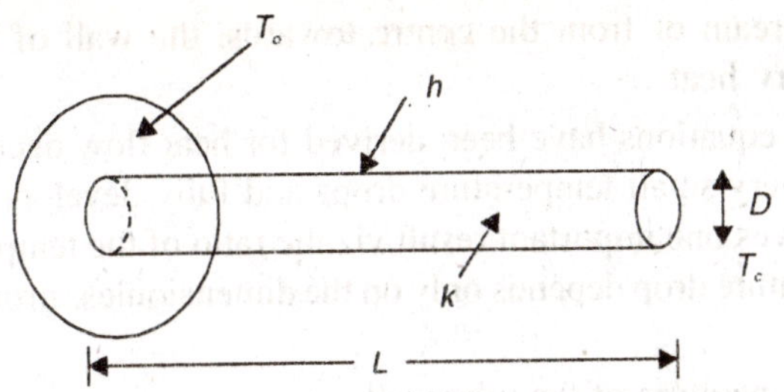

Fig. 5.11 *Heat transfer from a rod with an insulated end*

$$Q = \sqrt{k\,h\,P\,A}\,(T_o - T_c)\tanh(mL)$$

Here

$h = 25$ w/m² °C $P = 0.079$ m

$k = 45$ w/m °C $A = 0.00049$ m²

$T_o = 200$ °C $T_c = 40$°C

$$m = \sqrt{\frac{25 \times 0.079}{45 \times 0.00049}} \quad \frac{1}{m}$$

$$m = \sqrt{89.6} = 9.46 \text{ m}^{-1}$$

$$L = 15 \times 10^{-2} = 0.15 \text{ m}$$

$$mL = 1.419 \qquad \tanh(1.419) = 0.890$$

The heat transfer rate, on substitution of the above, is

$$Q = \sqrt{25 \times 0.079 \times 45 \times 0.00049}\ (200 - 40)\ 0.890$$

$$= 0.2087 \times 160 \times 0.890$$

$$= 29.7 \text{ w}$$

5.3 HEAT TRANSFER TO FLUIDS IN STREAMLINE FLOW INSIDE TUBES

Significance of Graetz Number (N_{GZ})

The mechanism of heat transfer between fluid in laminar flow and solid differs very much from that of heat flow to fluid, flowing in turbulent conditions. In laminar flow, eddies do not form and the heat must flow largely by the mechanism of conduction. We know in a fully developed flow, the velocity profile in the case of laminar flow is parabolic in shape. This will be true only if flow is isothermal so that viscosity is invariant with position in the cross section. The effect of viscosity gradients is to cause secondary flows either from the wall towards the centre of the stream or from the centre towards, the wall of the tube. These cross flows carry heat.

Theoretical equations have been derived for heat flow occurring under the assumption of very small temperature drops and fully developed laminar flow. This analysis gives one important result viz the ratio of the temperature range to the inlet temperature drop depends only on the dimensionless group called Graetz number

t_w = temperature of the tube wall

t_b = average fluid temperature at the tube outlet.

The mathematical expression is

$$\frac{t_b - t_a}{t_w - t_a} = \chi^1 \left(\frac{wc_p}{kL}\right) = \chi^1 \left(N_{G_z}\right) \quad \text{...(1)}$$

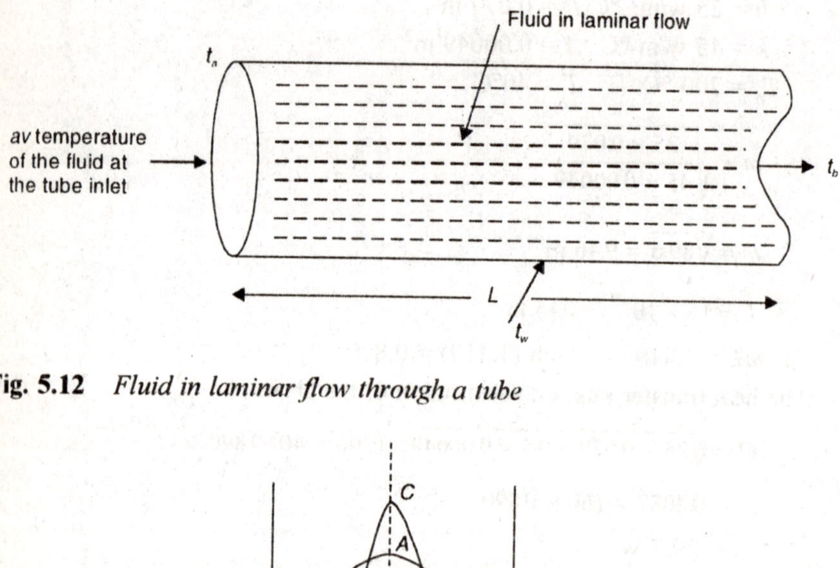

Fig. 5.12 *Fluid in laminar flow through a tube*

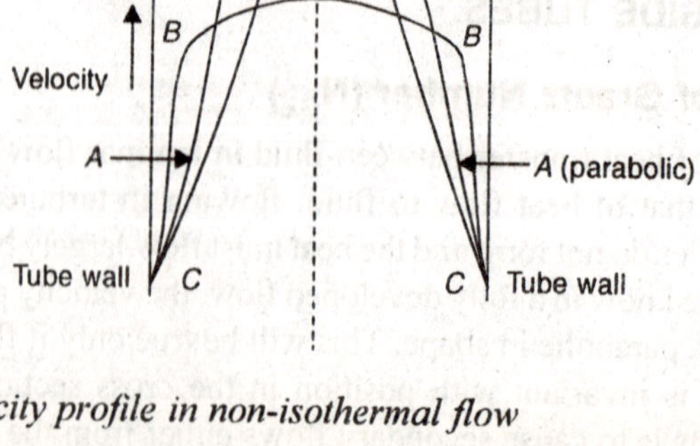

Fig. 5.13 *Velocity profile in non-isothermal flow*

t_a = average fluid temperature at the tube inlet

t_b = average fluid temperature at the tube outlet

t_w = temperature of wall in contact with fluid

BB = heating of liquid and cooling of gas

CC = cooling of liquid and heating of gas

w = mass flow rate of the fluid lbs/hr

L = heated length of the tube

c_p = specific heat of the fluid at constant pressure

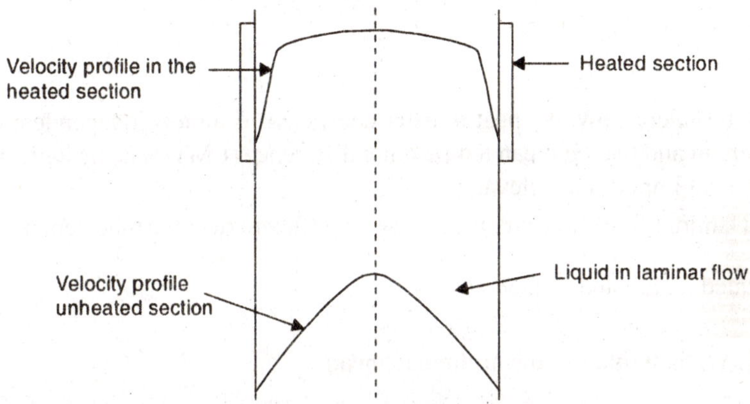

Fig. 5.14 *Velocity in isothermal and non-isothermal sections*

k = thermal conductivity of the fluid in question.

N_{Gz} = Graetz number (dimensionless)

The above equation does not contain the heat transfer coefficient h_i. An equation involving h_i can be derived as follows

$$h_i = \frac{q}{A_T \overline{\Delta t_i}} \qquad \text{...(2)}$$

h_i = average heat transfer coefficient over the entire tube

q = rate of heat transfer

A_T = heating surface area

$\overline{\Delta t_i}$ = average temperature drop over fluid, from tube wall to average fluid temperature

Striking an enthalpy balance

$$q = w\, c_p\, (t_b - t_a) \qquad \text{...(3)}$$

$$A_T = \pi D L \qquad \text{...(4)}$$

D = inside diameter of the tube

Elimating q and A_T from equations (2) (3) and (4), we obtain

$$h_i = \frac{w\, c_p\, (t_b - t_a)}{\pi D L \quad \overline{\Delta t_i}} \qquad \ldots(5)$$

The left side of equation (5) is expressed in the form of Nu

$$N_{N_u} = \frac{h_i\, D}{k} = \frac{1}{\pi} \frac{w\, c_p}{k\, L} \frac{(t_b - t_a)}{\overline{\Delta t_i}}$$

$$= \frac{N_{G_z}\, (t_b - t_a)}{\pi\, \overline{\Delta t_i}} \qquad \ldots(6)$$

In turbulent flow, the heat transfer coefficient is almost independent of the tube length and the log mean temperature difference (LMTD) is the logical form of practical importance/relevance

In laminar flow, however the average coefficient over the tube depends upon the heated length and in fact $h_i \propto \frac{1}{L}$

If $\overline{\Delta t_i}$ is written as the arithmetic mean

$$\overline{\Delta t_i} = \frac{(t_w - t_a) + (t_w - t_b)}{2}$$

$$= \overline{\Delta t_a}$$

we get

$$\frac{h_{ia}\, D}{k} = \frac{2\, N_{G_z}}{\pi} \frac{(t_b - t_a)}{(t_w - t_a) + (t_w - t_b)} \qquad \ldots(7)$$

h_{ia} = heat transfer coefficient, based on the arithmetic average temperature drop.

If equation (1) is sovled for t_b

$$\frac{t_b - t_a}{t_w - t_a} = \chi' \left(\frac{w\, c_p}{k\, L} \right) = \chi' \left(N_{G_z} \right)$$

and the resulting expression for t_b is substituted into the denominator of equation (7) the result is

$$\frac{h_{ia}\, D}{k} = \frac{2\, N_{G_z}}{\pi} \frac{(t_b - t_a)}{(t_w - t_a)\left[2 - \chi' \left(N_{G_z} \right)\right]} \qquad \ldots(8)$$

Now substituting for $\frac{t_b - t_a}{t_w - t_a}$ from the above equation (theoretical derivation)

we obtain,

$$\frac{h_{ia}\, D}{k} = \frac{2\, N_{G_z}}{\pi} \cdot \frac{\chi'\left(N_{G_z}\right)}{2 - \chi'\left(N_{G_z}\right)}$$

$\chi\left(N_{G_z}\right)$ = Some other function of Graetz number.

Experimentation indicates, for fluids of moderate viscosity

$$\chi\left(N_{G_z}\right) = 2\left(N_{G_z}\right)^{\frac{1}{3}}$$

and hence $\frac{h_{ia}\, D}{k} = N_{N_{ua}} \quad 2\left(N_{G_z}\right)^{\frac{1}{3}}$

$$\frac{h_{ia}\, D}{k} = 2\left(\frac{w\, c_p}{k\, L}\right)^{\frac{1}{3}}$$

For viscous fluids and large temperature drops, the above equation is modified to include the same viscosity correction term used in turbulent flow correlation.

$$\frac{h_{ia}\, D}{k} = 2\left(\frac{w\, c_p}{k\, L}\right)^{\frac{1}{3}} \left(\frac{\mu}{\mu_w}\right)^{0.14}$$

Extreme conditions

If the tube is very long

$$t_b \simeq t_w$$

An asymptotic pattern is reached, then

$$t_b - t_a = t_w - t_a$$

and $\quad t_w - t_b = 0$

Then for this situation, equation (6) becomes

$$\frac{h_{ia}\, D}{k} = \frac{2\, w\, c_p}{\pi\, k\, L} = \frac{2}{\pi} N_{G_z}$$

This condition is approached for values of N_{G_z} of 10 to 13.

The steps for the development of equations (8) are given below

$$\frac{h_{ia}\, D}{k} = N_{N_{ua}} = 2\left(\frac{w\, c_p}{k\, L}\right)^{\frac{1}{3}} \left(\frac{\mu}{\mu_w}\right)^{0.14}$$

$$h_{ia} = \frac{w\, c_p\, (t_b - t_a)}{\pi\, D\, L \quad \Delta t_a}$$

$$\frac{t_b - t_a}{t_w - t_a} = \chi'\left(N_{G_z}\right)$$

$$t_b - t_a = t_w \times \chi'\left(N_{G_z}\right) - t_a\, \chi'\left(N_{G_z}\right)$$

$$t_b = t_a + t_w\, \chi'\left(N_{G_z}\right) - t_a\, \chi'\left(N_{G_z}\right) \qquad \text{...(1)}$$

we know $\dfrac{h_{ia}\, D}{k} = \dfrac{2\, N_{G_z}}{\pi} \dfrac{(t_b - t_a)}{(t_w - t_a) + (t_w - t_b)}$

substituting for t_b from equation (1), in the denominator of the above expression we obtain

$$\frac{h_{ia}\, D}{k} = \frac{2\, N_{G_z}}{\pi} \frac{(t_b - t_a)}{(t_w - t_a) + t_w - t_a - t_w\, \chi'\left(N_{G_z}\right) + t_a\, \chi'\left(N_{G_z}\right)}$$

$$= \frac{2\, N_{G_z}}{\pi} \frac{(t_b - t_a)}{2t_w - 2t_a - t_w\, \chi'\left(N_{G_z}\right) + t_a\, \chi'\left(N_{G_z}\right)}$$

$$= \frac{2\, N_{G_z}}{\pi} \frac{t_b - t_a}{(t_w - t_a)\, 2 - \chi'\, N_{G_z}\, (t_w - t_a)}$$

$$= \frac{2\, N_{G_z}}{\pi} \frac{t_b - t_a}{(t_w - t_a)\left[2 - \chi'\, N_{G_z}\right]}$$

Problems on the Use of Graetz Number and Grashof Number and Nusselt Equation

1. A light motor oil is to be heated from 150 to 250°F in a 0.364″ pipe which is 15′ long. The pipe wall is at 350°F. How much oil can be heated in this pipe in lbs/hr? What heat transfer coefficient can be expected?

$k = 0.082$ Btu/hr ft °F

$c_p = 0.48$ Btu/lb °F

μ_{at} 150 °F = 6.0 cp

at 250 °F = 3.3 cp

at 350 °F = 1.37 cp

Assume laminar flow condition.

Solution

We have to solve for w and h_{ia} from the two equations

$$\frac{h_{ia}\, D}{k} = 2\left(\frac{w\, c_p}{k\, L}\right)^{\frac{1}{3}}\left(\frac{\mu}{\mu_w}\right)^{0.14}$$

and
$$h_{ia} = \frac{w\, c_p\, (t_b - t_a)}{\pi\, D\, L\, \Delta \bar{t}_a}$$

$$\mu = \frac{6 + 3.3}{2} = 4.5 \text{ c.p} \qquad \mu_w = 1.37 \text{ c.p}$$

$$D = \frac{0.364}{12} = 0.0303 \text{ ft} ; \quad \left(\frac{\mu}{\mu_w}\right)^{0.14} = 1.187$$

$$\frac{0.0303\, h_{ia}}{0.082} = 2 \times 1.187\left(\frac{0.48\, w}{0.082 \times 15}\right)^{\frac{1}{3}}$$

$$h_{ia} = 4.69\, w^{\frac{1}{3}}$$

$$\Delta \bar{t}_a = \frac{(350 - 150) + (350 - 250)}{2} = 150\ °\text{F}$$

$$L = 15' \quad D = 0.0303'$$

$$t_b - t_a = 250 - 150 = 100\ °\text{F}$$

$$\therefore\ h_{ia} = \frac{0.48 \times 100\, w}{\pi \times 0.0303 \times 15 \times 150} = 0.224 \text{ w}$$

i.e.
$$4.69\, w^{\frac{1}{3}} = 0.224 \text{ w}$$

or
$$w = \left(\frac{4.69}{0.224}\right)^{\frac{3}{2}} = 95.9 \text{ lbs/hr and } h_{ia} = 0.224 \times 95.9 = 21.5 \text{ Btu/hr ft °F}$$

2. Use of Grashof modulus/number

Atmosphere air is contained between two horizontal parallel plates separated by a distance of 1inch. The lower and upper plates are maintained at 140°F and 60°F respectively. Determine the rate of heat transfer across the air space per sq. ft. of the plate surface.

The simplified equations valid for this situation are

For laminar

$$N_{Nu_m} = 0.195 \times \left(N_{Gr_m}\right)^{\frac{1}{4}}$$

$$1 \times 10^4 < N_{G_{r_b}} < 4 \times 10^5$$

For turbulent condition

$$N_{N_{u_m}} = 0.068\left(N_{G_{r_b}}\right)^{\frac{1}{3}} \quad 4 < 10^5 < N_{G_{r_b}}$$

where $N_{N_{u_m}}$ and $N_{G_{r_b}}$ are based on the distance 'b' between the plates

The physical properties of air at the temperature 100°F $\left(\dfrac{140+60}{2}\right)$

$k = 0.0169$ Btu/hr ft °F $\rho = 0.070$ lbs/c ft

$\nu = 0.662$ ft²/hr $\beta = 1.79 \times 10^{-3}$ °F^{-1}

$N_{P_r} = 0.70$ $g = 4.17 \times 10^8$ ft/hr²

Solution

The Grashof number for a spacing $b = 1$ inch is calculated as follows

$$N_{G_r} = \frac{x^3 \rho^2 g \beta \Delta t}{\mu^2} \qquad x \to b$$

$$= \frac{g \beta (\Delta t) b^3}{\nu^2}$$

$$= \frac{\left(4.17 \times 10^8\right)\left(1.79 \times 10^{-3}\right)(140-60)\left(1^3\right)}{(0.662)^2 \times 12^3}$$

$$= 7.9 \times 10^4$$

Nusselt number, $N_{N_{u_m}} = 0.195 \times \left(N_{G_{r_b}}\right)^{\frac{1}{4}}$

$$= 0.195 \times \left(7.9 \times 10^4\right)^{\frac{1}{4}}$$

$$= 3.27$$

$$N_{n_{u_m}} = \frac{h D}{k} = \frac{q' b}{k(t_1 - t_2)} = 3.27$$

$$\therefore = q' = 3.27 \times 0.0169 \times (140-60) \times 12$$

$$= 53 \text{ Btu/hr ft}^2$$

Problem on Nusselt Equation Application

Water at a mean temperature of 37.8°C flows through a 2 cm i.d. smooth walled tube with a mean velocity of 152 cm/sec. The tube wall is 17°C above the water temperature. Determine the heat transfer coefficient and the total heat transfer rate to water for a 3 m length of the tube.

Physical properties at the mean temperature are

$$v = \frac{\mu}{\rho} = 0.687 \times 10^{-2} \text{ cm}^2/\text{sec}$$

$$k = 0.63 \text{ k cal/hr m °C}$$

$$N_{P_r} = 4.52$$

Solution

$$N_{R_e} = \frac{Du\rho}{\mu} = \frac{Du}{v} = \frac{2 \times 152}{0.687 \times 10^{-2}}$$

Flow is turbulent as $N_{R_e} > 10000$ we will use the Nusselt equation (Dittus-Boelter equation) without the viscosity correction term

$$\frac{hD}{k} = 0.023 \left(N_{R_e}\right)^{0.8} \left(N_{P_r}\right)^{0.4}$$

$$\frac{h \times \frac{2}{100}}{0.63} = 0.023\,(56300)^{0.8}\,(4.52)^{0.4}$$

$$h = 6900 \text{ k cal/hr m}^2 \text{ °C}$$

Heat transfer rate

$$q = h\,(\pi D L)\,(T_w - T_m)$$

$$= 6900 \left(\frac{22}{7} \times \frac{2}{100} \times 3\right)(17)$$

$$= 22110 \text{ k cal/hr}$$

Fluid Flow in a Tube: Solutions for Constants Wall Heat Flux and Constant Wall Surface Temperature

Flow through a conduct in a tube or pipe is the most common type and is wholly viscous in nature. There is no existence of boundary layer separation. In practice almost any variations or combinations of the boundary conditions can exist at a tube surface. The limiting cases are the situations where we have constant heat flux and constant surface temperature.

The first situation can be met experimentally by electrical heating through a heater sheath or by winding a tube uniformly with a heating wire or ribbon. The other situation that is one of constant temperature condition is most nearly realized in actual practice by allowing the condensation of a saturated vapour on the outside surface of the tube. The expressions corresponding to these conditions are developed below.

Specified Wall Heat Flux

The control volume shown below in Δx thick and bounded by the tube walls. This will be the basis for our considerations.

A first law of thermodynamics analysis for a steady-state condition will give

$$q_1 - q_2 + q_3 = 0 \qquad ...(1)$$

where q_1 = energy transported into the control volume by fluid flow

q_2 = energy transported out of the control volume by fluid flow

q_3 = heat transferred to the control volume by the condition at the wall

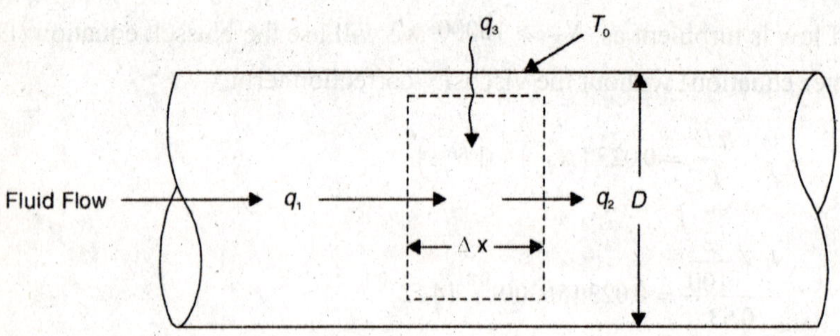

Fig. 5.15 *Fluid flow in a tube-heat transfer under specified wall heat flux condition*

The above terms may be evaluated one by one as

$$q_1 = \rho \frac{\pi D^2}{4} v_x c_p T\Big|_x \qquad ..(2)$$

$$q_2 = \rho \frac{\pi D^2}{4} v_x c_p T\Big|_{x+\Delta x} \qquad ..(3)$$

$$q_3 = \frac{q_r}{A} \pi D \Delta x \qquad ...(4)$$

Making these substitutions into the expression based on the first law of thermodynamics, we get

$$\rho \frac{\pi D^2}{4} c_p v_x \left(T\Big|_{x+\Delta x} - T\Big|_x\right) - \frac{q_r}{A} \pi D \Delta x = 0 \qquad ...(5)$$

For a fully developed flow, $v_x = v_x(x)$ therefore the above expression may be arranged as

$$\rho \frac{\pi D^2}{4} c_p v_x \frac{\left(T\Big|_{x+\Delta x} - T\Big|_x\right)}{\Delta x} - \frac{q_r}{A} \pi D = 0 \qquad ...(6)$$

Taking the limit as $\Delta x \to 0$. We have

$$\rho\, c_p\, v_x \frac{D}{4}\frac{dT}{dx} - \frac{q_r}{A} = 0 \qquad \ldots(7)$$

The temperature gradient may be solved for as

$$\frac{dT}{dx} = \frac{1}{\rho\, c_p\, v_x \dfrac{D}{4}} \cdot \frac{q_r}{A} \qquad \ldots(8)$$

and after separating the variables and integrating, we obtain an expression for the mean fluid temperature as a function of position along the tube

$$T - T_e = \frac{1}{\rho\, c_p\, v_x \dfrac{D}{4}} \int_o^x \frac{q_r}{A}\, dx \qquad \ldots(9)$$

which, for the case of constant wall flux, turns out to be

$$T - T_e = 4\,\frac{\dfrac{q_r}{A}}{\rho\, c_p\, v_x}\,\frac{x}{D} \qquad \ldots(10)$$

where T_e is the temperature at the entrance i.e. $x = 0$

Equation (8) may be employed with any known function for wall heat flux.

One quantity that is often of interest to an engineer is the surface temperature of a tube through which a coolant flows. In this case, with the heat flux known and the mean temperature available from equation (10), the surface temperature can be determined from

$$T_o = T_\infty + \frac{\dfrac{q_r}{A}}{h}$$

The convective heat transfer coefficient may be evaluated, using an appropriate expression.

Constant Wall Surface Temperature

The development of an appropriate analytical expression for the case of constant surface temperature, will be based on the following sketch.

As in the case considered earlier, at steady state condition

$$q_1 - q_2 + q_3 = 0 \qquad \ldots(1)$$

The individual evaluation of these q's yields the following

$$q_1 = \rho \frac{\pi D^2}{4} c_p\, v_x\, T\big|_x \qquad \ldots(2)$$

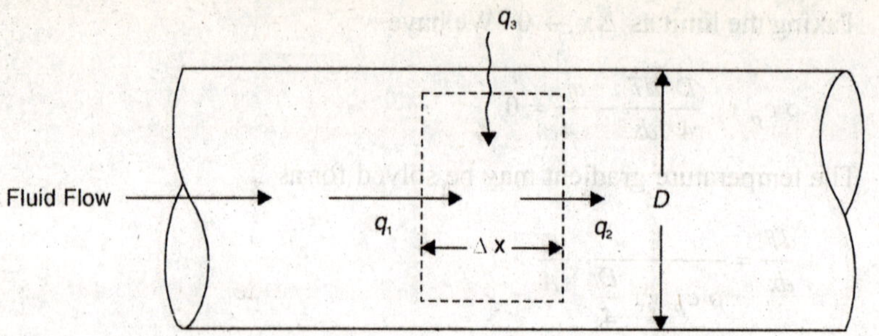

Fig. 5.16 *Fluid flow in a tube-heat transfer under constant wall surface temperature*

$$q_2 = \rho \frac{\pi D^2}{4} c_p\, v_x\, T\big|_{x+\Delta x} \qquad \text{...(3)}$$

$$q_3 = h\,\pi\, D\, \Delta x\,(T_o - T) \qquad \text{...(4)}$$

on substituting the above expressions for the q's in equation number (1), we obtain

$$\rho \frac{\pi D^2}{4} c_p \left(T_{x+\Delta x}\, v_x - v_x\, T\big|_x\right) - h\,\pi\, D\, \Delta x\,(T_o - T) = 0 \qquad \text{...(5)}$$

Taking the flow to be fully developed in which case $v_x = v_x(x)$ dividing by Δx and rearranging, we get

$$\rho\, c_p\, v_x \frac{D}{4} \frac{T\big|_{x+\Delta x} - T\big|_x}{\Delta x} + h\,(T - T_o) = 0 \qquad \text{..(6)}$$

In the limit $\Delta x \to 0$ this turns out to be

$$\rho\, c_p\, v_x \frac{D}{4} \frac{dT}{dx} + h\,(T - T_o) = 0 \qquad \text{...(7)}$$

The temperature gradient may be written as

$$\frac{dT}{dx} = -\frac{h}{\rho\, c_p\, v_x} \frac{4}{D} (T - T_o) \qquad \text{...(8)}$$

we can now separate the variables and perform the integration from the entrance to some position x, along the flow axis, as follows

$$\int_{T_e - T_o}^{T - T_o} \frac{dT}{T - T_o} = -\frac{h}{\rho\, c_p\, v_x} \frac{4}{D} \int_o^x dx \qquad \text{...(9)}$$

$$ln \frac{T - T_o}{T_e - T_o} = -\frac{h}{\rho c_p v_x} 4 \frac{x}{D} \quad \text{...(10)}$$

The variation in temperature with x may now be expressed as

$$\frac{T - T_o}{T_e - T_o} = e^{(-N_{st})\left(4\frac{x}{D}\right)} \quad \text{...(11)}$$

It is necessary that x/D be greater than be in order that the assumption of fully developed conditions be valid (specifically $h \neq h(x)$)

It may be noted that in equation (11) the Stanton number arose quite naturally from our analysis. The final determination of $T(x)$ calls for the evaluation of N_{s_t} for which an appropriate expression can be used.

Illustration

A closed container filled with a hot liquid is in a room whose walls and air are at a fixed temperature. Identify all the modes of heat transfer by which the hot liquid gets cooled. Comment on features that would contribute to a good container design.

Solution

The given situation is the hot liquid is separated from its cooler surroundings by a plastic flask, an air space and a plastic cover

A schematic representation of the given situation is presented below.

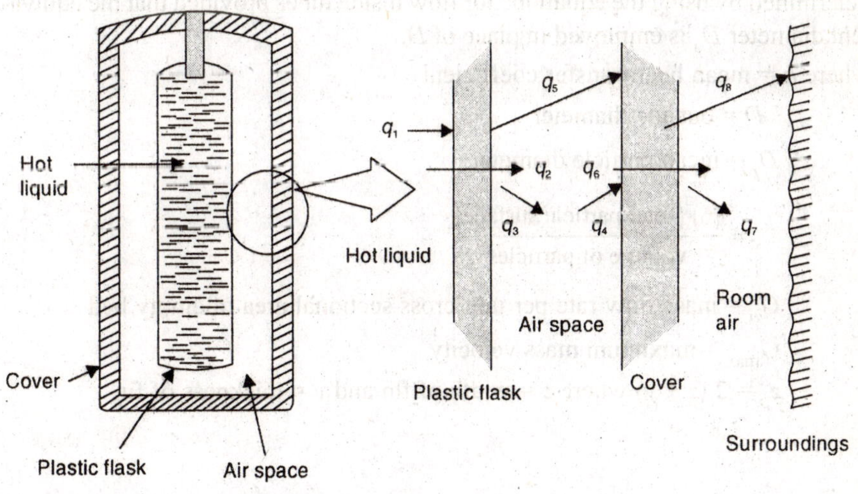

Fig. 5.17 *Heat transfer from a closed container*

Pathways for energy transfer from the hot liquid are as follows

q_1 = free (natural) convection from the hot liquid to the flask

q_2 = conduction through the flask

q_3 = natural convection from the flask to the air

q_4 = natural convection from the air to the cover.

q_5 = net radiation exchange between the outer surface of the flask and the inner surface of the cover.

q_6 = conduction through the cover

q_7 = natural convection from the cover to the room air.

q_8 = net radiation exchange between the outer surface of the cover and the surroundings.

Comments: Improvements in design can be effected by (1) using aluminized (low emissivity) surfaces for the flask and cover to reduce net radiation and (2) evacuating the air space or using a filler material to bring about a reduction in natural convection.

Flow Over Submerged Bodies

Correlations, which are normally used for laminar and turbulent flow normal to the outside surface of cylinders (laminar or turbulent flow over submerged bodies), spheres tube banks etc are summarized in the following tables. By normal flow is meant that the direction of flow is at right angles to the axis of the cylinder or tube. For flow parallel to the outside of tubes the value of h may be determined by using the equations for flow inside tubes provided that the equivalent diameter D_e is employed in place of D.

where h = mean heat transfer coefficient

D = outside diameter

D_p = mean particle diameter

$$= \frac{(6)\,(\text{total particle surface})}{\text{volume of particles}}$$

G_o = mass flow rate per unit cross sectional area of empty bed

G_{max} = maximum mass velocity

$z_p = 2\,(z + y)$ where z = width of fin and y = thickness of fin

Table 5.10 *Mean heat transfer coefficient h for flow over cylinders, spheres etc*

Fluid	*Flow geometry*	*Correlation*	*Range of application*
Air, water and hydrocarbon oils	Flow normal to single cylinder i.e. at right angles to axis of cylinder	$\frac{hD}{k_f}\left(p_r\right)^{-0.3} = 0.35 + 0.56\left(\frac{DG}{\mu_f}\right)^{0.52}$	$1 < \frac{DG}{\mu_f} < 1000$
Air	Flow normal to single cylinder	$\frac{hD}{k_f}\left(p_r\right)_f^{-0.3} = 0.26\left(\frac{DG}{\mu_f}\right)^{0.60}$	$1000 < \frac{DG}{\mu_f} < 50{,}000$
Gases	Flow normal to single sphere	$\frac{hD}{k_f} = 0.37\left(\frac{DG}{\mu_f}\right)^{0.60}\left[\left(p_r\right)_f\right]^{\frac{1}{3}}$	$17 < \frac{DG}{\mu_f} < 70{,}000$
Air	Flow normal to spheres	$\frac{hD}{k_f} = 2.0 + 0.6\left(p_{rf}\right)^{\frac{1}{3}}\left(R_{ef}\right)^{0.5}$	$17 < \frac{DG}{\mu_f} < 70{,}000$
Water and oil	Flow normal to single sphere	$\frac{hD}{k_f}\left(p_{rf}\right)^{-0.3} = 0.97 + 0.68\left(\frac{DG}{\mu_f}\right)^{0.5}$	$1 < \frac{DG}{\mu_f} < 1000$;
	Flow through packed bed of spheres	$\frac{hD_p}{k} = 0.8\left(\frac{D_p G_o}{\mu}\right)^{0.7}\left(p_r\right)^{\frac{1}{3}}$	7 mm $< D <$ 12.5 mm
Air	Flow across short fins consisting of strips or pins	$\frac{h}{(c_p)_b G_{max}}\left(\frac{c_p \mu}{k}\right)_f^{\frac{2}{3}} = \left[\frac{z_p G_{max}}{\mu_f}\right]^{-0.5}$	$260 < \frac{z_p G_{max}}{\mu_f} < 12000$

Table 5.11 *Mean heat transfer coefficient h for flow over tube banks*

Fluid	*Flow geometry*	*Correlation*	*Range of application*
Gases	Flow normal to tube banks 10 or more row deep (staggered tubes)	$\dfrac{hD}{k_f} = 0.33\left(\dfrac{D\,G_{max}}{\mu_f}\right)^{0.6}\left(N_{P_r}\right)^{\frac{1}{3}}$	$2000 < \dfrac{D\,G_{max}}{\mu_f} < 32{,}000$
	Flow normal to tube banks 10 or more rows deep (in line tubes)	$\dfrac{hD}{k_f} = 0.26\left(\dfrac{D\,G_{max}}{\mu_f}\right)^{0.6}\left(N_{P_r}\right)^{\frac{1}{3}}$ (G_{max} = maximum mass velocity) (D = outside diameter of tubes)	$2000 < \dfrac{D\,G_{max}}{\mu_f} < 32{,}000$
Air	Flow normal to banks of tubes; 10 or more rows deep (staggered or in-line)	$\dfrac{hD}{k_f} = b\left(\dfrac{D\,G_{max}}{\mu_f}\right)^{n}$ values of b and n are given in reference books	$2000 < \dfrac{D\,G_{max}}{\mu_f} < 40{,}000$

5.4 HEAT TRANSFER ACCOMPANIED BY CHANGE OF PHASE

In the preceding chapter the conditions of the system did not envisage any change in the chemical composition or that of phase of the fluid flowing. However, there are many industrial processes where the transfer of thermal energy is accompanied by change of phase. The process of heat transfer in such cases is not merely that of convection, but is quite different and complex. The phenomenon of transfer of heat associated with change of phase will be dealt with in this chapter.

The two important processes which fall under this category are those of boiling liquids and condensing vapours. These processes, in addition to all the variables associated with convection, depend also on those parameters which govern the change of phase from liquid to vapour in boiling and from vapour to liquid in condensation. These additional parameters like surface tension of fluid, latent heat and pressure make the phenomena so complex that it has not been possible to evolve detailed general equations describing the process or their analytical solutions so far.

Until recent past the processes involving heat transfer with change of phase were found primarily in steam generation and condensers of thermal power plants, evaporators and condensers of refrigerating systems and in chemical industry.

But the cooling problem in nuclear reactions or rocket motors has thrown a challenge to engineers to develop the knowledge of boiling heat transfer so as to dissipate heat energy at rates larger than 10^8 k cal/hr m^2. The large magnitude of this rate may be realised by comparing it to the maximum heat transfer rate in a modern boiler 2 × 10^5 k cal/hr-m^2

Phenomenon of Boiling

Boiling, though a familiar phenomenon is extremely complicated. In order to understand the heat transfer mechanism during boiling, we shall first discuss the phenomenon of boiling as such and the existence of various regimes of boiling.

Boiling is the process of vapour formation in a liquid receiving heat. Usually, the temperature of the boiling liquid is taken to be equal to the saturation temperature, t_s corresponding to the boiling pressure. However, experience shows that the liquid is at a slightly higher temperature that t_s.

If water is allowed to boil over a horizontal surface, the temperature profile may be as presented below. It is observed that the bulk of the water is at a fairly uniform temperature (slightly higher than t_s = 100°C). In a layer of 2.5 mm a steep gradient up to the surface of temperature of 109°C when the heating is done at the rate of 10,300 k cal/hr m^2.

In general, the boiling process which finds maximum engineering application is the one is which the heating surface is submerged beneath the free surface of the liquid in a container as discussed above.

Heat Transfer from Condensed Vapours

Condensation of a saturated vapour is accomplished by bringing the vapour into contact with a surface whose temperature is below the dew point of the vapour. The condensing vapour may comprise a single substance, a mixture of condensable and non-condensible substances or mixture of two or more condensing vapours. A common example of the condensation of one constituent from its mixture with a second non-condensible substance is the condensation of water from a mixture of steam and air.

Condensetes of organic subtances commonly wet metallic surfaces, forming a film of liquid on the cold surface which is called filmwise condensation. Water condensate usually wets metallic surfaces but under some conditions may not do so in such cases dropwise condensation occurs.

Filmwise Condensation

This type is more common that dropwise condensation. The liquid condensation forms a film or a continuous layer of liquid that flows over the surface of the tube under the action of gravity. It is the layer of liquid interposed between the vapour and the wall of the tube that provides the resistance to the flow of heat and hence controls the heat transfer coefficient of the condensing vapour.

The thickness of the condensate layer in filmwise condensation is governed by the configuration of the surface, the rate of condensation and the rate at which the liquid flows from the surface.

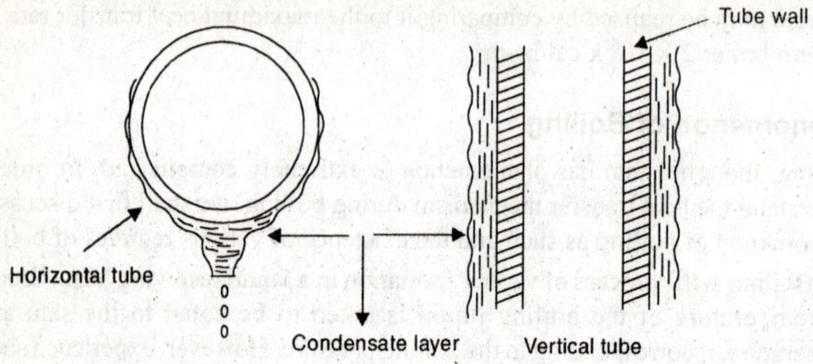

Fig. 5.18 *Filmwise condensation on a horizontal tube and on a vertical tube*

Dropwise Condensation

The condensate is formed as fine drops and these drops (fine) tend to coalesce into rivulets which flow down the tube under the action of gravity, sweep away the condensate and clear the surface for more droplets. During dropwise condensation large areas of the cold tube are bare and are directly exposed to vapour. Since the liquid film is absent, the resistance to the heat flow at these bare areas is very low and thereby the h value is also correspondingly very high.

h_{av} for dropwise > 5 h_{av} for filmwise condensation.

There are chemicals known as drop promoters which help in the formation of drops e.g mercaptans on copper alloys, fatty acids, oleic acids; another situation; Condenstion of steam on polished chromium covered with stearic acid.

A study of dropwise condensation on vertical and inclined surfaces has indicated the following mechanism.

Vapour condenses on the surface in small drops which are generally of uniform size and shape. The small drops grow by condensation on their surfaces and by coalescence until one or more drops reach the maximum size possible for adhering to the surface. These large drops then roll and grow rapidly by coalescing with other drops in the path. In this way the surface is cleared of condensate periodically by droplets sweeping down the surface, exposing a bare strip of metal. The cycle is then replaced starting with the condensation of vapour on the surface in small drops.

In can be as high as 70000 Btu/ft^2 hr °F

Heat Transfer Coefficients for Filmwise Condensation

The basic equations for the rate of heat transfer in filmtype condensation were first derived by Nusselt.

Assumpations Involved in the Derivation

1. Only resistance is by the layer of condensate.
2. Condensate flows in laminar condition under gravitational influence.
3. Velocity of liquid at wall = 0
4. Velocity of liquid at the outside of the film is not influenced by the velocity of the vapour.
5. Temperatures of wall and vapour are constant
6. Physical properties of the fluid are taken at t_f (mean film temperature)
7. The situation is heat transfer between a pure saturated vapour and a colder surface, with the condensate leaving the tube at the condensing temperature.

Vertical tubes

The average h, based on the entire tube is, from the Nusselt theory,

$$\bar{h}\left(\frac{\mu_f}{k_f^3\,\rho_f^2\,g}\right)^{\frac{1}{3}} = 1.47\left(\frac{4T}{\mu_f}\right)^{-\frac{1}{3}}$$

where μ_f, k_f and ρ_f are the physical properties of the condensate at the mean film temperature

$g = 4.17 \times 10^8$ ft/hr^2

$\bar{h}$ = average heat transfer coefficient

T = condensate loading/unit tube perimeter (lb/ft-hr) at the bottom of the tube

The mean film temperature is evaluated from the relation

$$t_f = t_h - \frac{3(t_h - t_w)}{4}$$

where t_f = mean film temperature

t_h = temperature of condensing vapour

t_w = temperature of outside surface of the tube wall

T is defined as condensate loading/unit perimeter at bottom

$$T = \frac{W}{\pi D_o}$$

W = condensate rate, for the entire tube lbs/hr

D_o = outside tube diameter

The term $\frac{4T}{\mu_f}$ is a N_{R_e} for the flow of liquid film down a vertical surface which can be shown as follows

$$N_{R_e} = \frac{4 r_H G}{\mu}$$

To get r_H, we take S_c = cross sectional area of the annular ring of the condensate flowing down the tube

Then $r_H = \dfrac{S_c}{\pi D_o}$

and $G = \dfrac{w}{S_c}$

substituting for r_H and G in the equation for N_{R_e} we get

$$N_{R_e} = 4 \frac{S_c}{\pi D_o} \cdot \frac{w}{S_c \mu} = \frac{4 w}{\pi D_o \mu} = \frac{4 T}{\mu}$$

$$\because \frac{w}{\pi D_o} = T$$

The Nusselt theory equation in another form

$$\bar{h}\left(\frac{\mu_f^2}{k_f^3 \rho_f^2 g}\right)^{\frac{1}{3}} = 1.47 \left(\frac{4T}{\mu_f}\right)^{\frac{-1}{3}}$$

The above equation could also be used in a different form in which the term T has been eliminated by means of the definition of $\bar{h}$

$$\bar{h} = \frac{q}{A_o (t_h - t_w)} = \frac{q}{A_o \Delta t_o}$$

q = total heat transferred Btu/hr

Δt_o = difference between t_h and t_w

since only the latent heat of condensation is transferred

$q = w\lambda$

λ = latent heat of condensation in Btu/lb

The heat transfer area $A_o = \pi D_o L$

$$\therefore \bar{h} = \frac{w \lambda}{\pi \Delta t_o D_o L} = \frac{T \lambda}{L \Delta t_o}$$

$$\boxed{T = \frac{w}{\pi D_o}}$$

Eliminating T from the first equation by means of the above equation

$$\bar{h}\left(\frac{\mu_f^2}{k_f^3 \rho_f^2 g}\right)^{\frac{1}{3}} = 1.47 \left(\frac{4 \bar{h} \cdot L \Delta t_o}{\lambda \mu_f}\right)^{\frac{-1}{3}}$$

solving for $\bar{h}$ gives

$$\bar{h} = 0.943 \left(\frac{k_f^3 \rho_f^2 g \lambda}{\Delta t_0 L \mu_f} \right)^{\frac{1}{4}}$$

For horizontal tubes the corresponding Nusselt expressions are

$$\bar{h} \left(\frac{\mu_f^2}{k_f^3 \rho_f^2 g} \right)^{\frac{1}{3}} = 1.51 \left(\frac{4 T'}{\mu_f} \right)^{\frac{-1}{3}}$$

$$\bar{h} = 0.725 \left(\frac{k_f^3 \rho_f^2 g \lambda}{\Delta t_o D_o \mu_f} \right)^{\frac{1}{4}}$$

where T' = condensate loading/unit length of the tube
i.e. mass flow rate per unit length lbs/ft hr –w/L

Use of Nusselt Equations (Filmwise Condensation)

The above equation shown for a single horizontal tube could be used also for filmwise condensation on a vertical stack of horizontal tubes, where the condensate falls cumulatively from tube to tube and the total condensate from the entire stack finally drops from the bottom tube. T' is redefined as the loading of the bottom tube

$$T' = \frac{W}{L N}$$

W = total condensate for the entire stack, lbs/hr

N = number of rows of tubes in a vertical tier i.e. the no. of horizontal tubes in a vertical tier

It is likely that a portion of the condensate splashes away from each individual tube instead of dripping completely to the tube below

T' is written as

$$T' = \frac{W}{L N^{\frac{2}{3}}}$$

The equation for a stack of horizontal tubes becomes

$$\bar{h} = 0.725 \left(\frac{k_f^3 \rho_f^2 g \lambda}{N^{\frac{2}{3}} \Delta t_o D_o \mu_f} \right)^{\frac{1}{4}}$$

In the case of vertical tubes, it has been observed that the coefficients measured experimentally are about 20% larger than those computed from the equations. This may be due to the effect of ripples (small waves). The coefficients are taken as

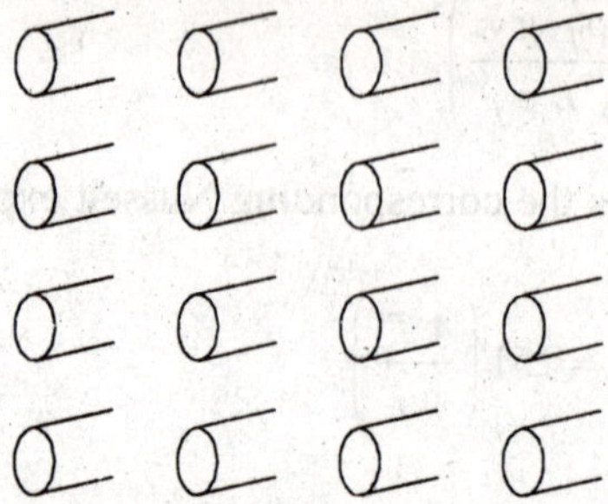

Fig. 5.19 *A stack of horizontal tubes*

$$\bar{h}\left(\frac{\mu_f^2}{k_f^3\,\rho_f^2\,g}\right)^{\frac{1}{3}} = 1.76\left(\frac{4\,T}{\mu_f}\right)^{\frac{-1}{3}}$$

↓ instead of 1.47

$$\text{Similarly } \bar{h} = 1.13\left[\frac{k_f^3\,\rho_f^2\,g\,\lambda}{\Delta t_o\,L\,\mu_f}\right]^{\frac{1}{4}}$$

↓(in lieu of the value 0.943)

For a given amount of liquid on the wall, the occurrence of ripples causes a variation in the thickness of the film and gives a larger average value of $\bar{h}$ than when no ripples occur.

To allow for the fact that the condensate is cooled from the saturation temperature to the film temperature, for both vertical and horizontal tubes, it has been shown that the latent heat of condensation λ should be multiplied by the term $\left(1 + 0.4\,\frac{c_p\,\Delta t}{\lambda}\right)^2$

This correction becomes important at high pressures where c_p increases and λ decreases.

Illustration

A square array of 400, $1/4''$ diameter tubes is used to condense dry saturated steam at 100 psia. The tubes are maintained at 322°F by a coolant flowing inside and are 30 inches long. Calculate the heat transfer rate and the amount of condensate produced—horizontal tubes and filmwise condensation. Assume no splashing away of the condensate.

Solution

From an appropriate set of steam tables, we have

$$t_h = t_{sat} = 327.8°F$$

$$\lambda = 890 \text{ Btu/lb}$$

The mean film temperature is

$$t_f = t_h - \frac{3(t_h - t_w)}{4}$$

$$= 327.8 - \frac{3}{4}(327.8 - 322)$$

$$= 327.8 - \frac{3}{4}(5.8)$$

$$= 323.3$$

Noting the condensate properties at this temperature

$$\rho_f = 56.45 \text{ lbs/ft}^3$$

$$\mu_f = 0.415 \text{ lb/hr ft}$$

$$k_f = 0.393 \text{ Btu/hr ft °F}$$

$$\Delta t_o = t_h - t_w$$

$$= 327.8 - 322 = 5.8 \text{ F}$$

The arrangement is a square array of 400 tubes

There will be 20 tubes directly over each other i.e. $N = 20$

$$\bar{h} = 0.725\left(\frac{k_f^3 \rho_f^2 g \lambda}{N(\Delta t_o) D_o \mu_f}\right)^{\frac{1}{4}}$$

(with out splashing)

$$= 0.725\left[\frac{(0.393)^3 (56.45)^2 (32.2)(890)(3600)^2}{(20)(5.8)\frac{0.25}{12}(0.415)}\right]^{\frac{1}{4}}$$

$$\bar{h} = 2105 \text{ Btu/hr ft}^2 \text{ °F}$$

The total heat transfer is then

$$q = \bar{h} A_{total} \Delta t_o$$

$$= (400)\, \bar{h}\, \pi d_o\, L\, \Delta t_o = (400)\,(2105)\left(\frac{0.25}{12}\right)\pi\left(\frac{30}{12}\right)(5.8)$$

$$= 8 \times 10^5 \text{ Btu/hr}$$

The mass flow of condensate is obtained from

$$q = w\,\lambda,\; w = \frac{q}{\lambda} = \frac{8 \times 10^5}{890} = 901 \text{ lbs/hr}$$

Problem

A shell and a tube condenser with vertical 3/4″ 16 BWG copper tubes has chlorobenzene condensing at atmosphere pressure in the shell. The tubes are. 5′ long. Cooling water at an average temperature of 75°F is flowing in the tubes. If the water side coefficient is 400 Btu/ft^2 hr°F (a) What is the coefficient of the condensing chlorobenzene? (b) What would be the coefficient in a horizontal condenser with the same number of tubes if the average number of tubes in a vertical stack is six? Neglect fouling factors and tube-wall resistance.

Effect of Non-condensible Gases on the Rate of Condensation

The presence of even small amounts of non-condensing gas in a condensing vapour seriously reduces the rate of condensation. A layer of gas accumulates on the film of the condensate and establishes an added thermal resistance which is not present during the condensation of pure vapour. Additional resistance lies in the fact that the vapour of the condensing substance must flow by diffusion through the layer of gas to reach the film of condensate and the resistance offered by the gas to diffusion further impedes the process of condensation.

An important example of condensation in the presence of small amounts of fixed or non-condensible gas is condensation of steam containing air. The reduction in the value of h could be seen from the figure.

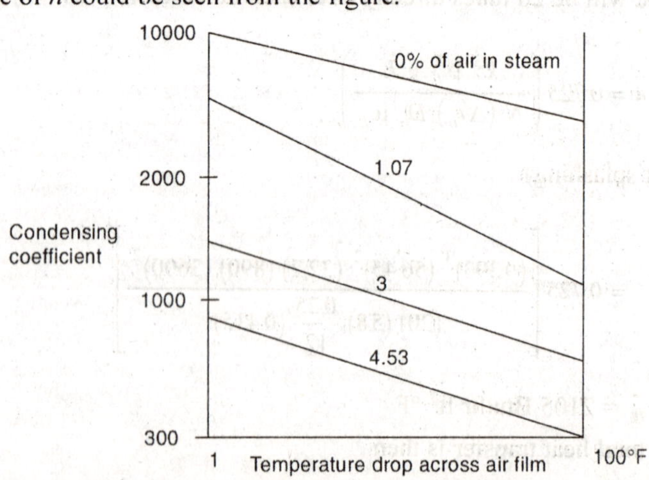

Fig. 5.20 *Effect of non-condensible gas on the value of 'h'*

For example, as little as 1.0 or 2 % non-condensable gases in a steam system may reduce the rate of transfer of heat by 75 to 80%. In practice, a steam condenser is operated with appropriate auxiliaries (steam traps or air vents) to eliminate air in the system.

Illustration on the Computation of '*h*'.

A condenser is fitted with 1 inch tubes 3′ long. If the condenser is mounted horizontally, the condensate will contact five tubes in its descent. If the tube surface temperature is 150°F and steam is condensing at 1 atm, recommend whether to mount the condenser vertically or horizontally. Also find out whether the flow of the condensate is laminar or turbulent in the case of vertical mounting. For horizontal mounting take into account the splashing effect.

Solution

Data: The fluid property data must be computed at t_f, according to the equation

$$t_f = t_h - \frac{3\,(t_h - t_w)}{4}$$

t_h = condensing vapour temperature

t_w = temperature of the outside surface of the tube wall

$t_f = 212 - 0.75\ (62)$

$= 165.5°F$

212 – 150 = 62 °F

$\mu_{165.5} = 0.37\ c_p \quad 0.37 \times 2.42 = 0.885$ lb/hr ft

$k_{165.5} = 0.38$ Btu/hr °F ft

$\rho_{165.5} = 61.0\ \text{lbs/ft}^3$

$\lambda = 970$ Btu/lb (steam tables)

For vertical position, the equation is

$$\bar{h} = 1.13\left[\frac{k_f^3\ \rho_f^2\ g\ \lambda}{\Delta t_o\ L\ \mu_f}\right]^{\frac{1}{4}}$$

L = heated length of the tube in feet

$$= 1.13\left[\frac{(0.383)^3\ (61.0)^2\ 32.2 \times 3600^2 \times 970}{(62)\ (0.37 \times 2.42)\ 3}\right]^{0.25}$$

$\bar{h} = 955$ Btu/hr °F sq/ft

The data can be tested for turbulence

We know

$$\bar{h} = \frac{w\,\lambda}{\pi\, D_o\, \Delta t_o\, L} = \frac{T\,\lambda}{L\,\Delta t_o}$$

or $$T = \frac{\bar{h}\, L\, \Delta t_o}{\lambda}$$

substituting the various values

$$T = \frac{(955)\,(3)\,(65)}{970}$$

$$= 178 \text{ lbs/ft hr}$$

$$N_{R_e} = \frac{4\,T}{\mu} = \frac{4 \times 178}{0.37 \times 2.42}$$

$$= 800$$

$$\frac{4\,T}{\mu_f} < 2100$$

For horizontally mounted condenser, the equation to be used is

$$\bar{h} = 0.725 \left(\frac{k_f^3\, \rho_f^2\, g\, \lambda}{N^{\frac{2}{3}}\, \Delta t_o\, D_o\, \mu_f} \right)^{\frac{1}{4}}$$

N = no. of tubes in the stack

$= 1045$ Btu/hr ft^2°F

$\bar{h}$ for horizontal mounting is larger than that for vertical mounting.

Hence horizontal mounting of the condenser is recommended.

Overall Coefficients in Condensers

Surface condenser—an equipment in which vapour is being condensed on the outer surface of a tube while a liquid, usually cooling water, flows in turbulent condition inside the tubes.

In majority of cases the tube temperatures have not been measured; hence only the overall coefficients of heat transfer are available.

$$q = U\,A\,(\Delta t) \qquad \frac{q}{A} = q' = \frac{\Delta t}{\frac{1}{U}} = \frac{\Delta t}{\Sigma R}$$

We know the overall resistance to the heat flow

$$\frac{1}{U_o} = \frac{D_o}{D_i\, h_i} + \frac{x_w}{k_m}\, \frac{D_o}{\overline{D}_L} + \frac{1}{h_o}$$

$$\Sigma R = \frac{1}{U_0} = \text{sum of the individual series resistance}$$

$$= R_v + R_w + R_d + R_L$$

R_v — Vapour side; R_w — Resistance of the tube wall; R_d — Dirt deposit; R_L — Water side

Assuming heat transfer area to the same the following approach is used

Wilson Plot

Wilson suggested a graphical analysis of the overall coefficients of heat transfer which facilitates the computation of the film coefficients for condensing vapours.

Let us consider that a series of runs made in condensing substantially air-free vapour at a given temperature, employing different water velocities

$$\text{We know } \frac{1}{U_o} = R_v + R_w + R_d + R_L$$

According to the theoretical equation of Nusselt, the resistance on the vapour side depends on the temperature difference and the temperature of the condensate film; therefore R_v should vary somewhat with the change in water velocity. The dirt resitance (scale on the heating surface) and the tube wall resistance also depend on their respective temperatures and upon the influence the temperature has on k_m. The thermal resistance of any scale on the heating surface might differ in various runs. However except where very high water velocities are used the water side resistance is usually the major resistance from condensing steam to water and under these conditions, serious error would not be introduced by assuming that the sum of the first three individual resistances $R_v + R_w + R_d$ is approximately constant. Wilson found in a study of steam condenser operations that the sum of the condensate, scale and tube-wall resistances remained substantially constant for varying cooling water rules.

As is well known, the water side resistance is an inverse function of the water velocity V through the tubes and neglecting the effects of changes in water temperature due to changes in water velocity, the water-side resistance R_L could be taken to be a function of the water velocity alone. As Wilson points out a plot of $\frac{1}{U}$ vs $\frac{1}{\phi(V')}$ should give a straight line when plotted to ordinary rectangular coordinates.

For turbulent flow of water in a given apparatus, $\phi(V')$ may be taken as $C_1 (V')^{0.8}$ and therefore the expression

$$\frac{1}{U_o} = R_v + R_w + R_d + R_L$$

can be rewritten as

$$\frac{1}{U_o} = R_v + R_w + R_d + \frac{1}{C_1 (V')^{0.8}}$$

where c_1 = empirical constant and may be considered as the apparent individual coefficient of heat transfer from tube to water based on the outside surface for water velocity of 1 ft/sec.

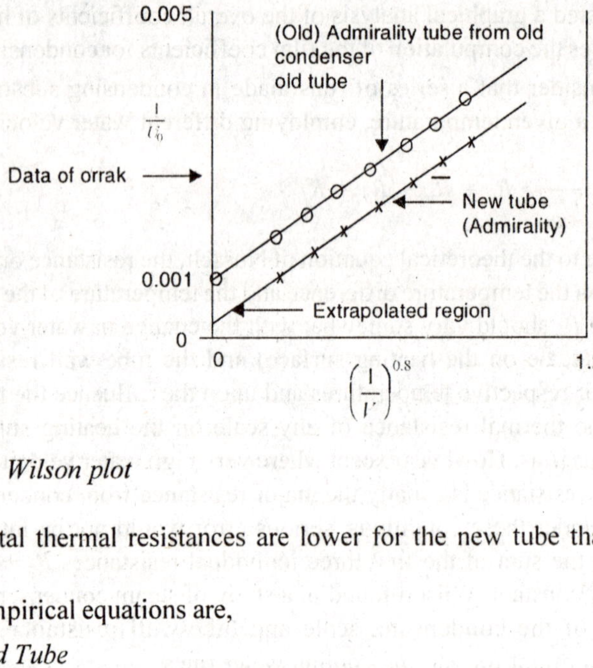

Fig. 5.21 *Wilson plot*

The total thermal resistances are lower for the new tube than for the old tube.

The empirical equations are,

For the Old Tube

$$\frac{1}{U_o} = 0.00092 + \frac{1}{268\,(V')^{0.8}}$$

For the Clean Tube

$$\frac{1}{U_o} = 0.00040 + \frac{1}{268\,(V')^{0.8}}$$

where U_0 is expressed as Btu/hr sq ft (on the steam side) °F over all temperature difference

V' = water velocity in the tubes in feet/sec

Use of Wilson's Plot to Evaluate h_0

The reciprocal slope c_1, for each set of tests was 268, indicating that the water-side coefficient at a water velocity of 1 ft/sec was268 Btu/hr sq ft (°F difference from tube to water)

The experimentation involved the use of tube of $D_0 = 1.00$ and $D_1 = 0.902''$ and hence the water-side coefficients per sq. ft of inside surface was

$$h_i = 268/0.902 = 297$$

$$x_w = 0.00408 \text{ ft}$$

and $\quad k_w = 63$ Btu/hr ft °F

for the metal tube in question

Therefore $R_w = 0.000068$

Now referring to the curve for the clean tube R_w may be deducted from the intercept of 0.00040 giving by difference ($R_d = 0$, because of clean tube)

$$R_v = 0.00040 - 0.00068$$

$$= 0.000332 = \frac{1}{3010}$$

That is, the vapour-side coefficient of heat transfer was 3010.

Thus by the aid of this graphical method, the observed over-all resistance has been subdivided into three component resistances

$$R_v = 0.000332 \quad R_w = 0.000068$$

$$R_L = \frac{1}{268}\frac{1}{(V')^{0.8}}$$

The same procedure for tubes with scale deposits yields the apparent coefficient of heat transfer through the scale as 1920.

Illustration

The effect of scale on heat flux was studied experimentally using a 2.3 cm *id* copper tube. Water flows inside the tube and steam condenses on the outside. Overall heat transfer coefficient U_0 in cals/sec cm^2°C was correlated by the following equation.

$$\frac{1}{U_o} = 2.21 + \frac{1}{0.56\, V^{0.8}} \text{ for clean tubes}$$

and
$$\frac{1}{U_o} = 5.90 + \frac{1}{0.56\, V^{0.8}} \text{ for fouled tubes}$$

where V is the velocity of water in the tube.. The wall thickness of the copper tube is 0.12 cm. Calculate the fouling coefficient assuming the steam film coefficient to be constant. If the thermal conductivity of copper at the average temperature of flow is 0.91 cals/sec cm° C, calculate the steam film coefficient

$$\frac{1}{U_o} = R_v + R_w + R_d + \frac{1}{0.56\,V^{0.8}} \text{ (fouled tube)}$$

$$\frac{1}{U_o} = R_v + R_w + R_d + \frac{1}{0.56\,V^{0.8}} \text{ for clean tube}$$

$$(R_d = 0) \quad \uparrow 0$$

i.e. $$= R_v + R_w + \frac{1}{0.56\,V^{0.8}}$$

The intercepts are given by the expressions in the problem statement, as
For dirty tube = 5.90
For clean tube = 2.21

$$R_d = 5.90 - 2.21 = 3.69$$

$$R_d = \frac{D_o}{D_i\,h_{di}} \qquad D_i = 2.30 \text{ cm}$$

$$D_o = 2.54 \text{ cm}$$

$$3.69 = R_d = \frac{2.54}{(2.30)\,h_{di}}$$

$$h_{di} = \frac{2.54}{(2.3)\,(3.69)}$$

$$R_w = \frac{x_w\,D_o}{k_w\,D_L}$$

R_w can be computed
Using the values of R_w and R_d thus obtained, the R_v value can be determined. From this the steam film coefficient value could be obtained.

Example in Condensation Heat Transfer

1. Saturated steam at the temperature 90°C is condensing on a single horizontal tube 2 cm in diameter. The surface temperature is 70°C. For this situation compute the average heat transfer film coefficient. The thermo-physical properties are given below at the average temperature of 80°C

$k = 0.674$ w/m °C

$\rho = 917.8$ kg/m^3

$\lambda = 2.283 \times 10^6$ J/kg

$\mu = 0.359 \times 10^{-3}$ kg/m s

Solution

We know $N_{N_u} = \dfrac{hD}{k} = 0.725\left(\dfrac{g\,\rho^2\,\lambda\,D^3}{\mu\,k\,\Delta T}\right)^{0.25}$

When this equation (Nusselt equation) is rearranged by solving for h and g is evaluated as 9.807 m/s² the following expression is got

$$h = 1.283\left(\frac{k^3\,\rho^2\,\lambda}{D\mu\,\Delta T}\right)^{0.25}$$

D = 2 cm = 0.020 m

ΔT = 20, i.e. (90 – 70)

The given values of the thermo-physical properties may now be substituted into the above expression

$$h = 1.283\left[\frac{(0.674)^3\,(971.8)^2\,\left(2.283\times10^6\right)}{(0.020)\left(0.359\times10^{-3}\right)20}\right]^{0.25}$$

$$= 1.283\left[\frac{(0.3021)\left(0.9444\times10^6\right)\left(2.283\times10^6\right)}{0.1436\times10^{-3}}\right]^{0.25}$$

$$= 1.283\left(0.4536\times10^{16}\right)^{0.25}$$

$$= 1.283\times0.821\times10^4$$

$$= 10.530 \text{ w/m}^2\ ^\circ\text{C}$$

2. **If steam were condensing under the conditions of the previous example but on a vetical array of eight tubes in a vertical column, what could be the magnitude of the average heat transfer coefficient?**

Solution

The thermo-physical properties are the same as of the preceding illustration.

In this situation we have a bank of horizontal tubes, the equivalent tube diameter (D_e) has to be substituted for the single tube diameter (D)

The equivalent tube diameter for computing the average heat transfer coefficient is the sum of the outside tube diameters of all the tubes in a vertical column of the tube pattern

$D_e = n\,D_T$

D_e = the equivalent tube diameter of the tube bank (m)

n = the number of tubes in a vertical column of the tube bank pattern

D_T = the diameter of a single tube in the bank (m)

The expression to be employed is

$$h = 1.283\left(\frac{k^3\,\rho^2\,\lambda}{D_e\,\mu\,\Delta T}\right) \text{ w/m}^2\ ^\circ\text{C}$$

$n = 8 \quad D_T = 0.020$ m

$D_e = (8)\,(0.02) = 0.16$ m

These quantities on substitution into the above expression along with the thermo-physical properties, yield the value of 'h'.

$$h = 1.283\left[\frac{(0.674)^3\,(971.8)^2\,\left(2.283 \times 10^6\right)}{(0.160)\left(0.359 \times 10^{-3}\right) 20}\right]^{0.25}$$

$$= 1.283\left[\frac{(0.3021)\left(0.9444 \times 10^6\right)\left(2.283 \times 10^6\right)}{1.149 \times 10^{-3}}\right]^{0.25}$$

$$= (1.283)\left(0.05669 \times 10^{16}\right)^{0.25}$$

$$= (1.283)\,(0.488 \times 10^4) = 6260 \text{ w/m}^2\ ^\circ\text{C}$$

3. If steam were condensing under conditions of the previous example but on a vertical flat plate 1.5 m high, what would be the magnitude of the average heat transfer film coefficient? The thermo-physical values needed in this computation are as given in the previous example.

Solution

We know $N_{N_u} = \frac{h L}{k} = 0.943\left(\frac{g\,\rho^2\,\lambda\,L^3}{\mu\,k\,\Delta T}\right)^{0.25}$ from Nusselt equation.

When the Nusselt equation is rearranged by solving for h and g is evaluated as 9.807 m/sec² the following expression results

$$h = 1.669\left(\frac{k^3\,\rho^2\,\lambda}{L\,\mu\,\Delta T}\right)^{0.25}$$

in which L = height of the vertical surface (m).

It may be noted that the numerical value in the above expression, 1.669 is the product of the dimensionless coefficient 0.943 and the acceleration of gravity (g) to the 0.25 power.

The thermo-physical properties can be substituted in the above expression

$$h = 1.669 \left[\frac{(0.674)^3 \, (971.8)^3 \, \left(2.283 \times 10^6\right)}{(1.5) \left(0.359 \times 10^{-3}\right) (20)} \right]^{0.25}$$

$$= 1.669 \left[\frac{(0.3021) \left(0.9444 \times 10^6\right) \left(2.283 \times 10^6\right)}{10.77 \times 10^{-3}} \right]^{0.25}$$

$$= 1.669 \left(60.48 \times 10^{12}\right)^{0.25}$$

$$= (1.669) \left(2.789 \times 10^3\right)$$

$$= 4655 \text{ w/m}^2 \text{ °C}$$

Examples in Heat Transfer with Change of Phase

Determine the heat transfer coefficient for steam at 120°C condensing on a vertical wall which is maintained at a constant temperature of 100°C. The length of the condensing wall is 100 mm. At the average temperature of 110°C $\left[\frac{(120 + 100)}{2}\right]$ the physical properties of saturated water are as under

$\rho = 951$ kg/m^3

$\mu = 26.4 \times 10^{-6}$ kg sec/m^2

$k = 58.9 \times 10^{-2}$ k cal/hr m °C

the latent heat of condensation of steam (λ) at 120°C is 526.1 k cal/kg.

Solution

The Nusselt number is computed from the expression

$$N_{N_u} = 0.943 \left\{ \frac{g \, \rho^2 \, \lambda \, L^2}{k \, \mu \, \Delta t} \right\}^{\frac{1}{4}}$$

$$= 0.943 \left\{ \frac{\left(127 \times 10^6\right) (951)^2 \, (526.1) \, (0.1)^3}{\left(26.4 \times 10^{-6} \times 9.81 \times 3600\right) \left(58.9 \times 10^{-2}\right) 20} \right\}^{\frac{1}{4}}$$

[$\because \mu = 26.4 \times 10^{-6}$ kg sec/m^2 $= 26.4 \times 10^{-6} \times 9.81 \times 3600$ kg/m hr and $g = 127 \times 10^6$ m^2/hr]

$$N_{N_u} = 0.943 \left(5.51 \times 10^{12}\right)^{\frac{1}{4}}$$

$= 1445$

hence the heat transfer coefficient $h = N_{N_u} \dfrac{k}{L}$

$$h = \frac{\left(1445 \times 58.9 \times 10^{-2}\right)}{0.1}$$

$= 8510$ k cal/hr m^2 °C

It may be noted that the magnitude of heat transfer coefficient in condensing vapours is comparatively larger that that of free convection and forced convection at low fluid velocities. In the above example, if the transfer area is 1 m^2, then the average heat flux 'q' is given by

$$q = h\left(t_f - t_\infty\right) = 8510 \times 20$$

$= 170{,}200$ k cal/hr m^2

and the quantity of water condensed per unit area is

$$m = \frac{q}{\lambda} = \frac{170{,}200}{526.1}$$

$= 324$ kg/hr m^2

The area of the condensing surface can thus be enhanced to the extent depending upon the quantity of vapour required to be condensed.

Illustration on Heat Transfer with Change in Phase

A ½ in *OD*, 5 ft long tube is to be employed to condense steam at 6 psia. Estimate the heat transfer coefficients for this tube in the (a) horizontal and (b) vertical positions. Assume that the average tube-wall temperature is 130°F. At the average temperature of the condensate film $\left(T_f = \dfrac{(170 + 130)}{2} = 150\ °\text{F}\right)$, the physical property values relevant to this problem are

$k = 0.383$ Btu/hr ft °F

$\rho = 61.2$ lbs/c ft

$\lambda = 996.3$ Btu/lb (from steam tables)

$\mu_f = 1.06$ lb/hr ft

$T_{s_v} = 170$°F

Solution

For the tubes in horizontal position the equation is

$$h = 0.725\left[\frac{\rho_l\,(\rho_l - \rho_v)\,g\,\lambda}{D\,\mu_l\left(T_{s_v} - t_s\right)}\right]^{0.25}$$

where T_{s_v} temperature of saturated vapour, °F

T_s = wall surface temperature °F subscript l = denotes liquid

sv = saturated vapour

$$h = 0.725\left[\frac{(0.383)^3\,(61.2)^2\left(4.17\times10^8\right)(996.3)}{\left(\frac{0.5}{12}\right)(1.06)\,(170-130)}\right]^{0.25}$$

= 1920 Btu/hr ft² F

(b) In the vertical position the tube may be treated as a vertical plate of area $\pi\,D\,L$ and according to the following equation, h is given by

$$h = 0.943\left[\frac{\rho_l\,(\rho_l - \rho_v)\,\lambda\,k^3}{\mu_l\,L\left(T_{s_v} - T_s\right)}\right]^{0.25}$$

$$h = 0.94\left[\frac{(61.2)^2\left(4.17\times10^8\right)(996.3)\,(0.383)^3}{(1.06)\,(5)\,(170-130)}\right]^{0.25}$$

= 730 Btu/hr ft² °F

Illustration on Heat Transfer in Condensing Vapours

Determine the heat transfer coefficient for steam at 142°F condensing on the outside of a horizontal 1-m diameter cylinder at 138°F. The latent heat of condensation and other physical properties of steam at the average temperature of 140°F are

ρ = 61.4 lb/ft³

μ = 1.14 lb/ft hr

k = 0.376 Btu/hr ft °F

λ = 1014 Btu/lb

Solution

The Nusselt equation, for condensation on horizontal tubes, is

$$N_{N_u} = \frac{hL}{k} = 0.725\left(\frac{g\,\rho^2\,\lambda\,L}{k\,\mu\,\Delta t}\right)^{0.25}$$

substituting the given values for this situation, we obtain

$$h = 0.725\left[\frac{(418)\,10^6\,(1.90)^2\,(32{,}600)\,(0.376)^3}{\left(\frac{1}{12}\right)(0.0354)\,(142-138)}\right]^{0.25}$$

$$= 0.725\ (3856)$$

$$= 2796 \text{ Btu/hr ft}^2\ {}^\circ\text{F}$$

Problems on Heat Transfer with Phase Change

1. Calculate the average heat flux from a 12 mm diameter tube to water boiling outside at 100°C. The tube is immersed in water and is held at 500°C by electrical heating.

2. A copper container with a flat bottom is used to boil water at 100°C. The temperature of the bottom of the container in 115°C. Calculate the heat transfer coefficient.

3. Freon-12 at 60°C condenses on the outer surface of a vertical tube 90 cm long and 2.5 cm in diameter held at 55°C. Calculate the rate of condenstion per hour assuming the following properties for Freon-12.

$\rho = 1280$ kg/m^3

$\mu = 0.9$ kg/m hr

λ (or) $L = 32$ k cal/kg

$k = 0.072$ k cal/hr m °C

4. A condenser is required to handle 10000 kg/hr of steam at 0.1 kg/cm^2 pressure. The heat transfer surface is made of horizontal brass tubes 3 cm ID, 1 mm thick and 3 m long. 400,000 kg/hr of cooling water at 25°C circulates through the tubes with a velocity of 21 m/sec. How many tubes are required in the condenser?

5. A horizontal tube 15 mm O.D. is exposed to saturated steam at 100°C. The outer tube wall is maintained at 99°C by passing a liquid metal through the tube. Calculate (i) the condensing coefficient and (ii) the mass of steam condensed per meter of tube length per minute.

6. Calculate '*h*' between a horizontal heating surface and boiling water if the heat flux is 30,000 k cal/hr m^2 and the pressure is 15 atm.

5.5 BOILING OR EVAPORATION HEAT TRANSFER (HEAT TRANSFER TO BOILING LIQUIDS)

When a surface is exposed to a liquid and is maintained at a temperature above the saturation temperature of the liquid, boiling may occur and the heat flux will depend on the difference in temperature between heat exchange surface and the saturation temperature. There are a few types of boiling. They are

Pool Boiling

This process of boiling is encountered when the heated surface is submerged below a free surface of the liquid.

Sub-cooled or Local Boiling

If the temperature of the liquid is below the saturation temperature, the process is called sub-cooled or local boiling. That is the bulk liquid temperature is below the saturation value. In this phenomenon, a liquid below its saturation temperature is in contact with a surface hotter than the saturation temperature of the liquid. Very high heat transfer rates are obtained as a result of the local generation and collapse of the bubbles on the surface.

Saturated or Bulk Boiling

If the liquid is maintained at the saturation temperature, the process is known as saturation level or saturated boiling.

Regimes of Boiling

Consider for example the boiling water on an electrically heated horizontal platinum wire. The heat flux q/A may be computed from the measured input of electrical energy and the measured surface area of the wire. The temperature difference Δt between the surface temperature t_w of the wire and the saturation temperature t_{sat} of the liquid may be obtained by the techniques of resistance thermometry with suitable correction for radial gradients in temperature within the wire.

The resulting graph of $\frac{q}{A}$ vs Δt is shown in the following sketch for water boiling at 1 atm. It is similar in shape to the boiling curves which are reported for other fluids and other pressures.

$A–B$ = Natural convection

$B–C$ = Nucleate boiling

$C–D$ = Partial film boiling

DEF = Film boiling

The temperature drop corresponding to point c is referred to as critical temperature drop

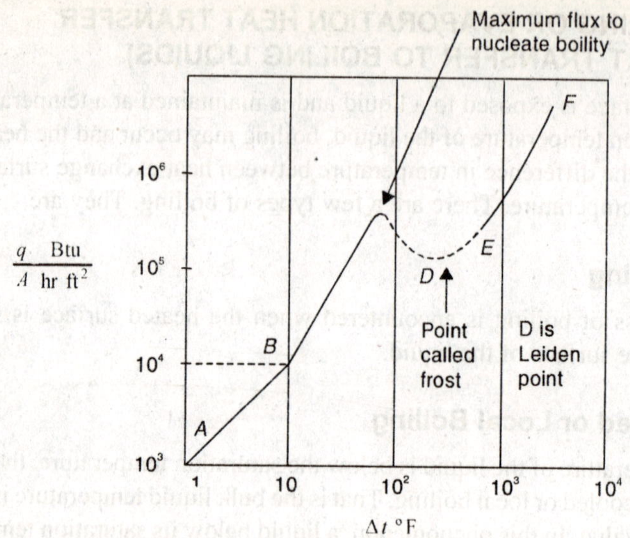

Fig. 5.22 *Plot of q/A versus* Δt *for water boiling at 1 atm pressure*

Since by definition $h = \frac{\left(\frac{q}{A}\right)}{\Delta t}$, the above plot is readily convertible into a plot of h vs Δt .

This plot is shown below. the different regimes are also indicates on the graph.

Region I

Free-convection currents are responsible for the motion of the fluid near the surface. In this region the liquid near the heated surface is superheated slightly and its subsequently evaporates when it rises to the surface.

Region II

Bubbles begin to form on the surface of the wire and are dissipated in the liquid after breaking away from the surface. This region indicates the beginning of nucleate boiling.

Region III

As the temperature excess is increased further, bubbles form more rapidly and rise to the surface of the liquid, where they are dissiputed.

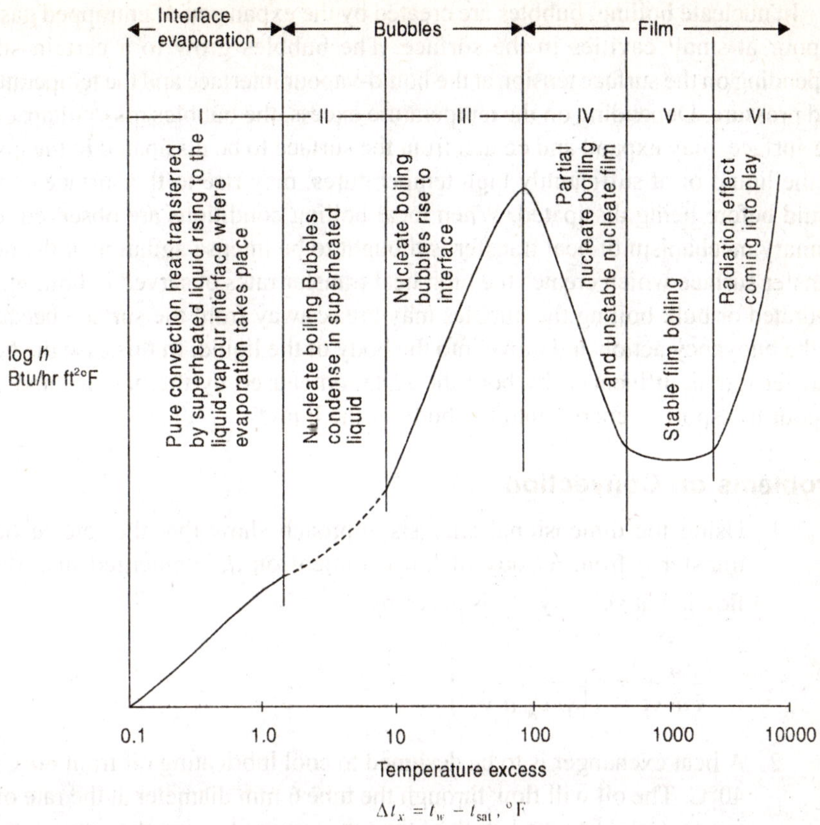

Fig. 5.23 *Heat flux data from an electrically heated platinum wire*

Region IV

Eventually bubbles are formed so rapidly that they blanket the heating surface and prevent the inflow of fresh liquid from taking their places. At this point the bubble coalesce and form vapour film which covers the surface. The heat must be conducted through the film before it can reach the liquid and effect the boiling process. The thermal resistance of this film causes a reduction in heat flux and this phenomenon is illustrated in region IV, the film-boiling region.

This region represents a transition from nucleate boiling to film boiling and is unstable. Stable film boiling is eventually encountered in region V.

The surface temperature required to maintain stable film boiling are high and once this condition is attained, a significant portion of heat lost by the surface may be the result of thermal radiation, as shown in region VI.

The temperature of the wire corresponding to point B usually exceeds the melting point of the wire resulting in the burnout of the wire.

In nucleate boiling, bubbles are created by the expansion of entrapped gas or vapour at small cavities in the surface. The bubbles grow to a certain size, depending on the surface tension at the liquid-vapour interface and the temperature and pressure. Depending on the temperature excess, the bubbles may collapse on the surface, may expand and detach from the surface to be dissipated in the body of the liquid or at sufficiently high temperatures, may rise to the surface of the liquid before being dissipated. When local boiling conditions are observed, the primary mechanism of heat transfer is thought to be intense agitation at the heat transfer surface which creates the high heat transfer rates observed in boiling. In saturated or bulk boiling the bubbles may break away from the surface because of the buoyancy action and move into the body of the liquid. In this case the heat transfer rate is influenced by both the agitation caused by the bubbles and the vapour transport of energy into the body of the liquid.

Problems on Convection

1. Using the dimensional analysis approach show that the rate of heat transfer q from a body of linear dimension d, submerged in a fluid flowing at velocity V_∞ is given by

$$\frac{q}{k\,d\,\Delta t} = c\left(\frac{k}{\rho\, c_p\, d\, v_\infty}\right)^n$$

2. A heat exchanger is to be designed to cool lubricating oil from 60°C to 40°C. The oil will flow through the tube 6 mm diameter at the rate of 1 m/sec. Calculate the length of the tube required, using the properties of oil which are given below.

   ρ = 800 kg/m^3 k = 0.12 k cal/m hr°C

   μ = 12 centipoise; c_p = 0.45 k cal/kg°C

3. 300 kg.hr of water is to be heated from 30°C to 45°C while flowing through a horizontal 2 cm diameter steel pipe. If the pipe is jacketed with steam condensing at 100°C what length of pipe is required. Hint The flow rate will be found to be laminar in pipe. calculate LMTD.

4. In a counter-flow heat exchanger lubricating oil is cooled from 40°C to 30°C by water entering at 15°C. The velocity of oil is 0.6 m/sec and it flows through the inside tube of 4.375 cm ID and 5.0 cm OD. If the film coefficient between the outside tube surface and water is $450\ \frac{\text{k cal}}{\text{hr m}^2\,°\text{C}}$ find the flow rate of oil and water for 9 m^2 of inside tube surface. The properties of oil are given below

   c_p = 0.46 k cal/kg°C

   μ = 81.5 kg/hr m

   k = 0.114 k cal/hr m C

$\rho = 896 \text{ kg/m}^3$

5. In a double pipe heat exchanger, liquid sodium flows through the inner pipe at the rate of 910 kg/hr while mercury flows in the annulus at the rate of 18,200 kg/hr. Sodium enters at 500°C and leaves at 300°C. Mercury enters at 50°C and leaves at 150°C. Determine the length of the exchanger. The dimensions of the pipe are

	ID	*OD*
Inner pipe	2.060 cm	2.625 cm
Outer pipe	4.025 cm	4.750 cm

The following average properties for mercury and sodium may be used.

	Mercury	*Sodium*
$\rho\ (\text{kg/m}^3)$	13,344	832
k (k cal/hr m°C)	8.7	63
c_p (k cal/kg°C)	0.032	0.32
μ (kg/hr m)	4.5	1.08

6. A steam pipe 8.75 cm OD is exposed to air. Wind at a speed of 50 kg/hr blows squarely on the pipe. The surface temperature of the pipe is 96°C and the air temperature is 10°C calculate the heat loss per m length of the pipe.
7. Calculate Grashof number with the following data

$d = 25 \text{ mm};\ t_\infty = 150°\text{C};\ t_w = 100°\text{C}$

$\nu = 2.31 \times 10^{-5} \text{ m}^2/\text{sec};\ N_{pr} = 0.69$

8. A horizontal cylinder 2.5 cm in diameter and 20 cm in length is suspended in still air which is at a temperature of 20°C. Calculate the rate of heat transfer from the cylinder surface if its temperature is 60° C.
9. A transformer is to be cooled by free convection. The sheet is 30 cm tall and 30 cm × 4 cm in cross section. The surface temperature is 95°C. The ambient air temperature is 40°C. Calculate the heat dissipation.
10. An enclosed vertical square layer of air 60 cm × 60 cm × 10 cm thick is heated from below at a rate of 50 k cal/hr while its upper surface is kept at 25° C by means of a cooling device. Calculate the temperature of bottom.
11. A horizontal cylinder 5 cm in diameter is placed in a fluid for which k = 0.024 k cal/hr m°C. If the Grashof and Prandtl numbers are 10,000 and 10 respectively, find the heat loss by natural convection per m length of the cylinder. The temperature difference between cylinder surface and the fluid or 70°C.
12. The temperature of the surface of a horizontal cylinder in quiet air is 250°C. The air temperature is 50°C. What should be the diameter of the cylinder if the Grashof number is 2×10^8.

6

Industrial Heat Exchange Equipment

Categorization of the Basic Types of Heat Exchanges

There are several ways of classifying an industrial heat exchanger designed for a particular duty. The method of classification presented below attempts to include all the known and possible methods of classification.

A heat exchanger may be classified according to:

1. The nature of heat transfer process
   (a) Recuperative heat exchanger.
   (b) Regenerative heat exchanger: Again with (b) the solid matrix may either be stationary or slowly rotating type
2. The terminal temperature difference of the fluids
   (a) Contact or mixing type (direct type)
   (b) Closed or non-mixing type
3. The nature of flow arrangement of the fluids (a) parallel (b) counter flow (c) cross flow (d) mixed flow
4. The number of passes that the heat transfer fluid makes
   (a) Single pass (b) Multi-pass units
5. The mechanical design of the heat exchanger surface
   (a) Shell and tube type
   (b) Plate type (c) extended surface with fins
   (c) Double pipe type
6. The physical state of heat exchanging fluids
   (a) Gas to gas (direct, e.g. air cooled i.c. engines)
   (b) Liquid to liquid (oil coolers)
   (c) Gas to liquid (economisers)

(d) Gas to a coupling liquid and then to gas (water cooled i.c. engines)
(e) Liquid or gas to condensing vapour (condensers)
(f) Gas to a boiling liquid (evaporators)

7. The operating pressure range
   (a) Low pressure units (0 to 8 atm)
   (b) Medium pressure units (8 to 35 atm)
   (c) High pressure units (35 atm and above)

Industrial Heat—Exchange Equipment

The catagorization can also be done as follows.

There are many varieties. One characteristic common to most of the heat exchangers is the transfer of heat from a hot phase to a cold phase with the two phases separated by a solid boundary. Other types of heat exchangers exist in which the fluids mix. These include cooling towers and jet condensers.

6.1 TYPES OF HEAT EXCHANGERS AND DEFINITIONS

Two basic type are there:

- In line or unidirectional flow exchangers
- Cross flow exchangers

Flow is along the same axis in the in-line exchangers but the two fluids may flow in the same or opposite directions giving rise to the names parallel and counter flow.

The in-line exchanger may consist simply of two concentric tubes with one fluid flowing in the inner tube and the other in the annulus. Alternatively, there may be a number of tubes within a large tube or shell and to increase the heat transfer, the shell fluid is made to flow partly across the tubes by means of baffles. Counter and parallel flows also occur in plate heat exchangers in which the fluids flow between closely spaced plates sealed at the edges.

The following sketch shows some simple in-line arrangements.

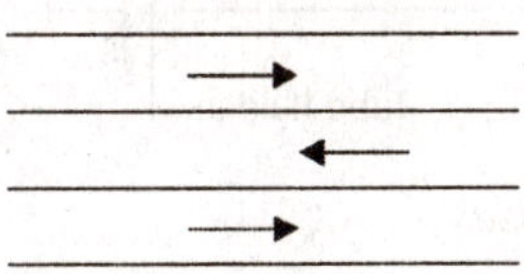

(a) Counter flow concentric tube exchanger

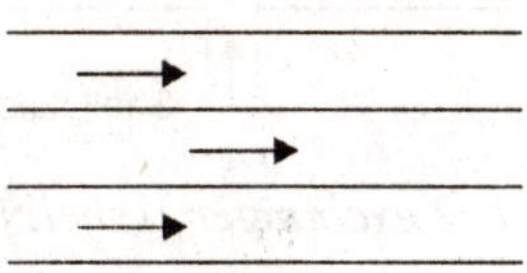

(b) Parallel flow concentric tube exchanger

Fig. 6.1 *Counter and parallel concentric tube exchanger*

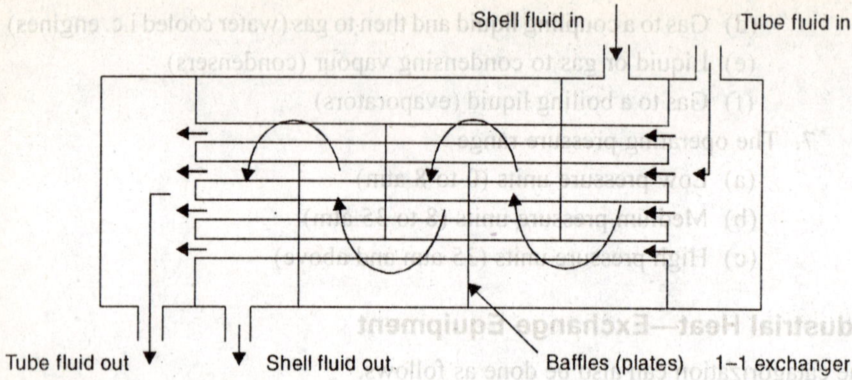

Fig. 6.1 (contd.) *Shell and tube heat exchanger in parallel flow (1 shell pass and 1 tupe pass)*

Tube and shell exchanger in parallel flow tube fluid may be reversed to give rise to counter flow—1 shell pass and 1 tube pass exchanger (Fig. 6.1).

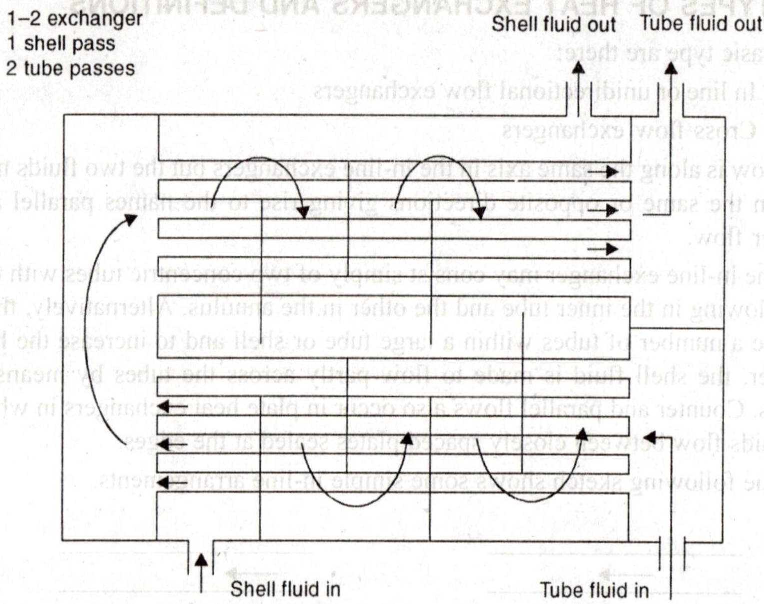

Fig. 6.2 *1–2 exchanger: 1shell pass, 2 tube passes*

An 1–2 heat exchanger implies that there are two tube passes and one shell pass. Flow is partly counter and partly parallel. (Fig. 6.2).

A heat exchanger having two shell passes and four tube passes is called a 2–4 exchanger.

The cross flow exchanger is, as its name implies, one in which the two fluids streams flow at right angles. Gas to gas heat exchangers are often of this type.

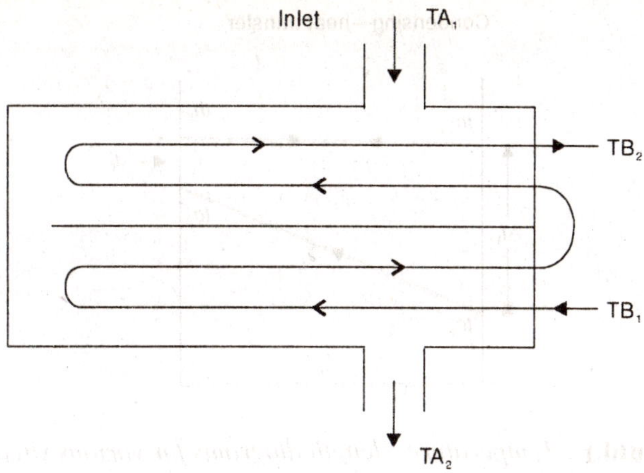

Fig. 6.3 *2–4 heat exchanger.—2shell passes, four tube passes*

The temperature variations of the fluids parallel and counter flow are shown in the flowing figures. Temperatures are plotted against length of the heat exchanger surface.

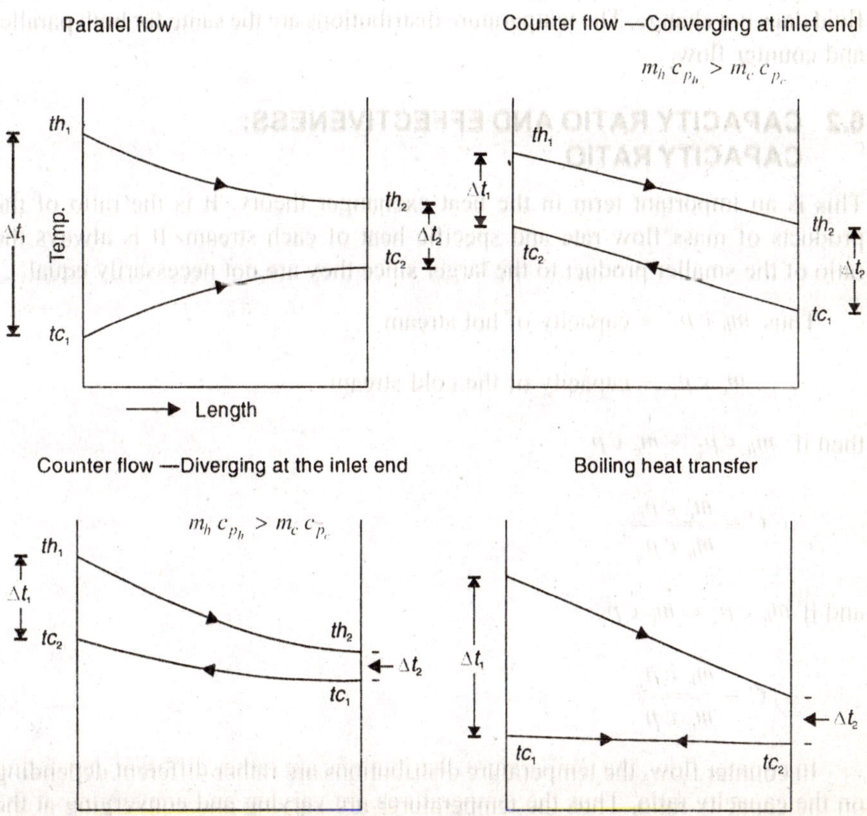

Fig. 6.4 *Temperature—length diagrams for various situations*

Condensing—heat transfer

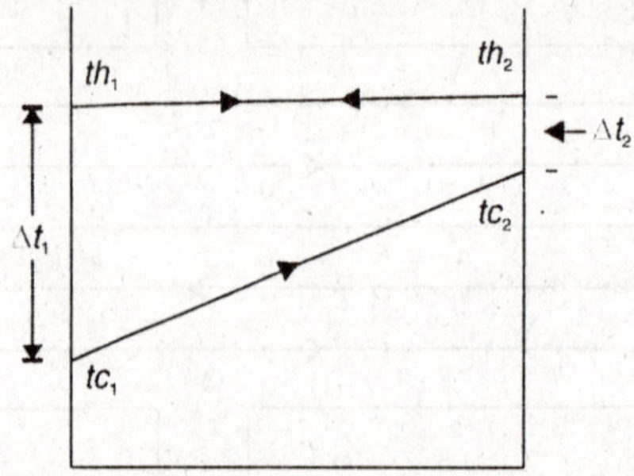

Fig. 6.4 (contd.) *Temperature—length diagrams for various situations*

Suffixes 1 and 2 denote inlet and outlet of the individual streams.

Temperature distributions with a change of phase are also shown in the above figures. These will occur in boiling and condensing. Only the phase change takes place in the heat exchanger so the temperature of the boiling or condensing fluid does not change. The temperature distributions are the same for both parallel and counter flow.

6.2 CAPACITY RATIO AND EFFECTIVENESS: CAPACITY RATIO

This is an important term in the heat exchanger theory. It is the ratio of the products of mass flow rate and specific heat of each stream. It is always the ratio of the smaller product to the larger since they are not necessarily equal.

Thus $m_h\ cp_h$ = capacity of hot stream

$m_c\ cp_c$ = capacity of the cold stream

then if $m_h\ cp_h > m_c\ cp_c$

$$C = \frac{m_c\ cp_c}{m_h\ cp_h}$$

and if $m_h\ cp_h < m_c\ cp_c$

$$C = \frac{m_h\ cp_h}{m_c\ cp_c}$$

In counter flow, the temperature distributions are rather different depending on the capacity ratio. Thus the temperatures are varying and converging at the inlet end when $m_h\ cp_h > m_c\ cp_c$ and the temperatures are diverging at the inlet end when $m_c\ cp_c > m_h\ cp_h$.

The appropriate capacity ratio equation has to be used.

In parallel flow it is obvious that tc_2 will approach th_2 for an infinitely long heat exchanger, but can never exceed th_2.

In counter flow, it is quite normal for tc_2 to exceed th_2 and consequently the counter flow exchanger is the more effective unit.

Effectiveness

It is the ratio of energy actually transferred to the maximum theoretically possible. Again the definition depends on the relative thermal capacities of the streams. The maximum theoretical transfer will take place in counter flow, in an exchanger of infinite length and in such a case $tc_2 \rightarrow th_1$ when $m_h\ c\,p_h > m_c\ c\,p_c$ and $th_2 \rightarrow tc_1$ when $m_h\ c\,p_h < m_c\ c\,p_c$.

To determine the maximum possible heat transfer for the heat exchanger, we first recognise that this maximum possible value could be attained if one of the fluids were to undergo a temperature change equal to the maximum temperature difference present in the heat exchanger $(th_1 - tc_1)$. The fluid which might undergo this maximum difference is the one having the minimum value of mc in accordance with the energy balance.

Thus the maximum transfers in the two cases are

$$q_{max} = (mc)_{min}\left(Th_{inlet} - Tc_{inlet}\right)$$

$$m_c\ c\,p_c\ (th_1 - tc_1)$$

when $m_h\ c\,p_h > m_c\ c\,p_c$ and $m_h\ c\,p_h\ (th_1 - tc_1)$ when $m_h\ c\,p_h < m_c\ c\,p_c$

The actual transfers in the two cases are $m_c\ c\,p_c\ (tc_2 - tc_1)$ and $m_h\ c\,p_h\ (th_1 - th_2)$ and hence E, the effectiveness becomes

For Parallel Flow

$$E_c = \frac{m_c\ c\,p_c\ (tc_2 - tc_1)}{m_c\ c\,p_c\ (th_1 - tc_1)} \quad \text{when } m_h\ c\,p_h > m_c\ c\,p_c$$

and $E_h = \dfrac{m_h\ c\,p_h\ (th_1 - th_2)}{m_h\ c\,p_h\ (th_1 - tc_1)}$ when $m_h\ c\,p_h < m_c\ c\,p_c$; these definitions may be used in either counter or parallel flow, but the value of E will be lower in parallel flow.

For Counter Flow

$$E_c = \frac{tc_2 - tc_1}{th_1 - tc_1} \text{ and } E_h = \frac{th_1 - th_2}{th_1 - tc_1}$$

the effectiveness equations may be used condensing and boiling situations.

For Condensing

$$E = \frac{tc_2 - tc_1}{th_1 - tc_1} \quad (m_h\, c\, p_h > m_c\, c\, p_c)$$

For Boiling

$$E = \frac{th_1 - th_2}{th_1 - tc_1} \,(m_h\, c\, p_h < m_c\, c\, p_c)$$

Problem

0.2 kg/sec of an alcohol is to be cooled from 75°C to 35°C in a counter flow heat exchanger. Cooling water enters the exchanger at 12°C and at the rate of 0.16 kg/sec. The c_p for alcohol is 2.52 kj/kg °k and for water 4.187 kj/kg °k. Calculate the capacity ratio and effectiveness.

For the hot stream i.e. alcohol

$m_h\, c\, p_h = 0.2 \times 2.52 = 0.504$ KJ/sec °K for the cold stream.

$m_c\, c\, p_c = 0.16 \times 4.187 = 0.671$ KJ/sec °K

from the equation $C = \dfrac{m_h\, c\, p_h}{m_c\, c\, p_c} \quad (m_h\, c\, p_h < m_c\, c\, p_c)$

$$= \frac{0.504}{0.671} = 0.75$$

To determine E,

$$E = \frac{th_1 - th_2}{th_1 - tc_1} \text{ if } m_h\, cp_h < m_c\, cp_c$$

$$= \frac{75 - 35}{75 - 12} = 0.635$$

6.3 DETERMINATION OF HEAT EXCHANGER PERFORMANCE

The primary purpose of a heat exchanger is to achieve the required heat transfer rate using the smallest possible transfer area and the fluid pressure drop. A large exchanger can mean unnecessary capital outlay and high pressure drop means reduced efficiency of the plant considered overall. Generally a smaller exchanger can be made by finning surfaces to increase the overall heat transfer coefficient.

However this leads to a higher fluid pressure drop and the best design is often a compromise between conflicting requirements. In fact a number of different designs for a given duty may be acceptable.

Double-pipe Heat Exchanger

This is the simplest type of a heat exchanger. This consists of two concentric pipes with one fluid flowing through the inner centre pipe while the other fluid moves co-currently or counter currently in the annular space. The length of each section is usually limited to standard pipe lengths so that if an appreciable heat transfer surface is required banks of sections are frequently used. If the required area is too large, a double pipe heat exchanger is not recommended.

This type of an exchanger can be used both for liquid-liquid and gas-liquid exchangers and also for gas to gas exchangers.

Either fluid may be sent through the tube or annulus at relatively high velocities, thereby aiding in the heat transfer process.

Shell and Tube Exchangers

When the required heat transfer surface is large, this type is preferred.

Large heat transfer surface can be achieved economically and practically by placing tubes in a bundle; the ends of the tubes are mounted in a tube sheet which is done by expanding the end of the tube into a close-fitting hole in the tube sheet. The tube bundle is then enclosed by a cylindrical casing (the shell) through which the second fluid flows around the tube bundle.

Parallel flow through all tubes at a low velocity gives a low heat transfer coefficient and a low pressure drop. For higher rates of heat transfer, multi pass operation may be employed. In this type of construction the fluid in the tubes is diverted by baffles, within the distribution header. The liquid passes back and forth through some fraction of the tubes at high velocities which give good heat transfer coefficient.

The number of tube passes employed depends on the economics of the design and operation and upon the space available. Complexity in design must be balanced against improved performance. Another disadvantage of the multi pass exchanger is the added frictional loss due to the higher linear velocities and entrance and exit losses in the headers.

Baffles are provided within the shell to divert the flow of the shell fluid into a path predominantly across the tubes of the exchanger. The constantly changing velocity of the shell fluid tends to impart turbulence to the stream, thereby improving heat transfer. Baffles consist of semicircular discs of sheet metal, pierced to accommodate the tubes. The baffles direct as much flow as possible normal to the tubes in the shell side as well as serve to keep the tubes from sagging.

Multi pass on the shell side may cause complexities in construction and higher friction losses-they are employed only in larger installations.

The difficulty with this type of construction is that the tube bundle cannot be removed for cleaning and no provision is made for differential expansion between the tubes and the shell (which will be encountered since large temperature difference between the two fluids exchanging heat are likely to exist).

In order to make the removal of the tube bundle possible and to allow for considerable expansion of the tube a floating heat exchanger is used. Here, one of the tube sheets is independent of the shell so that the entire tube bundle can be removed for cleaning of the shell and the outside of the tubes.

Selection of the type of shell and tube heat exchanger will depend on the factors such as:

Cost	Corrosion
Cleanabilty	Operating pressure
Temperatures	Pressure drop
and hazards	

Large-sized *h, e*'s are designed for a particular application while small units may be designed for multipurpose uses.

A fixed tube sheet is the cheapest.

The nature of shell side fluid will influence the type of heat exchanger.

Least corrosive and cleanest fluid should be placed in the shell; corrosive fluid should be placed in the tubes to effect savings on the cost of an expensive alloy shell.

Extended Surface Heat Exchanger

Finned Tube Units

When viscous liquids are heated in a concentric tube or tubular exchanger by condensing steam or by a hot fluid/liquid of low viscosity, the film coefficient for the viscous fluid will be much smaller than that on the other side and will hence control the rate of heat transfer. Examples are heating of oils in storage or in pipes or in air heating.

The relative magnitude of '*h*' is about 10 for the oil or air compared with 2000 for the steam. In order to compensate for the high resistance of oil or air, the heat exchanger (heat transfer surface) exposed to these fluids may be increased by extension of the surface as in the addition of fins to the outside of the tubes. The fins increase transfer area substantially in a given amount of space.

Fins may be longitudinal, i.e. along the length of the tube: or Fins may be traverse, i.e. radial.

6.4 THE LOG MEAN TEMPERATURE DIFFERENCE (LMTD)

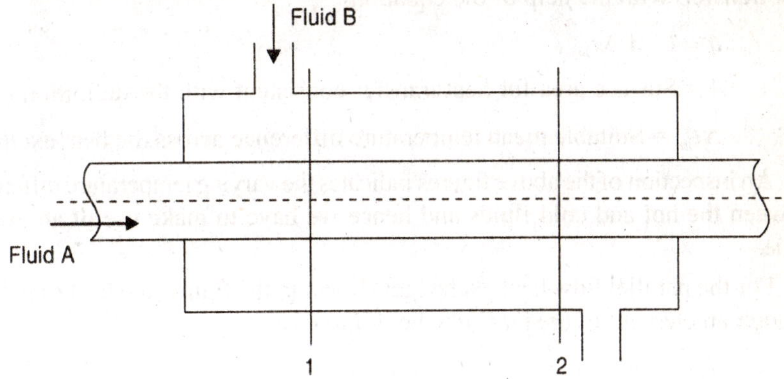

Fig. 6.5 *A double-pipe heat exchanger*

Consider a double pipe heat exchanger. The fluids may flow in either parallel or counter flow and the temperature profiles for these two cases are indicated below.

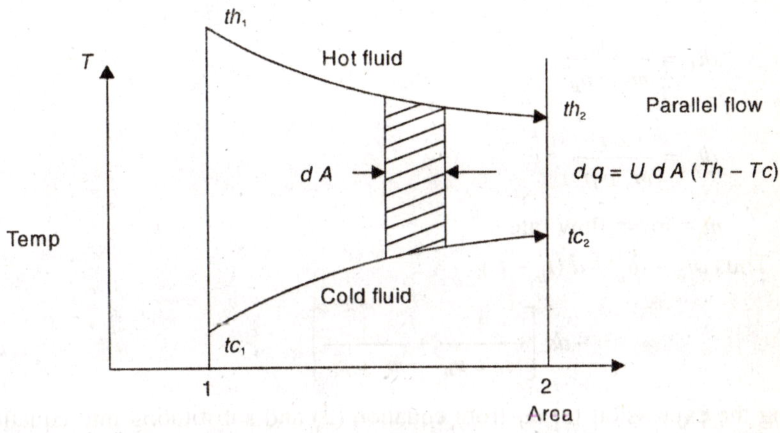

Fig. 6.6 *Temperature profile for parallel flow*

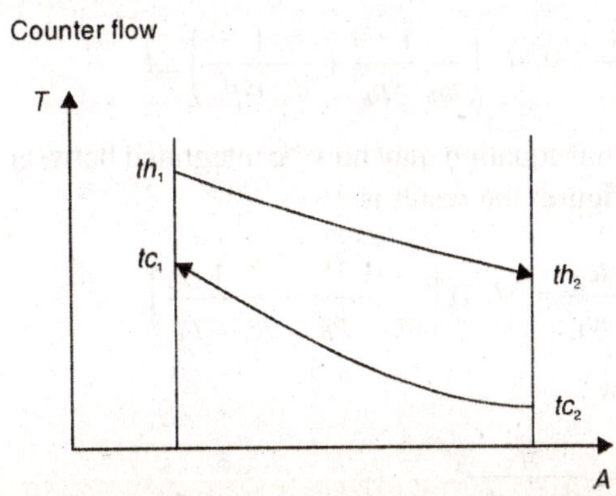

Fig. 6.7 *Temperature profile for counter flow*

We propose to compute in this double-pipe heat exchanger arrangement, the heat transfer, with the help of the equation

$$q = U\ A\ \Delta t_m$$

A = Surface area for heat transfer consistent with the definition of U.

Δt_m = Suitable mean temperature difference across the heat exchanger

An inspection of the above figures indicates the varying temperature differences between the hot and cold fluids and hence we have to make use of an average value.

For the parallel flow heat exchanger shown in the figure, the heat transferred through an element of area dA may be written as:

$$dq = -m_h\ c\,p_h\ dt_h = m_c\ c\,p_c\ dt_c \qquad ...(1)$$

we also know

$$dq = U\,(t_h - t_c)\ dA \qquad ...(2)$$

From equation (1) we get

$$dt_h = \frac{-dq}{m_h\ c\,p_h}$$

$$dt_c = \frac{dq}{m_c\ c\,p_c}$$

m = mass flow rate

Thus $dt_h - dt_c = d\,(t_h - t_c)$

$$= -dq\left[\frac{1}{m_h\ c\,p_h} + \frac{1}{m_c\ c\,p_c}\right] \qquad ...(3)$$

using the expression for dq from equation (2) and substituting into equation 3, we obtain

$$\frac{d\,(t_h - t_c)}{t_h - t_c} = -U\left(\frac{1}{m_h\ c\,p_h} + \frac{1}{m_c\ c\,p_c}\right) dA \qquad ...(4)$$

This differential equation may now be integrated between conditions 1 and 2 as shown in the figure; the result is

$$ln\,\frac{th_2 - tc_2}{th_1 - tc_1} = -U\,A\left(\frac{1}{m_h\ c\,p_h} + \frac{1}{m_c\ c\,p_c}\right) \qquad ...(5)$$

We also know

$$m_h\ c\,p_h = \frac{q}{th_1 - th_2}$$

$$m_c \, c\,p_c = \frac{q}{tc_2 - tc_1}$$

substituting these relations into equation (5) we get

$$q = U\,A \frac{(th_2 - tc_2) - (th_1 - tc_1)}{ln\left[\dfrac{(th_2 - tc_2)}{(th_1 - tc_1)}\right]} \quad \text{...(6)}$$

Equations $q = U\,A\,(\Delta t)_m$ and the above equation, on comparison show that

$$\Delta T_m = \frac{(th_2 - tc_2) - (th_1 - tc_1)}{ln\left[\dfrac{(th_2 - tc_2)}{(th_1 - tc_1)}\right]}$$

which is the LMTD

The LMTD approach to the heat exchanger analysis is useful when the inlet and the outlet temperatures are known or are easily determined. The LMTD is then easily computed and the heat flow, surface area or U may be determined. When the inlet or exit temperatures are to be evaluated for a given heat exchanger, the analysis frequently involves an iterative procedure because of the log function in the LMTD. In such cases the analysis is performed more easily with the aid of the method based on the effectiveness of the heat exchanger in transferring a given amount of heat. The effectiveness method also offers many advantages for analysis of problems in which a comparison between various types of heat exchangers must be made for purposes of selecting the most suitable type for accomplishing a particular heat transfer objective.

6.5 EFFECTIVENESS—NTU RELATIONSHIP

$$q_{max} = (mc_p)_{min}\,(th_{inlet} - tc_{inlet})$$

For parallel flow heat exchanger

$$E_h = \frac{th_1 - th_2}{th_1 - tc_1} \text{ based on } m_h\,c\,p_h < m_c\,c\,p_c \quad \text{...(1)}$$

or $$E_c = \frac{tc_2 - tc_1}{th_1 - tc_1} \quad (m_h\,c\,p_h > m_c\,c\,p_c) \quad \text{...(2)}$$

subscripts on E, designate the fluid which has the minimum value of mc_p

For counter flow exchanger

$$E_h = \frac{th_1 - th_2}{th_1 - tc_2} \quad \text{...(3)}$$

$$E_c = \frac{tc_1 - tc_2}{th_1 - tc_2} \qquad ...(4)$$

The denominator depends on the way the inlet and outlet temperatures are called.

We can derive an expression for effectiveness in parallel flow as indicated below.

We have already seen

$$ln \frac{th_2 - tc_1}{th_1 - tc_1} = -U A \left(\frac{1}{m_h \; c \, p_h} + \frac{1}{m_c \; c \, p_c} \right)$$

(Equation 5 of the previous section that is the section under LMTD)

$$ln \frac{th_2 - tc_2}{th_1 - tc_1} = -\frac{U A}{m_c \; c \, p_c} \left[1 + \frac{m_c \; c \, p_c}{m_h \; c \, p_h} \right] \qquad ...(5)$$

or
$$\frac{th_2 - tc_2}{th_1 - tc_1} = \text{Exp} \left[\frac{-U A}{m_c \; c \, p_c} \left(1 + \frac{m_c \; c \, p_c}{m_h \; c \, p_h} \right) \right] \qquad ...(6)$$

If the cold fluid is the minimum fluid i.e. $m_c \; c \, p_c < m_h \; c \, p_h$

$$E_c = \frac{tc_2 - tc_1}{th_1 - tc_1} \text{ (for parallel flow)}$$

Rewriting the temperature ratio in equation (5) above, by using the relation

$$m_c \; c \, p_c \, (tc_2 - tc_1) = +m_h \; c \, p_h \, (th_1 - th_2)$$

or
$$th_2 = -\frac{m_c \; c \, p_c}{m_h \; c \, p_h} (tc_2 - tc_1) + th_1$$

The above value of th_2 is now substituted in the relation $\frac{th_2 - tc_2}{th_1 - tc_1}$ (for th_2)

$$\frac{th_2 - tc_2}{th_1 - tc_1} = \frac{th_1 + \frac{m_c \; c \, p_c}{m_h \; c \, p_h} (tc_1 - tc_2) - tc_2}{th_1 - tc_1} \qquad ...(7)$$

The right side of equation (7) is now rewritten as:

$$= \frac{(th_1 - tc_1) + \left(\frac{m_c \; c \, p_c}{m_h c \, p_h} \right) (tc_1 - tc_2) + (tc_1 - tc_2)}{(th_1 - tc_1) \text{ (adding and subtracting } tc_1 \text{ in the numerator)}}$$

$$= 1 - \left(1 + \frac{m_c \, cp_c}{m_h \, cp_h}\right) \cdot E \qquad \ldots(8)$$

Inserting the above expression back in equation (6) i.e. equating the right sides of equations (6) and (8), gives for the effectiveness

$$E = \frac{1 - \text{Exp.}\left[\left(\frac{-U\,A}{m_c \, cp_c}\right)\left(1 + \frac{m_c \, cp_c}{m_h \, cp_h}\right)\right]}{\left(1 + \frac{m_c \, cp_c}{m_h \, cp_h}\right)} \qquad \ldots(9)$$

It can be proved that the same expression results for the effectiveness when $m_h \, cp_h < m_c \, cp_c$ (i.e. the hot fluid is the minimum fluid) except that $m_c \, cp_c$ and $m_h \, cp_h$ are interchanged.

Therefore, the effectiveness is usually written as:

$$E = \frac{1 - \text{Exp.}\left[\left(-\frac{U\,A}{c_{\min}}\right)\left(1 + \frac{c_{\min}}{c_{\max}}\right)\right]}{1 + \frac{c_{\min}}{c_{\max}}} \qquad \ldots(10)$$

C = capacity

A similar analysis, if applied to the case of counter flow would yield for E.

$$E = \frac{1 - \exp\left[\left(-\frac{U\,A}{c_{\min}}\right)\left(1 - \frac{c_{\min}}{c_{\max}}\right)\right]}{1 - \left(\frac{c_{\min}}{c_{\max}}\right)\exp\left[\left(-\frac{U\,A}{c_{\min}}\right)\left(1 - \frac{c_{\min}}{c_{\max}}\right)\right]}$$

The derivation of this expression is given later in the text.

The groupings of the terms $\frac{U\,A}{c_{\min}}$ is termed the number of transfer units (NTU) since it is indicative of the size of the heat exchanger.

Kay and London have presented effectiveness ratios for various heat exchanger arrangements like parallel flow, counter flow,

One shell pass, 2, 4, 6 tube passes

Two shell passes, 4, 8, 12, etc tube passes, cross flow heat exchangers.

Parallel Flow Exchanger Performance

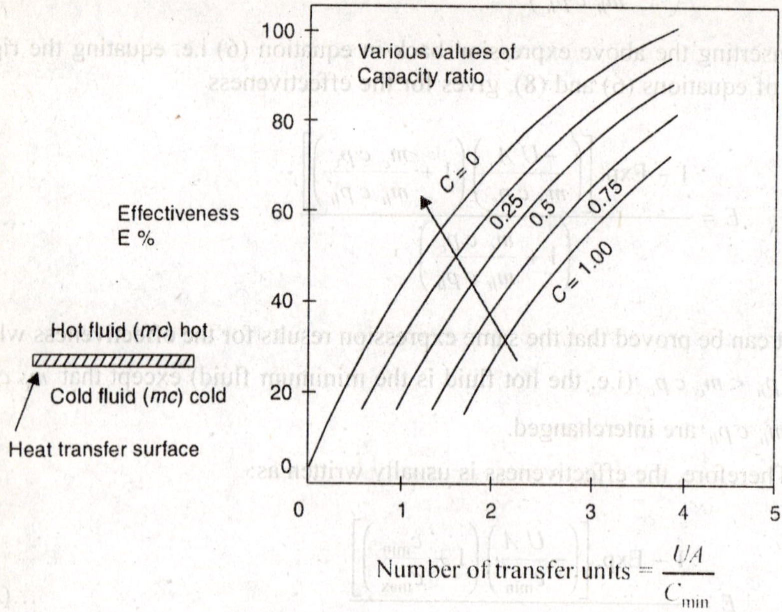

Fig. 6.8 *Plot of effectiveness versus number of transfer units*

For Boiling and Condensation

The capacity ratio C becomes 0 for both boiling and condensation since the larger thermal capacity is in each case infinite. This follows by definition of the heat capacity.

$$c_p = \frac{dh}{dt} = \infty \text{ when } dt = 0 \;\; c = 0$$

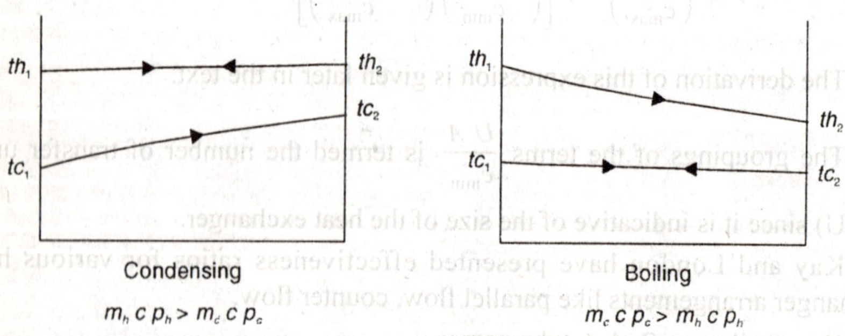

Fig. 6.9 *Temperature profile for condensation and boiling*

For effectiveness the expressions viz.

$$E_h = \frac{th_1 - th_2}{th_1 - tc_2}$$

$$E_c = \frac{tc_1 - tc_2}{th_1 - tc_2}$$

In boiling and condensing $C = 0$

For counter flow, the expression for E,

$$E = \frac{1 - \exp\left[\left(-\frac{UA}{c_{min}}\right)\left(1 - \frac{c_{min}}{c_{max}}\right)\right]}{1 - \left(\frac{c_{min}}{c_{max}}\right)\exp\left[\left(-\frac{UA}{c_{min}}\right)\left(1 - \frac{c_{min}}{c_{max}}\right)\right]}$$

reduces to $\because \frac{c_{min}}{c_{max}} = C = 0$

$$E = 1 - e^{-NTU}$$

Similarly, the expression for parallel flow i.e.

$$E = \frac{1 - \exp\left[\left(-\frac{UA}{c_{min}}\right)\left(1 + \frac{c_{min}}{c_{max}}\right)\right]}{1 + \frac{c_{min}}{c_{max}}}$$

$\because C = 0$

$E = 1 - \exp(-NTU)$

For the other limiting value of c, i.e. $c = 1$

$$m_h\, c\, p_h = m_c\, c\, p_c$$

for equal thermal capacities

In the case of parallel flow

$$E = \frac{1 - \exp\left[\left(-\frac{UA}{c_{min}}\right)(1 + 1)\right]}{1 + 1}$$

$$= \frac{1 - \exp\left[\left(-\frac{UA}{c_{min}} \cdot 2\right)\right]}{2}$$

or $$E = \frac{1 - e^{-2\text{NTU}}}{2}$$

For Counter Flow

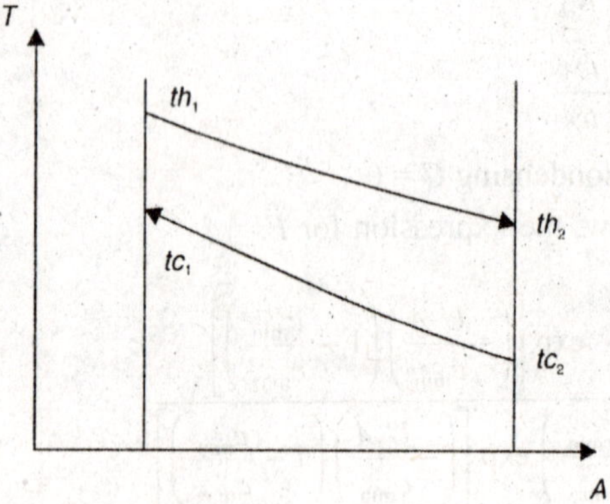

Fig. 6.10 *Temperature profile for counter flow*

$$E = \frac{1 - \exp\left[\left(-\frac{U\,A}{c_{\min}}\right)\left(1 - \frac{c_{\min}}{c_{\max}}\right)\right]}{1 - \left(\frac{c_{\min}}{c_{\max}}\right)\exp\left[\left(-\frac{U\,A}{c_{\min}}\right)\left(1 - \frac{c_{\min}}{c_{\max}}\right)\right]}$$

When $\frac{c_{\min}}{c_{\max}} = 1$, the above expression becomes indeterminate

It is hence necessary to do a fresh analysis

We know $E_h = \frac{th_1 - th_2}{th_1 - tc_2}$

and also $(th_1 - th_2) = (tc_1 - tc_2)$ or $th_1 - tc_1 = th_2 - tc_2$

$(\text{since } m_h\; c\,p_h = m_c\; c\,p_c)$

$(\text{also } q = U\,A\,(th_1 - tc_1) = m_h\; c\,p_h\;(th_1 - th_2))$

$$\therefore (th_1 - th_2) = \frac{U\,A}{m_h\; c\,p_h}(th_1 - tc_1)$$

$$= \text{NTU}\,(th_1 - tc_1) \quad \text{or} \quad \frac{th_1 - th_2}{th_1 - tc_1} = \text{NTU}$$

E may be written as:

$$E = \frac{th_1 - th_2}{(th_1 - th_2) - (tc_2 - th_2)} \text{ (adding and subtracting } th_2)$$

$$= \frac{\text{NTU}\,(th_1 - tc_1)}{(\text{NTU})\,(th_1 - tc_1) - (tc_2 - th_2)} : \text{Since } \frac{th_1 - th_2}{th_1 - tc_1} = \text{NTU}$$

but $(tc_2 - th_2) = -(th_1 - tc_1)$

$$\therefore\ E = \frac{\text{NTU}}{\text{NTU} + 1} \text{ when } c = 1.$$

6.6 ILLUSTRATION ON THE ECONOMICAL SIZE OF THE HEAT EXCHANGER (USE OF THE CHART NTU$_{MAX}$ US EFFECTIVENESS IN SOLVING THIS PROBLEM)

A counter flow double-pipe exchanger is used to heat 10000 lbs/hr of water from 100 to 190°F by cooling an oil (c_p = 0.5) from 300 to 200°F. The overall heat transfer coefficient is 150 Btu/hr ft^2°F. A similar arrangement is to be built at another plant location, but it is desired to compare the performance of a single counter flow heat exchanger with two smaller counter flow heat exchangers equal in area surface connected in series on the water side and in parallel on the oil side, as shown in the sketch. The oil flow is split equally between two heat exchangers and it may be assumed that the overall heat transfer coefficient for the smaller exchangers is the same as for the large exchanger. If the smaller exchangers cost 20% more per unit surface area which would be the most economical arrangement of the two?

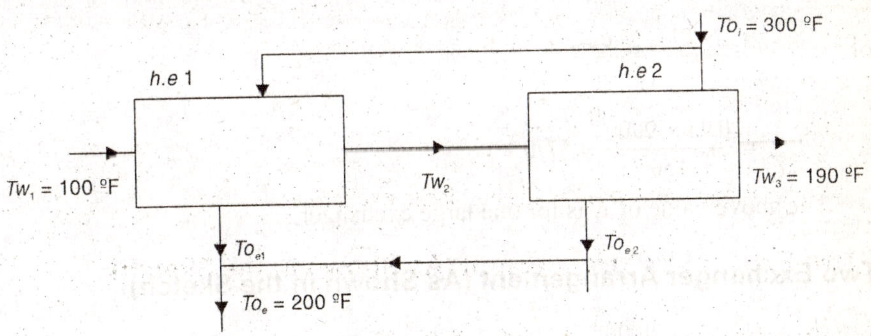

Fig. 6.11 *Two-exchanger arrangement with equal heat transfer areas—in series on water side and in parallel on the oil side*

Unknown temperatures are Tw_2, To_{e1} and To_{e2}

Solution

Let us calculate the surface area required for both the arrangements

Large Heat Exchanger

$$q = m_c\ c p_c\ \Delta T_c = m_h\ c p_h\ \Delta T_h$$
$$= (10000)\ (1)\ (190 - 100) = m_h\ cp_h\ (300 - 200)$$
$$= 9 \times 10^5 \text{ Btu/hr}$$
$$cp_h = 0.5$$
$$m_h = \frac{9 \times 10^5}{(0.5)\ (100)} = 18 \times 10^3$$
$$= 18000 \text{ lbs/hr}$$
$$m_c cp_c = 10000 \text{ Btu/hr°F}$$
$$m_h cp_h = 18000 \times 0.5 = 9000 \text{ Btu/hr°F}$$

and hence the oil is in the minimum fluid

$$E_h = \frac{\Delta T_h}{300 - 100} = \frac{100}{200} = 0.5$$

$$\text{cap. ratio} = \frac{c_{min}}{c_{max}} = \frac{9000}{10000} = 0.9$$

using the chart of NTU_{max} versus E, for case of counter flow, we read for an value of 0.5 and capacity ratio of 0.9, NTU_{max} as 0.95

$$NTU_{max} = \frac{U A}{c_{min}}$$

or
$$A = \frac{NTU_{max.}\ c_{min}}{U}$$
$$= \frac{0.95 \times 9000}{150} = 57 \text{ ft}^2$$

The above value of A is for one large exchanger.

Two Exchanger Arrangement (As Shown in the Sketch)

We have $m_h\ c p_h = \dfrac{9000}{2} = 4500$ BTU/hr °F

$$m_c cp_c = 10000$$

$$\frac{c_{min}}{c_{max}} = \frac{4500}{10000} = 0.45$$

The number of transfer units is the same for each exchanger since UA and c_{min} are the same. This requires that the effectiveness be the same for each exchanger. Thus

$$E_1 = \frac{To_i - To_{e1}}{To_i - Tw_1} = E_2 = \frac{To_i - To_{e2}}{To_i - Tw_2}$$

i.e.
$$\left. E_1 = \frac{300 - To_{e1}}{300 - 100} = E_2 = \frac{300 - To_{e2}}{300 - Tw_2} \right\} \quad ...(1)$$

where the nomenclature for the temperatures is as per the sketch.

Since the average exit oil temperature has to be 200°F, we may make use of the following relation.

$$\frac{To_{e1} + To_{e2}}{2} = 200 \quad ...(2)$$

An energy balance on the second exchanger yields

$$10000\left(Tw_3 - Tw_2\right) = 4500\left(To_i - To_{e2}\right)$$

i.e.
$$10000\left(190 - Tw_2\right) = 4500\left(300 - To_{e2}\right) \quad ...(3)$$

From the above three equations (1), (2) and (3) we can evaluate the three unknowns viz.

To_{e1}, To_{e2} and Tw_2

Thus $To_{e1} = 185°F$

$To_{e2} = 215°F$

and $Tw_2 = 152°F$

Then the effectiveness E, is computed as

$$E_1 = \frac{300 - 185}{300 - 100} = 0.575$$

$$E_2 = \frac{300 - 215}{300 - 152}$$

Using this value of E, i.e. 0.575

$\frac{c_{min}}{c_{max}} = 0.45$, we read from the plot of NTU_{max} us E, $NTU_{max} = 1.1$

$$A = NTU_{max} \frac{c_{min}}{U}$$

$$= \frac{(1.1)(4500)}{(150)} = 33 \text{ ft}^2$$

If is thus found that 33 ft^2 of area is required in each of the small exchangers that is a total of 66ft^2 that which is greater than that for a single large unit (57 ft^2). Moreover the cost/unit area is greater so that the most economical choice would be the single large exchanger. It may however be noted that the pumping costs for the oil would probably be less with two smaller exchangers, so that this could be resulting in a decision in favour of the smaller exchangers if the pumping costs represented a sizable economic factor.

Parallel Flow–When the Hot Fluid is the Minimum Fluid–Effectiveness–NTU Relationship

$$ln \frac{th_2 - tc_2}{th_1 - tc_1} = -U A\left(\frac{1}{m_h\ c\,p_h} + \frac{1}{m_c\ c\,p_c}\right) \quad ...(1)$$

Rewriting the above expression

$$ln \frac{th_2 - tc_2}{th_1 - tc_1} = -\frac{U A}{m_h\ c\,p_h}\left[1 + \frac{m_h\ c\,p_h}{m_c\ c\,p_c}\right]$$

or
$$\frac{th_2 - tc_2}{th_1 - tc_1} = \text{Exp}\left[-\frac{U A}{m_h\ c\,p_h}\left(1 + \frac{m_h\ c\,p_h}{m_c\ c\,p_c}\right)\right] \quad ...(2)$$

We know $E_h = \dfrac{th_1 - th_2}{th_1 - tc_1}$ when the hot fluid is the minimum fluid

$$m_c\ c\,p_c\ (tc_2 - tc_1) = m_h\ c\,p_h\ (th_1 - th_2)$$

or
$$tc_2 = \frac{m_h\ c\,p_h}{m_c\ c\,p_c}(th_1 - th_2) + tc_1$$

The above value of tc_2 is substituted for tc_2 in the relation $\dfrac{th_2 - tc_2}{th_1 - tc_1}$, we then get

$$\frac{th_2 - tc_2}{th_1 - tc_1} = \frac{th_2 - \dfrac{m_h\ c\,p_h}{m_c\ c\,p_c}(th_1 - th_2) - tc_1}{(th_1 - tc_1)} \quad ...(3)$$

Adding and subtracting th_1 in the numerator of the above expression we get

$$\frac{th_2 - tc_2}{th_1 - tc_1} = \frac{(th_2 - th_1) - \dfrac{m_h\ c\,p_h}{m_c\ c\,p_c}(th_1 - th_2) + (th_1 - tc_1)}{(th_1 - tc_1)}$$

$$= 1 - E - \frac{m_h\ c\,p_h}{m_c\ c\,p_c} \cdot E$$

or $$= 1 - E\left(1 + \frac{m_h\ c\,p_h}{m_c\ c\,p_c}\right) \qquad \ldots(4)$$

equations (2) and (4) on comparison yield

$$1 - E\left(1 + \frac{m_h\ c\,p_h}{m_c\ c\,p_c}\right) = \exp\left[-\frac{U\,A}{m_h\ c\,p_h}\left(1 + \frac{m_h\ c\,p_h}{m_c\ c\,p_c}\right)\right]$$

$$\text{Therefore } E = \frac{1 - \exp\left[-\dfrac{U\,A}{m_h\ c\,p_h}\left(1 + \dfrac{m_h\ c\,p_h}{m_c\ c\,p_c}\right)\right]}{\left(1 + \dfrac{m_h\ c\,p_h}{m_c\ c\,p_c}\right)} \qquad \ldots(5)$$

In general the above two expressions for E i.e. when the cold fluid is the minimum fluid and when the hot fluid is the minimum fluid can be written as.

$$E = \frac{1 - \exp\left[\left(-\dfrac{U\,A}{c_{\min}}\right)\left(1 + \dfrac{c_{\min}}{c_{\max}}\right)\right]}{1 + \dfrac{c_{\min}}{c_{\max}}} \qquad \ldots(6)$$

where c is the thermal capacity.

Effectiveness for a Counter Flow Heat Exchanger

The approach for developing an expression for the effectiveness, E, for a counter flow heat exchanger is similar to the one used for the parallel flow heat exchanger. Equating the rates of flow of heat as given by the LMTD formulation and the energy balance, we have the following expressions in which

Th_i = inlet temperature of the hot fluid

Th_o = outlet temperature of the hot fluid

Tc_i = inlet temperature of the cold fluid

Tc_o = outlet temperature of the cold fluid

$$q = U\,A\,\frac{(Th_i - Tc,o) - (Th,o - Tc,i)}{ln\left[\dfrac{(Th_i - Tc,o)}{(Th,o - Tc,i)}\right]}$$

$$= m_h\ c\,p_h\,(Th_i - Th,o)$$

Simplifying and rearranging, we get

$$ln\left(\frac{Th_i - Tc,o}{Th,o - Tc,i}\right) = \frac{U\,A}{m_h\,c\,p_h}\left(1 - \frac{Tc,o - Tc,i}{Th,i - Th,o}\right)$$

$$= \frac{U\,A}{m_h\,c\,p_h}(1 - R)$$

where $R = \dfrac{Tc,o - Tc,i}{Th,i - Th,o}$

or $$\frac{Th.o - Tc_i}{Th_i - Tc.o} = \exp\left[-\frac{U\,A}{m_h\,c\,p_h}(1 - R)\right] = \beta$$

or $$(Th,o - Tc,i) = \beta\,(Th_i - Tc,o)$$

We now substitute the expression for T_{c_o} obtained in the preceding derivation of E for parallel flow and assume that the hot fluid is the one with the minimum heat capacity rate. Thus, we have

$$(Th,o - Tc,i) = \beta\left[Th,i - Tc,i - R(Th_i - Th,o)\right]$$

where $$R = \frac{Tc.o - Tc.i}{Th_i - Th,o} = \frac{m_h\,c\,p_h}{m_c\,c\,p_c}$$

Rearrangement yields

$$Th,o\,(1 - R\beta) + R\beta\,Th,i = Tc,i\,(1 - \beta) + \beta\,Th,i$$

Subtracting Th,i from both sides of the above equation, multiplying by (–1) and regrouping the terms, one obtains

$$(1 - R\beta)\,(-Th,o + Th,i) = (1 - \beta)\,(-Tc,i + Th,i)$$

or $$\frac{Th_i - Th,o}{Th_i - Tc_i} = \frac{1 - \beta}{1 - R\beta}$$

Recognizing the left side as E and substituting for β, we finally get

$$E = \frac{1 - \exp\left[-\left(\dfrac{U\,A}{m_h\,c\,p_h}\right)(1 - R)\right]}{1 - R\exp\left[-\left(\dfrac{U\,A}{m_h\,c\,p_h}\right)(1 - R)\right]}$$

where $R = \dfrac{m_h\,c\,p_h}{m_c\,c\,p_c} < 1$

If the cold fluid were the one with the minimum value of heat capacity rate we would get exactly the same expression as above with R defined as $m_c\,c\,p_c/m_h\,cp_h$. Hence the effectiveness for a counter-flow exchanger can be written as:

$$E = \frac{1 - \exp\left\{-\left(\frac{UA}{c_{min}}\right)\left[1 - \frac{c_{min}}{c_{max}}\right]\right\}}{1 - \left(\frac{c_{min}}{c_{max}}\right)\exp\left\{-\left(\frac{UA}{c_{min}}\right)\left(1 - \frac{c_{min}}{c_{max}}\right)\right\}}$$

where $C = mc_p$

A closer look at the above equation reveals that it contains three dimensionless quantities, namely E $\left(\frac{c_{min}}{c_{max}}\right)$ and $\left(\frac{UA}{c_{min}}\right)$. The last group is called the number of transfer units or NTU as already mentioned. With a given overall heat transfer coefficient U and a known value of c_{min}, one of the objectives in heat exchanger calculations would be to determine the surface area A. This can be readily done once the number of transfer units NTU is determined.

Therefore, the quantity NTU is often considered as indicative of the size of a heat exchanger.

The LMTD and NTU approaches are really two sides of the same coin; depending on the nature of the data given, one can choose either the LMTD approach or the NTU approach. The above expression is more convenient to use when all the four are not prescribed.

6.7 MEAN TEMPERATURE DIFFERENCE IN MULTI PASS HEAT EXCHANGERS

The flow paths are so arranged that either or both fluids must reverse directions one or more times in passing through the exchanger. The combination of parallel and counter flow is called reversed flow.

Let us consider a 1–2 exchanger

1 shell pass and 2 tube passes, represented in the following sketch

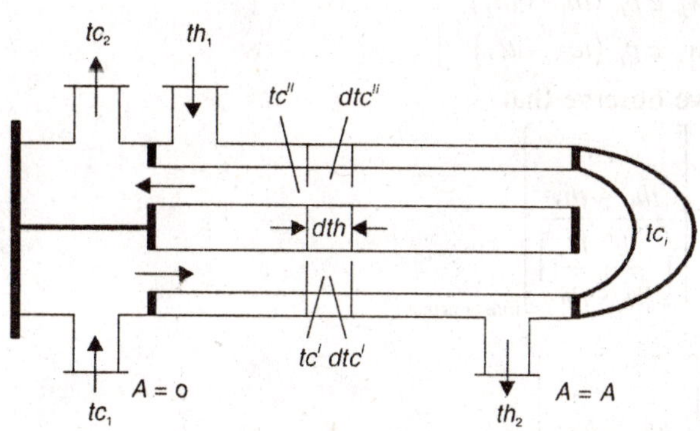

Fig. 6.12 *1–2 exchanger—one shell pass and 2 tube passes*

t_h = shell fluid temperature (hot)
t_c = tube side fluid temperature

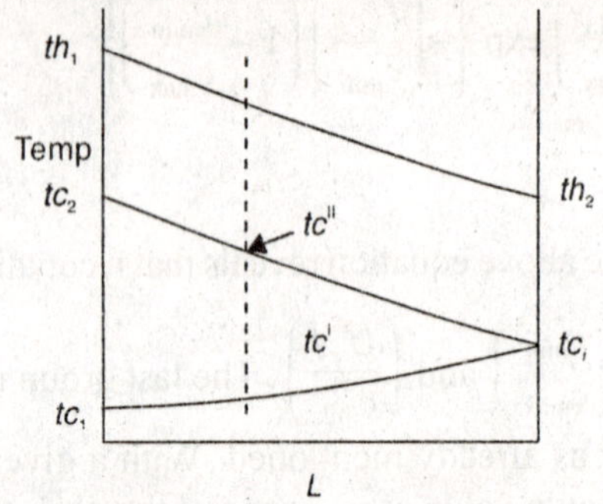

Fig. 6.13 *Temperature profile in a 1–2 exchanger*

The true mean temperature is determined by making use of the following assumptions:

1. Shell fluid temperature is an average isothermal temperature at any cross section.
2. The heating surface in each pass is the same.
3. *U* is constant.
4. Each fluid flows at a constant rate of flow.
5. The heat capacities are constant.
6. No phase change occurs—that is there is no evaporation or condensation.
7. The heat losses are negligible.

The overall heat balance where Δt is the true temperature difference is

$$\left.\begin{aligned} q &= U\,A\,\Delta t \\ &= w_h\,c\,p_h\,(th_1 - th_2) \\ &= w_c\,c\,p_c\,(tc_2 - tc_1) \end{aligned}\right\} \qquad \ldots(1)$$

from which we observe that

$$\Delta t = \left[\frac{th_1 - th_2}{\left(\dfrac{U\,A}{w_h\,c\,p_h}\right)}\right]_{\text{true or actual}}$$

$$= \left[\frac{tc_2 - tc_1}{\left(\dfrac{U\,A}{w_c\,c\,p_c}\right)}\right]_{\text{actual}} \qquad \ldots(2)$$

In the above sketch let t_h be the temperature of the hot or shell fluid at any cross section of the shell at $L = x$, between $L = o$ and $L = L$.

Let tc^{I} and tc^{II} represent the temperatures in the first and second tube passes respectively and at the same cross section as T.

Let a'' be the external surface per foot of length (heat transfer area per unit length)

In the incremental surface $dA = a'' \, dL$ the shell temperature changes by $-dt_h$.

Over the area dA

$$- w_h \, c p_h \, dt_h = U \frac{dA}{2}\left(th - tc^{I}\right) + U \frac{dA}{2}\left(th - tc^{II}\right) \quad \text{...(3)}$$

$$- w_h \, c p_h \, dt_h = U \, dA \left(th - \frac{tc^{I} + tc^{II}}{2}\right) \quad \text{...(4)}$$

$$-\int \frac{U \, dA}{w_h \, c p_h} = \int \frac{dth}{th - \dfrac{\left(tc^{I} + tc^{II}\right)}{2}} \quad \text{...(5)}$$

In equation (5), t_h, tc^{I} and tc^{II} are dependent variables

Similarly for 1–4 exchanger

$$-\int \frac{U \, dA}{w_h \, c p_h} = \int \frac{dth}{th - \dfrac{\left(tc^{I} + tc^{II} + tc^{III} + tc^{iv}\right)}{4}} \quad \text{...(6)}$$

and for 1–6 exchanger

$$-\int \frac{U \, dA}{w_h \, c p_h} = \int \frac{dth}{th - \dfrac{\left(tc^{I} + tc^{II} + tc^{III} + tc^{iv} + tc^{V} + tc^{iv}\right)}{6}} \quad \text{...(7)}$$

Considering again 1–2 exchanger, let us define

$$\frac{th_1 - th_2}{tc_2 - tc_1} = \frac{w_c \, c p_c}{w_h \, c p_h} = R$$

$\dfrac{tc_2 - tc_1}{th_1 - tc_1} = S$ (This is nothing but the effectiveness E in true counter flow).

From which the internal relationship of $w_c\ c\,p_c$ and $w_h\ c\,p_h$ can be seen i.e.

$$\left(\frac{U\,A}{w_h\ c\,p_h}\right)_{\text{actual}} = \left(\frac{U\,A\,R}{w_c\ c\,p_c}\right)_{\text{actual}}$$

Equation (5) has been solved for the right side and is got (in terms of R and S)

$$\left[\frac{U\,A}{w_c\ c\,p_c}\right]_{\text{actual}} = \frac{1}{\sqrt{R^2+1}}\, ln \frac{2 - S\left(R + 1 - \sqrt{R^2+1}\right)}{2 - S\left(R + 1 + \sqrt{R^2+1}\right)}$$

The above expression is for the true temperature difference for 1–2 parallel flow-counter flow.

Let us now see how this compares with the LMTD for counter flow, employing the same process temperatures

$$q = w_c\ c\,p_c\,(tc_2 - tc_1)$$

$$= U\,A\,\frac{(th_1 - tc_2) - (th_2 - tc_1)}{ln\left[\dfrac{(th_1 - tc_2)}{(th_2 - tc_1)}\right]}$$

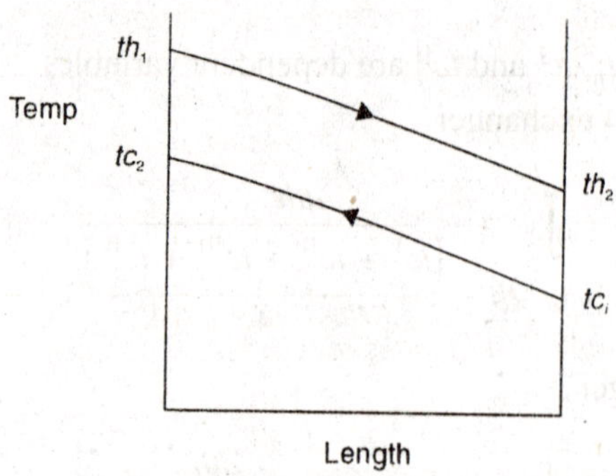

Fig. 6.14 *Temperature profile for counter flow*

from which

$$\left(\frac{U\,A}{w_c\ c\,p_c}\right)_{\text{counterflow}} = \frac{tc_2 - tc_1}{\dfrac{(th_1 - tc_2) - (th_2 - tc_1)}{ln\left(\dfrac{th_1 - tc_2}{th_2 - tc_1}\right)}}$$

$$= \frac{ln\left\{\dfrac{[(1-S)]}{(1-RS)}\right\}}{R-1}$$

Hence the ratio of true temperature difference to the LMTD is say F.

$$F = \frac{(R-1)\, ln \left[\dfrac{2 - S\left(R + 1 - \sqrt{R^2+1}\right)}{2 - S\left(R + 1 + \sqrt{R^2+1}\right)}\right]}{\left\{\sqrt{R^2+1}\; ln \dfrac{(1-S)}{(1-RS)}\right\}}$$

which is
$$\frac{\dfrac{tc_2 - tc_1}{\left(\dfrac{U\,A}{w_c\, c\, p_c}\right)_{\text{true}}}}{\dfrac{tc_2 - tc_1}{\left(\dfrac{U\,A}{w_c\, c\, p_c}\right)_{\text{counter flow}}}}$$

$$= \frac{\left(\dfrac{U\,A}{w_c\, c\, p_c}\right)_{\text{counter flow}}}{\left(\dfrac{U\,A}{w_c\, c\, p_c}\right)_{\text{true}}}$$

The Fourier equation for 1–2 exchanger can now be written

$$Q = U\,A\,\Delta t = U\,A\,\text{F} \quad \text{(LMTD)}.$$

Temperature-Length Curve for a 2–4 Heat Exchanger is given below

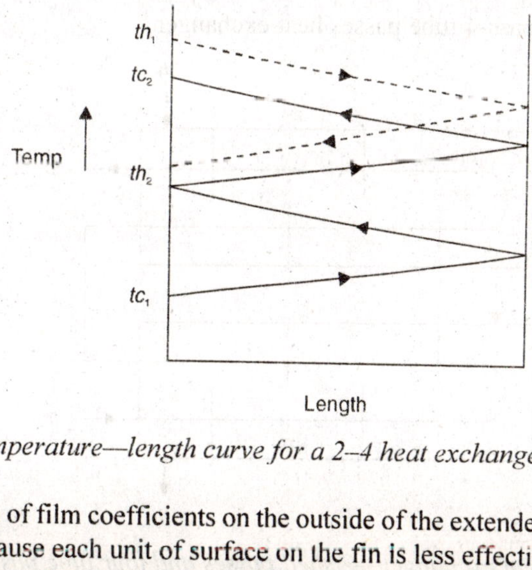

Fig. 6.15 *Temperature—length curve for a 2–4 heat exchanger*

The calculation of film coefficients on the outside of the extended surfaces (fins) is complex because each unit of surface on the fin is less effective than a unit of surface on the outside of the tube. When the fluid on the fin side is being cooled the heat must be transferred first to the fin and then along the fin to the tube wall. There will therefore be a temperature gradient along the fin from tip to the tube and the temperature difference between the fluid and the fin will therefore increase from tip to the base. The calculation of the true mean temperature of the surface either for cooling or heating the viscous fluid is therefore complex and although some equations have been obtained for these conditions it is generally more convenient and reliable to employ empirical relations obtained from measurements on the particular form of surface proposed.

Mean Temperature Difference in Multipass Exchangers

In an exchanger with one shell pass and several tube passes, the fluids in the tubes and shell will flow co-currently in some of the passes and counter currently in the others. The mean temperature difference for countercurrent flow is greater than for parallel flow, so that with the multi pass exchanger there is no easy way of finding the true mean temperature difference. A graphical method for finding a correction factor (to be applied to the log mean temperature difference for counter flow) has been suggested by Bowman Mueller and Nagle.

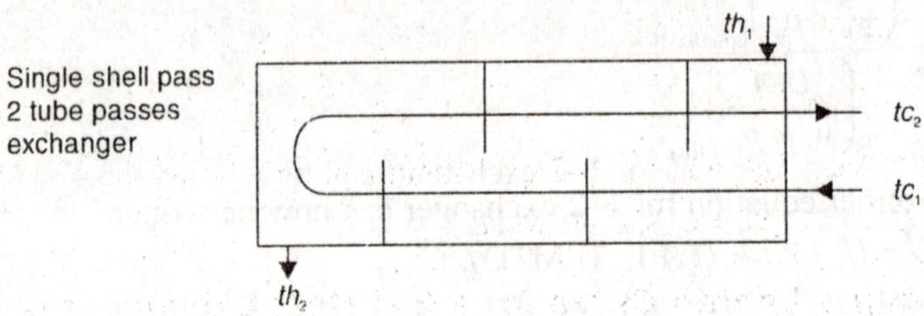

Fig. 6.16 *1–2 heat exchanger*

Double shell pass-4 tube passes heat exchanger

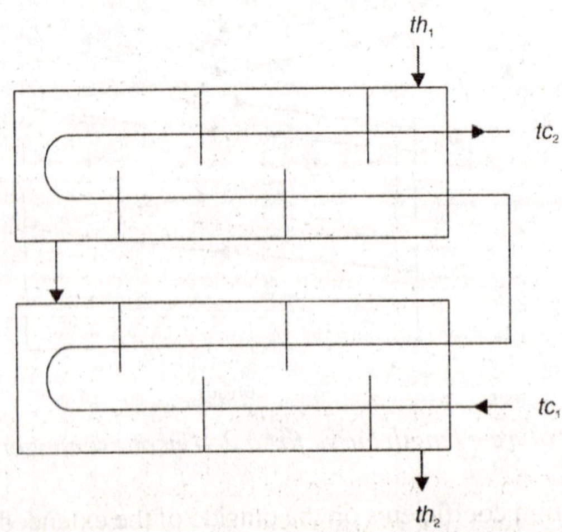

Fig. 6.17 *2–4 heat exchanger—2 shell passes and four tube passes*

Then for a unit with one shell side pass and an even number of passes on the tube side, the correction factor F is found from the following graph.

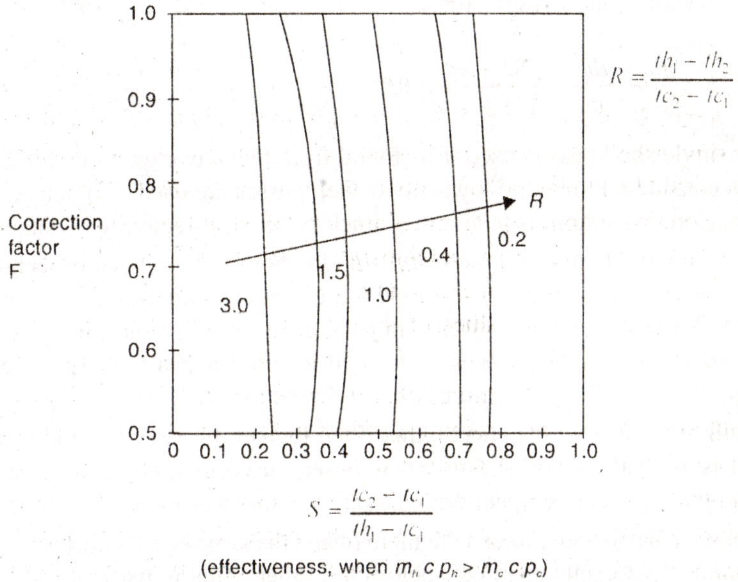

Fig. 6.18 *Correction factor plot for one shell pass and even number of tube passes*

$$R = \frac{th_1 - th_2}{tc_2 - tc_1} = \frac{\text{fall in temperature of the hot fluid}}{\text{rise in temperature of the cold fluid}}$$

F is given by

$$\Delta T_{\text{true mean}} = \Delta T_{\text{log mean}} \times F$$

This chart is developed on the assumption that an average temperature can be taken for the shell fluid at any cross section, that there is an equal transfer surface in each pass, that U is constant, that the c_p of each fluid is constant throughout the exchanger and that no phase change occurs in either fluid. If F has a value less than 0.75, it is generally desirable to use a different arrangement because the heat transfer area is not being used in the most effective manner.

There is a similar graph available for each of the other arrangements such as double shell pass etc.

Illustration on the Use of the Chart

Suppose an exchanger is required to operate over the temperature ranges

$th_1 = 182°C$, $th_2 = 99°C$

$tc_1 = 10°C$ and $tc_2 = 115°C$

$$S = \frac{tc_2 - tc_1}{th_1 - tc_1} = \frac{115 - 10}{182 - 10} = 0.6$$

$$R = \frac{th_1 - th_2}{tc_2 - tc_1} = \frac{182 - 99}{115 - 10} = 0.8$$

If a single shell pass is used, F is found from the above graph and it is 0.65 which is considered to be too low. If two shell passes are used, F obtained from the corresponding graph, is 0.95 and a much better heat recovery is obtained.

To obtain maximum heat recovery from the hot fluid, tc_2 must be as high as possible. When tc_2 exceeds th_2, a temperature cross in said to occur and with a single shell pass any further increase in tc_2 will result in a rapid fall in F. With more shell passes, higher values of tc_2 can be used and a better recovery is obtained.

It will be seen from the curves, that if the temperature of one fluid remains constant as for instance with a simple condenser, the correction factor F becomes unity and the log mean temperature is used.

Shell-side baffles are rather difficult to fit and the same effect may be obtained by mounting two exchangers one below the other. This is particularly useful where each unit is large and single exchanger would be unwieldy.

Examples

1. (a) Water at the rate of 150 lbs/min is heated from 100 to 160°F by an oil having a c_p of 0.45. The fluids are used in a counter flow double pipe heat exchanger and the oil enters the exchanger at 230°F and leaves at 150°F. The overall coefficient is 60 Btu/hr ft²°F. Calculate the heat transfer area.

The total heat transfer is,

$$q = m_w \, c_{p_w} \, (\Delta T)_w$$

$$= (150)\,(1)\,(160 - 100)$$

$$= 9000 \text{ Btu/m}$$

$$= 540{,}000 \text{ Btu/hr}$$

The LMTD is calculated as

$$\Delta T_m = \frac{(230 - 160) - (150 - 100)}{ln\left[\dfrac{(230 - 160)}{(150 - 100)}\right]}$$

$$= 59.6°\text{F}$$

$$q = U\,A\,\Delta T_m$$

$$A = \frac{5.4 \times 10^5}{60 \times 59.6} = 151 \text{ ft}^2$$

(b) Instead of a double-pipe heat exchanger of the above example, it is desired to use a shell and tube heat exchanger with the water making one shell pass and the oil making two tube passes. Calculate the area required for this exchanger assuming that the overall coefficient remains at 60 Btu/hr ft^2°F.

To solve this problem, we have to determine the correction factor F to be used with the LMTD. This is obtained from the graph for 1 shell and 2 tube pass heat exchanger.

The parameters of the figure are

$$S = \frac{tc_2 - tc_1}{th_1 - tc_1}; \quad R = \frac{th_1 - th_2}{tc_2 - tc_1}$$

$$tc_1 = 100°\text{F},\ th_1 = 230°\text{F}$$

$$tc_2 = 160°\text{F},\ th_2 = 150°\text{F}$$

$$S = \frac{60}{130} = 0.46$$

$$R = \frac{80}{60} = 1.33$$

F from the plot is 0.73 (plot for 1–2 exchanger)

$$A = \frac{q}{U\,\Delta T_m\,F}$$

$$= \frac{5.4 \times 10^5}{(60)\,(0.73)\,(59.6)} = 207 \text{ ft}^2$$

Illustration 2

Water at the rate of 30000 lbs/hr is heated from 100 to 130°F in a shell and tube exchanger. In the shell side one pass is used with water as the heating fluid flowing at the rate of 15000 lbs/hr. It enters the heat exchanger at 200°F. The U is 250 Btu/hr ft^2°F and the average water velocity in the 3/4″ ID tubes is 1.2 ft/sec. Because of space limitations the exchanger must not be longer than 8 feet. Consistent with this restriction, calculate the number of tube passes, the number of tubes per pass and the length of the tubes.

Solution

We first assume one tube pass and check to see if it satisfies the conditions specified in the problem. The exit temperature of the hot water is calculated from.

$$q = m_c\,cp_c\,\Delta T_c = m_h\,cp_h\,\Delta T_h$$

$$\Delta T_h = \frac{(30000)\,(1)\,(30)}{(15000)\,(1)} = 60$$

$$T_{h_{exit}} = 200 - 60 = 140\ °F$$

For counter flow

$$\text{LMTD} = \Delta T_m = \frac{(200-130)-(140-100)}{ln\left[\dfrac{(200-130)}{(140-100)}\right]} = 53.6°\text{ F}$$

$$q = U\,A\,\Delta T_m = m_c\,c\,p_c\,\Delta Tc$$

$$= (30000)\,(1)\,(30) = 9 \times 10^5 \text{ Btu/hr}$$

$$\therefore\ A = \frac{9 \times 10^5}{(250)\,(53.6)} = 67.3 \text{ ft}^2$$

Using the average water velocity in the tubes and the flow rate, we calculate the total flow area.

$$m_c = \rho\,A\,u$$

$$A_{\text{flow}} = \frac{30000}{(62.4)\,(1.2)\,(3600)} = 0.111 \text{ ft}^2$$

This area is the product of the number tubes and the flow area per tube.

$$0.111 = n\,\frac{\pi d^2}{4}$$

$$n = \frac{(0.111)\,(4)\,(144)}{\pi\,(0.75)^2}$$

$$= 36.1$$

or the number of tubes = 36

The surface area per tube per foot of length is

$$\pi d = \frac{\pi\,(0.75)}{12} = 0.196 \text{ ft}^2\text{/ft tube}$$

We recall that the total surface area required for a one-tube pass exchanger was 37.3 ft^2. We may thus compute the length of tube for this type of exchanger as

$$n\,\pi\,d\,L = 67.3$$

$$L = \frac{67.3}{(36)\,(0.196)} = 9.55 \text{ ft}$$

This length is greater than the allowable 8 ft so that we must use more than one tube pass, when we increase the number of passes we correspondingly increase the total surface area required because of the reduction in LMTD caused by the correction factor F.

We next try out, two-tube passes. From the appropriate graph

$F = 0.88$

$$\text{Thus } A_{\text{total}} = \frac{q}{F \cdot U \cdot \Delta Tm} = \frac{9 \times 10^5}{(250)(0.88)(53.6)}$$

$$= 76.3 \text{ ft}^2$$

The number of tubes per pass is still 36, because of the velocity requirement. For the two-tube pass system the total surface area is now related to the length by

$$A_{\text{total}} = 2n\,\pi d\,L$$

$$L = \frac{76.3}{(2)(36)(0.196)} = 5.41 \text{ ft}$$

This length is within the 8 ft stipulation so that the final design choice is
Number of tubes/pass = 36
Number of passes = 2
Length of tubes/pass = 5.4 feet

Problems

1. Hot water enters a counter flow heat exchanger at 210°F. It is used to heat a cool stream of water from 40 to 90°F m_c = 10000 lbs/hr; m_h'= 20000 lbs/hr

   U = 146.2 (in appropriate units)

   What is the area of the heat exchanger? Calculate the capacity ratio and the effectiveness of the heat exchanger.
2. It is desired to design a heat exchanger to cool continuously 110,000 lbs/hr of a solution from 150 to 103°F using 100,000 lbs/hr of cooling water available at a temperature of 50°F. The specific heat of the solution is 0.91 and the 'U' may be assumed to be 400. Calculate the heat transfer area required for each of the following proposals
   (a) Parallel flow
   (b) Counter flow
   (c) Reversed current exchanger with two shell passes and four tube passes with the hot solution flowing through the shell and the cold water flowing through the tubes.
3. A counter flow h.e. consists of a bundle of 20 mm diameter tubes contained in a shell. Oil flowing in the tubes is cooled by water flowing in the shell. The flow area within the tubes is 4.4×10^{-3} m². The flow of oil is 2.5 kg/sec, it enters at 65°C and leaves at 48°C. Water enters the shell at 20 kg/sec at 15°C. Calculate the area of the tube surface and the effectiveness of the heat exchanger. For the oil in the tubes take Nu_d = $0.023\,(Re_d)^{0.8}\,(Pr)^{0.33}$.

c_p = 2.15 kJ/kg °K

ρ = 880 kg/m^3

μ = 2.2 × 10^{-5} pascals

k = 190 × 10^{-6} kw/m °K

For water h = 1.2 kw/m^2 °K c_p = 4.19 kJ/kg °K

You may make any suitable assumptions.

Answer: Area = 2.23 m^2 effectiveness 34%.

Single Pass Heat Exchanger Analysis

Illustration

Benzene which is available as saturated liquid at 170°F is to be cooled to 115°F for the purpose of transporting to another place (location) in a chemical plant. The benzene is supplied at a rate of 8000 lb/hr and the cooling water is available at 55°C and its flow rate is 5000 lbs/hr. Determine the required heat exchanger areas for the following configurations.

(i) Single-pass, counter flow

(ii) Shell and tube with water making one shell pass, and benzene, four tube passes. '*U*' can be taken as 55 Btu/hr ft^2°F for both the flow arrangements.

Solution

The heat transfer rate is given by

$$q = m\,c_p\,\Delta T$$

For benzene c_p = 0.42 Btu/lb °F

$$q = (8000)\,(0.42)\,(170 - 115)$$
$$= 185{,}000 \text{ Btu/hr}$$

The exit temperature of the water stream can now be found

Again for water

$$q = m\,c_p\,\Delta T$$

$$185{,}000 = (5000)\,(1)\left(T_{H_2O_{exit}} - 55\right)$$

$$T_{H_2O_{exit}} = 55 + \frac{185{,}000}{5000}$$
$$= 55 + 37$$
$$= 92°\text{F}$$

(i) Single pass counter flow situation

Fig. 6.19 *Single pass counter flow configuration*

We can determine the heat transfer area

$$A = \frac{q}{U\,\Delta T_{lm}}$$

$$\Delta T_{lm} = \frac{(170-92)-(115-55)}{ln\left(\dfrac{170-92}{115-55}\right)}$$

$$= \frac{18}{ln\,\dfrac{78}{60}} = 68.6\ °F$$

$$A = \frac{185{,}000}{(55)\,(68.6)} = 49.0\ ft^2$$

(ii) Shell and tube heat exchanger

Water—shell fluid-one pass

Benzene—tube fluid-four passes

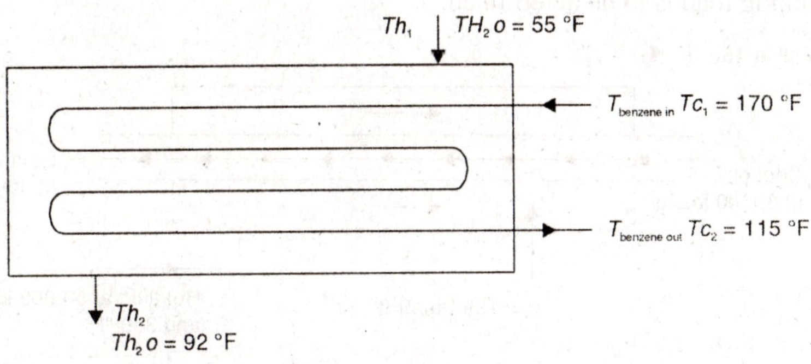

Fig. 6.20 *Shell and tube heat exchanger*

We need to apply the correction factor for the LMTD

$$S = \frac{Tc_2 - Tc_1}{Th_1 - Tc_1} = \frac{115 - 170}{55 - 170} = 0.478$$

$$R = \frac{th_1 - th_2}{tc_2 - tc_1} = \frac{55 - 92}{115 - 170} = 0.673$$

The correction factor is now read from the appropriate graph as 0.93 and the area becomes

$$A = \frac{q}{U\ F\ \Delta T_{lm}} = \frac{185{,}000}{(55)\ (0.93)\ (68)}$$

$$= 52.7 \text{ ft}^2$$

The heat transfer area is more for the second situation.

Problems on Heat Exchangers

1. Hot oil having a specific heat of 0.5 Btu/lb°F flows through a counterflow heat exchanger at a rate of 50,000 lbs/hr with an inlet temperature of 380°F and an outlet temperature of 150°F. Cold oil, whose specific heat is 0.4 Btu/lb°F, flows at a rate of 80000 lbs/hr and leaves at 300°F. Determine the area of the heat exchanger needed to handle the load if the U_i = 135 Btu/hr °F ft^2.

Solution

Systematic way of doing this problem is presented here.

Given: In a counter-flow heat exchanger hot oil flows at a rate of 50,000lbs/hr. The inlet and outlet temperatures of the hot fluid are 380°F and 150°F respectively. The hot oil is being cooled by cold oil whose specific heat is 0.4 Btu/lb. The flow rate of the cold oil is 80,000 lbs/hr. The outlet temperature of the cold fluid is 300°F. The U_i = 135 Btu/hr ft^2 °F.

Objective in this problem: The area of the heat exchanger necessary to handle the cooling load is to be determined.

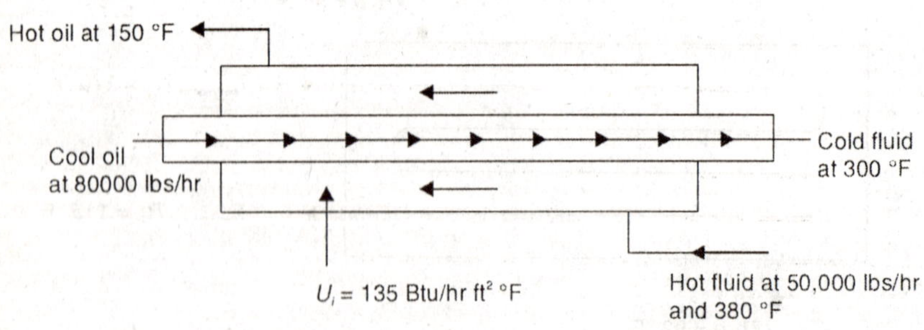

Fig. 6.21 *Counter current heat exchanger with the specified parameters of this situation*

Assumptions which normally hold good:

1. The tubes are clean and free of scale.
2. There are no heat losses.
3. The U is a constant over the length of the heat exchanger.
4. The specific heats of the oils remain constant.

Relevant principles: The nett heat lost by the hot oil equals the nett heat gained by the cold oil.

$$q = m_h \, c p_h \, (Th_i - Th_o) = m_c \, c p_c \, (Tc_o - Tc_i)$$

Also $q = U_i \; A_i \; \Delta T_{lm}$

Analysis: To determine the temperature Tc_i

$$m_c \, c p_c \, (Tc_o - Tc_i) = m_h \, c p_h \, (Th_i - Th_0)$$

substitution of the values gives

$$(80000)\,(0.4)\,(300 - Tc_i) = (50000)\,(0.5) \times (380 - 150)$$

$$Tc_i = 120°\text{F}$$

To determine LMTD

$$\text{LMTD} = \frac{(Th_o - Tc_i) - (Th_i - Th_o)}{ln \dfrac{Th_o - Tc_i}{Th_i - Th_o}}$$

$$= \frac{30 - 80}{ln \dfrac{30}{80}} = 51°\text{F}$$

Since $q = m_h \, cp_h \, (Tc_i - Th_o)$

$$= U_i \; A_i \; \Delta T_{lm}$$

or $\quad 50000\,(0.5)\,(380 - 150) = 135\,(A_i)\,51$

$$A_i = 835 \text{ ft}^2$$

i.e. the total area, based on the inside surface, needed to handle the load, is 835 ft^2.

2. A certain heat exchanger has a total outside surface area of 17.5 m^2. It is to be used for cooling oil at 200°C with a mass flow rate of 10000 kgs/hr having a specific heat of 1900 j/kg °k. Water at a flow rate of 3000 kgs/hr is available at 20°C as a cooling agent. If U_0 = 300 w/m^2 °K, estimate the temperature of the oil as it comes out of the heat exchanger, if the heat exchanger is operated

1. In a parallel-flow mode and
2. In a counter-flow mode.

Solution

Given: Hot oil at 200°C with a mass flow rate of 10000 kg/hr and a specific heat of c_p = 1900 J/kg °K is to be cooled by water. The flow rate of the water is 3000 kg/hr. The inlet temperature of the water is 20°C. The U_0 = 300 w/m^2 °K. The area A_0 = 17.5 m^2.

Objective: To estimate the temperature of the oil as it leaves if the exchanger is operated in a parallel flow mode and to estimate the temperature of the oil as it leaves if the exchanger is operated in a counterflow mode.

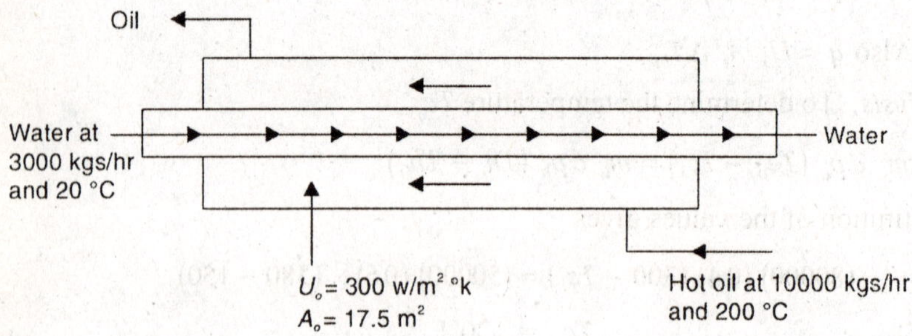

Fig. 6.22 *Counterflow arrangement*

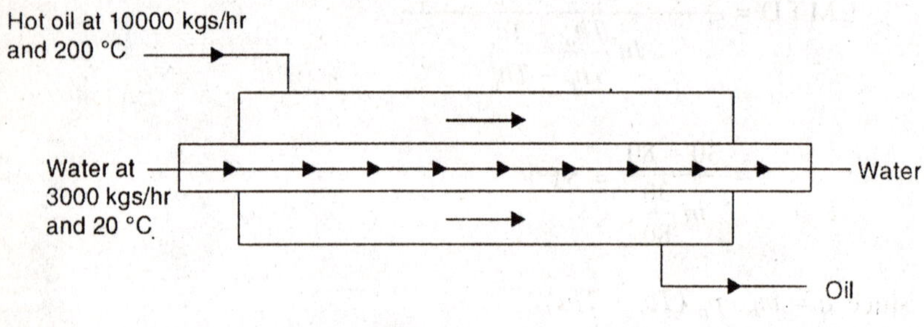

Fig. 6.23 *Parallel flow arrangment*

The assumptions involved in the solution:

1. The tubes are free of scale and are clean.
2. There are no heat losses.
3. The U is constant
4. The *cp's* of oil and water remain constant.

The principle/equations involved: The net loss of heat by the oil equals it net heat gained by the water which also equals the net heat transferred across the heat transfer area of the heat exchanger.

$$q = m_h \ c p_h \ (Th_i - Th_o) = m_c \ cp_c \ (Tc_o - Tc_i)$$

Also $q = U_0 \ A_0 \ \Delta T_{lm}$

We note that for a parallel flow heat exchanger

$$\text{LMTD}_{\text{P flow}} = \frac{(Th_i - Tc_i) - (Th_o - Tc_o)}{ln\left(\dfrac{Th_i - Tc_i}{Th_o - Tc_o}\right)}$$

and for a counterflow heat exchanger

$$\text{LMTD}_{\text{C flow}} = \frac{(Th_i - Tc_o) - (Th_o - Tc_i)}{ln\left(\dfrac{Th_i - Tc_o}{Th_o - Tc_i}\right)}$$

The properties/quantities are given as

$A_o = 17.5\ \text{m}^2$, $Tc_i = 20°\text{C}$

$m_h = 10000$ kg/hr, $cp_h = 1900$ j/kg °K

$m_c = 3000$ kg/hr, $U_o = 300$ w/m² °K

$Th_i = 200°\text{C}$, $cp_c = 4181.8$ j/kg °K (for water at 20°C)

Analysis of the situation:

$$q = c\,p_h\ m_h\ (Th_i - Th_o) = m_c\ c\,p_c\ (Tc_o - Tc_i)$$

$$(10000)\,(1900)\,(200 - Th_o) = 3000\,(4186) \times (Tc_o - 20)$$

$$Tc_o = 322.6 - 1.513\ Th,o$$

This equation holds good both for the parallel flow and the counterflow operation

1. For the parallel flow operation

$$Q = m_h\ cp_h\ (Th_i - Th_o) = U_o\ A_o\ \Delta T_{lm}$$

$$= U_o\ A_o \left\{ \frac{(Th_i - Tc_i) - (Th_o - Tc_o)}{ln\left[\dfrac{(Th_i - Tc_i)}{(Th_o - Tc_o)}\right]} \right\}$$

on substitution we get

$$(10000)\,(1900)\,(200 - Th,o) = 300\,(17.5) \times \left\{ \frac{(200-20) - (Th,o - Tc_o)}{ln\,\dfrac{(200-20)}{(Th_o - Tc_o)}} \right\} \times (3600)$$

substituting for Tc_o and simplifying we have

$$ln \frac{180}{2.513\ Th,o - 322.6} = 1.01 \frac{2.513\ (200 - Th,o)}{(200 - Th,o)}$$

$$= 2.54$$

or $$\frac{180}{2.513\ Th,o - 322.6} = e^{2.54} = 12.68$$

or $\quad 180 = 12.68\ (2.513\ Th,o - 322.6)$ and $Th,o = 134°C$

Since $Tc,o = 322.6 - 1.513\ Th,o$

We have $Tc,o = 322.6 - (1.513)\ (134)$

or $\quad Tc,o = 120°C$

2. For the counterflow operation

$$Q = m_h\ c\ p_h\ (Th_i - Th_o)$$

$$= U_o\ A_o\ (\Delta T)_{lm} = U_o\ A_o \times \left(\frac{(Th,i - Tc,o) - (Th,o - Tc_i)}{ln \dfrac{(Th_i - Tc_o)}{(Th,o - Tc,i)}} \right)$$

substituting the appropriate numerical values we obtain

$$= (10000)\ (1900)\ (200 - Th,o)$$

$$= 300\ (17.5) \times \left\{ \frac{(200 - Tc,o) - (Th,o - 20)}{ln \dfrac{(200 - Tc,o)}{(Th,o - 20)}} \right\} \times (3600)$$

substituting for Tc,o and rearranging we get

$$ln \left[\frac{1.513\ Th,o - 122.6}{Th,o - 20} \right] = \frac{0.513\ (Th,o - 200)}{0.986\ (200 - Th,o)}$$

$$= -0.518$$

This after simplification gives

$Th,o = 120.6°C$

Finally the energy equation yields

$Tc,o = 140.1°C$.

Comments/observations: It may be noted that in the parallel flow heat exchanger the temperature Th,o is greater than the temperature Tc,o. These two will be equal only when the area of the parallel flow heat exchanger is infinite. This is due to a progressively decreasing driving potential ($Th - Tc$) that is available for heat transfer as one moves from the inlet to the outlet. In the case of counterflow heat exchanger, the temperature of the existing cold fluid is observed to be greater than the temperature of the out going hot fluid. It can be seen from this example that a counterflow heat exchanger will transfer a greater quantity of heat than a parallel flow heat exchanger, operating under otherwise similar conditions.

Illustrations on Heat Exchangers Area Determination

1. Lubrication oil leaves an industrial engine at a temperature of 90°C with a mass flow rate of 0.025 kg/sec and is to be cooled to 50°C before returning to the engine. To accomplish this objective, the oil is to flow in the shell side of one-shell-pass, even number of tube passes heat exchanger made of tubes whose o.d is $D_o = 1.9$ cm. Water passes through the tubes at a mass flow rate of 0.04 kg/sec and with an inlet temperature of 20°C. The specific heat of water is 4.18 J/kg°C while that of the oil in question is 2100 J/kg°C. The overall heat transfer coefficient based on the outside area of the tubes is $U_0 = 250$ w/m^2°C (a) calculate the total outside heat transfer area A of the tubes (b) If the shell length is not to exceed 0.9 m, how may tubes passes are required?

Solution

We know $q = \text{U A F } \Delta T_{lm}$

It may be noted that the oil is the hotter fluid (h) and the water is the colder fluid (c) so that

$$m_h = 0.025 \text{ kg/sec}, \; cp_h = 2100 \text{ j/kg °C}$$

$$Th_i = 90°\text{C}, \; Th_0 = 50°\text{C}$$

$$m_c = 0.04 \text{ kg/sec}, \; cp_c = 4.18 \text{ j/kg °C}$$

and $Tc_i = 20°\text{C}$

$$c_h = m_h \, cp_h = (0.025)\, 2100 = 52.5$$

$$c_c = m_c \, cp_c = (0.04)\,(4180) = 167.2$$

The outlet temperature of the water stream is needed to find both ΔT_{lm} and the correction factor F.

This temperature is found through the energy balance equation

$$c_h \left(Th_i - Th_o\right) = c_c \left(Tc_o - Tc_i\right)$$

Thus $52.5 \times (90 - 50) = 167.2\,(Tc_o - 20)$

$$Tc_o = 32.6°\text{C}$$

The heat transfer rate q is found from an energy balance on the oil alone (or on water alone)

$$q = c_h\,(Th_i - Th_o) = 52.5 \times (90 - 50)$$

$$= 2100 \text{ w}$$

The value of LMTD is then evaluated as if the actual heat exchanger were pure counterflow exchanger

$$\Delta T_{lm} = \frac{(50 - 20) - (90 - 32.6)}{ln\left[\frac{50 - 20}{90 - 32.6}\right]}$$

$$= 42.2$$

The correction factor F, which takes into account the fact that the actual exchanger is not a pure counterflow but a single-shell-pass, even-multiple-tube pass exchanger is found from the appropriate graph.

Thus we have

$$S = \frac{32.6 - 20}{90 - 20} = 0.18$$

$$R = \frac{90 - 50}{32.6 - 20} = 3.17$$

F as read from the plot, corresponding to one-shell-pass, even number of tube passes is 0.93.

Using this value of F along with the previously calculated q, and ΔT_{lm} and the given value of U, the required heat transfer area (surface area) can be determined as follows.

$$A = \frac{q}{U F \Delta T_{ln}} = \frac{2100}{(250)(0.93)(42.2)}$$

$$= 0.214 \text{ m}^2$$

The total length of 1.9 cm diameter tubing needed is L and it is given by

$$A = \pi D L_{tube}$$

$$L_{tube} = \frac{0.214}{\pi(0.019)} = 3.95 \text{m}$$

Since the shell length L_{shell} is not to exceed 0.9 m, the number n of tube passes required is

$$h = \frac{L_{tube}}{L_{shell}}$$

$$= \frac{3.59}{0.9} \simeq 4$$

The number of tube passes calculated above, was predicted implicitly on the assumption of only a single tube pass from the inlet manifold to the exit manifold.

If, instead one desires to use two separate tubes, then only two tube passes of each of these separate tubes would be required in the shell length of 0.9 m.

2. Seventy-two thousand lbs/hr of engine oil whose specific heat $c_p = 0.5$ Btu/lb °F is to be cooled from 240°F to 123°F in a heat exchanger by cooling water available at 60°F whose mass flow rate is 40000 lbs/hr. The oil is to flow inside the exchange tubes and the '*U*' value based on outside tube area has been calculated to be $U = 33$ Btu/hr ft² °F. In this calculation it was observed that the relatively low oil side coefficient of heat transfer is the controlling resistance to heat transfer, so that the value of *U* remains about the same whether water flows along or across the tubes. Determine the required heat transfer surface area for the following heat exchanger configurations.

(a) Simple parallel flow exchanger

(b) Simple counterflow exchanger

(c) One-shell pass, two-tube pass exchanger

(d) Two shell-pass, four tube pass exchanger.

Solution

The outlet (exit) temperature of water is required to compute the LMTD appropriate to each of the four exchanger types. Therefore since oil is the hotter fluid and the water the cooler fluid

$$Th_i = 240°\text{F},\ Th_e = 123°\text{F},\ Tc_i = 60°\text{F}$$

$$c_h = m_h\, c p_h = 72000\ (0.5) = 36000$$

$$c_c = 40000\ (1) = 40000$$

The energy balance equation for all of the exchangers

$$c_h\,(Th_i - Th_e) = C_c\,(Tc_e - Tc_i)$$

$$36000\ (240 - 123) = 40000\ (Tc_e - 60)$$

Thus $Tc_e = 165°\text{F}$

It may be observed that the exit temperature of 165°F is greater than the required exit oil temperature of 123°F. This is not possible in a parallel flow where of course, both fluids exit at the same end of the heat exchanger. In such an exchanger both fluids also enter at the same end and then exchange energy as they flow in the same direction along the exchanger. Thus the II law of thermodynamics for this type of exchanger does not allow the original cooler fluid to ever reach a temperature which is above that of the hot fluid at the same axial position along the exchanger length. Hence a simple parallel flow exchanger cannot be used, since regardless of its surface area it cannot perform the required cooling job.

Next the LMTD for the pure counterflow exchanger will be calculated, since this is required in the calculations for all the remaining exchangers.

$$\text{LMTD} = \frac{(123 - 60) - (240 - 165.3)}{ln\left(\frac{123 - 60}{240 - 165.3}\right)}$$

$$= 68.7°\text{F}$$

The required energy transfer rate q, is determined by an energy balance on either fluid, hence

$$q = m_h\ c\,p_h\ (Th_i - Th_e)$$
$$= 72000 \times 0.5\ (240 - 123)$$
$$= 4{,}212{,}000 \text{ Btu/hr}$$

We know $q = U\ A\ \Delta\ T_{lm}$

$$4{,}212{,}000 = 33 \times A \times 68.7$$

and $$A = 1858 \text{ sq ft (for counterflow)}$$

For the one-shell pass, two-tube pass and two-shell pass four-tube pass exchangers, the appropriate rate equations is

$$A = \frac{q}{U\ F\ \Delta\ T_{lm}}$$

The value of the correction factor F can be found from the appropriate graphs. Oil is the tube-side fluid and water is the shell-side fluid.

$$Th_i = 240°\text{F},\ Th_e = 123°\text{F}$$
$$Tc_i = 60°\text{F and } Tc_e = 165.3°\text{F}$$

$$S = \frac{123 - 240}{60 - 240} = 0.65$$

$$R = \frac{60 - 165.3}{123 - 240} = 0.9$$

Using these values of S and R on the figure, for one-shell-pass, two-tube-pass exchanger, one notes that there is no intersection at $S = 0.65$ with $R = 0.9$ and therefore, there is no solution. Hence as was the case for the simple parallel flow exchanger, even a one-shell-pass, two-tube-pass exchanger of infinite area can not perform the task of the present requirement in this problem.

Turning next to the two-shell-pass, two-tube-pass exchanger, the corresponding value of F (S and R are 0.65 and 0.9 respectively), the correction factor is $F = 0.88$.

Using the equation $q = U\ A\ F\ \Delta\ T_{lm}$ the needed area is

$$A = \frac{4212000}{(33)\ (0.88)\ (68.7)}$$

$$= 2111 \text{ ft}^2$$

The required areas for the four configurations of the heat exchanger are given below:

(a) Parallel flow: Exchanger is unable to do the cooling job regardless of its area.

(b) Counterflow $A = 1858$ ft^2

(c) One-shell pass, two-tube-pass: Exchanger is unable to perform the cooling job irrespective of its area.

(d) Two-shell pass four-tube pass $A = 2111$ ft^2.

These results also reinforce what was discussed earlier in connection with the pure counterflow exchanger being the best or ideal exchanger even when it does not have infinite heat transfer area. Hence it is seen that two of the exchangers investigated were unable to perform the required cooling job, even if they could have infinite area. If the two remaining exchangers that could do the duty called for the counterflow exchanger requires less area than does the two-shell-pass four-tube-pass exchanger as long as both exchangers give essentially the same overall heat transfer coefficient U, as was the situation in this problem.

Illustration

A certain heat exchanger has a total outside area (surface) of 17.5 m^2. It is to be employed for cooling an oil at 200°C with a mass flow rate of 10000 kgs/hr having a heat capacity of 1900 J/kg°k. Water at a flow rate of 3000 kgs/hr is available at 20°C as a cooling agent. U = 300 w/m^2 K.

Calculate the effectiveness of the heat exchanger when it is used in (1) parallel mode (2) counter-flow mode.

Solution

Adopting a methodical way, we can proceed as follows.

Given: Hot oil at 200°C with a mass flow rate of 10,000 kg/hr and a heat capacity of 1900 J/kg°k is to be cooled. The inlet temperature of the water is $T_{c,i}$ =20°C. The U_0 = 300 w/m^2 °K, A_0 = 17.5 m^2

Objective: The effectiveness E, of the heat exchanger when used in both a parallel flow mode and a counterflow mode is to be determined.

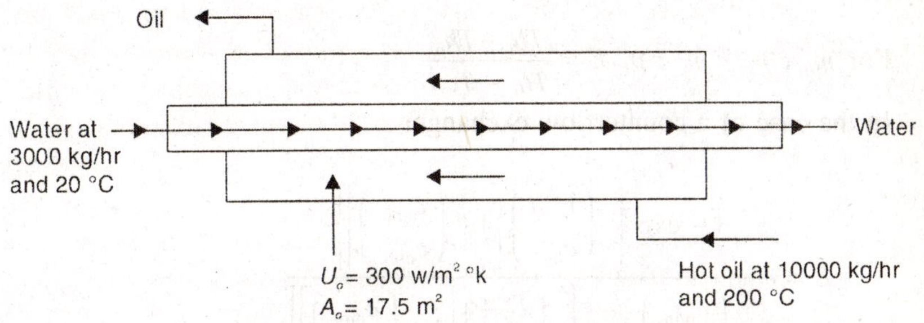

Fig. 6.24 *Counterflow arrangement*

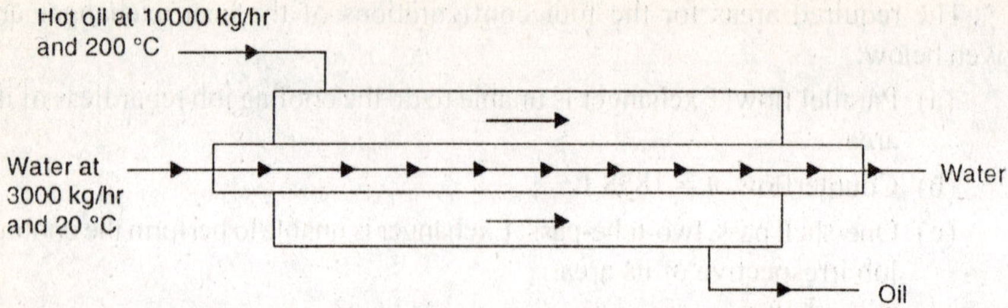

Fig. 6.25 *Parallel flow arrangement*

Assumptions to be used in the solution:

1. The tubes are clean and scale-free
2. There are no heat losses
3. U remains constant
4. The heat capacities of the oil a water remain constant.

Principle/equations involved: The net heat energy lost by the oil equals the heat energy absorbed by the water and the heat energy transferred by the heat exchanger

$$C = m\,c_p$$

where C = the heat capacity rate

m = mass flow rate

c_p = the heat capacity

For a parallel flow heat exchanger, the effectiveness is given by:

$$E = \frac{1 - \text{Exp}\left\{-\left(\frac{U\,A}{C_{min}}\right)\right\}\left[\left(1 + \frac{C_{min}}{C_{max}}\right)\right]}{1 + \frac{C_{min}}{C_{max}}}$$

For $m_h\, c\, p_h > m_c\, c\, p_c$ $E = \dfrac{Tc,o - Tc,i}{Th,i - Tc,i}$

For $m_h\, c\, p_h < m_c\, c\, p_c$ $E = \dfrac{Th_i - Th_o}{Th_i - Tc,i}$

In the case of a counterflow exchanger

$$E = \frac{1 - \exp\left\{-\left(\frac{U\,A}{C_{min}}\right)\left[1 - \left(\frac{C_{min}}{C_{max}}\right)\right]\right\}}{1 - \left(\frac{C_{min}}{C_{max}}\right)\exp\left\{-\left(\frac{U\,A}{C_{min}}\right)\left[1 - \left(\frac{C_{min}}{C_{max}}\right)\right]\right\}}$$

For $m_h\ c p_h > m_c\ c p_c$, $E = \dfrac{Tc,o - Tc,i}{Th,i - Tc,i}$

For $m_h\ c p_h < m_c\ c p_c$, $E = \dfrac{Th,i - Th,o}{Th;i - Tc,i}$

Properties:

$A_o = 17.5\ m^2$, $Tc,i = 20°C$

$m_h = 10{,}000$ kg/hr, $cp_h = 1900$ j/kg °C

$m_c = 3000$ kg/hr, $U_o = 300\ w/m^2$ °K

$Th,i = 200°C$, $cp_c = 4189$ j/kg °K (for water at 20°C)

Analysis:

$$C_c = m_c\ c p_c = (3000)\ (4189)$$
$$= 12.6 \times 10^6\ \text{j/°K hr}$$

$$c_h = m_h\ c p_h = (10000)\ (1900)$$
$$= 19 \times 10^6\ \text{j/°K hr}$$

$$c_{min} = 12.6 \times 10^6\ \text{j/°K hr}$$

and $$\frac{c_{min}}{c_{max}} = \frac{12.6 \times 10^6}{19 \times 10^6} = 0.66$$

$$\text{NTU} = \frac{U\ A}{c_{min}} = \frac{(300)\ (3600)\ (17.5)}{(12.6)\ (10^6)} = 1.5$$

1. For the parallel flow exchanger

$$E = \frac{1 - \exp\left[-1.5\ (1 + 0.66)\right]}{1 + 0.66} = 0.55$$

which is the same as read from the plot of effectiveness us NTU, for a parallel flow exchanger.

Since $m_h\ c p_h > m_c\ c p_c$

$$E = 0.55\ \frac{Tc,o - Tc,i}{Th,i - Tc,i}$$

$$(Th_i - Tc,i)\ (0.55) = (Tc,o - Tc_i)$$

$$Tc_o = Tc_i + (Th_i - Tc,i)\ (0.55)$$

$$Tc,o = 20 + (0.55)\ (200 - 20)$$

= 120°C.

2. For the counterflow heat exchanger

$$E = \frac{1 - \exp\left[-1.5\,(1 - 0.66)\right]}{1 - (0.66)\exp\left[1.5\,(1 - 0.66)\right]} = 0.67$$

which corresponds to the value of E read from the plot of effectiveness versus, NTU, for a counterflow exchanger.

Since $m_h\ c\,p_h > m_c\ c\,p_c$

$$E = 0.67 = \frac{Tc_o - Tc,i}{Th,i - Tc,i}$$

$$Tc,o - Tc,i = 0.67\,(Th,i - Tc,i)$$

$$Tc,o = Tc,i + 0.67\,(Th,i - Tc,i)$$

$$= 20 + 0.67 \times (200 - 20)$$

$$= 140°C$$

Remarks/observations: The outlet temperature values for the cooling water agree with those of the earlier problem on the same heat exchanger. From theoretical considerations, one expects to get identical answers to a heat exchanger problem regardless of whether the LMTD or the NTU approach is adopted. However, inaccuracies in reading the charts often bring about a discrepancy in the results obtained by the two approaches.

Illustration

A certain heat exchangers has one shell pass and four tube passes, with a total surface area of 100 ft^2. Oil with a heat capacity of 0.6 Btu/lb°F comes at temperature of 350°F at a rate of 6545 lbs/hr. Cooling water at a rate of 6000 lbs/hr is available at 60°F. Determine the exit temperatures of the oil and the water if the U value is 180 Btu/hr ft^2 F.

Solution

Given: A heat exchanger with one shell and four tube passes has a total surface area of 100 ft^2. U = 180 Btu/hr ft^2 F. The heat exchanger is used to cool oil. For the oil.

Th,i = 350°F, cp_h= 0.6 Btu/lb °F and the flow rate is 6545 lbs/hr. The coolant is water at 60°F with a mass flow rate of 6000 lbs/hr.

Objective: To determine the exit temperature of the oil and the water.

Assumptions:

1. The tubes are clean and scale free.
2. The heat capacities are constant in the temperature range under consideration.

3. The U remains constant and does not change with temperature.

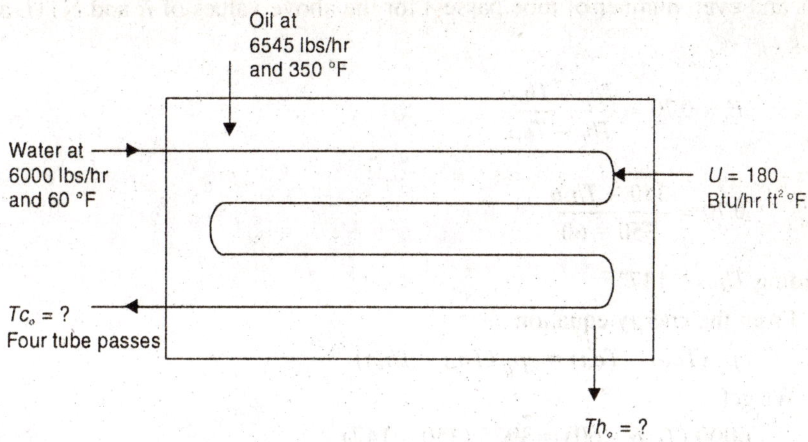

Fig. 6.26 *1–4 heat exchanger—1 shell pass and four tube passes*

Principles/expressions to be used: This problem involves the use of the plot E versus NTU, with R as the parameter

$$\left(\frac{c_{\min}}{c_{\max}} = R\right)$$

For $m_h \, c\,p_h < m_c \, c\,p_c$, $E = \dfrac{Th,i - Th,o}{Th_i - Tc,i}$

For $m_h \, c\,p_h > m_c \, c\,p_c$, $E = \dfrac{Tc,o - Tc,i}{Thi - Tci}$

The relevant energy equation is, $m_c \, cp_c \, (Tc, o - Tc_i) = m_h \, cp_h \, (Th_i - Th_o)$

Properties: Oil flow rate = 6545 lbs/hr

Tc_i = 60°F, m_c = 6000 lbs/hr

cp_h = 0.6 Btu/lb °F, Th_i = 350°F

cp_c = 1.00 Btu/lb °F for water at 60°F

Analysis: To determine the effectiveness

m_h = 6545 lbs/hr

$$c_c = m_c \, c\,p_c = (6000)\,(1) = 6000 \text{ Btu/hr °F}$$

$$c_h = m_h \, c\,p_h = (0.6)\,(6545) = 3925 \text{ Btu/hr °F}$$

$$R = \frac{c_{\min}}{c_{\max}} = \frac{4805}{6000} = 0.65$$

$$\text{NTU} = \frac{U\,A}{c_{\min}} = \frac{(180)\,(100)}{3925} = 4.59$$

From the appropriate plot of E vs NTU, with R as the parameter (single shell pass and even number of tube passes) for the above values of R and NTU, and since $c_h < c_c$

$$E = 0.70 = \frac{Th_i - Th,o}{Th_i - Tc,i}$$

or
$$0.70 = \frac{350 - Th,o}{350 - 60}$$

yielding $Th,o = 147°F$

From the energy equation

$$cp_c\,(Tc,o - Tc,i) = cp_h\,(Th,i - Th,o)$$

We get

$$6000\,(Tc,o - 60) = 3925\,(350 - 147)$$

or $Tco = 192.8°F$

Comments: From the analysis we observe that the outlet oil temperature is 147°C and the outlet water temperature is 192.8°F.

Illustration on Multipass Heat Exchanger

Determine the area (heat transfer) of a two-shell pass and four-tube pass heat exchanger, to heat oil with a mass flow rate of 10000 lbs/hr from 60°F to 150°F if hot water at a flow rate of 8000 lbs/hr is available at a temperature of 200°F. $cp_{oil} = 0.4$ Btu/lb °F The overall heat transfer coefficient is 95 Btu/hr ft^2 °F. Oil is the tube fluid.

Solution

Given: $m_c = 10000$ lbs/hr

$cp_c = 0.4$ Btu/lb °F, $Tc_i = 60°F$

$Tc_o = 150°F$

The heat exchanger uses hot water at 200°F and at a mass flow rate of 8000 lbs/hr. $U = 95$ Btu/hr-ft^2 °F

The oil is the tube fluid.

Objective: The surface area of the heat exchanger is to be determined.

Assumptions:

1. There are no heat losses.
2. The overall heat transfer coefficient is constant and uniform.
3. The heat capacity of the fluid is constant over the temperature range in question. This applies to both the fluids.
4. The tubes are scale free and clean.

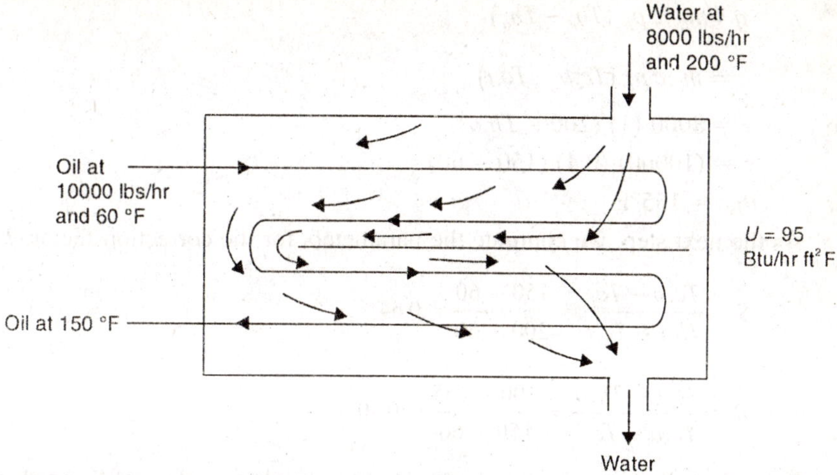

Fig. 6.27 *2–4 heat exchanger—two shell pass and four tube passes*

Principles employed: The nett heat lost by the water is equal to the net heat gained by the oil and it is also equal to the heat energy transmitted across the heat transfer surface.

$$q = m_h\, cp_h\,(Th_i - Th_o) = m_c\, cp_c\,(Tc,o - Tc,i)$$

$$= \text{UAF (LMTD)}$$

where F is the correction factor for a multi tube multi shell heat exchangers which can be read from the appropriate plot-F vs S, with R as the parameter.

$$\text{LMTD} = \frac{(Th_i - Tc_o) - (Th_o - Tc,i)}{ln\dfrac{(Th_i - Tc,o)}{(Th,o - Tc,i)}}$$

$$S = \frac{Tc_o - Tc_i}{Th_i - Tc_i}$$

$$R = \frac{Th,i - Th,o}{Tc,o - Tc,i}$$ Both are temperature ratio parameters

Properties:

$m_h = 8000$ lbs/hr

$m_c = 10000$ lbs/hr

$Tc_i = 60°\text{F}$, $Tc,o = 150°\text{F}$

$Th_i = 200°\text{F}$, $cp_c = 0.4$ Btu/lb°F

$cp_{\text{water}} = cp_h = 1$ Btu/lb °F

To find Th_o

$$q = m_h\, c\, p_h\, (Th_i - Th_o)$$

$$= m_c\, c\, p_c\, (Tc,o - Tc,i)$$

so $$= 8000\ (1)\ (200 - Th,o)$$

$$= (10000)\ (0.4)\ (150 - 60)$$

or $th,o = 155°F$

As the next step, we compute the parameters for the correction factor, F

$$S = \frac{Tc,o - Tc_i}{Th,i - Tc,i} = \frac{150 - 60}{200 - 60} = 0.64$$

$$R = \frac{Th,i - Th,o}{Tc,o - Tc,i} = \frac{200 - 155}{150 - 60} = 0.50$$

From the relevant correction factor plot, for the above values of S and R one can read the value of F

$$F = 0.96$$

$$\text{Also LMTD} = \frac{(200 - 150) - (155 - 60)}{ln\left\{\dfrac{(200 - 150)}{(155 - 60)}\right\}}$$

$$= 70.1°F$$

Since the amount of heat transferred is given by:

$$q = m_h\, c\, p_h\, (Th_i - Th,o)$$

$$= UAF\ (LMTD)$$

We have (8000) (1) (200 – 155) = 95 A (0.96) (70.1)

or $A = 56.3\ ft^2$.

Illustration on heat exchanger area calculation

Hot engine oil at 100°C with a mass flow rate of 0.03 kg/sec and a specific heat of 2100 J/kg°C is to be cooled to 45°C by making it pass through single-shell-pass, even number of tube passes exchanger in which the coolant is engine fuel whose specific heat is 2130 J/kg°C. The engine fuel enters the heat exchanger at 15°C and is to leave at 75°C. An U_0 of 200 $\frac{W}{m^2\,°C}$ is expected and is based on the outside tube area. Find the required heat transfer surface area by using the effectiveness—NTU approach.

Solution

The capacity rate C_c of the cooler engine fuel will be determined first by the overall energy balance

$$c_h\,(Th_i - Th_e) = c_c\,(Tc_e = Tc_i)$$

Since the ratio of c_{min} to c_{max} is required in the solution functions of the effectiveness-NTU approach

From the given information, we have

$$Th_i = 100°C \qquad Th_e = 45°C$$

$$c_h = m_h \, c p_h$$

$$= (0.03)\,(2100) = 63$$

$$Tc_i = 15°C \qquad Tc_e = 75°C$$

and $\quad cp_c = 2130$ j/kg °C

From the above equation, we can estimate c_c

$$c_c\,(Tc_e - Tc_i) = c_h\,(Th_i - Th_e)$$

$$c_c\,(75 - 15) = 63\,(100 - 45)$$

$$c_c = 57.75$$

Hence $c_{min} = 57.75 \quad c_{max} = 63$

and therefore $\dfrac{c_{min}}{c_{max}} = 0.917$

Then the effectiveness E of this exchanger can be determined from the effectiveness expression

$$E = \frac{Tc_e - Tc_i}{Th_i - Tc_i} = \frac{75 - 15}{100 - 15}$$

$$= 0.706$$

The plot of effectiveness-NTU for one shell pass, even-numbers (multiples) of tube passes exchanger has to be considered here. This plot with $E = 0.706$ and $\dfrac{c_{min}}{c_{max}} = 0.917$. does not yield any solution. In other words this exchanger with an infinite amount of surface area giving an infinite number of transfer units, can not have an effectiveness as high as 70.6%, for the capacity ratio in question.

Thus we will go to a two-shell tube even multiple of four tube passes heat exchanger to find out if it is capable of handling the required heat transfer load.

Referring to the appropriate plot, we can observe that

$$\text{NTU} \simeq 3.1 = \frac{U_o\,A_o}{c_{min}}$$

$$= 200 \times \frac{A_o}{57.75}$$

$$A_o = 0.895 \text{ m}^2$$

In this problem since all the temperatures (inlet and outlet) were known in advance, the correction factor method could also have been used with approximately the same amount of work, though some design engineers favouring the effectiveness—NTU method argue that the need to calculate the algebraically messy logarithmic mean temperature difference gives the effectiveness – NTU approach an advantage even in this type of design problem. The reader can contrast the relative amount of work in the two methods for the design type problem by comparing the problem just solved with the equivalent one solved by the $F - \Delta T_{lm}$ method.

Illustration

An already-designed oil cooler is a cross flow exchanger with the oil unmixed and the water coolant mixed and a heat transfer surface area of 190 ft^2. It is intended to use this heat exchanger, for a part of the day, at an off-design point where the oil mass flow rate would be doubled to 22500 lbs/hr entering the exchanger at 230°F while the water coolant mass flow rate remained at 20,000 lbs/hr entering at 60°F.

The U_o = 60 Btu/hr ft^2 °F, cp_{oil} = 0.45 Btu/lb °F

cp_{water} = 1.0 Btu/lb °F

Determine the exit temperature of both oil and water at this off-design condition.

Solution

This problem calls for a performance calculation, because both the exit temperatures are unknown and would necessitate, an iterative type of solution by the correction factor method.

The effectiveness—NTU method is the appropriate solution technique in this case. We have

$$m_h\, c\,p_h = c_h = 22500\,(0.45)$$

$$= 10{,}125$$

$$m_c\, c\,p_c = c_c = 20000\,(1.0) = 20000$$

$$\text{Therefore } c_{min} = c_{un\ mixed} = 10{,}125$$

$$c_{max} = c_{mixed} = 20{,}000$$

$$Th_i = 230°\text{F and } Tc_i = 60°\text{F; NTU} = \frac{UA}{c_{min}}$$

$$= \frac{(60)\,(190)}{10{,}125} = 1.126$$

$$\frac{c_{mixed}}{c_{un\ mixed}} = \frac{20000}{10{,}125} = 1.975$$

At NTU = 1.126 and the capacity rate ratio of mixed to unmixed fluid of 1.975, the plot of E vs NTU for the cross flow arrangement/configuration, yields a value of effectiveness $E \simeq 56\% = 0.56$. Next the oil exit temperature Th_e is found from the effectiveness expression $c_h = c_{min}$, in this problem

$$0.56 = \frac{10{,}125\,(230 - Th_e)}{10{,}125\,(230 - 60)}$$

Therefore $Th_e = 135°F$

Finally, the other exit fluid temperature Tc_e, can be determined from the overall energy balance or from the expression for effectiveness

$$E = \frac{c_c\,(Tc_e - Tc_i)}{c_{min}\,(Th_i - Tc_i)}$$

The latter yields

$$0.56 = \frac{20000\,(Tc_e - 60)}{10125\,(230 - 60)}$$

$Tc_e = 108°F$, which is the cooling water temperature.

The oil exit temperature, found earlier is $Th_e = 135°F$.

It may be noted that when the effectiveness vs NTU plot for the cross flow exchanger is used even though the capacity rate ratio is in terms of mixed to unmixed fluids rather than in terms of minimum to maximum capacity rates, the number of transfer units is still calculated using the capacity rate c_{min}.

Illustration

In a water-to-water heat exchange process, 750 kg/sec of pressurized liquid water is to be cooled from 288°C to 38°C by 1500 kg/sec of cooling water available at the temperature of 10°C. This is to be accomplished in a multi pass shell-and-tube heat exchanger where the initial estimate of the U_0 is900 w/m^2 °C. It is proposed to use a one-shell-pass, even number of tube passes exchanger if it can handle the heat transfer objective. It this is not feasible, an attempt will be made to use a two-shell-pass, even multiple tube pass unit. Finally if this still does not perform the job an attempt will be made to employ a number of equal size, equal heat transfer area, two shell-pass exchangers connected in series for the hot fluid and with equal mass flow rates of 10°C cold water entering each exchanger. For the colder water, $cp_c = 4180$ j/kg °C, while the hotter water has $cp_h = 4350$ j/kg °C. Find the required heat transfer area and the type of the heat exchanger or number of two-shell-pass exchangers to be used.

Solution

The capacity rate are determined first

$$c_c = m_c\, cp_c = (1500)\,(4180)$$

$= 6,270,000$

$c_h = m_h \; c p_h = 750\,(4350)$

$= 3,262,500$

Therefore $c_{\min} = c_h = 3,262,500$

and $\dfrac{c_{\min}}{c_{\max}} = 0.52$

The effectiveness required of the heat exchanger is determined next

$$E = \frac{c_h\,(Th_i - Th_e)}{c_{min}(Th_i - Tc_i)}$$

$$= \frac{3,262,500\,(288 - 38)}{3,262,500\,(288 - 10)} = 0.9$$

Referring to the plot of effectiveness vs NTU for one shell pass and even number of tube passes with these values of E and $c_{\min}/c_{\max}$, we can observe that there is no single-pass, even-multiple tube pass exchanger which can handle the heat transfer requirement, as there is no intersection of $E = 0.90$ with a curve of constant $c_{\min}/c_{\max} = 0.52$.

Likewise an examination of the plot pertaining to the two-shell-pass, even multiple of four tube passes configurations yields no solution, unless this intersection occurs beyond NTU = 5.5 where the curve $c_{\min}/c_{\max}$ is practically horizontal. Therefore, as a next step we will try to find a solution employing two-shell-pass, even multiple of four passes exchangers, identical in area through which the hotter water passes in series and half the coolant mass flow rate 750 kg/sec enters each of the two heat exchangers at 10°C.

Let us designate the first exchanger as '*a*' and the second exchanger as '*b*'.

The inlet temperatures to exchanger '*a*' are $Thi\,a = 288$°C and $Tci\,a = 10$°C. The outlet temperatures Th_{ea} and Tc_{ea} are unknown. The outlet temperature of the hot fluid from a, is the inlet temperature of the hot fluid for exchanger *b*, so $Th_{ib} = Th_{ea}$. Also $Tc_{ib} = 10$°C and $Th_{eb} = 38$°C while Tc_{eb} is not known.

$c_{ca} = m_{ca}\; c\,p_c = c_{ch} = m_c\; c\,p_c$

$= 750\,(4180)$

$= 3,135,000 = c_{\min}$

$$\left(\frac{c_{\min}}{c_{\max}}\right)_a = \left(\frac{c_{\min}}{c_{\max}}\right)_b$$

$$= \frac{3,135,000}{3,262,500}$$

$= 0.96$

For both exchangers $c_c = c_{min}$ and the effectiveness, can be found out as follows

$$E_a = \frac{Tc_{ea} - Tc_{ie}}{Th_{ia} - Tc_{ia}} = \frac{Tc_{ea} - 10}{288 - 10} = \frac{Tc_{ea} - 10}{278}$$

$$E_b = \frac{Tc_{eb} - Tc_{ib}}{Th_{ib} - Tc_{ib}} = \frac{Tc_{eb} - 10}{Th_{ea} - 10}$$

However since both exchangers have the same area (unknown), the same U and the same c_{min}, they have the same NTU. In addition since c_{min}/c_{max} is also the same, it follows that their effectiveness is also same, that is $E_a = E_b$. Thus equating the two preceding expression for E_a and E_b, yields

$$\frac{Tc_{ea} - 10}{278} = \frac{Tc_{eb} - 10}{Th_{ea} - 10} \qquad ...(1)$$

Equation (1) contains three unknown temperatures but Tc_{ea} and Tc_{eb} can be expressed in terms of Th_{ea} by overall energy balances on exchangers a and b separately.

Thus for the heat exchanger a

$$c_h\,(288 - Th_{ea}) = c_{ca}\,(Tc_{ea} - 10)$$

$$3{,}262{,}500\,(288 - Th_{ea}) = 3{,}135{,}000\,(Tc_{ea} - 10)$$

or $$Tc_{ea} - 10 = 299.7 - 1.04\,Th_{ea} \qquad ...(2)$$

The energy balance for the exchanger b gives

$$c_h\,(Th_{ea} - 38) = c_{cb}\,(Tc_{eb} - 10)$$

or $$Tc_{eb} - 10 = 1.04\,Th_{ea} - 39.5 \qquad ...(3)$$

Inserting equations (2) and (3) into equation (1) and solving for the resulting quadratic equation for Th_{ea}, yields

$Th_{ea} = 98.3°C$

The value Tc_{ea} is needed in the equation given earlier for E_a with Th_{ea} now known, Tc_{ea} is found next from equation (2), to be

$Tc_{ea} = 207.4°C$

$$\text{Thus } E_a = \frac{Tc_{ea} - 10}{278} = \frac{207.4 - 10}{278} = 0.71$$

with $E_a = 0.71$ and $\left(\dfrac{c_{min}}{c_{max}}\right)_a = 0.96$

One finds from the expression for NTU

$$(\text{N T U})_a = 3.3 = \frac{U A_a}{c_{\min a}}$$

$$= \frac{900\ A_a}{3{,}135{,}000}$$

$$A_a = 11495 \text{ m}^2$$

Since exchanger '*b*' has the same area as exchanger '*a*', two identical heat exchangers of the two-shell-pass even multiple of four tube passes type are required with total exchanger surface of nearly 23,000 m^2.

It is also of interest to see what area would be required in a pure counterflow exchanger if economy and compactness did not dictate a multi-shell, multi-tube-pass heat exchanger. Earlier it was observed that the effectiveness of a single exchanger would have to be 0.9 at a capacity rate ratio $c_{\min}/c_{\max} = 0.52$.

From the effectiveness-NTU plot for a counter current heat exchanger

$$\text{N T U} = 3.55 = \frac{AU}{c_{min}}$$

$$= \frac{900 \cdot A}{3{,}262{,}500}$$

$$A = 12869 \text{ m}^2$$

With this first estimate of the required heat transfer surface area for the two, required two-shell-pass, multi-tube-pass exchangers or the single pure counterflow exchangers along with the multitude of other design criteria/requirements such as size, allowable pressure drop, cost etc, a design engineer or a consultant should decide which of these two options is a better one. Details of such design decisions are discussed in length by Taboreh in his article in viz "Evolution of heat exchanger design techniques" in the journal "Heat Transfer Engineering" vol. 1 No. 1 July–September 1979, pp 15–29.

Illustration on the Calculation of Exit Temperature of the Fluid in a Heat Exchanger

Eleven thousand lbs/hr of a saturated steam vapour at 8 psia leaves a small steam turbine and it is to be condensed to a saturated liquid. A small shell-and-tube heat exchanger is to be employed with city water entering the tubes at 50°F and leaving at 160°F. The estimated U_0 (on clean conditions) is 225 Btu/hr ft^2°F. The enthalpy of vaporization of the 8 psia steam is $\Delta H_v = 988.4$ Btu/lb and its saturation temperature is 182.8°F (from steam tables). (a) Determine the mass flow rate of the coolant needed and the required heat transfer area (b) Subsequent to a certain period of operation, the heat exchanger becomes fouled or dirty and is still required to completely condense the 11,000 lbs/hr of the steam. Determine the required mass flow rate of the coolant and the exit coolant temperature under this condition. The inside fouling resistance due to city water is 0.0015 and that due to steam is 0.0005.

Solution

The required mass flow rate of the coolant in the initial clean condition of the exchanger is found out by an overall energy balance on the exchanger

$$m_c\ cp_c\ (Tc_e - Tc_i) = m_h\ \Delta H_v$$

$$m_c\ (1)\ (160 - 50) = (11{,}000)\ (988.4)$$

$$m_c = 98,\ 840 \text{ lbs/hr}$$

The heat transfer surface area is determined by equating the heat transfer rate q as given by the basic rate equation.

The expression to be employed is

$$U_o\ A_o\ F\ \Delta T_{lm} = m_h\ \Delta H_v$$

However in the problem statement, the type of shell-and-tube exchanger, number of shell passes and number of tube passes were not specified. However, since one of the fluids (steam) is undergoing a phase change at constant temperature, then the LMTD with $F = 1$ is to be used irrespective of the exchanger type and the form of this temperature difference appropriate to the phase change process is given by the expression

$$\Delta T_{\text{LMTD}} = \frac{Te - Ti}{ln\left(\dfrac{Ti - T_p}{T_e - T_p}\right)}$$

In this problem $T_p = 182.8°\text{F}$, $T_i = 50°\text{F}$ and $T_e = 160°\text{F}$

$$\Delta T_{\text{LMTD}} = \frac{160 - 50}{ln\left(\dfrac{50 - 182.8}{160 - 182.8}\right)} = 62.4°\text{F}$$

using this value in the expression

$U_o\ A_o\ F\ \Delta T_{\text{LMTD}} = m_h\ \Delta H_v$, we get

$(250)\ A_o\ (1)\ (62.4) = 11{,}000\ (988.4)$

$A_o = 774.4$ sq. ft

also $m_c = 98{,}840$ lbs/hr

In (b), it is required to study the exchanger performance after it has been in service for a long period of time to entail the formation of scale deposits on the surface. It is well known that the fouling layers on both the inside and the outside of the tubes add resistance to heat transfer and hence bring about a reduction in the magnitude of 'U' relative to its "clean" value of 225 Btu/hr ft^2 °F.

The inside fouling resistance due to the city water and the outside fouling resistance due to the steam are given in the problem statement as 0.0015 and 0.0005 respectively.

The U_{of} (for this changed situation) is calculated to be

U_{of} = 155.2 Btu/hr ft^2 °F (assuming that the tube is very thin)

Using the equations $U_o\ A_o\ F\ \Delta T_{\text{LMTD}} = m_h\ \Delta H_v$ and

$$\Delta T_{\text{LMTD}} = \frac{Te - Ti}{ln\left[\dfrac{Ti - T_p}{Te - T_p}\right]}$$, we can get the following

$$(155.2)\,(774.4)\,(1) = \frac{Te - 50}{ln\left[\dfrac{182.8 - 50}{182.8 - T_e}\right]} = 11{,}000\,(988.4)$$

$$ln\left[\frac{132.8}{182.8 - T_e}\right] = \frac{T_e - 50}{90.46}$$

This equation can now be solved by iteration for T_e, by employing an initial estimate of T_e on the right side and calculating T_e on the left, then using this value as the new estimate on the right until two successive iterates for T_e agree closely enough. As an initial estimate of T_e, one can use the T_e value which the arithmetic average temperature difference

$$182.8 - \frac{50 + Te}{2}$$

would give if it were in the equation $U_o\ A_o\ F\ \Delta T_{lm} = m_h\ \Delta H_v$ instead of the correct ΔT_{LMTD}.

Doing this gives an initial estimate of T_e = 135°F, which when used on the right side of the above equation gives T_e = 130.9°F. This when used on the right side of the above equation, yields 128.5°F. Continuation of this process leads to the new coolant exit temperature of T_e = 124.6°F

Next, the energy balance on the exchanger yields the new mass flow rate of the coolant m_c

$$m_c\ cp_c\ (Tc_e - Tc_i) = m_h\ \Delta H_v$$

$$m_c\ (1)\,(124.6 - 50) = 11{,}000\,(988.4)$$

m_c = 145,740 lbs/hr

Also T_e = 124.6°F

It is thus seen that the new coolant mass flow rate of 145,740 lbs/hr is needed to bring about complete condensation of the steam after the exchanger becomes fouled and this leads to a new exit coolant temperature of 124.6°F.

Examples on the Use of the Effectiveness Method to Calculate *U* and *A* (Their Product)

1. A counterflow heat exchanger is to be designed to cool 2000 lbs/hr of oil from a temperature of 140°F to 90°F. Water at 60°F is available for the cooling medium. The specific heat of the oil is assumed to be 0.5 Btu/lb°F. The maximum allowable exit temperature is 80°F because of possible scale formation at higher temperatures. Determine the value of the product of the total area and the overall heat transfer coefficient for this situation.

Solution

The heat to be transferred per hour to cool the oil equals

$$q = 2000\ (0.5)\ (140 - 90) = 50{,}000 \text{ Btu/hr}$$

From this the water rate may be obtained assuming the maximum allowable temperature rise, hence

$$50{,}000 = m_c\ (1)\ (80 - 60)$$

$$m_c = 2500 \text{ lbs/hr}$$

The value of for $\dfrac{m_h\ c\,p_h}{m_c\ c\,p_c}$ is

$$\frac{m_h\ c\,p_h}{m_c\ c\,p_c} = \frac{2000\ (0.5)}{2500\ (1)} = 0.4$$

and the effectiveness ratio for this case equals

$$E = \frac{t'_a - t'_b}{t'_a - t_a} = \frac{140 - 90}{140 - 60} = \frac{50}{80} = 0.625$$

where t'_a = inlet temperature shell side

t'_b = outlet temperature, shell side

t_a = inlet temperature tube side.

It may be mentioned in this context that,

t_b = outlet temperature, tube side

From the appropriate figure E vs $\dfrac{U\,A}{m_h\ c\,p_h}$

$$\frac{U\,A}{m_h\ c\,p_h} = 1.15$$

Therefore the value for UA would be

$$UA = 1.15\ (2000)\ (0.5)$$

$$= 1150 \text{ Btu/hr °F.}$$

In order to compare this procedure with that using the log mean temperature, a further calculation will be made. The value for the log mean temperature difference equals

$$(\Delta t)_{lm} = \frac{(140 - 80) - (90 - 60)}{ln \dfrac{140 - 80}{90 - 60}}$$

$$= 43.3$$

The value of UA equals

$$q = 50{,}000 = U\,A\,(\Delta t)_{lm}$$

$$U\,A = \frac{50{,}000}{43.3}$$

$$= 1150 \text{ Btu/hr °F}$$

2. Assume that the heat exchanger mentioned in the previous example has been built. If for some reason, it is desired to operate it at an oil flow rate of 1000 lb/hr what would be the expected terminal temperatures of the oil and water? The inlet oil temperature for the new condition will be 160°F. The water rate and entering water temperature remain the same.

Solution

The ratio of the products of the flow rates and specific heats is

$$\frac{m_h\ c\,p_h}{m_c\ c\,p_c} = \frac{(1000)\ (0.5)}{(2500)\ (1)} = 0.2$$

and the value for $\dfrac{U\,A}{m_h\ c\,p_h}$ is

$$\frac{U\,A}{m_h\ c\,p_h} = \frac{1150}{(1000)\ (0.5)} = 2.3$$

From the appropriate plot of E vs $\dfrac{U\,A}{m_h\ c\,p_h}$ for a counterflow heat exchanger, the value for the effectiveness ratio is approximately

$$E = \frac{t'_a - t'_b}{t'_a - t_a} = 0.86$$

and $\quad t'_a - t'_b = 0.86\ (160 - 60) = 86°\text{F}$

Therefore $t'_b = t'_a - 86 = 160 - 86 = 74°F$

The heat transfer rate is

$$q = 1000\ (0.5)\ (160 - 74)$$
$$= 43{,}000 \text{ Btu/hr}$$

The terminal water temperature may be found as follows

$$43{,}000 = 2500\ (1)\ (t_a - t_b)$$
$$t_b = 17.2 + 60 = 77.2°F.$$

Illustration on Heat Exchanger Area Computation

Benzene is obtained from a fractionating column as saturated vapour at 176°F. Determine the heat transfer area required to condense and subcool 8000 lb/hr of benzene to 115°F if the coolant is water with the flow rate of 40,000 lbs/hr available at 55°F. Compare the areas needed for (a) counterflow and (b) parallel flow configurations. An overall heat transfer coefficient of 200 Btu/hr ft^2 °F may be used for this situation. The latent heat of condensation for benzene is 169.6 Btu/lb.

Solution

The temperature length diagrams for the counter and parallel flows are indicated below.

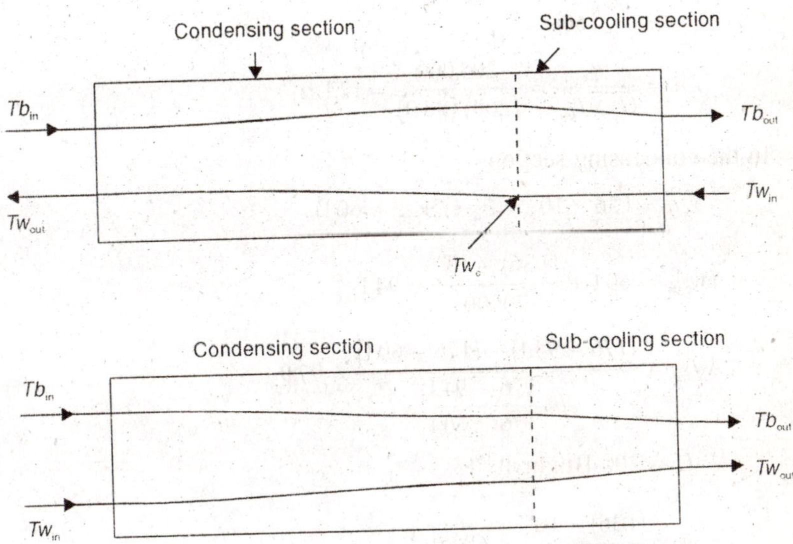

Fig. 6.28 *Temperature—length diagram for counter flow and parallel configurations*

Determining the total heat to be transferred we have, for the condensing section

$$q = (8000)\ (169.6) = 1.36 \times 10^6 \text{ Btu/hr}$$

and in the sub cooling section

$$q = (8000)\,(176 - 115)\,(0.42)$$

$$= 205{,}000 \text{ Btu/hr}$$

The above sketches (Temperature – Length) will be referred to in the analysis of the counterflow and parallel flow cases.

For the counterflow configuration, in the sub-cooling section

$$q = 205{,}000 = mc_{p_w}\,(Tw_c - 55)$$

$$Tw_c = 55 + \frac{205{,}000}{40{,}000 \times 1}$$

$$= 55 + 5.12 = 60.1$$

In the sub cooling section for (a) then

$$\Delta T_{lm} = \frac{(176 - 60.1) - (115 - 55)}{ln \dfrac{176 - 60.1}{115 - 55}}$$

$$= \frac{115.9 - 60}{ln \dfrac{115.9}{60}} = \frac{55.9}{0.658} = 84.9$$

$$A = \frac{q}{U\,\Delta T_{lm}} = \frac{205{,}000}{(200)\,(84.9)} = 12.1 \text{ ft}^2$$

In the condensing section

$$q = 1.36 \times 10^6 = c_w\,(Tw_{out} - 60.1)$$

$$Tw_{out} = 60.1 + \frac{(1.36) \times 10^6}{40000} = 94.1$$

$$\Delta T_{lm} = \frac{(176 - 94.1) - (176 - 60.1)}{ln \dfrac{176 - 94.1}{176 - 60.1}} = 97.9$$

$$U = 200 \text{ Btu/hr ft}^2\text{°F}$$

$$A = \frac{(1.36) \times 10^6}{(200)\,(97.9)} = 69.5 \text{ ft}^2$$

The required area for the counterflow case is the sum of the two parts or 81.6 ft^2.

For the parallel flow case, in the condensing section

$$q = 1.36 \times 10^6 = c_w\,(Tw_c - 55)$$

$$Tw_c = 55 + \frac{1.36 \times 10^6}{40{,}000} = 89°F$$

$$\Delta T_{lm} = \frac{(176 - 55) - (176 - 89)}{ln \dfrac{176 - 55}{176 - 89}} = 103.1$$

$$A = \frac{1.36 \times 10^6}{(200)\,(103.1)} = 66.0 \ \text{ft}^2$$

In the sub-cooling section

$$q = 205{,}000 \text{ Btu/hr} = c_w\,(Tw_{out} - 89)$$

$$Tw_{out} = 89 + \frac{205{,}000}{40{,}000} = 94.1°F$$

$$\Delta T_{lm} = \frac{(176 - 89) - (115 - 94.1)}{ln \dfrac{176 - 89}{115 - 94.1}}$$

$$= 46.3°F$$

$$A = \frac{205{,}000}{(200)\,(46.3)} = 22.1 \text{ ft}^2$$

The total area for the parallel flow configuration is 88.1 ft^2.

As seen from the above results, counterflow configuration is more efficient (it requires less heat transfer area) than the parallel one.

The latter requires 8% more area in this case.

Illustration on the Use of Correction Factor Chart in the Heat Exchanger Area Compution

Benzene, is available as saturated liquid at 170°F and it is to be cooled to 115°F for transporting. The benzene is supplied at a rate of 8000 lbs/hr. The cooling water is available at 55°F and a flow rate of 5000 lbs/hr. Determine the required heat exchanger area for the following configurations:

(i) Single pass counterflow

(ii) Shell-and-tube with water making one shell pass and benzene four tube passes

An overall heat transfer coefficient of 55 Btu/hr ft^2 °F may be used in each case.

Solution

Striking an energy balance

$$q = m\, c_p\, \Delta T_{\text{Benzene}}$$

$$= (8000)\,(0.42)\,(170 - 115)$$

$$= 185{,}000 \text{ Btu/hr}$$

and the required exiting water temperature

$$q = m\, c_p\, \Delta T_{\text{H20}}$$

$$185{,}000 = (5000)\,(1)\,(T_{\text{H}_2\text{O exit}} - 55°\text{F})$$

$$T_{\text{H}_2\text{O exit}} = 55 + \frac{185{,}000}{5000}$$

$$= 55 + 37 = 92°\text{F}$$

For single pass counterflow, configuration referring to the sketch given below,

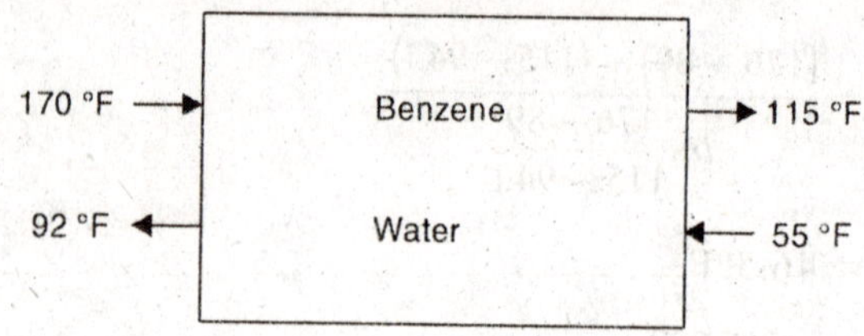

Fig. 6.29 *Single-pass counter flow arrangement*

$$\Delta T_{lm} = \frac{(170 - 92) - (115 - 55)}{ln \dfrac{170 - 92}{115 - 55}}$$

$$= \frac{18}{ln \dfrac{78}{60}} = 68.6°\text{F}$$

Therefore the heat transfer area is

$$A = \frac{q}{U\, \Delta T_{lm}} = \frac{185{,}000}{(55)\,(68.6)}$$

$$= 49.0 \text{ ft}^2$$

For the shell-and-tube configuration

The situation is represented in the sketch given below:

For this case, the correction factor chart has to be used to calculate the actual temperature difference. The correction factor found from the chart (by using the appropriate temperature difference ratios, the parameters in the chart) is 0.93.

The heat transfer area is hence

$$A = \frac{q}{U\ F\ \Delta T_{lm}} = \frac{185{,}000}{(55)\,(0.93)\,(68.6)}$$

$$= 52.7\ \text{ft}^2.$$

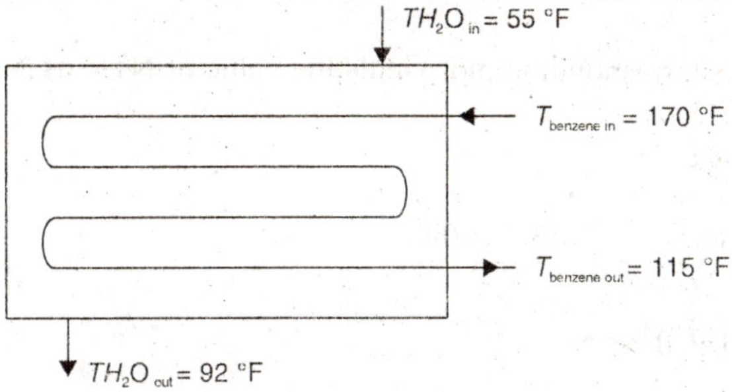

Fig. 6.30 *Shell and tube arrangement*

Illustration on the Application of the Effectiveness Concept

1. Rework the above example using the heat exchanger effectiveness approach to determine the heat transfer area required.

Solution

From previous calculations (the ones done in the above problem) the following values have been established.

$$q = 185{,}000\ \text{Btu/hr},\ C_{H_2O} = 5000\ \text{Btu/hr °F}$$

$$T_{H_2O\,(exit)} = 92°\text{F},\ C_{Benzene} = 3360\ \text{Btu/hr °F}$$

Benzene turns out to be the minimum fluid

The required value of 'E' may be found out from the equation

$$q = E\,c_{min}\,(Th_{in} - Tc_{in})$$

$$E = \frac{q}{c_{min}\left(Th_{in} - Tc_{in}\right)}$$

$$= \frac{185{,}000}{(3360)\,(170 - 55)} = 0.479$$

This value and the numerical value $\frac{c_{min}}{c_{max}} = \left(\frac{3360}{5000}\right) = 0.672$ allow NTU to be determined for each of the configuration is question

For the counterflow, single-pass case $NTU = \dfrac{UA}{c_{min}} = 0.8$ and the area required is determined as:

$$A = \frac{0.8\, c_{min}}{U} = \frac{(0.8)\,(3360)}{55} = 48.9 \text{ ft}^2$$

For the situation where in the heat exchanger is with one shell pass and four tube-passes, the corresponding plot yields the value of NTU as $NTU = \dfrac{UA}{c_{min}}$

$$= 0.85$$

and $$A = \frac{0.85\, c_{min}}{U} = \frac{(0.85)\,(3360)}{55}$$

$$= 51.9 \text{ ft}^2$$

2. A shell and tube heat exchanger with one shell pass and four tube passes has 52 ft² of heat transfer area. The effective overall heat transfer coefficient of this unit is 55 Btu/hr ft² °F. This exchanger was designed for use with water and benzene. It is now intended to cool a stream of oil (cp = 0.53 Btu/lb °F) at 250°F, flowing at 12000 lbs/hr with cooling water available at 55°F and a flow rate of 5000 lbs/hr. In this application what will be the exiting temperatures of the two fluid steams'.

Solution

We can use the following steps in the solution of this problem

$$q = c_c\,(Tc_{out} - Tc_{in})$$
$$= (5000)\,(T_{H2o\ out} - 55)\,(1)$$
$$= 5000\,(T_{H2o\ out} - 55) \text{ Btu/hr} \qquad \ldots(1)$$
$$q = c_H\,(Th_{in} - Th_{out})$$
$$= (12000)\,(0.53)\,(250 - T_{oil\ out}) \qquad \ldots(2)$$

we know $q = U A F\, \Delta T_{lm}$

$$= (55)\,(52)\left[\frac{(T_{oil} - 55) - (250 - T_{H2o})}{\left[ln\,\dfrac{T_{oil} - 55}{250 - T_{H2o}}\right]}\right]$$

We also know

$$q = E\, c_{min}\,(T_{oil\ in} - T_{H2o\ in})$$
$$q = (5000)\,(250 - 50)$$
$$= E\,(10^6) \text{ Btu/hr} \qquad \ldots(4)$$

Equations (1), (2), (3) and (4) must all be satisfied: among them they contain three unknowns: they are q, $T_{H2o\ out}$ and $T_{oil\ out}$.

Were equations (1), (2) and (3) used to solve this problem, a trial-and-error procedure would be required. The unknown temperatures are tied up in the argument of a logarithm in equation (3); moreover, the factor F is a function of the unknown temperatures as well and is determined from graphs while the procedure to solve the problem this way is not extremely difficult such a prospect is not overly attractive.

Equations (1) and (2) may be used along with equation (4) in a straight forward fashion involving no trial-and-error methods. This approach is definitely more attractive than the one discussed above.

Carrying out the necessary calculations

$$\frac{c_{min}}{c_{max}} = \frac{5000}{6360} = 0.786$$

$$\frac{U\,A}{c_{min}} = \frac{(55)\,(52)}{5000} = 0.572$$

$E = 0.36$ (from the appropriate plot)

The heat transferred is now evaluated for equation (4) as

$$q = (0.36)\ 10^6 \text{ Btu/hr}$$
$$= 360{,}000 \text{ Btu/hr}$$

and the sought-for exit temperatures may be evaluated from equations (1) and (2) yielding

$$T_{H2o\ out} = 55 + \frac{360{,}000}{5000}$$
$$= 55°F + 72 = 127°F$$

$$T_{oil\ out} = 250 - \frac{360{,}000}{6360} = 193°F$$

Some comments may be made in respect of the plots of heat exchanger effectiveness. In every case, the value of E reaches a maximum at some modest value of NTU, on the order of 2 to 5. For greater heat transfer areas, the value of NTU will increase however, the heat exchanger effectiveness, once the maximum value is approached, will not show any significant increase. Thus the idea that any increase in heat transfer area in a exchanger is accompanied by a corresponding increase in exchanger capacity is false.

The reader may also note that certain values of E may never be achieved for certain equipment configurations and operating conditions. For example, in a single-pass counterflow situation with $\frac{c_{min}}{c_{max}} = 1$, a value of $E = 0.90$ is impractical.

Example on the Heat Exchanger Area Computation

A heat exchanger is required to cool 5000 kgs/hr of oil at 70°C to 40°C with water entering at 15°C. The exit temperature difference between the two fluids should be 10°C. Find the ratio of areas required for a 1-4 exchanger (one shell-pass and four tube passes) and a 2-4 exchanger. The hot fluid flows through the shell and the cold fluid through the tube.

Solution

$$th_1 = 70°C = ts_1$$

$$th_2 = 40°C = ts_2$$

$$tc_2 = th_2 - 10 = 40 - 10 = 30°C = t_2$$

Therefore $R = \dfrac{th_1 - th_2}{tc_2 - tc_1} = \dfrac{70 - 40}{30 - 15} = 2$

$$S = \frac{tc_2 - tc_1}{th_1 - tc_1} = \frac{30 - 15}{70 - 15} = 0.273$$

For these values of R and S the correction factors for the 1–4 and 2–4 configurations form the appropriate LMTD correction factor plots, are

$$F_{1\text{-}4} = 0.92$$

and $F_{2\text{-}4} = 0.98$

Since q, U and $(\Delta t)_m$ are the same for both the cases we can write

$$q = U\left(A_{1-4}\right)[\Delta t]_m \; F_{1-4}$$

$$= U\left(A_{2-4}\right)[\Delta t]_m \; F_{2-4}$$

So $\dfrac{A_{1-4}}{A_{2-4}} = \dfrac{F_{2-4}}{F_{1-4}} = \dfrac{0.98}{0.92}$

$$= 1.065.$$

2. Calculate the surface area required for a heat exchanger which has to cool 55,000 kg/hr of alcohol from 66°C to 40°C using 40,000 kg/hr of water entering at 5°C. Assume that U based on the outer tube area is 500 k cal/hr m^2°C, specified heat of alcohol 0.9. Consider the following configurations:

(i) Counterflow tube and shell

(ii) Parallel-flow tube and shell

(iii) Reversed-current exchanger with 2 shell pass and 72 tube passes. Alcohol flows through the shell.

Solution

Let us first find the outlet temperature of water, tc_2

$$m_h\ c p_h\ (th_1 - th_2) = m_c\ c p_c\ (tc_2 - tc_1)$$

or $\quad 55{,}000 \times 0.9 \times (66 - 40) = 40000 \times 1 \times (tc_2 - 5)$

or $\quad tc_2 = 37.2°\text{C}$

$$q = m_h\ cp_h\ (th_1 - th_2)$$
$$= 55{,}000 \times 0.9 \times (66 - 40)$$
$$= 1{,}285{,}000 \text{ kg/hr}$$

(i) $$(\Delta tm)_{\text{counter flow}} = \frac{(\Delta t)_2 - (\Delta t)_1}{ln \left\{ \dfrac{\Delta t_2}{\Delta t_1} \right\}}$$

$$(\Delta t)_2 = th_1 - tc_2 = 66 - 37.2 = 28.8°\text{C}$$

$$(\Delta t)_1 = th_2 - tc_1 = 40 - 5 = 35°\text{C}$$

Therefore $$(\Delta t)_{m \text{ counter flow}} = \frac{28.8 - 35}{ln \dfrac{28.8}{35}}$$

$$= 31.1°\text{C}$$

or $$(A)_{\text{counterflow}} = \frac{q}{U\ (\Delta t_m)_{\text{counterflow}}}$$

$$= \frac{1{,}285{,}000}{500 \times 31.1}$$

$$= 82.5 \text{ m}^2$$

(ii) $$(\Delta t)_2 = th_2 - tc_2 = 40 - 37.2 = 2.8°\text{C}$$

$$(\Delta t)_1 = th_1 - tc_1 = 66 - 5 = 61°\text{C}$$

or $$(\Delta t)_m = \frac{2.8 - 61}{ln \dfrac{2.8}{61}}$$

$$= \frac{58.2}{3.08} = 18.9°\text{C}$$

so $$(A)_{\text{parallel flow}} = \frac{1285000}{(500)\ (18.9)}$$

$$= 136 \text{ m}^2$$

(iii) $(\Delta tm)_{2-72} = (F_{2-72})(\Delta tm)_{\text{counter flow}}$

$$\text{Now } R = \frac{ts_1 - ts_2}{t_{t_2} - t_{t_1}} \text{ or } \frac{th_1 - th_2}{tc_2 - tc_1} = \frac{66 - 40}{37.2 - 5} = 0.81$$

$$S = \frac{t_{t_2} - t_{t_1}}{ts_1 - t_{t_1}} = \frac{tc_2 - tc_1}{th_1 - tc_1} = \frac{37.2 - 5}{66 - 5} = 0.528$$

From the appropriate LMTD correction factor plot (2 shell passes and even number of tube passes)

$$F_{2-72} = 0.965$$

$$\text{Therefore } A_{2-72} = \frac{q}{U(F_{2-72})(\Delta t)_{m \text{ counterflow}}}$$

$$= \frac{1285000}{(500)(0.965)(31.1)}$$

$$= 85.7 \text{ m}^2.$$

3. Determine the length of counterflow heat exchanger to heat 4,000 kg/hr of oil from 10°C to 20°C using hot water at 70°C. The hot water flows with a velocity of 75 cm/sec through a copper pipe of Inner Diameter = 1.8 cm and Outer Diameter = 2.1 cm. The oil flows through the annulus between this pipe and a 3 cm Inner Diameter steel pipe. Neglect the tube wall resistance and scale resistance. The following properties of oil and water may be used

	Water	*Oil*
cp (k cal/kg °C)	1.0	0.45
k (k cal/hr m °C)	0.57	0.12
ρ (kg/m^3)	1000	850
v (m^2/sec)	0.405×10^{-6}	0.72×10^{-5}

Solution

$$m_h = \text{(area) (velocity) (density)}$$

$$= \left(\frac{\pi}{4}\right)^2 \left(\frac{1.8}{100}\right)^2 \left(\frac{75}{100} \times 3600\right)(1000)$$

$$= 690 \text{ kg/hr}$$

We know $m_h \, cp_h \, (th_1 - th_2) = m_c \, cp_c \, (tc_2 - tc_1)$

or $$690 \times 1 \times (70 - th_2) = 4000 \times 0.45 \times (20 - 10)$$

$$70 - th_2 = \frac{4000 \times 0.45 \times 10}{690}$$

$$= 26.1$$

So $th_2 = 70 - 26.1 = 43.9°C$

The LMTD is given by

$$(\Delta tm) = \frac{(th_1 - tc_2) - (th_2 - tc_1)}{ln\left(\frac{th_1 - tc_2}{th_2 - tc_1}\right)}$$

$$= \frac{(70 - 20) - (43.9 - 10)}{ln\left(\frac{50}{33.9}\right)}$$

$$= \frac{16.1}{ln\,(1.48)} = \frac{16.1}{0.4}$$

$$= 40.25°C$$

The next step is to compute U, To do this, we need to have the values of h_1 and h_0 which can be got by the first determining N_{Re} for water-flow and oil-flow and then using appropriate convection equations

N_{Re} for water-flow is

$$(N_{Re})_{\text{water}} = \left(\frac{Du}{v}\right)_{\text{water}}$$

$$= \frac{(0.75 \times 3600)\left(1.8 \times 10^{-2}\right)}{(0.405)\,10^{-6} \times 3600}$$

$$= 33{,}400$$

N_{Re} for oil-flow is

$$(N_{Re})_{\text{oil}} = \left(\frac{Du}{v}\right)_{\text{oil}}$$

The velocity of oil is given by

$$u_{\text{oil}} = \frac{\text{volume flow per hour}}{\text{area of flow}}$$

$$= \frac{\dfrac{(4000)}{(850)}}{\dfrac{\pi}{4}\left(3^2 - 2.1^2\right) \times 10^{-4}}$$

$$= \frac{4.72}{3.61 \times 10^{-4}}$$

$$= 1.31 \times 10^4 \text{ m/hr}$$

The equivalent diameter of the annulus is (3 – 2.1) = 0.9 cm

$$\text{Therefore } (N_{Re})_{oil} = \frac{1.31 \times 10^4 \times 0.9 \times 10^{-2}}{0.72 \times 10^{-5} \times 3600}$$

$$= 4550$$

Since the flow of both water and oil is turbulent, we can determine h, from the colburn equation

$$\frac{h}{\rho\, u\, c_p}\left(\frac{\mu\, c_p}{k}\right)^{\frac{2}{3}} = \frac{0.023}{(N_{Re})^{0.2}}$$

$$h_{oil} = h_o = \frac{(0.023)\, \rho\, u\, c_p}{(N_{Re})^{0.2}\left(\frac{c_p\, \mu}{k}\right)^{\frac{2}{3}}}$$

$$= \frac{0.023 \times 850 \times 1.31 \times 10^4 \times 0.45}{(4550)^{0.2}\left(0.72 \times 10^{-5} \times 850 \times 3600 \times \frac{45}{0.12}\right)^{\frac{2}{3}}}$$

$$= 870 \text{ k cal/hr m}^2\ {}^\circ\text{C}$$

Similarly

$$h_{water} = h_i = \frac{0.023 \times 1000 \times 0.75 \times 3600 \times 1.0}{(33400)^{0.2}\left[\left(0.405 \times 10^{-6}\right)(1000)(3600)\frac{1.0}{0.57}\right]^{\frac{2}{3}}}$$

$$= 4{,}150 \text{ k cal/hr m}^2\ {}^\circ\text{C}$$

$$\text{Thus } U = \frac{1}{\left(\frac{r_o}{r_i}\right)\left(\frac{1}{h_i}\right) + \frac{1}{h_o}}$$

$$= \frac{1}{\frac{2.1}{1.8}\left(\frac{1}{4150}\right) + \frac{1}{870}}$$

$$= \frac{1}{10^{-4}\,(2.8 + 11.5)} = \frac{10^4}{14.3}$$

$$= 700 \text{ k cal/hr m}^2\ °\text{C}$$

The rate of heat transfer is

$$q = m_c\, cp_c\,(tc_2 - tc_1)$$

$$= (4000) \times (0.45) \times (20 - 10)$$

$$= 18000 \text{ k cal/hr}$$

So $A = \dfrac{q}{U\,(\Delta tm)}$

$$= \frac{18000}{760 \times 40.25}$$

$$= 0.638 \text{ m}^2$$

or $\quad L = \dfrac{0.638}{\pi\,(2.1)\,10^{-2}} = 9.67 \text{ m}$.

4. 8000 kg/hr of air at 105°C is cooled by passing if through a counterflow heat exchanger. Determine the exit temperature of air if water enters at 15°C and flows at a rate of 7500 kg/hr U = 125 k cal/hr m² °C. The heat transfer area is 20 m².

Solution

Since the exit temperature of both air and water are not known, we are constrained to solve the problem by an iterative procedure. Let us assume that the exit temperature of the cold water i.e. water is 35°C

Then

$$q = 7500\,(35 - 15) \times 1 = 150{,}000 \text{ k cal/hr}$$

so $\quad (th_1 - th_2) = \dfrac{150{,}000}{(8000)\,(0.24)} = 78.3°\text{C}$

or $\quad th_2 = th_1 - 78.3 = 105 - 78.3 = 26.7°\text{C}$

thus, the LMTD is

$$\Delta t_m = \frac{(th_1 - tc_2) - (th_2 - tc_1)}{ln\left(\dfrac{th_1 - tc_2}{th_2 - tc_1}\right)}$$

$$\Delta t_m = \frac{(105 - 35) - (26.7 - 15)}{ln\left(\dfrac{70}{11.7}\right)}$$

$$= \frac{70 - 11.7}{ln\,(5.98)}$$

$$= \frac{58.3}{1.79} = 32.7°C$$

or $\quad q = U\,A\,\Delta tm$

$$= (125)\,(20)\,(32.7) = 81{,}700 \text{ k cal/hr}$$

This value of q does not check with the value of 150,000 on the basis of assumed tc_2. Hence tc_2 must be lower than 35°C. Let us make a second trial with $tc_2 = 30°C$

$$q = 7500\,(30 - 15) = 112{,}500 \text{ k cal/hr}$$

$$th_1 - th_2 = \frac{112500}{(8000)\,(0.24)} = 58.8$$

or $\quad th_2 = 100 - 58.1 = 42.2°C$

The LMTD is

$$\Delta t_m = \frac{(105 - 30) - (41.2 - 15)}{ln\left(\frac{75}{26.2}\right)}$$

$$= \frac{75 - 26.2}{ln\,(2.86)} = 46.4°C$$

or $\quad q = U\,A\,\Delta tm$

$$= (125)\,(20)\,(46.4)$$

$$= 116{,}000 \text{ k cal/hr}$$

The value of q is slightly higher than the assumed value of 112,500. This implies that the exit temperatures of water will be little higher than 30°C. A third trial will yield $tc_2 = 30.8°C$. Hence th_2 is found to be $th_2 = 43.3°C$.

5. A horizontal water heater is required to heat 15000 kg/hr of water from 15°C to 85°C by means of steam condensing at 110°C. The heater is consist of tubes having 1.9 cm Outer Diameter and 1.65 cm Inner Diameter. Determine the number of tubes required and the length of each tube.

Solution

The rate of heat transfer is given by

$$q = 15000 \times 1.0 \times (85 - 15)$$

$$= 1050{,}000 \text{ k cal/hr}$$

Since the latent heat of steam at 110°C is 533 k cal/kg, the steam flow rate is

$$m = \frac{1050{,}000}{533} = 1970 \text{ kg/hr}$$

The LMTD is given by

$$\Delta t_m = \frac{(110-85)-(110-15)}{ln\left(\dfrac{110-85}{110-15}\right)}$$

or $\qquad \Delta t_m = 52.5°C$

Therefore we may take the mean or bulk temperature of water to be (110 – 52.5) = 57.5°C. Note that the arithmetic mean temperature of water will be

$\dfrac{(15+85)}{2}$ = 50°C. For water at 57.5°C.

$$\mu = 1.88 \text{ kg/m-hr}$$

$$\rho = 985 \text{ kg/m}^3$$

$$c_p = 1.0 \text{ k cal/kg °C}$$

$$k = 0.57 \text{ k cal/m hr °C}$$

Let us assume that the velocity of water inside the tubes should not exceed 0.75 m/sec in order to keep the pressure drop reasonably low. This would require that the cross-sectional area of the tubes should be (assuming that ρ = 1000 kg/ m³ for water)

$$A_f = \frac{15000}{1000} \times \frac{1}{0.75 \times 3600} = \frac{1}{180} \text{m}^2$$

cross-sectional area of 1 tube.

$$= \frac{\pi}{4}(1.65)^2 \times 10^{-4} - 2.14 \times 10^{-4} \text{ m}^2$$

Therefore, the number of tubes required in this situation

$$= \left(\frac{1}{180}\right) \div (2.14) \times 10^{-4}$$

$$= 26$$

The value of N_{Re} for the flow of water is

$$N_{Re} = \frac{D\,u\,\rho}{\mu}$$

$$= \frac{(0.75 \times 3600)\,(0.0165)\,(985)}{1.88}$$

=23,400

It should be remembered that we have used the properties of water at the bulk temperature of 57.5°C.

Since the water flow is turbulent, we have may evaluate the waterside heat transfer coefficient h, by the equation

$$\frac{hD}{k} = 0.023\left(\frac{Du\rho}{\mu}\right)^{0.8}\left(\frac{c_p\,\mu}{k}\right)^{0.4}$$

$$\frac{Du\rho}{\mu} = 23{,}400$$

$$\left(\frac{c_p\,\mu}{k}\right) = \frac{1.88 \times 1.0}{0.57} = 3.3$$

So $$\frac{hD}{k} = (0.023)\,(23400)^{0.8}\,(3.3)^{0.4}$$

$$= (0.023)\,(3125)\,(1.613) = 116$$

or $$h = \frac{116 \times 0.57}{0.0165} = 4000 \text{ k cal/hr m}^2\,°\text{C}$$

The steam-side heat transfer coefficient h_2 and may be determined by using the equation.

$$h = 0.725\left\{\frac{\rho^2\, g\, L\, k^3}{\mu\, D\,(t_{sat} - tw)}\right\}^{\frac{1}{4}}$$

The above equation is valid for 1 tube. When there above are N tubes placed one above the other, the above equation is modified to the form

$$h = 0.725\left\{\frac{\rho^2\, g\, L\, k^3}{N\,\mu\, D\,(t_{sat} - tw)}\right\}^{\frac{1}{4}}$$

where L is the latent heat of condensation

In the above equation, the value of L is taken at t_{sat} while the other properties are evaluated at $t_f = \dfrac{(t_{sat} + t_w)}{2}$. Since t_w is not known we have to employ an iterative procedure. Let us assume that tw = 90°C.

Then $$t_f = \frac{110 + 90}{2} = 100°\text{C}$$

The properties of water at 100°C are

$\rho = 956$ kg/m^3

$c_p = 1.01$ k cal/kg °C

$k = 0.614$ k cal/hr m °C

$\mu = 1.03$ kg/hr-m

Therefore if assume $N = 6$, we obtain

$$h_2 = 0.725 \left\{ \frac{(956)^2 \times 9.81 \times 3600 \times 3600 \times 533 \times 0.614^2}{6 \times 1.03 \times 0.019 \times (110 - 90)} \right\}^{\frac{1}{4}}$$

$$= 6400 \text{ k cal/hr m}^2 \text{ °C}$$

The value of t_w can now be calculated by making an energy balance and this value can be checked with the assumed value.

$$h_1 A_1 (tw - 57.5) = h_2 A_2 (110 - tw)$$

$$(tw - 57.5) = \frac{h_2 A_2}{h_1 A_1} (110 - tw)$$

$$= \frac{6400}{4000} \left(\frac{1.9}{1.65} \right) (110 - tw)$$

$$= 1.84 (100 - t_w)$$

So $\quad t_w = 92°C$

Since the value of t_w is slightly different from the assumed value of 90°C, a second trial can be made by taking tw as 92°C. Note that a two-degree difference in tw will change $(t_{sat} - tw)$ from 20°C to 18°C and will thus change the value of h_2. The new value of h_2 is

$$h_2 = 0.725 \left\{ \frac{956^2 \times 9.81 \times 3600 \times 3600 \times 533 \times 0.614^3}{6 \times 1.03 \times 0.019 \times (110 - 92)} \right\}^{\frac{1}{4}}$$

$$= 6700 \text{ k cal/hr m}^2 \text{ °C}$$

The corresponding value of t_w then

$$(tw - 57.5) = \frac{6700}{4000} \left(\frac{1.9}{1.65} \right) (110 - tw)$$

or $\quad t_w = 92.1°C$

Since this value is very nearly equal to the assumed value of 92°C no further iteration is required. Note that we have neglected the resistance of the tube wall in above calculations.

The outside surface area required is

$$A_2 = \frac{q}{h_2\,(t_{sat} - tw)}$$

$$= \frac{1050{,}000}{6700\,(110 - 92)} = 8.7 \text{ m}^2$$

Therefore, the length of one tube will be

$$= \frac{8.7}{26 \times \pi \times 0.019} = 5.6 \text{ m}$$

Since the length is too much, we can use 2 passes so that the tube length is limited to 2.8 m. Thus there will be 2 passes, each pass having 26 tubes, each 2.8 m long.

6. 3000 kg/hr of furnace oil is to be heated from 10°C to 90°C in a shell and tube type heat exchanger. The oil is to flow inside the tubes while steam at 120°C is to flow through the shell. The tubes are to be 1.9 cm Outer Diameter and 1.65 cm Inner Diameter. Design the heat exchanger.

Solution

The bulk temperature of the oil may be taken as $t_b = \dfrac{10 + 90}{2} = 50°\text{C}$

The furnace oil's properties at 50°C

ρ = 900 kg/m^3

μ = 54 kg/hr m

c_p = 0.47 k cal/kg °C

k = 0.112 k cal/hr m °C

The required rate of heat transfer is given by

$$q = 3000 \times 0.47 \times (90 - 10)$$

$$= 113{,}000 \text{ k cal/hr}$$

In order to keep the pressure drop low, let us limit the velocity of oil to 5 cm/sec.

Then, we may use the following equation to determine the oil-side heat transfer coefficient h_1

$$\frac{hD}{k} = 1.86\left[\left(\frac{D\,u\,\rho}{\mu_b}\right)\left(\frac{c\,p\,\mu}{k}\right)_b\left(\frac{D}{L}\right)\right]^{\frac{1}{3}}\left(\frac{\mu_b}{\mu_w}\right)^{0.14}$$

Let us assume that the tube surface temperature will be nearly 100°C and that $D/L \simeq 150$. Then (μ_w = 12 kg/hr m)

$$\frac{h\,D}{k_b} = 1.86\left[\left(\frac{0.05 \times 3600 \times 0.019 \times 900}{30}\right)\left(\frac{54 \times 0.47}{0.112}\right)\left(\frac{1}{150}\right)\right]^{\frac{1}{3}} \times \left(\frac{54}{12}\right)^{0.14}$$

$$= (1.86)\left[(103)\,(228)\left(\frac{1}{150}\right)\right]^{\frac{1}{3}} (1.234)$$

$$= 1.86 \times 5.37 \times 1.234$$

$$= 12.34$$

or $$h_1 = \frac{12.34 \times 0.112}{0.019} = 73 \text{ k cal/hr m}^2\,°\text{C}$$

The value of the steam-side heat transfer coefficient may be obtained by the equation

$$h = 0.725\left\{\frac{\rho^2\; g\; L\; k^3}{\mu\; D\,(t_{sat} - t_w)}\right\}^{\frac{1}{4}}$$

Since h_2 will be much larger than h_1 and U will depend mainly on h_1, it is not necessary to know h_2 very accurately. Therefore we may use the assumed value of $t_w = 100°C$ and evaluate the properties of water at $t_f = \dfrac{100 + 120}{2} = 110°C$

At 110°C

$\rho = 950 \text{ kg/m}^3$

$k = 0.63$ k cal/hr m °C

$c_p = 1.015$ k cal/kg °C

$\mu = 0.97$ kg/m-hr

At = 120°C, the latent heat of condensation L is 527 k cal/kg. Hence h_2 is from the above equation (taking $N = 10$)

$$h_2 = 0.725\left\{\frac{\rho^3\; g\; L\; k^2}{\mu\; N\; D\,(t_{sat} - tw)}\right\}^{\frac{1}{1}}$$

$$= 0.725\left\{\frac{950^2 \times 9.81 \times 3600^2 \times 527 \times 0.63^2}{0.97 \times 10 \times 0.019 \times (120 - 100)}\right\}^{\frac{1}{4}}$$

$$= 6400 \text{ k cal/hr m}^2\,°\text{C}$$

The overall heat transfer coefficient U, based on the outside surface area is then (neglecting the tube wall resistance)

$$\frac{1}{U} = \frac{1}{h_2} + \frac{A_2}{A_1}\frac{1}{h_1}$$

$$= \frac{1}{6400} + \frac{1.9}{1.65} \times \frac{1}{73}$$

or $U = 62.3$ k cal/hr m² °C

$$\Delta t_m = \frac{(120 - 90) - (120 - 10)}{ln\left(\frac{30}{110}\right)} = \frac{80}{1.3} = 61.6°C$$

Therefore, the outside surface area required is

$$A_2 = \frac{q}{U\,\Delta t_m}$$

or $$A_2 = \frac{113000}{62.3 \times 61.6} = 29.5 \text{ m}^2$$

Total length of tubing required

$$= \frac{29.5}{\pi \times 0.019} = 495 \text{ m}$$

The number of tubes per pass is

$$= \left(\frac{3000}{900}\right) \div \left\{\frac{\pi\,(0.0165)^2}{4} \times 0.05 \times 3600\right\}$$

$$= 87$$

Length of each tube $= \frac{493}{87} = 5.69$ m

Since this tube length is too large we may use 2 passes

Therefore,

The length of each tube = 2.85 m

Number of tubes per pass = 87

Number of tube passes = 2.

7. A heat exchanger is to be designed. It is assumed that condensing vapour on the shell side will be held at constant pressure. Gas will flow through the tubes in a single pass. The total gas flow rate '*m*' and the mass flow rate *G* will be kept fixed. After selecting a tube diameter. *D*, the value of the number of tubes *N* in parallel, the gas film coefficient *h*, the total heat transfer area *A* and the tube length *L*, have been fixed. It is desired to explore the effect of larger tube size on the design.

Consider that *G*, *m*, *q* and other physical properties are constant and determine the effect of increase in tube diameter on each of the following

(i) Number of tubes needed
(ii) Heat transfer coefficient obtained
(iii) Total heat transfer surface needed and
(iv) Length of tubes required.

Solution

(i) Both G and m for gas, are constant so m/G = Area of flow = Constant

or $\frac{\pi D^2}{4} \times N = \text{constant}$

Hence if D increases, N decreases

(ii) $h = \frac{k}{D} \times 0.023 \times \left(\frac{G D}{\mu}\right)^{0.8} \left(\frac{c_p \mu}{k}\right)^{0.4}$

Since k, G, μ and c_p are constant

Hence h varies as $\left(\frac{1}{D}\right)^{0.2}$ or h decreases if D increases.

(iii) $q = U A \Delta t_m$

Since q and Δt_m are constant. UA is constant. Also $U \simeq h$ and therefore it decreases as D increases

Hence A should increase if D increases.

(iv) $A = N(\pi D L)$

Let prime indicate the value corresponding to increased D.

$$\frac{A'}{A} = \frac{N' \pi D' L'}{N \pi D L} = \frac{\dfrac{q}{U' \Delta tm}}{\dfrac{q}{U \Delta tm}} = \frac{U'}{U} = \frac{h'}{h}\left(\frac{D'}{D}\right)^{0.2}$$

or $\frac{L'}{L} = \frac{D}{D'} \times \frac{N}{N'} \times \left(\frac{D'}{D}\right)^{0.2}$

Now $N = \frac{4 \times A_{\text{flow}}}{\pi D^2}$ so $N' = \frac{4 \times A_{\text{flow}}}{\pi D'^2}$

or $\frac{N}{N'} = \left(\frac{D'}{D}\right)^2$

$$\text{So } \frac{L'}{L} = \frac{D}{D'}\left(\frac{D'}{D}\right)^2 \left(\frac{D'}{D}\right)^{0.2} = \left(\frac{D'}{D}\right)^{1.2}$$

Therefore the length L, increases if D increases.

Illustration on the Calculation of Heat Exchanger Area

1. 5.795 kg/sec of an oil flows through the shell side of a two-shell-pass, four-tube-pass oil cooler. It enters at 181°C and leaves at 38°C. Water flows in the tubes entering at 32°C and leaving at 49°C. cp_{oil} = 2282 J/kg °C U=416 w/m^2 °C. Find how much area the heat exchanger must have to accomplish this heat transfer.

Solution

$$\text{LMTD} = \frac{(Th_{\text{in}} - Tc_{\text{out}}) - (Th_{\text{out}} - Tc_{\text{in}})}{ln \dfrac{Th_{\text{in}} - Tc_{\text{out}}}{Th_{\text{out}} - Tc_{\text{in}}}}$$

$$= \frac{(181 - 49) - (38 - 32)}{ln \dfrac{181 - 49}{38 - 32}}$$

$$R = \frac{181 - 38}{49 - 32} = 8.412$$

$$S = \frac{49 - 32}{181 - 32} = 0.114$$

Referring to the correction factor graph for even number of shell and tube passes we observe that, F =0.92

It follows that

$Q = UAF$ (LMTD)

(5.795) (2282) (181 – 38) = 416 (A) (0.92) (40.76)

A = 121.1 m^2.

2. Consider the following parallel-flow heat exchange specification

cold flow enters at 40°C, c_c = 20000 w/ °C

hot flow enters at 150°C, c_h = 10000 w/ °C

A =30m^2, U = 500 w/m^2 °C

Determine the heat transfer and the exit temperatures by the effectiveness approach.

Solution

In this case we do not know the exit temperatures, so it is not possible to calculate the LMTD. Instead we can either go to the parallel flow effectiveness chart or to the equation

$$E = \frac{1 - \exp\left[-\left(1 + \frac{c_{min}}{c_{max}}\right)\text{NTU}\right]}{1 + \frac{c_{min}}{c_{max}}}$$

$$= fn\left(\frac{c_{min}}{c_{max}}, \text{NTU only}\right)$$

using $\text{NTU} = \frac{UA}{c_{min}} = \frac{(500)(30)}{10{,}000} = 1.5$

$$\frac{c_{min}}{c_{max}} = 0.5$$

and we obtain $E = 0.596$

Now from the equation $Q = E\, c_{min}(Th_{in} - Tc_{in})$

We find that

$$Q = 0.596\ (10{,}000)\ (110)$$

$$= 655{,}600, w = 655.6 \text{ kw}$$

Finally from energy balances, we get

$$Th_{out} = Th_{in} - \frac{Q}{c_h} = 150 - \frac{655{,}600}{10{,}000}$$

$$= 84.4°\text{C}$$

$$Tc_{out} = Tc_{in} + \frac{Q}{c_c} = 40 + \frac{655{,}600}{20{,}000}$$

$$= 72.8°\text{C}.$$

3. Suppose that we had the some kind of exchanger as we considered in the previous example, but that the area remained unspecified as a design variable. For this situation calculate the heat transfer area that would bring the hot fluid out at 90°C.

Solution

Once the exit cold fluid temperature is known, the problem can be solved with equal ease by either the LMTD or the effectiveness approach

$$Tc_{out} = Tc_{in} + \frac{c_h}{c_c}(Th_{in} - Th_{out})$$

$= 40 + 0.5\,(150 - 90) = 70°C$

Then using the effectiveness method

$$E = \frac{c_h\,(Th_{in} - Th_{out})}{c_{min}\,(Th_{in} - Tc_{in})}$$

$$= \frac{10000\,(150 - 90)}{10000\,(150 - 40)} = 0.5455$$

From the effectiveness versus NTU, plot for a parallel flow heat exchanger,

we read NTU $\simeq 1.15 = \dfrac{UA}{c_{min}}$

Thus $A = \dfrac{10000\,(1.15)}{500} = 23.00\ m^2$

we could also have calculated the LMTD

$$\text{LMTD} = \frac{(150 - 40) - (90 - 70)}{ln\left(\dfrac{110}{20}\right)}$$

$$= 52.79°C$$

Therefore from $Q = UA\,\Delta T_{ln}$, we obtain $A = \dfrac{10{,}000\,(150 - 90)}{500\,(52.79)}$

$$= 22.73\ m^2$$

It may be observed that the answers differ by 1% which reflects the graph-reading inaccuracy.

Example on the Evaluation of LMTD

Water enters the tubes of a small single-pass heat exchanger at 20°C and leaves at 40°C. On the shell side, 25 kg/min of steam condenses at 60°C. Calculate the overall heat transfer coefficient and the required flow rate of water if the area of the exchanger is 12 m². The latent heat of steam h_{fg} =2358.7 kj/kg at 60°C.

Solution

$$Q = m_{condensate}\, h_{fg\ 60°C}$$

$$= \frac{(25)\,(2358.7)}{(60)} = 983\ \text{kj/sec}$$

Since the condensate temperature is constant we can calculate the LMTD without naming the exchanger "parallel" or "counterflow"

$$\text{LMTD} = \frac{(60-20)-(60-40)}{ln\dfrac{60-20}{60-40}} = 28.85°\text{C}$$

Then $U = \dfrac{Q}{A\,\Delta T_{ln}}$

$$= \frac{(983)\,(1000)}{12\,(28.85)} = 2839 \text{ w/m}^2{}°\text{C}$$

and the flow rate of water is

$$m_{H_2o} = \frac{Q}{c_p\,(\Delta T)} = \frac{983{,}000}{(4174)\,20} = 11.78 \text{ kg/sec}$$

Limitations on the Characterization of Heat Exchangers with a LMTD

There are two basic limitations on characterizing heat exchangers with a LMTD. The first is that it is restricted to the single-pass parallel or counterflow configurations. This restrictions can be over come by adjusting the LMTD for other configurations.

The second limitation is more serious—that is one of constant value of *U*. *U* must be negligibly dependent on *T* to complete the integration of the equation

$$\int_o^A \frac{U}{m\,c\,p_c}\,dA = \int_{Tc_{in}}^{Tc_{out}} \frac{d\,T_c}{(\text{temp function})}$$

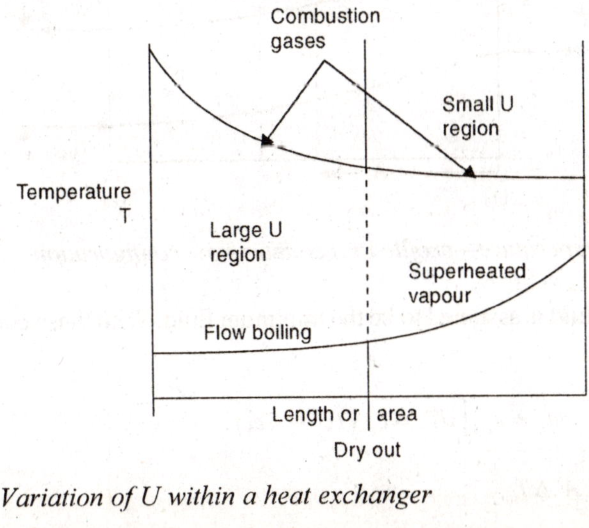

Fig. 6.31 *Variation of U within a heat exchanger*

If $U \neq f(T)$, the changing flow configuration and the variation of temperature can still give rise to serious variations of U within a given heat exchanger. The above figure shows a typical situation in which the variation of U within a heat exchanger might be very great. In this case the mechanism of heat exchange on the water side is completely altered when the liquiid is finally boiled away. If we had constant values of U for the two portions of the heat exchanger, then we could treat it as two different exchangers in series.

However, the more common difficulty that we face is that of designing heat exchangers in which U varies continuously with position within it. The problem is most severe in large industrial shell-and-tube configurations and less serious in compact heat exchangers with less surface area. If U depends on the locations, analyses of heat exchanger problems must be done using an average U defined by expression.

$$\int_o^A \frac{U\,dA}{A}.$$

Heat Exchanger Effectiveness for a Counterflow Operation—Derivation of the Expression

The analysis to follow is based upon the equation

$$q = E\,C_{\min}\,(Th_{in} - Tc_{in}) \qquad \ldots(1)$$

Counterflow operation of a single-pass exchanger is considered. The temperatures profile is shown in the sketch given below

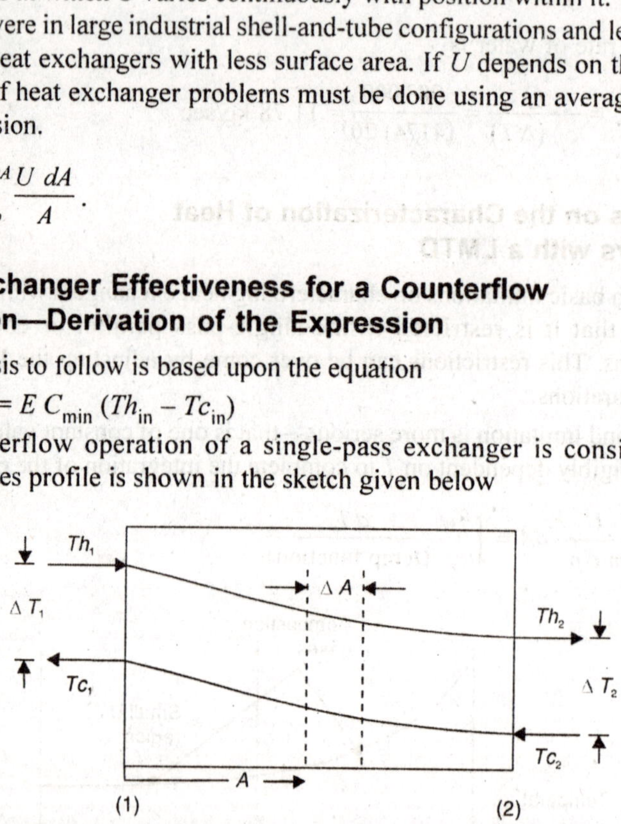

Fig. 6.32 *Temperature—profile in a counter flow configuration*

The cold fluid is assumed to be the minimum fluid. With these conditions the equations

$$q = \int_1^2 dq_c = c_c \int_1^2 dT_c = c_c\,(Tc_2 - Tc_1)$$

$$q = U\,A\,\Delta T_{lm}$$

are written as:

$$q = c_c\,(Tc_1 - Tc_2) = U\,A\,\frac{(Th_1 - Tc_1) - (Th_2 - Tc_2)}{ln\left[\frac{(Th_1 - Tc_1)}{(Th_2 - Tc_2)}\right]} \qquad ...(2)$$

The entering temperature of the hot fluid Th_1 may be expressed in terms of E, the effectiveness, using equation (1) as

$$Th_1 = Tc_2 + \frac{q}{E\,c_{min}}$$

$$= Tc_2 + \frac{Tc_1 - Tc_2}{E} \qquad ...(3)$$

The temperature difference at (1) between hot and cold fluid streams thus becomes

$$Th_1 - Tc_1 = Tc_2 - Tc_1 + \frac{Tc_1 - Tc_2}{E}$$

$$= \left(\frac{1}{E} - 1\right)(Tc_1 - Tc_2) \qquad ...(4)$$

The other temperature difference $Th_2 - Tc_2$ needed for substitution into equation (2) is got by solving the equations

$$\frac{c_h}{c_c} = \frac{(Tc_2 - Tc_1)}{(Th_2 - Th_1)}$$

for Th_2 yielding

$$Th_2 = Th_1 - \frac{c_{min}}{c_{max}}(Tc_1 - Tc_2)$$

thus $$Th_2 - Tc_2 = Th_1 - Tc_2 - \frac{c_{min}}{c_{max}}(Tc_1 - Tc_2) \qquad ...(5)$$

Employing equation (3), the above expression can be rearranged as

$$Th_2 - Tc_2 = \frac{Tc_1 - Tc_2}{E} - \frac{c_{min}}{c_{max}}(Tc_1 - Tc_2)$$

or $$Th_2 - Tc_2 = \left(\frac{1}{E} - \frac{c_{min}}{c_{max}}\right)(Tc_1 - Tc_2) \qquad ...(6)$$

Equations (4) and (6) are now substituted into equation (2).

With some rearrangement we get

$$ln \frac{\frac{1}{E} - \frac{c_{min}}{c_{max}}}{\frac{1}{E} - 1} = \frac{U\,A}{c_{min}}\left(1 - \frac{c_{min}}{c_{max}}\right)$$

The effectiveness E can be evaluated by taking the antilog of both sides of the above expression.

$$E = \frac{1 - \exp\left[-\frac{U\,A}{c_{min}}\left(1 - \frac{c_{min}}{c_{max}}\right)\right]}{1 - \frac{c_{min}}{c_{max}} \exp\left[-\frac{U\,A}{c_{min}}\left(1 - \frac{c_{min}}{c_{max}}\right)\right]} \qquad ...(7)$$

If may be recalled that equation (7) was derived assuming the cold fluid as minimum. In a similar way it can be shown that the same expression is obtained for the hot fluid as the minimum fluid. Thus with $\frac{U\,A}{c_{min}}$ designated NTU, the number of transfer units, we have, as a general expression for counterflow operation, the equation

$$E = \frac{1 - \exp\left[-\text{NTU}\left(1 - \frac{c_{min}}{c_{max}}\right)\right]}{1 - \frac{c_{min}}{c_{max}} \exp\left[-\text{NTU}\left(1 - \frac{c_{min}}{c_{max}}\right)\right]} \qquad ...(8)$$

In the case of parallel flow, it may be recalled that an analysis similar to that just gone through yields, for E. the expression

$$E = \frac{1 - \exp\left[-\text{NTU}\left(1 + \frac{c_{min}}{c_{max}}\right)\right]}{1 + \frac{c_{min}}{c_{max}}} \qquad ...(9)$$

The above two equations for single-pass heat exchangers relate the effectiveness E in terms of the two parameters NTU and $\frac{c_{min}}{c_{max}}$. It stands to reason that these expressions may be represented graphically for easy application. As mentioned earlier Kays and London have prepared plots of heat exchanger effectiveness for these single-pass cases and for various shell and tube and crossflow arrangements.

Illustration on the Computation of Exit Temperature in a Double-pipe Heat Exchanger

Steam passes through a turbine into a condenser. Liquid water from the condensed steam is used to heat ethylene glycol. The water is available at 195°F and with a mass flow rate of 5000 lb/hr. The ethylene glycol has a temperature of 85°F and a mass flow rate of 12000 lbs/hr. It is proposed to use a double pipe exchanger made of 2 × 1¼ standard type *M* copper tubing, with soldered fittings that is 20 ft long. Determine the outlet temperature of the ethylene glycol for counterflow configuration.

The properties of water are (at 140°F)

$\rho = 0.985\ (62.4)$ lbs/ft^3, $k_f = 0.376$ Btu/hr ft °R

$c_p = 0.9994$ Btu/lb °R, $\alpha = 6.02 \times 10^{-3}$ ft^2/hr

$\nu = 0.514 \times 10^{-5}$ ft^2/sec, $N_{Pr} = 3.02$

Also $m_h = 5000$ lb/hr and $T_1 = 195$°F

For ethylene glycol (at 140°F)

$\rho = 1.087\ (62.4)$ lbs/ft^3, $k = 0.150$ Btu/hr ft °R

$c_p = 0.621$ Btu/lb °R, $\alpha = 3.61 \times 10^{-3}$ ft^2/hr

$\nu = 5.11 \times 10^{-5}$ ft^2/sec, $N_{Pr} = 51$

Also $m_c = 12000$ lbs/hr and $t_1 = 85$°F

For seamless copper water tubing subscripts

a = annulus	p = inner pipe or tube we get
2 standard type *M*	Inner Diameter$_a$ = 0.1674 ft
1¼ standard type *M*	Inner Diameter$_p$ = 0.1076 ft

$$\text{Outer Diameter}_p = \frac{1.375}{12} = 0.1146 \text{ ft}$$

(i) *Flow areas*

$$A_p = \pi \frac{(0.1076^2)}{4} = 0.00909 \text{ ft}^2$$

$$A_a = \pi \frac{(0.1674^2 - 0.1146^2)}{4} = 0.0117 \text{ ft}^2$$

Because $A_a > A_p$, route the ethylene glycol (which has the higher flow rate) through the annulus

(ii) *Annulus Equivalent Diameters*

$$D_h = 0.1674 - 0.1146 = 0.0528 \text{ ft (friction)}$$

$$D_e = \frac{(0.1674^2 - 0.1146^2)}{0.1146} = 0.1299 \text{ ft (heat transfer)}$$

(iii) *Reynolds Numbers*

$$\text{Water } N_{Re_p} = \frac{5000}{3600}\,\frac{1}{0.985\,(62.4)\,(0.00909)}\,\frac{0.1076}{(0.514)\,10^{-5}}$$

$$= 5.2 \times 10^4$$

$$\text{Ethylene glycol } N_{Re_a} = \frac{12000}{3600}\,\frac{1}{1.087\,(62.4)\,(0.0117)}\,\frac{0.1299}{5.11 \times 10^{-5}}$$

(iv) *Nusselt Numbers*

$$\text{Water } N_{N_{u_p}} = 0.023\left(5.2 \times 10^4\right)^{0.8}(3.02)^{0.3} = 190$$

$$\text{Ethylene gylcol } N_{N_{ua}} = 0.023\left(1.07 \times 10^4\right)^{0.8}(51)^{0.4} = 185$$

(v) *Convection coefficients*

$$\text{Water } h_i = \frac{190\,(0.376)}{0.1076} = 664 \text{ Btu/hr ft}^2\ ^\circ\text{R}$$

$$h_{io} = \frac{664\,(0.1076)}{0.1146} = 623 \text{ Btu/hr ft}^2\ ^\circ\text{R}$$

$$\text{Ethylene glycol } h_o = \frac{185\,(0.150)}{0.1299} = 214 \text{ Btu/hr ft}^2\ ^\circ\text{R}$$

(vi) *Exchanger coefficient*

$$\frac{1}{U_o} = \frac{1}{623} + \frac{1}{214} \qquad U_o = 159 \text{ Btu/hr ft}^2\ ^\circ\text{R}$$

Outlet temperature

Subscripts c = cooler fluid or fluid being heated

h = warmer fluid or fluid being cooled

$$\frac{m_c\, c\,p_c}{m_h\, c\,p_h} = \frac{12000\,(0.612)}{5000\,(0.9994)} = 1.47$$

$$A_o = \pi\,(0.1146)\,(20) = 7.2 \text{ ft}^2$$

$$T_1 = 195^\circ\text{F} \qquad t_1 = 85^\circ\text{F}$$

$$T_2 = \frac{195\,(0.47) - 1.47\,(85) \times \left\{1 - \exp\left[\dfrac{159\,(7.2)\,(0.47)}{(12000)\,(0.612)}\right]\right\}}{1.47 \exp\left[\dfrac{159\,(7.2)\,(0.47)}{(12000)\,(0.612)}\right] - 1}$$

$T_2 = 174°F$ (water)

$$t_2 = 85 + \frac{195 - 174}{1.47} = 99.3°F \text{ (ethylene glycol)}$$

Heat balance (as a check on the calculations)

Water $q_h = 5000\ (0.9994)\ (195 - 174)$

$= 1.05 \times 10^5$ Btu/hr

Ethylene glycol

$$q_c = m_c\ cp_c\ \Delta T$$

$= 12000\ (0.612)\ (99.3 - 85)$

$= 1.05 \times 10^5$ Btu/hr

(vii) *Summary of the requested information*

Outlet temperature of ethylene glycol for counterflow = 99.3°F

If the outlet temperature of the ethylene glycol in not high enough, it is possible to use bigger tube diameters to increase the surface area; or to increase the mass flow rates thereby increasing the Reynolds numbers and correspondingly the Nusselt numbers: or to use several of this size exchanger in series. The decision will depend in part on the pressure drop encountered by each fluid stream.

The Effectiveness—NTU Relationship for a Counterflow Exchanger

Consider a counterflow exchanger. For this exchanger, we can start with the equation

$$\frac{d\,(th - tc)}{th - tc} = U\left[\frac{1}{m_h\ c\,p_h} - \frac{1}{m_c\ c\,p_c}\right] dA \qquad ...(1)$$

or
$$\frac{d\,(th - tc)}{th - tc} = \frac{U}{m_h\ c\,p_h}\left[1 - \frac{m_h\ c\,p_h}{m_c\ c\,p_c}\right] dA \qquad ...(2)$$

For brevity, let $m_h\ c\,p_h = c_h$ and $m_c\ c\,p_c = c_c$

Then Equation (2), becomes

$$\frac{d\,(th - tc)}{(th - tc)} = \frac{U}{c_h}\left(1 - \frac{c_h}{c_c}\right) dA \qquad ...(3)$$

Integrating equation (3), from $A = 0$, to $A = A$, we get

$$ln\left(\frac{th_1 - tc_2}{th_1 - tc_1}\right) = \frac{U\,A}{c_h}\left(1 - \frac{c_h}{c_c}\right)$$

or $$\left(\frac{th_1 - tc_2}{th_2 - tc_1}\right) = e^{\frac{UA}{c_h}\left(1 - \frac{c_h}{c_c}\right)} \qquad \ldots(4)$$

Inverting equation (4), we obtain

$$\frac{th_2 - tc_1}{th_1 - tc_2} = e^{-\left(\frac{UA}{c_h}\right)\left(1 - \frac{c_h}{c_c}\right)} \qquad \ldots(5)$$

Recall that the effectiveness of the heat exchanger is given by

$$E = \frac{tc_2 - tc_1}{th_1 - tc_1} \text{ for } c_c < c_h$$

$$E = \frac{th_1 - th_2}{th_1 - tc_1} \text{ for } c_h < c_c$$

Let us consider the two cases separately

Case I: ($c_h < c_c$ or $c_h = c_{min}$)

We know $E = \dfrac{c_h\,(th_1 - th_2)}{c_{\min}\,(th_1 - tc_1)}$

From the above equation we get

$$(th_1 - th_2) = E\,(th_1 - tc_1) \qquad \ldots(6)$$

or $$th_2 = th_1 + E\,(tc_1 - th_1) \qquad \ldots(7)$$

Since $q = c_c\,(tc_2 - tc_1) = c_h\,(th_1 - th_2)$

$$tc_2 = \frac{q}{c_c} + tc_1 = \frac{c_h}{c_c}\,(th_1 - th_2) + tc_1 \qquad \ldots(8)$$

Substituting for $(th_1 - th_2)$ from equation (6) into equation (8) we get

$$tc_2 = \frac{c_h}{c_c} \cdot E \cdot (th_1 - tc_1) + tc_1 \qquad \ldots(9)$$

Equations (7) and (9) permit us to eliminate the exit temperatures of the two fluids th_2 and tc_2 from equation (5). Substituting for th_2 and tc_2 from equations (7) and (9) into equation (5), we get

$$\frac{th_1 + E\,(tc_1 - th_1) - tc_1}{th_1 - \frac{c_h}{c_c}\,E\,(th_1 - tc_1) - tc_1} = e^{-\left(\frac{UA}{c_h}\right)\left(1 - \frac{c_h}{c_c}\right)}$$

or $$\frac{th_1 - tc_1\,(1 - E)}{(th_1 - tc_1)\left(1 - \frac{c_h}{c_c}\,E\right)} = e^{-\left(\frac{UA}{c_h}\right)\left(1 - \frac{c_h}{c_c}\right)}$$

or $$\frac{1-E}{\left(1-\frac{c_h}{c_c}E\right)} = e^{-\left(\frac{UA}{c_h}\right)\left(1-\frac{c_h}{c_c}\right)} \qquad ...(10)$$

Solving equation (10)

$$E = \frac{1-e^{-\left(\frac{UA}{c_h}\right)\left(1-\frac{c_h}{c_c}\right)}}{1-\frac{c_h}{c_c}e^{-\left(\frac{UA}{c_h}\right)\left(1-\frac{c_h}{c_c}\right)}} \qquad ...(11)$$

Case II: $c_c < c_h$

For this situation it is observed by following similar steps as given above, that

$$E = \frac{1-e^{-\left(\frac{UA}{c_c}\right)\left(1-\frac{c_c}{c_h}\right)}}{1-\left(\frac{c_c}{c_h}\right)e^{-\left(\frac{UA}{c_c}\right)\left(1-\frac{c_c}{c_h}\right)}} \qquad ...(12)$$

where c_{min} represents the smaller and c_{max} the larger of the two quantities $m_c\ c\,p_c$ and $m_h\ c\,p_h$.

The dimensionless ratio $\frac{UA}{c_{min}}$ is called the number of transfer units, NTU.

Thus for a counterflow heat exchanger

$$E = \frac{1-exp\left[(-NTU)\left(1-\frac{c_{min}}{c_{max}}\right)\right]}{1-\frac{c_{min}}{c_{max}}\ exp\left[(-NTU)\left(1-\frac{c_{min}}{c_{max}}\right)\right]} \qquad ...(13)$$

Illustration on the Use of NTU Method

1. 8000 kg/hr of air at 105°C is cooled by passing it through a counterflow heat exchanger. Find the exit temperature of air if water enters at 15°C and flows at a rate of 7500 kg/hr

U = 125 k cal/hr m^2 °C. The heat transfer area is 20 m^2. Solve this problem using the NTU method. Assume that the plot E is NTU is available for use.

Solution

$c_h = 8000 \times 0.24 = 1920$

$c_c - 7500 \times 1.0 = 7500$

or $\quad c_{min} = 1920 = c_h$

$c_{max} = 7500 = c_c$

$$\frac{c_{min}}{c_{max}} = \frac{1920}{7500} = 0.256$$

The number of transfer units

$$\text{NTU} = \frac{U\,A}{c_{min}} = \frac{125 \times 20}{1920} = 1.3$$

From Equation (13)

$$E = \frac{1 - e\left[-NTU\left(1 - \frac{c_{min}}{c_{max}}\right)\right]}{1 - \frac{c_{min}}{c_{max}}\,e\left[-NTU\left(1 - \frac{c_{min}}{c_{max}}\right)\right]}$$

$$e^{-\text{NTU}\left(1 - \frac{c_{min}}{c_{max}}\right)} = e^{1.3(1 - 0.256)} = 0.38$$

So $E = \dfrac{1 - 0.38}{1 - 0.256 \times 0.38} = \dfrac{0.62}{0.903} = 0.687$

From the equation

$$E = \frac{c_h\,(th_1 - th_2)}{c_{min}\,(th_1 - tc_1)}$$

$(E = 0.687)$

If can be seen,

or $\quad \dfrac{th_1 - th_2}{th_1 - tc_1} = 0.687$

so $\quad (th_1 - th_2) = 0.687\ (105 - 15) = 61.83$

or $\quad th_2 = 105 - 61.83 = 43.17°C$

Note that the value of E could also be directly obtained from the plot E vs NTU (with $\frac{c_{min}}{c_{max}}$ as the parameter) for $\frac{c_{min}}{c_{max}} = 0.256$ and NTU = 1.3. we find that $E = 0.69$

2. A parallel flow heat exchanger has to cool 1,000 kg/hr of oil from 60°C to 35°C. The cooling water enters the exchanger at 15°C and leaves at 25°C. Determine the value of UA. The heat capacity of oil is 0.5 k cal/kg °C.

Solution

The rate of flow of water which is the cold fluid, is given by

$$m_c\ cp_c\ (tc_2 - tc_1) = m_h\ cp_h\ (th_1 - th_2)$$

$$m_c = \frac{1000 \times 0.5\,(60 - 35)}{1 \times (25 - 15)} = 1250 \text{ kg/hr}$$

Now $c_h = 1000 \times 0.5 = 500$

$c_c = 1250 \times 1 = 1250$

thus $c_{min} = c_h = 500$

and $\dfrac{c_{min}}{c_{max}} = \dfrac{500}{1250} = 0.4$

$$E = \frac{c_h}{c_{min}}\left(\frac{th_1 - th_2}{th_1 - tc_1}\right) = \frac{60 - 35}{60 - 15} = \frac{25}{45}$$

$$= 0.55$$

From the appropriate plot of E vs NTU for $\dfrac{c_{min}}{c_{max}} = 0.4$ and $E = 0.55$,

$$\frac{UA}{c_{min}} = 1.1$$

or $UA = 1.1 \times 500 = 550$ k cal/hr °C

Illustration on Heat Exchanger Area Computation

Determine the heat transfer surface area required for a heat exchanger constructed from 1-inch Outer Diameter tube to cool 55,000 lbs/hr of a 95 percent ethyl alcohol solution (whose *cp* is 0.91 Btu/lb °F) from 150°F to 103°F, using 50,000 lbs/hr of water available at 50°F. Take that the over all heat transfer coefficient based on the outer-tube area is 100 Btu/hr ft^2 °F. The heat transfer surface area is required for the following arrangements:

(i) Parallel-flow tube and shell

(ii) Counter flow tube and shell and

(iii) Reversed current exchanger with two shell passes and 72 tube passes with the alcohol flowing through the shell and the water flowing through the tubes.

Solution

(i) The outlet temperature of the water for any of the three configurations can be got from an over all energy balance, assuming that the heat loss to the atmosphere is negligible.

Writing the energy balance is

$$m_h\ c\,p_h\ (Th_{in} - Th_{out}) = m_c\ c\,p_c\ (Tc_{out} - Tc_{in})$$

on substitution of the data in the above equation

$$(55{,}000)\ (0.91)\ (150 - 103) = (50{,}000)\ (1.0)\ (Tc_{out} - 50)$$

from which the outlet temperature of the water is found to be 97°F. The rate of heat transfer from the alcohol to the water is therefore

$$q = m_h\ c\,p_h\ (Th_{in} - Th_{out})$$

$$= (55{,}000)\ (0.91)\ (150 - 103) = 2{,}350{,}000 \text{ Btu/hr}$$

The LMTD for parallel flow is

$$\text{LMTD} = \frac{\Delta Ta - \Delta Tb}{ln\left(\dfrac{\Delta Ta}{\Delta Tb}\right)} = \frac{100 - 6}{ln\left(\dfrac{100}{6}\right)} = 33.4°\text{F}$$

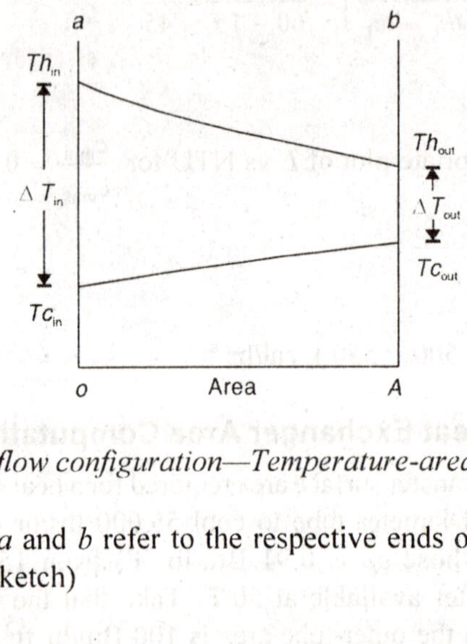

Fig. 6.33 *Parallel flow configuration—Temperature-area diagram*

Subscripts a and b refer to the respective ends of the heat exchanger (refer the sketch)

We know

$$q = U\ A\ \overline{\Delta T}$$

From the above equation, the heat transfer area is

$$A = \frac{q}{(U)\ (\text{LMTD})} = \frac{2{,}350{,}000}{(100)\ (33.4)} = 703 \text{ ft}^2$$

The length of the exchanger for a 1 inch. Outer Diameter tube would be too great, to be practical.

(ii) For the counterflow configuration, the appropriate mean temperature difference is 150 – 97 = 53°F because $m_h\ c\,p_h = m_c\ c\,p_c$

The required heat transfer surface area is

$$A = \frac{q}{(U)\,(\text{LMTD})} = \frac{2{,}350{,}000}{(100)\,(53)} = 444 \text{ ft}^2$$

Which is 40 percent less than that needed for parallel flow.

(iii) For the third arrangement *viz*: reversed current one. We estimate the appropriate mean temperature difference by applying the correction factor found from the correction factor chart corresponding to this configuration, to the mean temperature for counterflow

The temperature difference ratio is`

$$\frac{Tc_{\text{out}} - Tc_{\text{in}}}{Th_{\text{in}} - Tc_{\text{in}}} = \frac{90 - 50}{150 - 50} = 0.47$$

and the capacity ratio is

$$\frac{m_c\ cp_c}{m_h\ cp_h} = 1$$

From the appropriate correction factor chart, the correction factor F is obtained as 0.97 and the heat transfer surface area is

$$A = \frac{444}{0.97} = 460 \text{ ft}^2$$

The length of the heat exchanger for 72 1-inch Outer Diameter tubes in parallel would be

$$L = \frac{\frac{A}{72}}{\pi D} = \frac{6.4}{(\pi)\left(\frac{1}{12}\right)} = 24.4 \text{ ft}$$

It may be seen that this length is not unreasonable, but if it is desirable to shorten the exchanger, more tubes could be used.

Illustration on the Computation of Heat Exchanger Effectiveness

Based on a performance test on a well-baffled single shell, two-tube-pass heat exchanger, the following data are made available. Oil (cp = 0.5 Btu/lb °F) in turbulent flow inside the tubes entered at 160°F at the rate of 5000 lbs/hr and left at 100°F; water flowing on the shell side entered at 60°F and left at 80°F. A change in service conditions requires the cooling of a similar oil from an initial temperature of 200°F but at three fourths of the flow rate used in the performance test. Calculate the exit temperature of the oil for the same water rate and the inlet temperature as before.

Solution

The data obtained during the test can be used to determine the hourly heat capacity of the water and the over all conductance of the exchanger. The hourly heat capacity of the water can be calculated from the equation

$$\frac{c_c}{c_h} = \frac{Th_{out} - Th_{in}}{Tc_{out} - Tc_{in}}$$

$$c_c = c_h \frac{Th_{in} - Th_{out}}{Tc_{out} - Tc_{in}}$$

$$= (5000)\,(0.5)\left(\frac{160 - 100}{80 - 60}\right) = 7500 \text{ Btu/hr °F}$$

The dimensionless temperature ratio is

$$\frac{Tt_{out} - Tt_{in}}{Ts_{in} - Tt_{in}} = \frac{Th_{in} - Th_{out}}{Th_{in} - Tc_{in}} = \frac{60}{100} = 0.6$$

$$\frac{Ts_{in} - Ts_{out}}{Tt_{out} - Tt_{in}} = \frac{m_t\, c\, p_t}{m_s\, c\, p_s} = \frac{20}{60} = 0.33$$

Subscripts s and t, denote the shell and tube respectively.

From the appropriate correction factor chart F can be found

$F = 0.94$

The mean temperature difference is

$$\overline{\Delta T} = F \times \text{LMTD} = 0.94\,\frac{80 - 40}{ln\left(\frac{80}{40}\right)}$$

$$= 54.2\text{°F}$$

From the equation

$$q = U\,A\,\overline{\Delta T}$$

The UA term is

$$U\,A = \frac{q}{\overline{\Delta T}}$$

$$= \frac{(7500 \times 20)}{54.2} = 2760 \text{ Btu/hr °F}$$

Since the thermal resistance on the oil side is controlling, a decrease in velocity to 75 percent of the original value will increase the thermal resistance roughly by the velocity ratio raised to the 0.8 power. Under the new conditions the UA, the NTU and the hourly heat capacity ratio will therefore be approximately.

$$UA \simeq (2760)\,(0.75)^{0.8} = 2190 \text{ Btu/hr °F}$$

The number of transfer units

$$\text{NTU} = \frac{U\,A}{c_{oil}} = \frac{2190}{(0.75)\,(2500)} = 1.17$$

and $$\frac{c_{oil}}{c_{water}} = \frac{c_{min}}{c_{max}} = \frac{(0.75)\,(2500)}{7500} = 0.25$$

From an appropriate Effectiveness vs NTU chart (that is the one corresponding to one shell pass and two or multiples of two tube passes), the effectiveness (E) is found to be 0.63. Hence form the definition of E, the oil outlet temperature is

$$T_{oil\ out} = T_{oil\ in} - E\,\Delta T_{max}$$

$$= 200 - (0.63)\,(140) = 112\text{°F}.$$

Illustration on Heat Exchanger—Exit Temperature Calculation

In a double pipe heat exchanger, water flows through the inner pipe maintaining its outer surface at a mean temperature of 60°C. Steam at 1.4 kg/cm^2 flows through the annular space between the two pipes at the rate of 110 kg/hr. The temperature of steam at entrance to the exchanger is 205°C. Using the colburn equation determine the èxit temperature of steam if the pipes are 2.5 m long. The Outer Diameter of the inner pipe is 3.35 cm and the Inner Diameter of the outer pipe which is insulated is 5.25 cm. For steam, use

c_p = 0.48 k cal/kg °C, ρ = 0.688 kg/m^3 and k and μ values at 1 atm pressure.

Solution

The area of flow A_f is:

$$A_f = \frac{\pi}{4}\left(D_2^2 - D_1^2\right) = \frac{\pi}{4}\left(5.25^2 - 3.35^2\right)$$

$$= 12.9 \text{ cm}^2 = 12.9 \times 10^{-4} \text{ m}^2$$

$$G = \frac{w}{A_f} = \frac{110}{12.9 \times 10^{-4}} = 8.53 \times 10^4 \text{ kg/hr m}^2$$

$$D_E = D_2 - D_1 = 5.25 - 3.35$$

$$= 1.9 \text{ cm or } 0.019 \text{ m}$$

Let us assume (as a first approximation) that the steam leaves at 145°C, i.e. it undergoes a temperature drop of 60°C. Then

$$t_b = \frac{205 + 145}{2} = 175\text{°C}$$

$$t_w = 60\text{°C}$$

$$t_f = \frac{tb + tw}{2} = \frac{175 + 60}{2} = 117.5°\text{C}$$

At 117.50°C, the properties of steam are

$$\mu = 0.045 \text{ kg/hr m}$$

$$k = 0.0216 \text{ k cal/hr m °C}$$

Therefore

$$N_{pr_f} = \left(\frac{c_p \mu}{k}\right)_f = \frac{0.045 \times 0.48}{0.0216} = 1.0$$

$$N_{Re} = \frac{G D_{eq}}{\mu_f} = \frac{8.53 \times 10^4 \times 0.019}{0.045}$$

$$= 3.6 \times 10^4$$

We know

$$\frac{h}{G c_{p_h}} = \frac{0.023}{(N_{Re})^{0.2} (N_{pr_f})^{\frac{2}{3}}}$$

$$= \frac{0.023}{(36000)^{0.2} (1)^{\frac{2}{3}}}$$

$$\frac{h}{8.53 \times 10^4 \times 0.48} = \frac{0.023}{8.12}$$

or $\qquad h = 116 \text{ k cal/hr m}^2 \text{ °C}$

Now, if we denote the inlet and exit temperature of steam by t_1 and t_2, a heat balance gives

$$q = w c_p (t_1 - t_2) = h A \Delta t_m$$

where Δt_m = mean temperature difference between steam and pipe surface

Hence

$$110 \times 0.48 (205 - t_2) = 116 \left(\pi \times \frac{3.35}{100} \times 2.5\right) \Delta t_m$$

or $\qquad (205 - t_2) = 0.58 \Delta t_m \qquad \text{...(i)}$

and Δt_m as we know is

$$\Delta t_m = \frac{\Delta t_2 - \Delta t_1}{\log\left(\frac{\Delta t_2}{\Delta t_1}\right)}$$

In which Δt_2 and Δt_1 are the temperature differences at exit and entrance, respectively.

Hence

$$\Delta t_m = \frac{(t_2 - 60) - (t_1 - 60)}{\log_e \left(\dfrac{t_2 - 60}{t_1 - 60}\right)} = \frac{t_2 - 205}{\log_e \dfrac{t_2 - 60}{205 - 60}}$$

$$\Delta t_m = \frac{205 - t_2}{\log_e \left(\dfrac{145}{t_2 - 60}\right)} \quad \text{...(ii)}$$

From the expressions (1) and (2)

$$(205 - t_2) = 0.58 \frac{205 - t_2}{\log e \left(\dfrac{145}{t_2 - 60}\right)}$$

$$\log_e \left(\frac{145}{t_2 - 60}\right) = 0.58$$

$$\frac{145}{t_2 - 60} = 1.786$$

or $$t_2 = 141.3°C$$

Note that this value of t_2 agrees fairly well with our assumed value of 145°C. This means that our value s of t_b and t_f were close enough and the property values of k and μ were taken at the correct temperature. Hence there is no need for the second trial. And, outlet temperature of steam = 141.3°C.

Illustration on Heat Exchanger Area Computation

Water in two tube passes is used to cool oil in one shell pass. As shown in the following sketch, the water enters at 20°C and at the flow rate of 3 kg/sec and leaves at 70°C. The oil enters the exchanger at 180°C and leaves at 80°C. Find (a) the oil flow rate (b) the quantity of heat exchanged between oil and water and (c) the required heat exchanger area. U can be taken as 350 w/m^2 °C. For water an average c_p value of 4180 J/kg °C can be assumed. For oil at a mean temperature of 130°C, a c_p value of 2350 J/kg °C can be taken

Solution

$$q = m_c \, c p_c \left(Tc_{\text{out}} - Tc_{\text{in}}\right)$$

$$= m_h \, c\,p_h \left(Th_{in} - Th_{out}\right)$$

$$= 3 \text{ (kg/sec) } 4180 \times (70 - 20)$$

$$= m_h \,(2350)\,(180 - 80)$$

$$= 627{,}000 \text{ watts (the heat exchanged)}$$

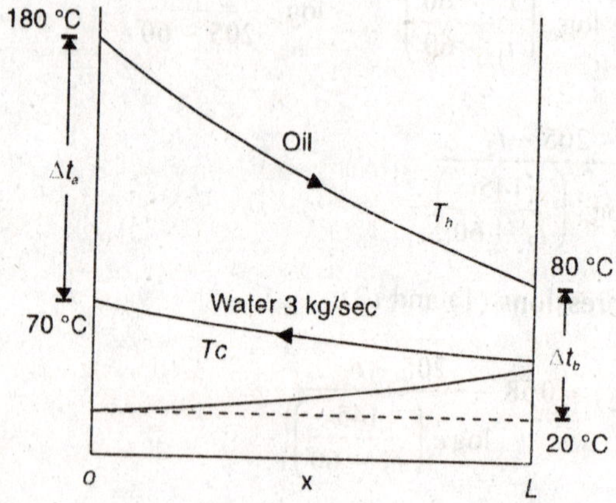

Fig. 6.34 *Temperature profile in an 1–2 exchanger*

Solving for m_h the oil flow rate

$$m_h = 2.67 \text{ kg/sec}$$

with reference to the figure.

$$\Delta Ta = 180 - 70 = 110°\text{C and } \Delta T_b = 80 - 20 = 60°\text{C}$$

The LMTD therefore is,

$$\text{LMTD} = \frac{110 - 60}{ln\dfrac{110}{60}} = 82.5°\text{C}$$

From the correction factor chart corresponding to this configuration, using the R value of 2.0 $\left(R = \dfrac{180 - 80}{70 - 20}\right)$ and S value of 0.3125 $\left(S = \dfrac{70 - 20}{180 - 20}\right)$, the correction factor F is read as, $F = 0.86$

The required heat exchanger area can then be determined

$$q = 627000 = \text{UAF}\,(\Delta T)_{\text{LM}}$$

$$= (350)\,(\text{A})\,(0.86)\,(82.5)$$

or $$A = 25.3 \text{ m}^2.$$

Illustration on the Use of NTU Concept

To illustrate the NTU method, the above example can be worked backward by assuming known area but unknown outlet temperatures. Then we are given oil in one shell pass and water in two tube passes with U = 350 w/m^2 °C and A = 25.3 m^2. The oil enters at 180°C and the water enters at 20°C. Determine the outlet temperatures if m_{oil} = 2.67 kg/sec and m_{water} = 3 kg/sec (the mass flow rates). The c_p values for oil and water can be taken as 2350 J/kg °C and 4180 J/kg °C respectively

$$c_{cold} = (3)\,(4180) = 12540 \text{ w/°C} = c_{max}$$

$$c_{hot} = (2.67)\,(2350) = 6270 \text{ w/°C} = c_{min}$$

$$\frac{c_{min}}{c_{max}} = \frac{6270}{12540} = 0.5$$

The NTU for this situation is

$$\text{NTU} = \frac{U\,A}{c_{min}} = \frac{(350)\,(25.3)}{6270}$$

$$= 1.41$$

The effectiveness for a 1–2 (1 shell pass and two tube passes) is given by the expression

$$E = 2\left[1 + r + B\,\frac{1 + \exp(-\text{NB})}{1 - \exp(-\text{NB})}\right]^{-1}$$

where $B = \left(1 + r^2\right)^{\frac{1}{2}}$

$$r = \frac{c_{min}}{c_{max}}$$

$$N = \text{NTU} = \frac{U\,A}{c_{min}}$$

As a matter of fact, the above expression for E, is applicable to the configurations one shell pass and (2, 4, 6, ...) tube passes

$$B = \left[(1 + 0.5)^2\right]^{\frac{1}{2}} = 1.118$$

$$E = 2\left\{1 + 0.5 + 1.118\,\frac{1 + \exp\left[-1.41\,(1.118)\right]}{1 - \exp\left[-1.41\,(1.118)\right]}\right\}^{-1}$$

$$= 0.625$$

From the definition of effectiveness

$$E = 0.625\,\frac{(6270)\,(180 - Th_{out})}{(6270)\,(180 - 20\ ^\circ C)}$$

$$= \frac{(12540)\,(Tc_{out} - 20)}{6270\,(180 - 20)}$$

Solving for Th and Tc, we get

$Th_{out} = 80.0°C$

$Tc_{out} = 70°C.$

Illustration on the LMTD Computation

In a single-pass heat exchanger, the hot fluid enters at 300°F and leaves at 220°F while the cold fluid enters at 75°F and exits at 200°F. Calculate the LMTD in the cases of parallel and counter flows and comment on the difference.

Solution

Referring to the temperature profile sketch, for a parallel flow situation, we can write that

$$\Delta Tb = Th_{out} - Tc_{out} = 220 - 200 = 20\ ^\circ F$$

$$\Delta Ta = Th_{in} - Tc_{in} = 300 - 75 = 225\ ^\circ F$$

$$\text{Then } \Delta T_{LM}\ (\text{parallel}) = \frac{225 - 20}{ln\left(\dfrac{225}{20}\right)}$$

$$= 84.7°F$$

For the case of the counter flow, referring to the appropriate temperature-length diagram, we can see that

$$\Delta Ta = Th_{in} - Tc_{out} = 300 - 200 = 100\ ^\circ F$$

$$\Delta Tb = Th_{out} - Tc_{in} = 220 - 75 = 145\ ^\circ F$$

from which

$$\Delta T_{LM}\ (\text{counterflow}) = \frac{100 - 145}{ln\left(\dfrac{100}{145}\right)}$$

$$= 121.1°F$$

For the same condition in this case, the LMTD for counter flow is 43 per cent higher; so a counter flow design will require only $\frac{1}{1.43}$ or 70 per cent of the exchanger area needed for a parallel flow system.

Illustration on the Computation of Tube Length in Heat Exchangers

1. A counter flow, concentric tube heat exchanger is employed to cool the lubricating oil for a large industrial gas turbine engine. The rate of flow of cooling water through the inner tube (D_i = 25 mm) is 0.2 kg/sec while that of oil through the outer annulus (D_o = 45 mm) is 0.1 kg/sec. The oil and water enter at temperatures of 100 and 30°C respectively. How long should the tube be made if the outer temperature of the oil is to be 60°C?

Solution

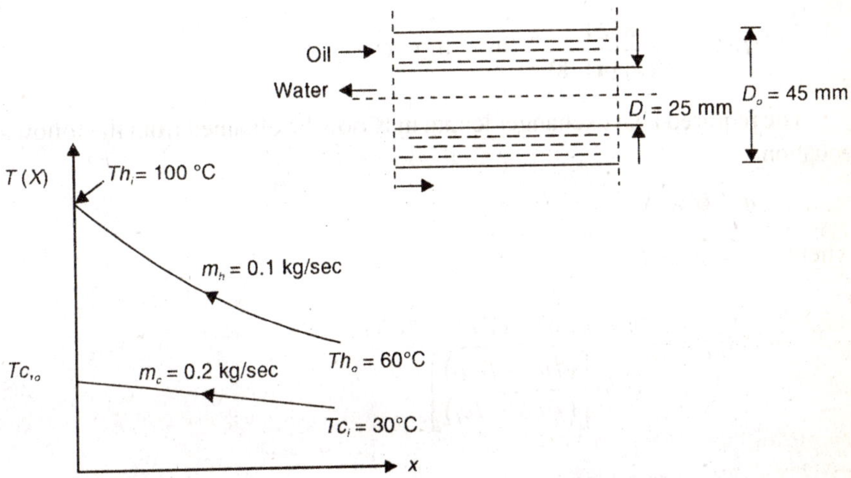

Fig. 6.35 *Temperature—length diagram in a counterflow exchanger*

We can make the following assumptions:

1. Heat loss to the surroundings is negligible
2. Constant properties
3. Negligible tube wall thermal resistance and fouling factors
4. Fully developed conditions for the water and oil (U independent of x)

The oil properties taken from literature at the temperature of T_h = 80°C, are

c_p = 2130 j/kg °C

$\mu = 3.25 \times 10^{-2}$ Ns/m²

k = 0.138 w/m °K

The temperatures of water are (at $Tc \approx 35\,°C$)

c_p = 4178 j/kg °K

$\mu = 725 \times 10^{-6}$ Ns/m²

k = 0.625 w/m °K

$N_{pr} = 4.85$

The required heat transfer rate may be obtained from the overall energy balance for the hot fluid.

$$q = m_h \; c p_h \; (Th_i \; - Th_o)$$

$$= 0.1 \times 2131 \times (100\text{-}60) = 8524 \text{ watts}$$

The outlet temperature of water can be calculated from the following equation

$$Tc,o = \frac{q}{m_c \; c p_c} + Tc_i$$

$$Tc,o = \frac{8524}{(0.2)\,(4178)} + 30 = 40.2\ °\text{C}$$

The required heat exchanger length may now be obtained from the following equation

$$q = U\,A\,\Delta T_{lm}$$

where $A = \pi\ D_i\ L$

$$\Delta T_{lm} = \frac{(Th_i - Tc,o) - (Th,o - Tc_i)}{ln\left[\dfrac{(Th_i - Tc,o)}{(Th,o - Tc_i)}\right]}$$

$$\Delta T_{lm} = \frac{59.8 - 30}{ln\left(\dfrac{59.8}{30}\right)} = 43.2\ °\text{C}$$

The overall heat transfer coefficient can be calculated from the equation (negligible tube wall thermal resistance and fouling factors)

$$U = \frac{1}{\left(\dfrac{1}{h_i}\right) + \left(\dfrac{1}{h_o}\right)}$$

For water flow through the tube

$$N_{Re} = \frac{4\,\text{m}}{\pi\ D_i\ \mu} = \frac{4 \times 0.2}{\pi\ (0.025\,\text{m})\ 725 \times 10^{-6}}$$

$$= 14050$$

The flow is seen to be turbulent (based on the magnitude of N_{Re}).

The heat transfer coefficient can be computed from the equation

$$N_{Nu} = 0.023\ N_{Re}^{0.8}\ N_{pr}^{0.4}\ (N_{pr} = 4.85)$$

$$= (0.023)\,(14050)^{0.8}\,(4.85)^{0.4} = 90$$

$$h_i = \frac{N_{Nu}\,k}{D_i} = \frac{90 \times 0.625}{0.025} = 2250 \text{ w/m}^2\ {}^\circ\text{K}$$

For the flow of oil through the annulus, the hydraulic diameter is

$$D_h = D_o - D_i = 0.02 \text{ m}$$

and the Reynolds number is

$$N_{Re} = \frac{\rho\,u\,D_h}{\mu} = \frac{\rho\,(D_o - D_i)}{\mu} \times \frac{m}{\dfrac{\rho\,\pi\left(D_o^2 - D_i^2\right)}{4}}$$

$$N_{Re} = \frac{4\text{m}}{\pi\,(D_o + D_i)\,\mu} = \frac{4 \times 0.1}{\pi\,(0.045 + 0.025) \times 3.25 \times 10^{-2}}$$

$$= 56$$

The flow through the annulus is laminar. The Nusselt number calculated in a similar way is 5.56.

$$N_{Nu} = \frac{h_o\,D_h}{k} = 5.56$$

$$h_o = 5.56 \times \frac{0.138}{0.020} = 38.4 \text{ w/m}^2\ {}^\circ\text{K}$$

The overall convection coefficient is then

$$U = \frac{1}{\left(\dfrac{1}{2250}\right) + \left(\dfrac{1}{38.4}\right)} = 37.8 \text{ w/m}^2 \text{ K}$$

and from the rate equation, it follows that

$$L = \frac{q}{U\,\pi\,D_i\,\Delta T_{lm}} = \frac{8524}{(37.8)\,\pi\,(0.025)\,(43.2)}$$

$$= 66.5 \text{ m}$$

The hot side convection coefficient controls the rate of heat transfer between the two fluids and the low value of h_o is responsible for the large value of L. A spiral tube arrangement would be needed.

Because $h_i >> h_o$, the tube wall temperature will follow closely that of the coolant water. Accordingly, the assumption of uniform wall temperature used to obtain h_o, is reasonable.

2. A shell and tube heat exchanger is to be designed to heat 2.5 kg/sec of water from 15°C and 85°C. It is desired to accomplish the heating by passing hot engine oil which is available at 160°C through the shell side of the exchanger. The oil is known to provide an average heat transfer coefficient (h_o) of 400 w/m² °k on the outside of the tubes. Ten tubes pass the water through the shell. Each tube is thin walled of diameter $D = 25$ mm and makes eight passes through the shell. If the oil leaves the exchanger at 100°C, what is its flow rate? How long should the tubes be to accomplish the desired heating?

Solution

The situation is represented schematically below,

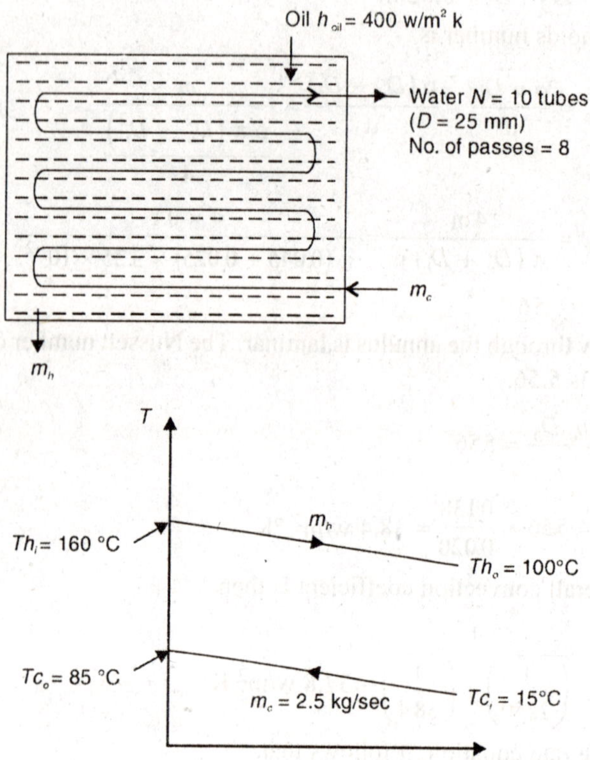

Fig. 6.36 *Schematic diagram of a shell and tube heat exchanger with the temperature values of the streams*

The usual assumptions made are:

1. Water flow in the tubes is fully developed
2. The tube wall thermal resistance and fouling effects are negligible
3. The properties are constant.

They are given below

The c_p of the engine oil 2350 J/kg °k at 130°C, the average temperature of the oil

Likewise for water, c_p = 4181 J/kg °k at an average temperature of 50°C

$\mu_{water} = 548 \times 10^{-6}$ N/s.m²

$k_{water} = 0.643$ w/m °K

$N_{pr} = 3.56$

The heat transfer required of the heat exchanger is

$$q = m_c \, cp_c \, (Tc,o - Tc,i)$$

$$= 2.5 \times 4181 \times (85 - 15) = 7.317 \times 10^5 \text{ watts}$$

Therefore, m_h can he computed

$$m_h = \frac{q}{cp_h \, (Th_i - Th,o)} = \frac{7.317 \times 10^5}{(2350)\,(160 - 100)} = 5.20 \text{ kg/sec}$$

The length of the tubes can be calculated as follows

$$q = \text{U A F } \Delta T_{lm} \text{ (cold fluid)}$$

and $$U = \frac{1}{\left(\frac{1}{h_i}\right) + \left(\frac{1}{h_o}\right)}$$

In the above equation h_i can be obtained by first computing N_{Re}

If m_1 is the water flow rate per tube

$$m_1 = \frac{m_c}{N} = 0.25 \text{ kg/sec}$$

N = no. of tubes

$$N_{Re} = \frac{4\,m_1}{\pi\, D\, \mu} = \frac{4 \times 0.25}{\pi\,(0.025)\,548 \times 10^{-6}}$$

$$= 23{,}234$$

Therefore the water flow is turbulent and

$$N_{Nu} = 0.23\, N_{Re}^{0.8}\, N_{pr}^{0.4}$$

$$= (0.023)\,(23{,}234)^{0.8}\,(3.56)^{0.4} = 119$$

$$h_i = \frac{k}{D} N_{Nu} = \frac{0.643}{0.025} \times 119$$

$$= 3061 \text{ w/m}^2 \text{ °K}$$

Therefore $$U = \frac{1}{\left(\frac{1}{400}\right) + \left(\frac{1}{3061}\right)} = 354 \text{ w/m}^2 \text{ °K}$$

$$\text{If } R = \frac{T_i - T_o}{t_o - t_i} \text{ and } S = \frac{t_o - t_i}{T_i - t_i}$$

where T_i and T_o are the inlet and outlet temperatures of the shell fluid and t_i and t_o are the inlet and outlet temperatures of the tube fluid.

For the situation

$$R = \frac{160 - 100}{85 - 15} = 0.86 \text{ and } S = \frac{85 - 15}{160 - 15} = 0.48$$

Using these values of *R* and *P* in the correction factor graph one can find out the value of the correction factor *F*.

F turns out be 0.87

Therefore

$$\Delta T_{lm \text{ cold fluid}} = \frac{(Th_i - Tc,o) - (Th,o - Tc,i)}{ln\left[\dfrac{(Th_i - Tc,o)}{(Th,o - Tc,i)}\right]}$$

$$= \frac{75 - 85}{ln\left(\dfrac{75}{85}\right)} = 79.9°\text{C}$$

Hence, since $A = N\pi\, DL$

where *N*, the no. of tubes is 10

$$L = \frac{q}{U N \pi D \cdot F \,\Delta T_{lm\,\text{CF}}}$$

$$= \frac{7.317 \times 10^5}{(354) \times 10 \cdot \pi\,(0.025)\,(0.87)\,(79.9)}$$

$$= 37.9 \text{ m}$$

It may be observed that with $\frac{L}{D} = \frac{37.9}{0.025} = 1516$, the assumption of fully developed flow conditions through out the tube is justified.

With eight passes, the shell length is approximately $\frac{L}{M} = 4.7$ m.

3. In a large steam power plant the condenser is a heat exchanger in which steam is condensed to liquid water. Assume the condenser to be a shell-and-tube heat exchanger comprising a single shell and 30,000 tubes each executing two passes. The tubes are of thin wall construction with $D = 25$ mm and steam condenses on their outer surface with an associated heat transfer coefficient of h_o = 11,000 w/m^2 k. The heat transfer rate that must be brought about by the exchanger is $q = 2 \times 10^9$ w and this is accomplished by passing cooling water through the tubes at a rate of 3×10^4 kg/sec (the flow rate per tube is therefore 1 kg/sec). The water enters at 20°C while the steam condenses at 50°C. What is the temperature of the cooling water coming out of the condenser? What is the required tube length L per pass. For water the properties at an average temperature $Tc \simeq 27$°C are c_p = 4179 j/kg °K, $\mu = 855 \times 10^{-6}$ Ns/m^2

k = 0.613 w/m °K and N_{pr} = 5.83.

Solution

The situation can be represented schematically as shown below

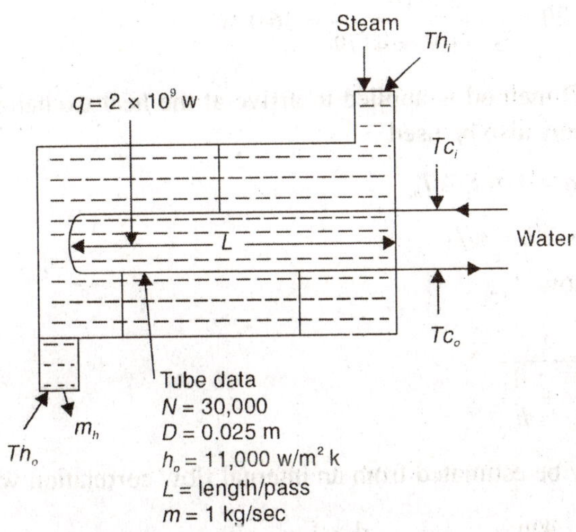

Fig. 6.37 *Shell and tube heat exchanger*

The temperature profile is

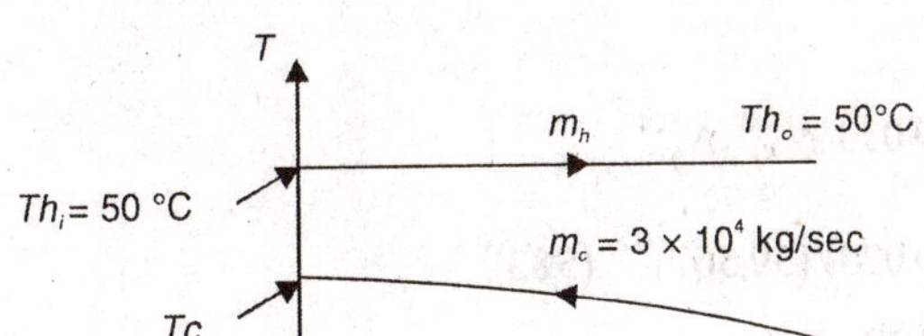

Fig. 6.38 *Temperature profile for the given situation*

The assumptions made are:

1. The thermal resistance of tube material and fouling factors are negligible.
2. The properties are constant.
3. Tube internal flow and thermal conditions are fully developed.

The cooling water outlet temperature can be calculated from the overall energy balance.

Accodingly

$$Tc,o = Tc,i + \frac{q}{m_c\, c\, p_c}$$

$$= 20 + \frac{2 \times 10^9}{3 \times 10^4 \times 4179} = 36.0\ °C$$

The LMTD method is applied to arrive at the heat exchanger design. The NTU method may also be used

We know $q = UAF\, \Delta T_{lm}$

where $A = N \times 2L \times \pi D$

We also know

$$U = \frac{1}{\frac{1}{h_i} + \frac{1}{h_o}}$$

in which h_i may be estimated from an internal flow correlation with

$$N_{Re} = \frac{4\,\mathrm{m}}{\pi D \mu} = \frac{4 \times 1}{\pi\,(0.025)\,(855) \times 10^{-6}}$$

$$= 59{,}567$$

Which implies that the flow is turbulent.

We know

$$N_{Nu} = 0.023\, N_{Re}^{0.8}\, N_{pr}^{0.4}$$

$$= (0.023)\,(59{,}567)^{0.8}\,(5.83)^{0.4}$$

$$= 308$$

Therefore,

$$h_i = N_{Nu}\frac{k}{D} = (308)\frac{0.613}{0.025}$$

$$= 7552\ \mathrm{w/m^2\ °K}$$

$$U = \frac{1}{\left(\frac{1}{7552}\right) + \left(\frac{1}{11000}\right)} = 4478 \text{ w/m}^2 \text{ °K}$$

The LMTD value is

$$\Delta T_{lm} = \frac{(Th,i - Tc,o) - (Th,o - Tc,i)}{ln\left[\frac{(Th,i - Tc,o)}{(Th,o - Tc,i)}\right]}$$

$$= \frac{(50 - 36) - (50 - 20)}{ln\left(\frac{14}{30}\right)} = 21 \text{ °C}$$

The correction factor F may be obtained from the appropriate graph. $F = 1$ for this situation.

It follows that the length of the tube per pass is

$$L = \frac{q}{U\,(N\,2\pi\,D)\,F\,\Delta T_{lm}}$$

$$= \frac{2 \times 10^9}{(4478)\,(30000 \times 2\pi \times 0.025) \times 1 \times 21}$$

$$= 4.51 \text{ m}$$

Therefore the total tube length is 9.02 (two passes)

Using the NTU method

$$c_h = c_{max} = \infty$$

and $$c_{min} = m_c\; c\,p_c$$

$$= 3 \times 10^4 \times 4179$$

$$= 1.25 \times 10^8 \text{ w/°K}$$

It follows that $q_{max} = c_{min}\,(Th,i - Tc,i) = 3.76 \times 10^9$ w

Hence $E = 0.53$ (effectiveness)

From the NTU-E chart, it is seen that NTU = 0.75 from which it may be shown that $L = 4.46$ per pass.

Illustration on Heat Exchanger Design

Water is available as a coolant at 60°F with a flow rate of 150 lb/min. It is to attain a temperature of 140°F by exchanging heat with an oil ($cp = 0.45$ Btu/lb °F). The oil will enter the exchanger at 240°F and leave at 80°F. The overall heat transfer coefficient is 50 Btu/hr ft² F. Determine

(i) The required area in a single-pass counter flow exchanger.

(ii) The required heat transfer area if a shell-and-tube exchanger is to be used with the water making one shell pass and the oil two tube passes.

(iii) The exit temperature if, for the exchanger of (i), the flow rate of water is reduced to 120 lb/min.

Solution

To solve this problem, we first determine the heat transfer rate using the water flow rate

$$q = m\,c_p\,(\Delta T)_{H_2o}$$

$$= (150)\,(1)\,(80) = 12{,}000 \text{ Btu/min}$$

Either the LMTD or the effectiveness approach may be used to solve for the heat transfer area in (i)

Let us use the NTU/effectiveness approach the required quantities are determined as

$$q = m\,c_p\,\Delta T_{H_2o} = m\,c_p\,\Delta T_{\text{oil}}$$

$$\downarrow c_{H_2o} \qquad\qquad \downarrow c_{\text{oil}}$$

$$c_{H_2o} = 150 \text{ Btu/min °F}$$

$$c_{\text{oil}} = \frac{12000}{160} = 75 \text{ Btu/min °F}$$

Therefore $c_{\text{oil}} = c_{\text{min}}$

$$\text{NTU} = \frac{U\,A}{c_{\text{min}}} = \frac{(50)\,(A)}{(75)\,(60)} = 0.0111\,A$$

$$E = \frac{q_{\text{actual}}}{q_{\text{max}}} = \frac{12000}{c_{\text{min}}\,(Th_{\text{in}} - Tc_{\text{in}})}$$

$$= \frac{12000}{(75)\,(240 - 60)} = 0.889$$

$$\frac{c_{\text{min}}}{c_{\text{max}}} = 0.5$$

From the appropriate plot, NTU = 3.3 and the required are (heat transfer) is

$$A = \frac{3.3}{0.0111} = 297 \text{ ft}^2$$

For (ii), the NTU/effectiveness approach will involve the same values for the parameters NTU, E and $c_{\text{min}}/c_{\text{max}}$.

From the appropriate plot for the shell-and-tube exchanger, the conditions of this problem are seen to be impossible as far as a shell-and-tube arrangement (configuration) is concerned. In order to use a shell-and-tube exchanger for the streams with flow rates and entering temperatures specified, less heat transfer would be accomplished and the temperature changes of the two fluid streams would be less than that specified in the problem statement.

In the case of (iii) of this illustration, we have

$A = 297\ \text{ft}^2$

$c_{H_2O} = (120)(1) = 120$ Btu/min °F

$c_{oil} = 75$ Btu/min °F, thus c_{oil} is c_{min}

That is, the oil is the minimum fluid

$$\text{NTU} = \frac{U\,A}{c_{min}} = 3.3$$

$$\frac{c_{min}}{c_{max}} = \frac{75}{120} = 0.625$$

and referring to the appropriate plot and reading the value of the effectiveness we see that $E = 0.86$

We may now determine q as

$$q = E\,c_{min}\,(Th_{in} - Tc_{in})$$
$$= (0.86)\,(75)\,(240 - 60)$$
$$= 11610 \text{ Btu/min}$$

Each of the existing stream temperatures determined easily from this result

$$q = c_{H_2O}\left(T_{H_2O\text{ out}} - T_{H_2O\text{ in}}\right)$$

$$= c_{oil}\left(T_{oil\text{ in}} - T_{oil\text{ out}}\right)$$

$$T_{H_2O\text{ out}} = 60 + \frac{11610}{120} = 157\,°\text{F}$$

$$T_{oil\text{ out}} = 240 - \frac{11610}{75} = 85.2\,°\text{F}$$

In this illustration, (iii) was a relatively simple evaluation, using the NTU/effectiveness approach, while a laborious trial-and-error procedure would have been required had the LMTD approach been adopted.

Example on Shell and Tube Heat Exchanger

In a electricity-generating facility, steam leaves a turbine and is piped to a condensing unit. After condensation occurs, it is desired to further cool the (distilled) water by means of a shell-and-tube exchanger. The water enters the heat exchanger at 110°F with a flow rate of 170,000 lbs/hr. The heat will be transferred to raw water from a nearby river. The raw water is available at 65°F and the mass flow rate is 150,000 lbs/hr. It is proposed to use a heat exchanger that has a 17 ¼ in ID shell and ¾ in OD 18 BWG tubes that are 16 feet long. The tubes are laid out on a 15/16 inch, triangular pitch.

The tube fluid will make two passes. The shell contains baffles that are spaced 1 ft apart. Determine the outlet temperature of the distilled water.

The properties of distilled water, and raw water are given below. Also other pertaining data are furnished below.

The properties of distilled water at 104°F are

$\rho = 0.994\ (62.4)$ lb/ft³ $k_f = 0.363$ Btu/hr ft °R

$c_p = 0.998$ Btu/lb °R $\alpha = 5.86 \times 10^{-3}$ ft²/hr

$\nu = 0.708 \times 10^{-5}$ ft/sec $N_{pr} = 4.34$

Also $T_1 = 110$°F $m_h = 170{,}000$ lb/hr

The properties of raw water are (at 68°F)

$\rho = 62.4$ lb/ft³ $k_f = 0.345$ Btu/lb ft °R

$c_p = 0.9988$ Btu/lb °R $\alpha = 5.54 \times 10^{-3}$ ft²/hr

$\nu = 1.083 \times 10^{-5}$ ft²/sec $N_{pr} = 7.02$

Also $t_1 = 65$°F and $m_c = 150{,}000$ lb/hr

For ¾ inch OD 18 BWG tubes

OD_t = ¾ inch = 0.0625 ft

ID_t = 0.652 inch = 0.0543 ft

Also, according to the tabulated values of the tube counts for shell-and-tube equipment for ¾ OD tubes on 15/16 triangular pitch, the number of tubes is 224.

No. of tube passes = 2.

Solution

Shell dimensions

D_s = 17.25 inch = 1.438 ft

No. of baffles = 15

Baffle spacing is 1 ft

The following calculations are performed in steps.

Subscript t denotes tubes

Subscript s denotes shell.

(i) *Flow areas*

$$A_t = 224\,\pi\,\frac{(0.0543)^2}{4\ (2)} = 0.259\ \text{ft}^2$$

$$A_s = 1.438\,\frac{(0.0156)\ (1)}{0.0781} = 0.287\ \text{ft}^2$$

Place the raw water through the tubes because it has the lower flow rate

(ii) *Shell equivalent diameter*

$$D_E = \frac{3.44\,(0.0781)^2 - \pi\,(0.0625)^2}{\pi\,(0.0625)}$$

$$= 0.044 \text{ ft}$$

(iii) *Reynolds numbers*

$$\text{Raw water } N_{Re_t} = \frac{150{,}000}{(3600)\,(62.4)\,(0.259)}\;\frac{0.0543}{(1.083)\times 10^{-5}}$$

$$= 1.29 \times 10^4$$

$$\text{Distilled water } N_{Re_s} = \frac{170{,}000}{(3600)\,(0.994)\,(6.2)\,(0.287)}\;\frac{0.0444}{0.708 \times 10^{-5}}$$

$$= 1.66 \times 10^4$$

(iv) *Nusselt numbers*

$$\text{Raw water } N_{Nu_t} = 0.023\left(1.29 \times 10^4\right)^{0.8}(7.02)^{0.4}$$

$$= 97.4$$

$$\text{Distilled water } N_{Nu_s} = 0.36 \times N_{Re}^{0.55}\, N_{pr}^{0.33}$$

$$N_{Nu_s} = 0.36 \times \left(10^4 \times 1.66\right)^{0.55}(4.34)^{0.33}$$

$$= 123$$

(The Nusselt number for the shell side is given by the above empirical equation and this equation has been verified on a number of commercial heat exchangers).

(v) *Convection coefficients*

$$\text{Raw water } h_i = \frac{(97.4)\,(0.345)}{0.0543} = 619 \text{ Btu/hr ft}^2\,°\text{R}$$

$$h_{io} = \frac{(619)\,(0.0543)}{0.0625} = 538 \text{ Btu/hr ft}^2\,°\text{R}$$

$$\text{Distilled water } h_o\; \frac{123\,(0.363)}{0.0444} = 1005 \text{ Btu/hr ft}^2\,°\text{R}$$

(vi) *Exchanger coefficient (clean or new)*

$$\frac{1}{U_o} = \frac{1}{538} + \frac{1}{1005}$$

$$U_o = 350 \text{ Btu/hr ft}^2\,°\text{R}$$

outlet temperatures $R = \dfrac{T_1 - T_2}{t_2 - t_1}$ $S = \dfrac{t_2 - t_1}{T_1 - t_1}$

$$R = \frac{150{,}000\ (0.9988)}{170{,}000\ (0.998)} = 0.883$$

$$A_o = 224\ \pi\ (0.0625)\ (16) = 703.7\ \text{ft}^2$$

$$\frac{U_o\ A_o}{m_c\ c\,p_c} = \frac{350\ (703.7)}{150{,}000\ (0.9988)} = 1.64$$

From the appropriate graph of S vs $\dfrac{U\,A_o}{m_c\ c\,p_c}$ with R as the parameter, S can be found out

$$S = 0.58 = \frac{t_2 - t_1}{T_1 - t_1}$$

$t_2 = 0.58\ (110 - 65) + 65$: $t_2 = 91.1°\text{F}$ (raw water)

$T_2 = 110 - 0.883\ (91.1 - 65)$: $T_2 = 86.9°\text{F}$ (distilled water)

Summary of requested information outlet temperatures

Raw water = 91.1°F

Distilled water = 86.9°F.

Examples on Double Pipe Exchangers

A bank of four diesel engines is used with alternators for the generation of electricity. The exhaust from the engines is discharged to the atmosphere. It is proposed to use all or part of the exhaust for heating air which can then be used for space heating to reduce costs. From measurements of velocity it is determined that the mass flow rate of exhaust available from the engines is 90 kg/hr. The exhaust-gas temperature is 600°k. The air a available at 20°C and is not of much use unless it can be heated to at least 80°C at a mass flow rate of 100 kg/hr. There are a number of 4 × 3 double pipe exchangers that are 2m long and made of type k copper tubing with compression fittings. Determine (a) how many of them are required and (b) the overall coefficient of all the exchangers. Assume counter flow arrangement.

The properties of air are (at 323°K)

$\rho = 1.088\ \text{kg/m}^3$ $k_f = 0.02814$ w/m °K

$c_p = 1.007$ j/kg °K $\alpha = 0.26 \times 10^{-4}\ \text{m}^2/\text{sec}$

$\nu = 18.2 \times 10^{-6}\ \text{m}^2/\text{sec}$ $N_{pr} = 0.703$

Also $m_c = 100$ kg/hr, $t_1 = 20°\text{C} = 293°\text{K}$

and $t_2 = 80°C = 353°K$

The properties of carbon dioxide gas at 500°k

$\rho = 1.0732$ kg/m³ $\quad k_f = 0.03352$ w/m °K

$c_p = 1.013$ j/kg °K $\quad \alpha = 0.3084 \times 10^{-4}$ m²/sec

$\nu = 21.67 \times 10^{-6}$ m²/sec $N_{pr} = 0.702$

Also $m_h = 90$ kg/hr and $T_1 = 600°K$

For seamless copper tubing.

4 standard type k $\quad ID_a = 9.8$ cm

3 standard type k $\quad ID_p = 7.384$ cm

$OD_p = 7.938$ cm.

Solution

The air is heated by the diesel exhaust with air inlet and outlet temperatures known. The diesel exhaust properties are not known.

Assumptions to be Used

1. Air properties are constant and evaluated at the average of the inlet and outlet temperatures.
2. Diesel exhaust has the same properties as carbon dioxide gas.
3. Diesel exhaust properties are constant. For lack of better information (i.e. outlet temperature of the exhaust gas) evaluate the exhaust properties at the average of the inlet temperatures $\frac{[600 + (20 + 273)]}{2} = 480 \simeq$ 500°K
4. The system is under steady state conditions.

   Composing the calculations in a methodical way, we proceed as follows.

(i) *Flow areas*

$$A_p = \pi \frac{(0.07384)^2}{4} = 4.28 \times 10^{-3} \text{ m}^2$$

$$A_a = \pi \frac{(0.098^2 - 0.07938^2)}{4} = 2.59 \times 10^{-3} \text{ m}^2$$

In view of the fact that $A_p > A_a$, let air be directed through the pipe that is the inner tube.

(ii) *Heat balance*

$$\text{Air; } q = \frac{100}{3600} (1007)(80 - 20) = 1.68 \times 10^3 \text{ w}$$

$$Co_2;\ T_2 = T_1 - \frac{q}{m_h\ c\,p_h}$$

$$= 600 - \frac{1.68 \times 10^3\ (3600)}{90\ (1013)} = 534\,°K$$

(iii) *Log-mean temperature difference*

$$LMTD = \frac{(600 - 353) - (534 - 293)}{ln\left(\frac{247}{241}\right)}$$

$$= 244°k$$

(iv) *Annulus equivalent diameter*

$$D_h = 0.098 - 0.07938 = 0.01862\,m$$

$$D_e = \frac{\left(0.098^2 - 0.07938^2\right)}{0.07938}$$

$$= 0.0416m$$

(v) *Reynolds numbers*

$$Air;\ N_{Re_p} = \frac{100}{3600}\ \frac{1}{(1.088)\ (0.00428)}\ \frac{0.07384}{18.2 \times 10^{-6}} = 2.42 \times 10^4$$

$$Co_2;\ N_{Re_a} = \frac{90}{3600}\ \frac{1}{(1.073)\ (0.00259)}\ \frac{0.0416}{21.67 \times 10^{-6}} = 1.73 \times 10^4$$

(vi) *Nusselt numbers*

$$Air;\ N_{Nu_p} = 0.023\left(2.42 \times 10^4\right)^{0.8}\ (0.703)^{0.4}$$

$$= 64.2$$

$$Co_2;\ N_{Nu_a} = 0.023\left(1.73 \times 10^4\right)^{0.8}\ (0.702)^{0.3}$$

$$= 50.8$$

(vii) *Film coefficients*

$$Air;\ h_i = \frac{64.2\ (0.02814)}{0.07384} = 24.5$$

$$h_{io} = 24.5\,\frac{0.07384}{0.07938} = 22.8\ w/m^2\ °K$$

$$Co_2;\ h_o = \frac{50.8\ (0.03352)}{0.0416} = 40.9 \text{ w/m}^2\ ^\circ K$$

(viii) *Fouling factors*

Air; $Rd_i = 0.0004 \text{ m}^2\ ^\circ K/w$

Co_2; $Rd_o = 0.002 \text{ m}^2\ ^\circ K/w$

(ix) *Exchanger coefficients*

$$\frac{1}{U_o} = \frac{1}{22.8} + \frac{1}{40.9}$$

$$U_o = 14.6 \text{ w/m}^2\ ^\circ K$$

$$\frac{1}{U} = \frac{1}{14.6} + 0.0004 + 0.002$$

$$U = 14.1 \text{ w/m}^2\ ^\circ K$$

(x) *Heat transfer area required*

$$q = U A_o \text{ LMTD}$$

$$A_o = \frac{1.68 \times 10^3}{(14.1)\ (244)} = 0.49 \text{ m}^2$$

One exchanger has a surface area of $A = \pi D_o L$

so $$L = \frac{0.49}{\pi\ (0.07938)} = 1.96 \text{ m}$$

Each available exchanger is 2m in length. Therefore only one exchanger is required

Summary of requested information

(a) One exchanger is required

(b) $U_o = 14.6$, $U = 14.1$ w/m² °K

Because the length of each exchanger is longer (more) than that required by the conditions of the problem more heat than specified will be exchanged. This implies that the outlet temperature of the air will be greater than 80°C and the outlet temperature of the diesel exhaust will be lower than 583°k.

Illustration on the Application of the Effectiveness—NTU Method to Calculated the Outlet Temperatures in a Heat Exchanger

In a electricity—generating facility, steam leaves a turbine and is piped to condensing unit. After condensation occurs, it is desired to further cool the (distilled) water by means of a shell-and-tube heat exchanger. The water enters the heat exchanger at 110°F with a flow rate of 170,000 lbs/hr. The heat will be transferred to raw water from a nearby river. The raw water is available at 65°F and the mass flow rate is 150,000lbs/hr. It is proposed to use a heat exchanger that has a 17 ¼ inch ID shell and ¾ inch OD 18 BWG tubes that are 16 feet long. The tubes are laid out on a 15/16 inch triangular pitch. The tube fluid will make two-passes. The shell contains baffles that are spaced 1 ft apart. Determine the outlet temperature of the distilled water using the effectiveness NTU method for the calculation, U_o = 350 Btu/hr ft² °k and A_o = 703.7 ft².

Solution

Data given are

	Distilled water (shell fluid)	*Raw water (tube fluid)*
m lbs/hr	170,000	150,000
c_p Btu/lb °R	0.998	0.9988
Inlet temperature °F	110	65

The effectiveness—NTU approach is employed when the overall heat transfer coefficient is known. We first determine the values of $m\ c_p$ for the two fluids

Raw water $(m\ c_p)$ = 150,000 (0.9988) = 149.820 Btu/hr °k

Distilled water $(m\ c_p)$ = 170,000 (0.998) = 169,660 Btu/hr °k

The raw water has the minimum value of $m\ c_p$

$$\frac{(m\ c_p)_{min}}{(m\ c_p)_{max}} = \frac{149{,}820}{169{,}660} = 0.885$$

$$\frac{U_o\ A_o}{(mc_p)_{min}} = \frac{350\ (703.7)}{149{,}820} = 1.64$$

With these values, one can find out the value of E from the appropriate plot

$E = 0.58$

The maximum heat that can be transferred

$$q_{max} = (m\ c_p)_{max}\ (T_1 - t_1)$$
$$= 149{,}820\ (110 - 65)$$

or $\quad q_{max} = 6.74 \times 10^6$ Btu/hr

The actual heat transferred by the exchanger then is,

$$q = E\ q_{max} = 0.58\ (6.74 \times 10^6)$$
$$= 3.91 \times 10^6 \text{ Btu/hr}$$

The outlet temperatures are now calculated with the heat-balance equations

$$q = m_c\ cp_c\ (t_2 - t_1)$$

Substituting

$$3.91 \times 10^6 = 149{,}820\ (t_2 - 65)$$

or $\quad t_2 = 91.1\ °F$ (distilled water)

Also, for the warmer fluid

$$3.91 \times 10^6 = 169{,}660\,(110 - T_2)$$

or $\quad T_2 = 86.9°F$ (raw water)

These results will be identical with the ones obtained by using the

$$S = \left(\frac{t_2 - t_1}{T_1 - t_1}\right) \text{ vs } \frac{U A_o}{c p_c\, m_c} \text{ plot.}$$

Illustration on the Computation of LMTD

1. A hot fluid at 100°C enters a double-pipe exchanger and is cooled to 75°C. A cooler fluid at 5°C enters the exchanger and is warmed to 50°C. Determine the LMTD for both counter flow and parallel-flow configurations.

Solution

$$T_1 = 100°C \qquad T_2 = 75°C$$

$$t_1 = 5°C;\ t_2 = 50°C$$

For counter flow

$$\text{LMTD} = \frac{(T_1 - t_2) - (T_2 - t_1)}{ln\left[\dfrac{(T_1 - t_2)}{(T_2 - t_1)}\right]}$$

$$= \frac{(100 - 50) - (75 - 5)}{ln\,\dfrac{50}{70}}$$

$$= 59.4°C \text{ (counter flow)}$$

For parallel flow

$$\text{LMTD} = \frac{(T_1 - t_1) - (T_2 - t_2)}{ln\left[\dfrac{(T_1 - t_1)}{(T_2 - t_2)}\right]}$$

$$= \frac{(100 - 5) - (75 - 50)}{ln\left(\dfrac{95}{25}\right)}$$

$$\text{LMTD} = 52.4°C \text{ (parallel flow).}$$

2. A hot fluid at 250°F enters a double-pipe exchanger and is cooled to 150°F. A cooler fluid enters the exchanger at 100°F and is to be warmed to 150°F. Determine the LMTD for both counter flow and parallel flow configurations.

Solution

Both fluids are to have the same outlet temperature

Given $T_1 = 250°F \quad t_1 = 100°F$

$T_2 = 150°F \quad t_2 = 150°F$

$$\text{For counter flow LMTD} = \frac{(250 - 150) - (150 - 100)}{ln\left(\frac{100}{50}\right)} = 72.1°F$$

$$\text{For parallel flow LMTD} = \frac{(250 - 100) - (150 - 150)}{ln\left(\frac{150}{0}\right)} = 0°F$$

(parallel flow)

For a finite heat-transfer rate and a finite overall heat-transfer coefficient, if parallel flow is to given equal outlet temperatures then the area needed must be infinite. This is not feasible economically.

Illustration on Effectiveness Ratio—NTU Method in Heat Exchangers

1. A counter flow oil heat exchanger is to be designed to accomplish an U of 225 k cal/hr m^2 °k. The area of heat transfer surface can be taken as 2 m^2. The mass flow rate of the cooling fluid (water) is 600 kg/hr with an inlet temperature of 30°C. It is now intended to employ the equipment for cooling oil, which is also flowing at the rate of 600 kg/hr at an initial temperature of 100°C. If the heat capacity of the oil is 0.5 k cal/kg °C calculate the estimated exit temperatures of the oil and water (ii) what would be the corresponding exit temperatures in case the heat transfer area is doubled? (iii) if the heat capacity ratios of both the fluids are made equal as 600 k cal/hr °C. Compute the exit temperatures of both the fluids for a surface area of 2m^2.

Solution

The capacity (ratios) of both the fluids can be calculated

$$m_h \, cp_h = 600 \times 0.5 = 300 \text{ k cal/hr °C}$$

$$m_c \, cp_c = 600 \times 1.0 = 600 \text{ k cal/hr °C}$$

Hence the hot fluid is the minimum fluid.

We know

$$\text{NTU} = \frac{U\,A}{c_{min}} = \frac{225 \times 2}{300} = 1.5$$

$$\text{and the capacity ratio } \frac{c_{min}}{c_{max}} = \frac{300}{600} = 0.5$$

The effectiveness of the heat exchanger E may by got by calculation using the expression for it for counter flow configuration as well as the chart for NTU-effectiveness in the case of counter flow arrangement.

By calculation $E = \dfrac{1 - e^{0.75}}{1 - 0.5\left(e^{-0.75}\right)}$

$$= \frac{0.528}{0.764} = 0.691$$

From the appropriate plot for NTU = 1.5 and $\dfrac{c_{min}}{c_{max}} = 0.5$, the corresponding value of E is 0.69.

In view of the calculated value being more precise, let us take the value of E as 0.691.

$$\therefore \quad E_h = \frac{th_1 - th_2}{th_1 - tc_1}$$

$$0.691 = \frac{100 - th_2}{100 - 30}$$

$$\therefore \ th_2 = (100 - [0.691) \times 70]$$
$$= 51.63°\text{C}$$

$\therefore$ The exit temperature of the oil, $th_2 = 51.63°\text{C}$.

The exit temperature of the water can be worked out from the energy balance for the heat exchanger.

$$m_h \, c\,p_h \left(th_2 - th_1\right) = m_c \, c\,p_c \left(tc_2 - tc_1\right)$$

from the above expression, $(tc_2 - tc_1) = (0.5)\ (48.37)$

$$\therefore \ tc_2 = 30 + 24.2 = 54.2°\text{C}$$

Therefore the exit temperature of the cold fluid, $tc_2 = 54.2°\text{C}$

(ii) If the heat transfer area is doubled then only the NTU value will be doubled in accordance with the equation $\dfrac{U\,A}{c_{min}} = \text{NTU}$.

While the capacity ratio will remain the same

For this situation $\text{NTU} = \dfrac{U\,A}{c_{min}} = \dfrac{225 \times 4}{300} = 3$ with $\dfrac{c_{min}}{c_{max}} = 0.5$, then E according to the equation

$$E = \frac{1 - e^{-\text{NTU}\left\{1 - \frac{c_{min}}{c_{max}}\right\}}}{1 - \frac{c_{min}}{c_{max}} e^{-\text{NTU}\left\{1 - \frac{c_{min}}{c_{max}}\right\}}}$$

$$= \frac{1 - e^{-1.5}}{1 - (0.5)\, e^{-1.5}}$$

$$= 0.875$$

The value of E can be got from the appropriate plot of NTU vs E with $\frac{c_{min}}{c_{max}}$ as the parameter.

In this case since NTU = 3 and $\frac{c_{min}}{c_{max}} = 0.5$. E from the plot is read as 0.87.

So for this case $0.875 = \frac{100 - th_2}{100 - 30}$ and the exit temperature of the oil,

$th_2 = 38.75°C$.

From the energy balance equation

$$tc_2 - tc_1 = \frac{m_h\, c\, p_h}{m_c\, c\, p_c}(th_1 - th_2) = \frac{1}{2} \times 61.25$$

$\therefore\ tc_2 = 30 + 30.625 = 60.625°C$

$\therefore$ The exit temperature of the water $tc_2 = 60.625°C$.

It will be interesting to note the effect of enhancing the area on the exit temperature. In the limiting case of very large area the effectiveness being unity,

$th_2 = 100 - (1 \times 70) = 30°C$

Which is the temperature of the cold fluid entering the heat exchanger and the corresponding exit temperature of water would be 65°C.

(iii) If the term $m\, c_p$ for the two fluids are made equal as 600 k cal/hr °C. Then

$$\text{NTU} = \frac{225 \times 2}{600} = 0.75$$

and $$\frac{c_{min}}{c_{max}} = \frac{m_h\, c\, p_h}{m_c\, c\, p_c} = 1$$

For this special case the expression

$$E = \frac{1 - e^{-\text{NTU}\left\{1 - \frac{c_{min}}{c_{max}}\right\}}}{1 - \frac{c_{min}}{c_{max}} e^{-\text{NTU}\left\{1 - \frac{c_{min}}{c_{max}}\right\}}}$$

becomes indeterminate

By the energy balance rate equation analysis the final result is

$$E = \frac{\text{NTU}}{1 + \text{NTU}} = \frac{0.75}{1.75} = 0.428$$

From the E vs NTU plot for NTU = 0.75 and $\frac{c_{min}}{c_{max}} = 1$,

the value of $E = 0.43$.

Taking the value of 0.428

$$0.428 = \frac{100 - th_2}{100 - 30}$$

or $\quad th_2 = 100 - (0.428 \times 70) = 70.04°C$

and $\quad tc_2 = 30 + [1 \times 29.96] = 59.96°C$

It is obvious that the effectiveness of the heat exchanger decreases appreciably when the dimensionless capacity ratio of the heat transfer fluids approaches unity. For the best performance this dimensionless ratio should approach zero, as in the case of boiler or condenser.

If the heat capacity-mass flow rate terms i.e. $m_h\, cp_h$ and $m_c\, cp_c$ are equal for the parallel flow arrangement, it can be seen that the expression

$$E = \frac{1 - e^{\frac{-UA}{c_{min}}\left\{1 + \frac{c_{min}}{c_{max}}\right\}}}{1 + \frac{c_{min}}{c_{max}}}$$

boils down to $E = \frac{1 - e^{-2\,\text{NTU}}}{2}$.

2. It is necessary to design a high temperature gas cooled reactor for a gas turbine power plant with helium as the reactor coolant and air as the working fluid for the gas turbine power plant. The terminal temperature across the heat exchanger for both the fluids are as follows $th_{in} = 775°C$; $th_{out} = 410°C$: $tc_{in} = 205°C$; $tc_{out} = 540°C$. The heat exchanger of the plant handles 1200 kg of air per sec with an average specific heat of 0.26 k cal/kg °C while the hot fluid flows through the heat exchanger at an average flow rate of 430 kg/sec and its average heat capacity is 1.4 k cal/kg °C. The overall heat transfer coefficient U is 300 k cal/hr m^2 °C. Determine the heat transfer area in the case of both parallel and counter flow arrangements.

Solution

Let us first determine the heat capacity ratio for the fluids

For the hot fluid Helium, $m_h\ c\,p_h$

$= 430 \times 1.3 \times 3600$

$= 2010 \times 10^3$ k cal/hr °C

For the cold fluid, air

$$m_c\ c\,p_c = 1200 \times 0.26 \times 3600$$

$$= 1122 \times 10^3 \text{ k cal/hr °C}$$

The cold fluid is seen to be the minimum fluid and the term $\dfrac{m_c\ c\,p_c}{m_h\ c\,p_h}$ becomes

$$\frac{m_c\ c\,p_c}{m_h\ c\,p_h} = \frac{1122}{2010} = 0.557$$

From the terminal temperatures of the fluids the heat exchanger effectiveness can be determined taking the air as the minimum fluid.

$$E_c = \frac{tc_{out} - tc_{in}}{th_{in} - tc_{in}} = \frac{540 - 205}{775 - 205} = 0.587$$

$$= 0.587$$

This magnitude of the effectiveness will be applicable for the cases of both the parallel as well as counter flow configurations

Referring to the NTU us E plot for the parallel flow arrangement for the values of $E = 0.587$ and $\dfrac{c_{min}}{c_{max}} = 0.557$

$$\text{NTU for the parallel flow} = 1.7 = \frac{U\,A}{c_{min}}$$

$\therefore$ Area of the heat transfer surface, A is

$$A = \frac{1.7 \times 1122 \times 10^3}{300} = 6360\,\text{m}^2$$

Like wise, for the case of counter flow arrangement, referring to the corresponding NTU vs E chart, with the same values of E and $\dfrac{c_{min}}{c_{max}}$.

NTU for the counter flow

$$= 1.2 = \frac{U\,A}{c_{min}}$$

$\therefore$ Area of the heat transfer surface

$$= \frac{1.2 \times 1122 \times 10^3}{300}$$

$$= 4490 \text{ m}^2$$

From the above, it is obvious that the requirement of heat transfer surface area is approximately 1.4 times larger in the case of parallel flow arrangement than the counter flow arrangement. Since the initial cost of the heat exchange equipment (first cost) is very much dependent on surface area the cost of a parallel flow heat exchanger under identical operating conditions is likely to be more.

Illustration on the Computation of Area and LMTD in Heat Exchangers

1. A steam condenser is transferring 240 kw of thermal energy at a condensing temperature of 60°C. The cooling water enters the condenser at 20°C, with a mass flow rate of 8000 kg/hour.

(a) Calculate the LMTD (log mean temperature difference)

(b) In case the U for the condenser surface is 1000 w/m^2 C what is the required heat transfer surface area?

$cp = 4.2$ kj/kg °K.

Solution

Fig. 6.39 *Schematic representation of a steam condenser*

$$\Delta T_{ln} = \frac{\Delta T_a - \Delta T_b}{ln\left(\frac{\Delta T_a}{\Delta T_b}\right)}$$

$Th = 60°\text{C} \qquad Tc_{\text{in}} = 20°\text{C}$

$\Delta T_a = Th - Tc_{\text{in}}$

$\Delta T_a = 40°\text{C}$

$\Delta T_b = Th - Tc_{\text{out}} \qquad Th = 60°\text{C}$

$Q = m\, c_p\, \Delta T$ where $\Delta T = Tc_{\text{out}} - Tc_{\text{in}}$

$Q = 240$ kw $\quad c_p = 4.2$ kj/kg °K

$m = 8000$ kg/hr $= 2.22$ kg/sec

$$\Delta T = \frac{240}{(2.22)(4.2)} = 25.7°\text{C}$$

$$\therefore Tc_{out} = Tc_{in} + \Delta T$$
$$= 20 + 25.7 = 45.7°\text{C}$$

$$\Delta Tb = Th - Tc_{out}$$
$$= 60 - 45.7$$
$$= 14.3°\text{C}$$

$$ln\left(\frac{\Delta Ta}{\Delta Tb}\right) = ln \frac{40}{14.3} = ln\, 2.80$$

$$= 1.0296$$

$$\Delta T_{ln} = \frac{40 - 14.3}{1.0296}$$

$$= \frac{25.7}{1.0296}$$

$$\Delta T_{ln} = 24.9°\text{C}$$

To obtain the heat transfer surface area, we can employ the basic expression

$$Q = U A \Delta T_{ln}$$

or
$$A = \frac{Q}{U \Delta T_{ln}}$$

Q = 240,000 watts

U = 1000 w/m^2 °C

ΔT_{ln} = 24.9°C

Therefore $A = \dfrac{240{,}000}{1000 \times 24.9}$

$= 9.64$ m^2.

2. Compute the mean temperature difference in a double-pipe parallel flow heat exchanger where 50 kgs/hour of water enters at the temperature of 80°C, flows-through the outer annulus and comes out at 50°C. Inside the inner pipe, water at 1000 kgs/hour enters at the temperature of 20°C and comes out at 35°C.

Solution

This situation can be represented by the following schematic diagram

Fig. 6.40 *Schematic representation of a double-pipe parallel flow heat exchanger*

$$\Delta T_{ln} = \frac{\Delta Ta - \Delta Tb}{ln\left(\dfrac{\Delta Ta}{\Delta Tb}\right)}$$

$\Delta Ta = Th_{in} - Tc_{in}$ $\qquad Th_{in} = 80°C \qquad Tc_{in} = 20°C$

$\Delta Ta = 80 - 20°C$

$= 60°C$

$\Delta Tb = Th_{out} - Tc_{out}$ $\qquad Th_{out} = 50°C \qquad Tc_{out} = 35°C$

$\Delta Tb = 50 - 35 = 15°C$

$$\Delta T_{ln} = \frac{60 - 15}{ln\dfrac{60}{15}} = \frac{45}{ln\ 4.0} = \frac{45}{1.3862}$$

$\Delta T_{ln} = 32.5°C$.

3. Determine the log mean temperature difference in double-pipe counter flow heat exchanger in which 500 kgs/hour of water enters at 80°C, flows through the outer annulus and exits at 50°C. As for the cold stream of water the situation is as follows. Inside the inner pipe 1000 kgs/hour of this water enters at 20°C and comes out at 35°C.

Solution

This situation is schematically represented as under

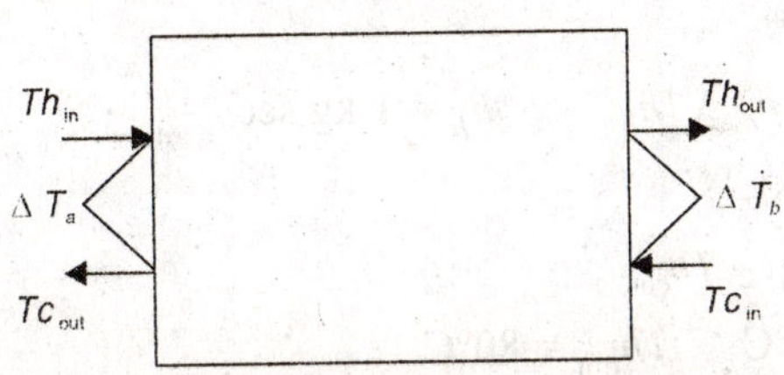

Fig. 6.41 *Schematic representation of a doble-pipe counter flow heat exchanger*

$Th_{in} = 80°C \qquad Tc_{out} = 35°C$

$\Delta Ta = 80 - 35 = 45°C$

$\Delta Tb = Th_{out} - Tc_{in} \qquad Th_{out} = 50°C \qquad Tc_{in} = 20°C$

$\Delta Tb = 50 - 20 = 30°C$

$$\text{Therefore } \Delta T_{ln} = \frac{45 - 30}{ln\left(\frac{45}{30}\right)}$$

$$= \frac{15}{ln\,(1.5)} = \frac{15}{0.4055}$$

$\Delta T_{ln} = 37.0°C$.

4. If a hot oil flowing at the rate of 1 kg/sec were cooled from 150°C to 80°C by water entering at 25°C and leaving at 90°C in a counter flow heat exchanger, what would be the heat transfer area needed if the value of the over all heat transfer coefficient were 500 w/m² °C. The heat capacity of the oil is 2 kj/kg °C.

Solution

The situation is depicted by the sketch given below

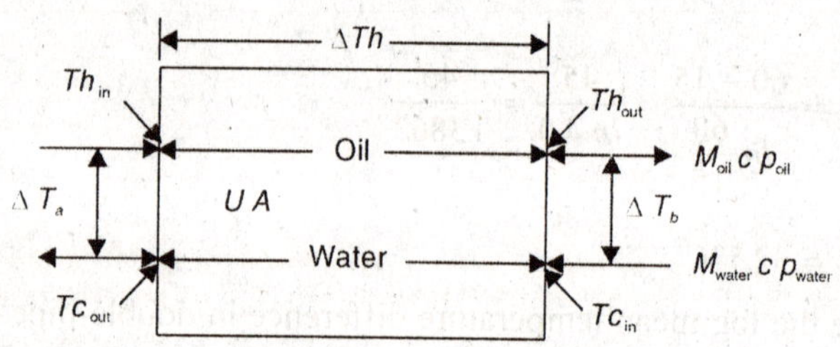

Fig. 6.42 *Schematic representation of counter flow heat exchanger*

The expression for the heat transfer area is

$$A = \frac{Q}{U\,\Delta T_{ln}}$$

$Q_h = m_h\, c\, p_h\, \Delta Th \qquad m_h = 1$ kg/sec

$cp_h = 2$ kj/kg °C

$\Delta Th = Th_{in} - Th_{out}$

$Th_{in} = 150°C \qquad Th_{out} = 80°C$

$\Delta Th = 150 - 80 = 70°C$

The heat flow transferred through the heat exchanger structure $Q = Q_h$

$$\therefore\ Q = (1)\,(2)\,(70) = 140 \text{ kw}$$
$$= 140{,}000 \text{ watts}$$
$$U = 500 \text{ w/m}^2\ ^\circ\text{C}$$

$$\Delta T_{ln} = \frac{\Delta Ta - \Delta Tb}{ln \dfrac{\Delta Ta}{\Delta Tb}}$$

$$\Delta Ta = Th_{\text{in}} - Tc_{\text{out}}$$
$$Th_{\text{in}} = 150^\circ\text{C}$$
$$Tc_{\text{out}} = 90^\circ\text{C}$$
$$\Delta Ta = 150 - 90 = 60^\circ\text{C}$$

$$\Delta Tb = Th_{\text{out}} - Tc_{\text{in}}$$
$$Th_{\text{out}} = 80^\circ\text{C}$$
$$Tc_{\text{in}} = 25^\circ\text{C}$$
$$\therefore\ \Delta Tb = 80 - 25 = 55\ ^\circ\text{C}$$

$$\Delta T_{ln} = \frac{60 - 55}{ln \dfrac{60}{55}} = \frac{5}{ln\ 1.091} = \frac{5}{0.0870}$$
$$= 57.47^\circ\text{C}$$

$$\text{Therefore } A = \frac{140{,}000}{(500)\,(57.47)}$$
$$= 4.87 \text{ square metre.}$$

5. A heat exchanger is used to heat water by an oil. The inlet oil temperature is 150°C, the mass flow rate of oil is 1 kg/sec and the overall heat transfer coefficient is 500 w/m² °C. To what temperature would the oil be cooled by doubling the cooling water mass flow rate of the previous problem. The inlet temperature of the water stream is 25°C cp_h = 2000 j/kg °C cp_c = 4200 j/kg °C.

Solution

Let us first find the mass flow rate of water corresponding to the conditions specified in the previous problem.

The value of Q as obtained in the previous problem is 140,000 w

$$cp_c = 4200 \text{ j/kg }^\circ\text{C}$$
$$Tc_{\text{out}} = 90^\circ\text{C (previous problem)}$$
$$Tc_{\text{in}} = 25^\circ\text{C}$$

$$\text{Therefore } m_c = 2\left[\frac{140{,}000}{4200\,(90-25)}\right] = 1.026 \text{ kg/sec}$$

In this problem, we need to determine both Th_{out} and Tc_{out}. This can be accomplished by formulating two expressions involving these temperatures.

Substituting the given quantities in the energy balance equation

$$(1)\,(2000)\,(150 - Th_{out}) = (1.026)\,(4200)\,(Tc_{out} - 25) \quad \ldots(1)$$

$$\left[m_h\, c\, p_h\, (Th_{out} - Th_{in}) = m_c\, c\, p_c\, (Tc_{out} - Tc_{in})\right]$$

Using the expression

$$Q = m_h\, c\, p_h\, (Th_{out} - Th_{in}) = U\, A\, \Delta T_{ln}$$

We can formulate the second equation i.e.

$$m_h\, c\, p_h\, (Th_{in} - Th_{out}) = U\, A\left[\frac{(Th_{in} - Tc_{out}) - (Th_{out} - Tc_{in})}{ln\dfrac{(Th_{in} - Tc_{out})}{(Th_{out} - Tc_{in})}}\right]$$

Substituting the given values of the relevant quantities, we obtain.

$$(1)\,(2000)\,(150 - Th_{out}) = 500\,(4.87)\left[\frac{(150 - Tc_{out}) - (Th_{out} - 25)}{ln\dfrac{(150 - Tc_{out})}{(Th_{out} - 25)}}\right]$$

$$Th_{out} + 1.218\left[\frac{(150 - Tc_{out}) - (Th_{out} - 25)}{ln\left(\dfrac{150 - Tc_{out}}{Th_{out} - 25}\right)}\right] = 150 \quad \ldots(2)$$

Equations (1) and (2) are two equations with the two unknowns Th_{out} and Tc_{out}, which conform to the mathematical requirements for solution. However equations (1) and (2) do not lend themselves readily to a direct mathematical solution. Consequently the practical method for solution is to solve by iteration, as illustrated below.

First Trial

Assume $Th_{out} = 70°C$

Let us solve equation (1) for Tc_{out} and substitute Th_{out} and Tc_{out} in equation (2) to see if it is valid

$Tc_{out} = 94.62 - 0.464\ Th_{out}$

$Tc_{out} = 94.62 - (0.464)(70)$

$= 94.62 - 32.48 = 62.14°C$

Now substituting Th_{out} and Tc_{out} in equation (2)

$$= 70 + 1.218 \left[\frac{(150 - 62.14) - (70 - 25)}{ln\left(\frac{150 - 62.14}{70 - 25}\right)} \right]$$

$$= 70 + 1.218 \left[\frac{87.86 - 45}{ln\ 1.952} \right]$$

$$= 70 + 1.218 \left[\frac{42.86}{0.669} \right]$$

$= 70 + 78.03$

$= 148.03 \neq 150$.

Second Iteration

Take $Th_{out} = 72°C$

$Tc_{out} = 94.62 - (0.464)(72) = 94.62 - 33.41$

$Tc_{out} = 61.21°C$ ...(1)

$$= 72 + 1.218 \left[\frac{(150 - 61.21) - (72 - 25)}{ln\left(\frac{150 - 61.21}{72 - 25}\right)} \right] \quad ...(2)$$

$$= 72 + 1.218\ \frac{88.79 - 47}{ln\ 1.889}$$

$$= 72 + 1.218 \left(\frac{41.79}{0.636} \right)$$

$= 72 + 80.03 = 152.03 \neq 150$.

Third Iteration

Take Tc_{out}, half way between 70°C and 72°C; it will satisfy the second equation i.e. equation (2)

The temperature $Th_{out} = 71°C$.

Illustration on LMTD/area Computation in Heat Exchangers

In a high temperature gas cooled nuclear reactor system for a 330 Mw installed capacity, the coolant employed is helium gas. The thermal output of the reactor is 850 Mw. This has to be transferred to the coolant gas ultimately. In the primary loop, the temperature of the helium at the outlet of the reactor is 775°C and in flowing through the steam generating heat exchanger it is cooled to a temperature of 410°C. In the secondary loop in which the working fluid to the cycle is steam water, the required amount of steam is generated in helium cooled steam generator, which is composed of the following four major sections.

1. Economiser where the feed water is heated from a temperature of 205°C to 350°C while the helium is cooled from a temperature of 520°C to 410°C. The flow arrangement is counter current here.

2. Evaporator in which the steam is generated at 350°C while the helium is cooled from 650°C to 520°C.

3. In the superheater section the temperature of the steam is increased from the saturation temperature of 350°C to the final temperature of 540°C while the corresponding temperature drop that the helium is subjected to is from 720°C to 650°C. In this section of the helium steam generator parallel flow is used.

4. A counter flow reheater section in which the expanded steam from the h.p. section of the turbine is reheated by the helium from a temperature of 360°C to 540°C while the helium is cooled from 775°C to 720°C.

(a) Compute the LMTD for each section of the helium cooled steam generator (b) estimate the heat transfer surface area for all the sections of the steam generator assuming the overall heat transfer and the percentage distribution of total heat transfer energy to be distributed between the different sections of the steam generator as given below.

Economiser 1600 $\frac{\text{k cal}}{\text{hr m}^2\ {}^\circ\text{C}}$ and 28 % of total energy

Evaporator 2100 $\frac{\text{k cal}}{\text{hr m}^2\ {}^\circ\text{C}}$ and 27 % of total energy

Superheater 1500 $\frac{\text{k cal}}{\text{hr m}^2\ {}^\circ\text{C}}$ and 30 % of total energy

Reheater 1400 $\frac{\text{k cal}}{\text{hr m}^2\ {}^\circ\text{C}}$ and 15 % of total energy.

Solution

(a) Economiser: For the counter flow arrangement

$$(\Delta t)_y = 410 - 205 = 205^\circ\text{C}$$

$$(\Delta t)_x = 520 - 350$$
$$= 170°C$$

Hence

$$\text{LMTD} = \frac{205 - 170}{ln\left(\frac{205}{170}\right)}$$

$$= \frac{35}{ln\,(1.205)} = 187.5°C$$

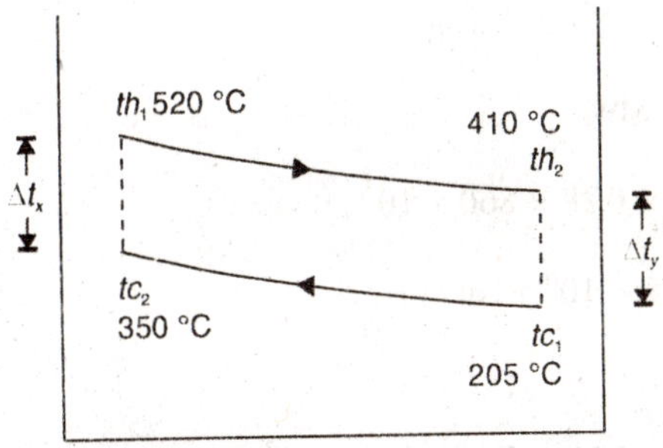

Fig. 6.43 *Temperature profile for counter flow configuration*

Evaporator

$$(\Delta t)_y = 650 - 350 = 300°C$$

$$(\Delta t)_x = 520 - 350 = 170°C$$

$$\text{LMTD} = \frac{300 - 170}{ln\,\frac{300}{170}} = \frac{130}{ln\,(1.763)} = 229°C$$

Superheater: For parallel flow arrangement

$$(\Delta t)_y = 720 - 350 = 370°C$$

$$(\Delta t)_x = 650 - 540 = 110°C$$

$$\text{LMTD} = \frac{370 - 110}{ln\,\frac{370}{110}} = \frac{260}{ln\,(3.36)} = 214°C$$

Reheater; for counter flow arrangement

$$(\Delta t)_y = 720 - 360 = 360°C$$

$$(\Delta t)_x = 775 - 540 = 235°C$$

$$\text{LMTD} = \frac{360 - 235}{ln \frac{360}{235}} = \frac{125}{ln\ 1.532} = \frac{125}{0.426} = 293°C$$

(b) To calculate the transfer area for each section, the expression $Q = U\,A\,(\Delta t)_{lm}$ can be employed.

First the individual proportion of the total energy to be transferred in each section can be determined as follows

$$Q_{\text{Economiser}} = Q_{\text{total}} \times \frac{28}{100}$$

where $Q_{\text{total}} = 850$ Mw

$$= 850 \times 0.28 \times 860 \times 10^3$$

$$= 204.5 \times 10^6 \text{ k cal/hr}$$

Similarly

$$Q_{\text{Evaporator}} = 850 \times \frac{27}{100} \times 860 \times 10^3$$

$$= 197.3 \times 10^6 \text{ k cal/hr}$$

$$Q_{\text{Superheater}} = 850 \times \frac{.30}{100} \times 860 \times 10^3$$

$$= 219.0 \times 10^6 \text{ k cal/hr}$$

$$Q_{\text{Reheater}} = 850 \times \frac{15}{100} \times 860 \times 10^3$$

$$= 109.5 \times 10^6 \text{ k cal/hr}$$

Then, the area of the Economiser

$$\frac{Q_{\text{Eco}}}{U_{\text{Eco}}\,(LMTD)_{\text{Eco}}} = \frac{204.5 \times 10^6}{1600 \times 187.5}$$

$$= 681 \text{ m}^2$$

The area of the Evaporator is

$$= \frac{197.3 \times 10^6}{2100 \times 229} = 410 \text{ m}^2$$

The area of the Superheater is

$$= \frac{219 \times 10^6}{1500 \times 214} = 682\,\text{m}^2$$

The area of the Reheater is

$$= \frac{109.5 \times 10^6}{1400 \times 293} = 267\,\text{m}^2$$

Total steam generator heat transfer area with which the coolant gas helium must be in contact = 2040 m^2.

2. An oil water is to be designed for cooling the oil from a temperature of 85°C to 38°C by a large body of stagnant water which may be assumed to remain at a constant temperature of 30°C. The heat transfer surface of the cooler consists of 30 metre length straight tube of 25 mm inside diameter. The average velocity of flow of the oil inside the cylindrical tube may be taken as 500 mm/sec. The specific heat and the specific gravity of the oil may be assumed as 0.6 k cal/kg°C and 0.8 respectively. Calculate the over all heat transfer coefficient for this oil cooler.

Solution

$$th_1 = 85\text{ °C}\ th_2 = 38\text{°C}$$

$$tc_1 = tc_2 = 30\text{°C}$$

Hence the LMTD of the heat exchanger is given by

$$\text{LMTD} = 24.4\text{°C}$$

The mass rate of flow of oil through the heat transfer tube is

$$m = \rho\, A\, u$$

$$\rho = 0.8 \times 1000 = 800\ \text{kg/m}^3$$

A is the cross sectional area for the flow of oil, which is equal to

$$\frac{\pi}{4} \times 25^2 \times 10^{-6} = 492 \times 10^{-6}\ \text{m}^2$$

'u' is the velocity of flow is 0.5 m/sec.

The mass flow rate is

$$m = 800 \times 492 \times 10^{-6} \times 0.5 \times 3600\ \text{kg/hr}$$

$$= 710\ \text{kg/hr}$$

and the total energy transferred is

$$Q = m_h\, c\, p_h\, (th_1 - th_2)$$

$$= 710 \times 0.6 \times (85 - 38)$$

$$= 20000\ \text{k cal/hr}$$

The corresponding area through which the energy transfer takes place is $A_{h\cdot t} = \pi d L$

$$= \pi \times 25 \times 10^{-3} \times 30 = 2.355 \text{ m}^2$$

Therefore applying the expression $Q = U A$ LMTD

$$U = \frac{20000}{2.355 \times 24.4} = 348 \text{ k cal / hr m}^2 \text{ °C}$$

3. In a heat exchanger, an oil is cooled from a temperature of 230°C to 160°C while the cold fluid's temperature rises from 30°C to 100°C as it flows past the heat exchanger. The hot and cold fluids enter the heat exchanger at opposite ends and the cold fluid makes two passes across the heat transfer surface. Calculate the log mean temperature difference for this situation. Assume that the hot fluid makes one shell pass. The heat exchanger is obviously a shell-and-tube one.

Solution

The non-dimensional parameters S and R are respectively

$$S = \frac{tc_2 - tc_1}{th_1 - tc_1} \quad \text{and} \quad R = \frac{th_1 - th_2}{tc_2 - tc_1}$$

$$S = \frac{100 - 30}{230 - 30} = 0.35 \text{ and } R = \frac{230 - 160}{100 - 30} = 1.0$$

The correction factor for the LMTD can be read from the appropriate correction factor plot for the values of $S = 0.35$ and $R = 1.0$ (the plot is for one shell pass and two tube passes)

The value of F found from the chart is $F = 0.95$

Uncorrected LMTD can be computed for the case of counter flow configuration

$$\text{LMTD} = \frac{(230 - 100) - (160 - 130)}{ln\left(\frac{130}{130}\right)}$$

$$= \text{indeterminate}$$

This can be solved by employing an energy balance-rate equation.

But for this situation as-an alternative, the LMTD for parallel flow configuration may be calculated

$$\text{LMTD} = \frac{(230 - 30) - (160 - 100)}{ln \dfrac{200}{60}}$$

$$= \frac{140}{ln\, 3.34}$$

$$= \frac{140}{1.205} = 116°C$$

Therefore the corrected LMTD with two tube passes for the cold fluid, for parallel flow

$$(LMTD)_{corrected} = F \times LMTD$$

$$= 0.95 \times 116$$

$$= 110°C.$$

4. In case the heat exchanger referred in the previous example is modified to have two shell passes for the hot fluid and four tube passes for the cold fluid calculate the LMTD for this situation. Can the modification be considered to be economical with respect to the initial cost of the heat exchanger?

Solution

The new value of F i.e. for 2 shell pass 4 tube passes (2–4) exchange can be read from the corresponding plot for the same values of R and S

$$F = 0.98$$

$$LMTD_{corrected} = 0.98 \times 116 = 113.5°C$$

As regards the economic aspect of this modification, the improvement in LMTD with the result of reduction of heat transfer area, has to be compared with additional material and fabrication cost of one more shell and two more tube passes. In this situation since as F for one shell two tube passes in fairly close to unity, the intended modification may not be justified economically.

Double-pipe Heat Exchangers—Suggested Calculation Methodology

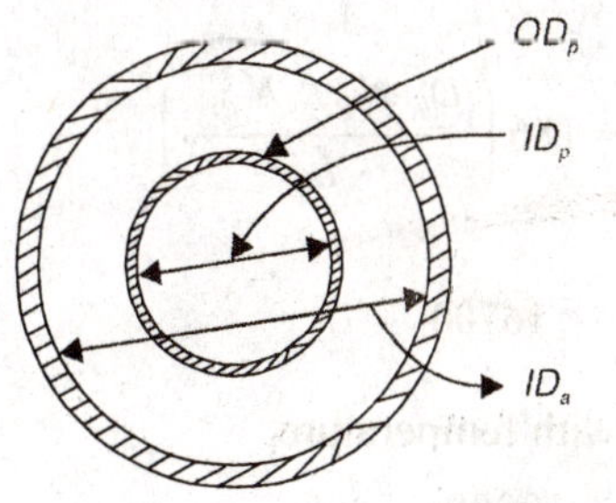

Fig. 6.44 *Double pipe heat exchanger*

1. Discuss the problem.
2. List assumptions

3. Obtain fluid properties
4. Obtain dimensions of the tubes: OD_p, ID_p, ID_a
5. Perform the calculations for both the tube and the annulus (subscripts p = pipe: a = annulus)

(i) *Flow areas*

$$A_p = \pi \frac{(ID_p)^2}{4}$$

$$A_a = \pi \frac{(ID_a^2 - OD_p^2)}{4}$$

(ii) *Annulus equivalent diameters*

$$D_e = \frac{(ID_a^2 - OD_p^2)}{OD_p}$$

(iii) *Reynolds numbers*

$$N_{Re_p} = \frac{u_p\, ID_p}{v_p} = \frac{m_p}{\rho_p\, A_p}\frac{ID_p}{v_p}$$

$$N_{Re_a} = \frac{u_a\, D_e}{v_a} = \frac{m_a}{\rho_a\, A_a}\frac{D_e}{v_a}$$

(iv) *Nusselt numbers*

Laminar flow $N_{Re} < 2200$

$$N_{Nu_p} = \frac{h_i ID_p}{k_p} = 1.86\left(\frac{ID_p\, N_{Re_p}\, N_{Pr_p}}{L}\right)^{\frac{1}{3}}$$

$$N_{Nu_a} = \frac{h_o\, D_e}{k_p} = 1.86\left(\frac{D_e\, N_{Re_a}\, N_{Pr_a}}{L}\right)^{\frac{1}{3}}$$

$$0.48 < N_{Pr} = \frac{v}{\alpha} < 16700$$

μ moderately changes with temperature.

Turbulent flow $N_{Re} > 10000$

$$N_{Nu_p} = \frac{h_i\, ID_p}{k_p} = 0.023\, N_{Re_p}^{0.8}\, N_{Pr_p}^{n}$$

$$N_{Nu_a} = \frac{h_o D_e}{k_a} = 0.023 \ N_{Re_a}^{0.8} \ N_{Pr_a}^{n}$$

$n = 0.4$, if fluid is being heated

$n = 0.3$, if fluid is being cooled

$$0.7 < N_{Pr} = \frac{\nu}{\alpha} < 160$$

$$\frac{L}{D} > 60$$

(v) *Convection coefficients*

$$\text{Pipe } h_i = \frac{N_{Nu_p} \ k_p}{ID_p} : h_{io} = h_i \frac{ID_p}{OD_p}$$

$$\text{Annulus } h_o = \frac{N_{Nu_a} \ k_a}{D_e}$$

(vi) *Exchanger coefficient*

$$\frac{1}{U_o} = \frac{1}{h_{io}} + \frac{1}{h_o} \text{ (neglecting resistance of inner pipe wall)}$$

Once the above calculations are carried out, the problem can be completed depending upon what information is sought that is, outlet temperatures or a heat balance, etc.

Outlet temperatures

Subscripts c = cooler fluid or fluid being heated;

h = warmer fluid or fluid being cooled

$$R = \frac{m_c \ c p_c}{m_h \ c p_h} : A_o = \pi \left(OD\right)_p \ L$$

$$T_2 = \frac{T_1 (R-1) - R t_1 \left\{1 - \exp\left[U_o \ A_o \ (R-1)/m_c \ c p_c\right]\right\}}{R \exp\left[U_o \ A_o \ (R-1)/m_c \ c p_c\right] - 1} \text{ (counter flow)}$$

$$t_2 = t_1 + \frac{T_1 - T_2}{R}$$

Heat balance

$$\text{LMTD} = \frac{(T_1 - t_2) - (T_2 - t_1)}{ln\left[\frac{(T_1 - t_2)}{(T_2 - t_1)}\right]} \text{ (counter flow)}$$

$$q = m_c\, c\, p_c\,(t_2 - t_1) = m_h\, c\, p_h\,(T_1 - T_2)$$

$$q = U_o A_o \text{ LMTD (clean)}$$

Area required to transfer heat

$$\frac{1}{U} = \frac{1}{U_o} + Rd_i + Rd_o$$

R_d = fouling factor

$$A_o = \frac{q}{U\,(\text{LMTD})}$$

$$L = \frac{A_o}{\pi\,(OD)_p}.$$

Shell and Tube Heat Exchangers

Suggested calculation methodology

1. Discuss the problem
2. List out assumptions
3. Obtain fluid properties
4. Obtain tube and shell dimensions

Tube dimensions

Take OD_t, ID_t from the appropriate tables

N_t = number of tubes or tube count from the appropriate table

N_p = number of passes the tube fluid makes

Shell data

D_s = shell inside diameter

B = baffle spacing

N_b = number of baffles

S_t = tube pitch

$C = S_T = OD_t$ = clearance between adjacent tubes.

5. Perform the calculations for the tube and the shell fluids (subscripts: t = tube s = shell)

(i) *Flow areas*

$$A_t = N_t\, \pi\, \frac{\left(ID_t^2\right)}{4\, N_p}$$

$$A_s = D_s\, \frac{CB}{S_T}$$

(ii) *Shell equivalent diameter*

$$D_e = \frac{4\left[S_T^2 - \pi\frac{(OD_t)^2}{4}\right]}{\pi(OD_t)} = \frac{4S_T^2 - \pi(OD)_t^2}{\pi(OD_t)} \quad \text{(square pitch)}$$

$$D_e = \frac{4\left[0.86\frac{S_T^2}{2} - \pi\frac{(OD_t)^2}{8}\right]}{\frac{\pi(OD_t)}{2}} = \frac{3.44S_T^2 - \pi(OD)_t^2}{\pi(OD_t)} \quad \text{(triangular pitch).}$$

(iii) *Reynolds numbers*

$$N_{Re_t} = \frac{u\ ID_t}{v} = \frac{m_t}{\rho_t\ A_t}\ \frac{ID_t}{v_t}$$

$$N_{Re_s} = \frac{u_s\ D_e}{v_s} = \frac{m_s}{\rho_s\ A_s}\ \frac{D_e}{v_s}.$$

(iv) *Nusselt numbers*

Tube side

Laminar flow $N_{Re_t} < 2200$

$$N_{Nu_t} = \frac{h_i\ ID_t}{k_t} = 1.86\left(\frac{ID_t\ N_{Re}\ N_{pr}}{L}\right)^{\frac{1}{3}}$$

$$0.48 < N_{pr} = \frac{v}{\alpha} < 16700$$

μ moderately changes with temperatures

Turbulent flow $N_{Re_t} > 10000$

$$N_{Nu_t} = \frac{h_i\ ID_t}{k_t} = 0.023\ N_{Re_t}^{0.8}\ N_{pr}^{n}$$

$n = 0.4$ if fluid is being heated

$n = 0.3$ if fluid is being cooled

$$0.7 < N_{pr} = \frac{v}{\alpha} < 160\text{:}\ \frac{L}{D} > 60$$

Shell side

$$N_{Nu_s} = \frac{h_o\ D_e}{k_s} = 0.36\ N_{Re_s}^{0.55}\ N_{Pr_s}^{\frac{1}{3}}$$

$$2 \times 10^3 < N_{Re_s} = \frac{u_s D_e}{v_s} < 1 \times 10^6$$

$$N_{Pr_s} = \frac{v}{\alpha} \geq 0$$

μ varies moderately with temperature.

(v) *Convection coefficients*

$$h_i = \frac{N_{Nu_t} k_t}{ID_t} : h_{io} = h_i \frac{ID_t}{OD_t}$$

$$h_o = \frac{N_{Nu_s} k_s}{D_e}.$$

(vi) *Exchanger coefficient*

$$\frac{1}{U_o} = \frac{1}{h_{io}} + \frac{1}{h_o}$$

Once the aforementioned calculations are made the problem can be completed, depending on what information is sought (i.e outlet temperatures or a heat balance etc.)

Outlet temperatures = subscripts; c = cooler fluid or fluid being heated; h = warmer fluid or fluid being cooled

$$R = \frac{m_c \, c p_c}{m_h \, c p_h} : A_o = N_t \, \pi \, (OD_t) \, L$$

Calculate $\frac{U_o A_o}{m_c \, c p_c}$: use the appropriate graphs read $S = \frac{t_2 - t_1}{T_1 - t_1}$

$t_2 = S(T_1 - t_1) + t_1$

$T_2 = T_1 - R(t_2 - t_1)$

Heat balance

$$\text{LMTD} = \frac{(T_1 - t_2) - (T_2 - t_1)}{ln\left[\frac{(T_1 - t_2)}{(T_2 - t_1)}\right]} \text{ (counter flow)}$$

Read the correction factor F_t from the appropriate plot

$$q = m_c \, c p_c \, (t_2 - t_1) = m_h \, c p_h \, (T_1 - T_2)$$

$$= U_o A_o F_t \, \text{LMTD}$$

Area required to transfer heat

Rd_i and Rd_o can be found from the appropriate table

$$\frac{1}{U} = \frac{1}{U_o} + Rd_i + Rd_o$$

$$A_o = \frac{q}{U F_t \text{ LMTD}}$$

$$L = A_o / \left[N_t \pi (OD)_t\right].$$

Efficiency of Heat Exchanger

It is of interest to establish a standard for comparison in heat exchangers. In particular it is important to have a standard of maximum performance against which heat exchangers can be compared. Sometimes the effectiveness is used for this purpose.

$$E = \frac{\text{Actual heat transferred}}{\text{Maximum heat that might be removed}}$$

It may be recalled, however that the denominator of the above expression is associated with an exchanger that has an infinite area. This seems somewhat objectionable in view of the fact that an infinite area is neither practical nor realistic, even though using the effectiveness as a measure of exchanger efficiency has merit.

Alternatively, if it has been shown that maximum heat can be transferred in a real exchanger arranged in counter flow. Consequently there is merit also in using a counter flow exchanger as a basis for comparison. Thus we can define the efficiency of an exchanger as

$$\eta = \frac{\text{Temperature difference in the exchanger}}{\text{Temperature difference for counter flow}}$$

$$= F_t$$

which is recognized immediately as the correction factor F_t. The objectionable aspect of this definition is that a counter flow exchanger is assigned an efficiency of 100%.

This is not in keeping with the traditional thermodynamic idea of efficiency. In which a value of 100% is related to the concept of entropy. Exchanger efficiency is still an unresolved and somewhat controversial issue.

Effectiveness—NTU Equation for Multipass Heat Exchanger

Effectiveness—NTU equations have been derived for shell-and-tube exchangers having one or more shell passes and one or more tube passes.

For a shell-and-tube exchanger with one shell pass and any integral multiple of two tube passes, we have

$$E = 2\left\{1 + \frac{(mc_p)_{\min}}{(mc_p)_{\max}} + \left[1 + \frac{(mc_p)^2_{\min}}{(mc_p)^2_{\max}}\right]^{\frac{1}{2}} \times \left(\frac{1 + \exp\left\{-\text{NTU}\left[1 + \frac{(mc_p)^2_{\min}}{(mc_p)^2_{\max}}\right]^{\frac{1}{2}}\right\}}{1 - \exp\left\{-\text{NTU}\left[1 + \frac{(mc_p)^2_{\min}}{(mc_p)^2_{\max}}\right]^{\frac{1}{2}}\right\}}\right)\right\}^{-1}$$

Plot of this equation is available in the form E vs NTU, with $(m\,c_p)_{\min} / (m\,c_p)_{\max}$ as an independent variable.

Some Points on Heat Exchanger Design

The LMTD and NTU methods of exchanger analysis are very useful concepts, but they do not normally result in complete "designs" of a system. In particular they are based on two stringent assumptions:

1. That U remains constant through the heat exchangers and
2. That U may be predicted from the correlations proposed for forced-convection heat a transfer.

In reality, U is rarely constant along the heat exchanger surface. The closest to a constant-U case is a double-pipe parallel or counter flow system or a compact exchanger with straight uniform flow passages.

Prediction of U from theory and dimensional analysis can give rise to large uncertainty in its value, except perhaps for circular double-pipe designs. Fortunately, there is a large amount of published data from compact heat exchanger designs. Accurate correlations are rather lacking for large complex, baffled shell designs. Manufacturers commonly combine extensive test data into a large computer program for predicting the overall heat transfer coefficient of complicated geometries. If the spatial variation of U can be predicted, it is quite clear that for the expression

$$ln\left(\frac{Th_{out} - Tc_{out}}{Th_{in} - Tc_{in}}\right) = -\frac{U\,A}{C_{\min}}\left(1 + \frac{c_{\min}}{c_{\max}}\right)$$

that the proper "constant" U to be used in the LMTD and NTU methods is the area-average.

$$\overline{U} = \frac{1}{A}\int U\, dA$$

Even with this refinement, the final estimate can be subject to considerable uncertainty.

While the average engineer may only be called upon to design a small moderately efficient compact exchanger, specialized manufacturers require an extremely detailed design processes. There is constant redesign and reconfiguring to optimize the system among heat transfer, pressure drop, total size and volume, cost, reliability and maintenance.

J. Taborek [(Evolution of heat exchanger design techniques. Heat transfer engineering vol. 1, no. 1 (1979), pp. (5–29)] presents an eloquent discussion of the shell-and-tube exchanger design process. The points made by Taborek are the following.

1. Cost questions relate not only to the fixed and operating costs of the device but also to the cost of the engineering design effort and the cost of poor design that is the penalty that results if the expected performance is not achieved.
2. A given design can be either heat transfer limited or pressure-drop-limited. In turbulent flow, heat transfer rises approximately as $m^{0.8}$ (mass flow rate) while pressure drop rises as $m^{1.8}$ and pumping power as $m^{2.8}$. It is thus simple to evolve a design that bumps up against a maximum power limit such a design in usually a poor one and should be reconfigured. Further if a design is severely heat transfer limited, it is usually well worth the effort to redesign for augmenting heat transfer.
3. A crucial decision is the allocation of one fluid to the shell side and one fluid to the tube side. Generally the tube side should receive the fluid that is at higher pressure, is more corrosive or has a high fouling tendency. The shell side should receive the more viscous fluid or the fluid that presents more pressure-drop difficulties. Finned tubes should be viewed with caution if heavy fouling is expected.
4. Even the best available heat transfer correlations result in an overall uncertainty of $\pm$ 30% in U. The design engineer must be ready to compensate for such uncertainties.
5. If a design results in a correction factor F that falls below 0.5, it is almost always a poor design and can be improved.
6. It is good to make sketches of temperature and flow distributions as one progresses with the design. One can learn much from such graphical representations.
7. Commercial heat exchangers design are making increasing use of massive computer programmes. One can safely predict that future designs may be performed entirely by computers including plotter-generated shop

drawings. This is an ominous prospect since no human being can completely check such results. It will also lead to a decrease in the number of experienced human designers.

In a way a good compromise in future heat exchanger work will be a computer-aided design (CAD) backed up by experienced engineers who can choose proper parameters and options. It is good that one is encouraged to study the use of computers to organize and optimize design procedures. Stoecker (W.F Stoecker, design of thermal systems, McGraw Hill. New York, 1980) has written an excellent guide for computer modeling simulation curve fitting, optimization search procedures and dynamic/geometric/ linear programming techniques which are applicable for the heat exchanger case. As a matter of fact all of Stoecker's examples relate to heat transfer problems.

Additional Considerations in Heat Exchanger Analysis and Design

Heat Exchanger Fouling

When a heat exchanger has been in operation over some extended period of time, there will likely to be build-up of scale on tube surfaces or a deterioration of the surface itself due to corrosion. These effects in time will alter the performance of the heat exchanger. A heat transfer surface which has been affected in such a manner is said to be fouled.

A fouled surface is normally considered to be one which presents some additional heat transfer resistance due to the build up of foreign material or scale. This additional thermal resistance will naturally cause the heat transfer to be less than in the case of no fouling resistance.

The prediction of scale build-up or the corresponding effect on heat transfer is a most difficult task. The actual performance of a heat exchanger can be evaluated after some period of service and the fouling resistance determined with a clean surface

$$q_o = U_o \; A \, \Delta T_{lm} = \frac{\Delta T_{lm}}{\Sigma R_{to}}$$

Similarly, for a fouled surface

$$q_f = U_f \; A \, \Delta T_{lm} = \frac{\Delta T_{lm}}{\Sigma R_{tf}}$$

where ΣR_{to} and ΣR_{tf} represent the total thermal resistance for clean and fouled surfaces respectively.

In the case of a clean surface

$$U_o = \frac{1}{\dfrac{1}{h_o} + \dfrac{A_o \ln \dfrac{r_o}{r_i}}{2\pi k} + \dfrac{A_o}{A_i h_i}}$$

where the terms in the denominator are thermal resistances due to convection on the outside tube surface, conduction through the tube wall and convection at the inside tube surface respectively.

In the case of a fouled surface, the expression for U is

$$U_f = \frac{1}{\dfrac{1}{h_o} + R_o + \dfrac{A_o \ln \dfrac{r_o}{r_i}}{2\pi k} + R_i + \dfrac{A_o}{A_i h_i}}$$

with the additional terms R_o and R_i representing the fouling resistances on the outside and inside tube surfaces respectively. Some typical values of fouling resistance to be employed in the above expression are given in the following table. These value of the fouling resistance have been suggested by the Tubular Exchanges Manufactures Association

Table 6.1 *Fouling resistance for certain fluids in heat exchangers*

Fluid	*Fouling resistance Hr. ft² °F/Btu*
Distilled water	0.0005
Sea water, below 125°F	0.0005
Above 125°F	0.001
Boiler feed water treated	0.001
City or well water, below 125°F	0.001
Above 125°F	0.002
Refrigerating liquids	0.001
Refrigerating vapours	0.002
Liquid gasoline and organic vapours	0.0005
Fuel oil	0.005
Quenching oil	0.004
Steam non-oil-bearing	0.0005
Industrial air	0.002

In the following table, some representative values of the overall heat transfer coefficient for different fluid combinations are given. These values have been suggested in the literature. These values are not to be taken as exact, but merely representative of the magnitude to be expected in a heat exchanger with the fluid combinations listed

Tabel 6.2 *Overall heat transfer coefficients—Approximate values*

Fluid combination	*U Btu/hr ft² °F*
Water to compressed air	10–30
Water to water, jacket water coolers	150–275
Water to brine	100–200
Water to gasoline	60–90
Water to gas oil or distillate	35–60
Water to organic solvents, alcohol	50–150
Water to condensing alcohol	45–120
Water to lubricating oil	20–60
Water to condensing oil vapours	40–100
Water to condensing or boiling Freon-12	50–150
Water to condensing ammonia	150–250
Steam to water, instantaneous heater	400–600
Storage tank water	175–300
Steam to oil, heavy fuel	10–30
Light fuel	30–60
Light petroleum distillate	50–200
Steam to aqueous solutions	100–600
Steam to gases	5–50
Light organics to light organics	40–75
Medium organics to medium organics	20–60
Heavy organics to heavy organics	10–40
Heavy organics to light organics	10–60
Crude oil to gas oil	30–55

Some Guidelines for the Heat Exchanger Design

To understand better the course of the design process, faced with such an array of trade-offs of advantanges and penalties. It is better to follow Taborek's (J. Taborek, "Evolution of Heat Exchanger Design Techniques" Heat Transfer Engineering vol. 1, no. 1. 1979, pp 15–29) list to design considerations for a large shell and-tube exchanger.

1. Decide which fluid should flow on the shell side and which should flow in the tubes. Normally, this decision will be made to minimize the pumping cost. If for example water is being used to cool an oil, the more viscous oil would flow in the shell. Corrosion behaviour, fouling and the problems of cleaning fouled tubes also weigh heavily in this decision.

2. Early in this process, the designer should assess the cost of the calculation in comparison with
   (a) The converging accuracy of the computation
   (b) The investment in the exchanger
   (c) The cost of miscalculation.
3. Make a rough estimate of the size of the heat exchanger using, for example U values from the tables given and/or anything else that might be known from experience. This serves to circumscribe the subsequent trial and error calculations: it will help to size flow-rates and to anticipate temperature variations and it will help to avoid subsequent errors.
4. Evaluate the heat transfer, pressure drop and cost of various exchanger configurations which appear reasonable for the application. This is usually done with large-scale computer programmes that have been developed and are constantly being improved as new research is included in them.

The computer runs suggested by this methodology are normally very complicated and might typically involve 200 successive redesigns, even when relatively efficient optimization procedures are employed.

However most students of heat transfer will not have to deal with such designs. Many if not most will be called upon frequently to design smaller exchangers in the range 0.1 to 10m^2. The heat transfer calculation can usually be done effectively with the methods described earlier in the text. Some useful sources of guidance in pressure drop calculation are Kern's classical treatment of Process Heat Transfer, the TEMA design book and Perry's chemical Engineer's Hand book.

In such a computation exercise we start off with one fluid to heat and one to cool perhaps we know the flow heat capacities (the C's), certain temperatures and/or the amount of heat that is to be transferred. The problem can be annoyingly wide open and nothing can be done until it is somehow delimited. The normal starting point is the specification or an exchanger configuration and to make this choice one needs experience. The descriptions covered in this section of the text provide a kind of first level of experience. For a second level involvement one has to refer to manufacture's catalogue.

Once the exchanger configuration is set, U will be approximately set and the heat transfer area becomes the basic design variable. The design can then proceed along the lines of the previous sections. If it is possible to begin with a complete specification of inlet and outlet temperatures

$$\underbrace{Q}_{(mc_p\,\Delta t)} = \underbrace{U}_{\text{known}} \underbrace{A\,F\,\text{LMTD}}_{\text{calculable}}$$

Then A can be calculated and the design exercise completed. Usually a revaluation of U and some iteration of the computation is required.

More often, we begin without full knowledge of the outlet temperatures. In such cases we normally have to invent an appropriate trial-and-error method to get the area and a more complicated sequence of trials if we seek to optimize pressure drop and cost by varying the configuration as well. If the C's are design variables then U will change significantly because the h's are generally velocity-dependent and more iteration will be needed.

Matching Heat Transfer Principles to Our Exchanger Needs

In this section we will apply the basic principle developed in earlier sections on conduction and convective heat transfer. We will demonstrate how these principles are used to match process requirements to the exchangers described in the manufacturer's bulletins. Limiting our discussion to tubular units, we find a list that includes categories such as shell and tube exchangers, tube banks double pipe sections and U tubes.

In the following paragraphs we discuss the basic characteristics of the many types of heat exchange equipment. In our summary we discuss criteria affecting service construction type and mechanical features of heat exchangers as well as factors affecting routing of fluid streams, flow rates and fouling coefficients.

From here on, let us assume that the type of exchanger, diameter and thickness of the tubes, tube lay out, fluid routing and all other factors which are described under the leading of how to select exchangers, have been specified. We will also assume that terminal temperatures and flow rates of the fluids have been given and that our problem is to determine the number of tubes length of tubes, pass arrangement and from these shell diameter and baffle spacing.

Although the details may differ the general procedure that we will follow is applicable to almost all exchanger problems including condensers and reboilers. These methods are shown in the following example.

Example

A hydrocarbon stream flowing at 50,000 lbs/hr through a tubular exchanger loses sensible heat and cools from 175 F to 100°F. Physical properties of the hydrocarbon stream may be taken as those of n-heptane given in any standard handbook. Cooling water at 80°F is available and it leaves the exchanger at 125°F. Because of its pressure, the hydrocarbon is routed through the tubes.

For safety reasons leakage is not permissible. Hence we must use a removable bundle, floating heat exchanger. Tubes are copper and are ¾ inch diameter 16 gauge 12 ft long and arranged on 1-in triangular centres. Assuming maximum allowable pressure drop of 10 psi for each fluid and using fouling factors given in Table 6.1 let us determine the stock-size heat exchanger necessary for this service.

Step 1

Since the exchanger involves only sensible heat and since the flow is countercurrent for half of the passes and parallel for the other, the effective temperature difference is less than it would be if countercurrent flow existed throughout. Hence

$$\Delta T_{\text{effective}} = F \times \Delta T_{lma}$$

The value of F depends on the flow arrangement. In this example, if the water leaves at 125°F, calculate the temperature efficiency (P) and the ratio of temperature change R in the two fluids to find values of F

$$P = \frac{T_1 - T_2}{T_1 - t_1} = \frac{175 - 100}{175 - 80} = 0.79$$

or
$$\frac{t_2 - t_1}{T_1 - t_1} = 0.47$$

$$R = \frac{t_2 - t_1}{T_1 - T_2} = \frac{125 - 80}{175 - 100} = 0.60$$

or
$$\frac{T_1 - T_2}{t_2 - t_1} = 1.67$$

Either set of consistent values 0.79 and 0.60 or 0.47 and 1.67 shows the same results.

For the 1–2 exchanger, F is less than 0.5 showing it to be inoperable for these conditions. For the two-pass shell arrangement F is 0.85 which is satisfactory.

Thus a two shell can be specified or two separate one-pass shells can be used in series to given the same effective ΔT. In this problem, we will use more water until F for a single shell pass has a value of 0.75 or more. We can find this value by assuming various water outlet temperatures

t_2 °F	P	R	F
125	0.79	0.60	<0.5
120	0.79	0.53	<0.5
115	0.79	0.47	<0.5
110	0.79	0.40	0.58
105	0.79	0.33	0.71
100	0.79	0.27	0.80

Using water outlet temperatures t_2 equal to 100°F, we find

$$q = 50{,}000 \times 0.561 \times (175 - 100)$$
$$= 2{,}110{,}000 \text{ Btu/hr}$$

The amount of water necessary to remove this heat is

$$2{,}110{,}000 = w \times 1.0 \times (100\text{–}80)$$

$$w = 105{,}500 \text{ lbs/hr}$$

For terminal water temperature of 100°F and a value of F equal to 0.80, the effective temperature difference is 33.2°F.

Step 2

Estimate an overall coefficient of 150 Btu/hr ft²°F, based on typical values expected for the individual coefficients. Use this value to find total area required which is $\dfrac{2{,}110{,}000}{(150)(33.2)}$ or 424 ft². Since the surface area of one 3/4 in by 12 ft long tube is 2.36 ft², the number of tubes becomes 424/2.36 or 180. From a table of tube counts available in the literature an 18 inch shell contains 214 tube in an inside packed, floating head bundles with two tube side passes or 196 if four tube passes. Use this for the first calculation.

Step 3

If two tube passes are used, the velocity at the average temperature is

$$U_{\text{tubes}} = \frac{50000}{40.5\left(\dfrac{0.302}{144}\right)\left(\dfrac{214}{2}\right)3600} = 1.53 \text{ feet / sec}$$

If four tube passes are used, the velocity is 3.35 feet/sec. Both seem reasonable. The higher velocity gives a better coefficient so let us try it first to see if the tube side pressure drop is satisfactory.

Tubeside pressure drop equals the sum of the pressure drops through the tubes and headers. For a Reynolds number of 37,400 at an average temperature of 137.5°F friction factor f is 0.0055. Pressure drop through the tubes is 1.00 psi from the equation $\Delta P_{\text{tubes}} = 4 f L u^2 P/2gc\, D$ and pressure drop through headers equals 0.78 from $\Delta p_{\text{headers}} = \dfrac{4 P u^2}{2\, gc} D$ per pass. Hence the total pressure drop is safely within the specified limit.

For a non-viscous fluid in turbulent flow, we calculate the heat transfer coefficient which equals 268 from

$$h = 0.023 \left(N_{Re}\right)^{0.8} \left(N_{Pr}\right)^{\frac{1}{3}} \frac{k}{D}.$$

Step 4

On the shell side flow area S at the central line equals inside diameter of shell less projected width of tubes baffle spacing. As a first guess set baffle spacing so that $u_{av} = 3$ feet/sec. Hence $w = ups$

$$105{,}500 = 3 \times 3600 \times 62.1\ (17.25 - 16 \times 0.75)\ \frac{B}{144}$$

Baffle spacing equals 4.3 inch or 33.5 sections are needed for the 12 ft length. Arbitrarily adjust B to 6 inch to give 24 sections and u equal to 2.14 feet/sec. For this spacing check shell side pressure drop. N_{Re} in the shell is 15700 and f is 0.13.

$\Delta P_s = 4\, f\, s$ (number of sections) length of cross flow path/tube pitch × $u^2\, P\, /\, 2g_c$.

Substituting in above equation yields a shell side pressure drop of 4.7 psi.

Next calculate shell side heat transfer coefficient from appropriate j_H factors. For shell side N_{Re} of 15,700, $j_H = 100$ use the following relation to calculate h, which for this problem equals 1000.

$j_H = h_s\, D_o\, /\, k\, (N_{Pr})^{0.33}$ (subscript S denotes shell side)

Since the shell side coefficients is large as compared with the tube side coefficient there is little incentive for increasing the velocity to the maximum permitted by the available pressure drop.

Step 5

Evaluate the overall heat transfer coefficients as follows.

$$\frac{1}{U_s} = \frac{1}{1000} + \frac{0.62}{268 \times 0.75} + \frac{(0.065) \times 0.75}{12 \times 220 \times 0.685} + 0.001 + 0.001$$

Since, U_s equals 163, area required becomes 2,110,000/(163) × (33.2) or 389 ft^2. Hence 165 tubes are required compared to 196 actually used.

Since the overall heat transfer coefficient is higher than assumed, a smaller exchanger can be tried. However, the next smaller shell is 16 inch. and contains 144 tubes in four passes.

Normal design	*Velocities* *Tube side ft/sec*	*Shell side ft/sec*
Water	3–6	2–4
Non viscous liquids	2–5	2–3
Viscous liquids	2–4	1–2
Gases low density	50–150	30–60
Gases, high density	20–80	20–40

To simplify the selection problem. We can group and classify heat exchangers in terms of three criteria.

1. *Function or service*: The apparatus must perform. Common terminology is to call the unit an exchanger if the transfer of sensible heat between the two process streams is involved. The unit may be called a heater if a service fluid provides the heat or a cooler if water or similar cooling medium absorbs heat from the major stream.

Similarly it is designated a condenser if the major process stream gives up latent heat and a reboiler or vapourizer if it receives latent heat. It is known as on evaporator if water is the substance vapourized.

2. *Construction type*: The shell and tube design is the most commonly specified unit because it is very economical and can be modified to meet a variety of flow rates and process conditions.

The double-pipe section has only one tube per shell and may be considered as the limiting case of a shell and type exchanger. While its capacity is extremely limited, it has the major advantage that any process requirement can be satisfied by using units in series or in parallel on both.

Tube banks differs from the other types in that they do not have a compartment to contain the service fluid. Use of this type is restricted to air-or water service. Where applicable, they are considerably cheaper per square foot of heating area than are shell-and-tube exchangers or double pipe sections.

3. *Mechanical features*: Primary considerations are: (1) the method of providing for thermal expansion and (2) the measures used to obtain reasonable velocities and practical tube lengths. Alternatives in the first case are fixed tube sheet, floating head and *U*-tube exchangers. Major characteristics and comparative features of these exchangers are given in literature. The number of times the fluids—traverse the length of the unit gives the pass arrangement of the exchanger and fulfills the second requirement.

Planning the Exchanger

After the type of exchanger has been selected, several other quantities and operating variables must be specified before the size of the exchanger can be determined. Major factors are:

1. Operating pressures of the fluids. In general the fluid under the higher pressure is put inside the tubes. All other factors being equal, this is the most economical arrangement as it reduces shell thickness.
2. Corrosivity of the fluids. Again, as an economy measure, if either fluid requires a special material of construction it is assigned to the tube side. In this way only the tubes need to be of the more costly material.
3. Scaling or fouling tendencies. Because it is easier to clean the inside of the tubes the worst offender is put inside the tubes. Another reason for this routing is that scale or dirt buildup appears to be lower at high velocities.
4. Permissible pressure drops. Pressure drops of the fluid passing through the shell can be regulated more easily by adjustment of the baffle spacing. Consequently the fluid is usually routed through the shell if this is a limiting factor.

5. Heat transfer considerations: Overall heat transfer coefficient tends to be higher if the more viscous fluid is put on the shell side where turbulence is induced by flow around the tubes. Since the relative magnitudes of the two fluid coefficients is not always predictable, it is frequently necessary to consider both routings quantitatively and to choose the one which gives the higher rate of heat transfer and thus gives a small exchanger.

Since routing is an economic problem, alternative exchanger designs must be made and evaluated costwise if the foregoing considerations do not indicate one obvious routing.

The exchanger should be designed and operated at the maximum available pressure drop if all other factors are equal. Overall thermal resistance is lower and a small exchanger can be selected than is necessary if a lower pressure drop is used. However caution must be exercised in applying this concept in as much as (1) pressure drop increases as the exchanger becomes fouled and allowance must be made so that the design flow rate will be maintained: (2) too high a velocity will result in erosion: (3) while fewer tubes or fewer tubes per pass are required, their length or number of passes required to obtain a given heat transfer may become unreasonable.

Also, it should be recognized that pumping energy must be provided. Hence, it is an economics problem whether to spend more for pumping and less for the exchanger or vice versa. Subject to these considerations normal design flow rates may fall in the ranges shown in the above table.

The majority of stock exchangers have 3/4, 7/8 or 1 in OD tubes. This choice reflects the influence of diameter or pressure drop and anticipated method of cleaning the tubes. Smaller diameter tubes give a higher heat transfer coefficient, all other factors being the same. However the smaller size has less area per foot of length and carries less fluid per tube. Thus more tubes are needed.

Normally, exchangers have tubes with heating lengths from 8 to 20 ft and with from one to eight tube passes. Length is determined by practical considerations such as availability of standard tubes and the floor space allocated to the exchanger.

Design Procedure

Major steps in selecting an exchanger include:

1. A tentative number of shell side passes is set. This is based on the effective temperature difference.
2. Approximate area requirement of the exchanger is determined by estimating the overall heat transfer coefficient and applying the equation $A = \frac{q}{U} \Delta Tm$. Normally, a stock exchanger with a definite number and length of tubes and a corresponding shell diameter is selected to provide this area.

3. A reasonable velocity through the tubes is set. This determines the number of tubes is parallel in each pass. In turn, this fixes the number of tube side passes required for the total number of tubes assumed in step 2. Tube side pressure drop is next calculated and compared with the available. Any adjustments necessary are then made in steps 2 and 3. Appropriate heat transfer correlation can than be used to determine the coefficient for the established tube side velocity.
4. Step 3 is repeated for the shell side fluid. The primary variable here is baffle spacing.
5. Individual heat transfer coefficients so found are combined with the tube wall resistances and the fouling factors, if any have been specified, into a new design coefficient. If comparison with the value assumed in stop 2 shows a significant difference in the coefficient or area, steps 2 to 5 are repeated with the new values.

Although no exact standards have been formulated, values of the fouling factor are normally taken such that the exchanger will stay on stream a reasonable length of time before it must be cleaned. Few data are available on the rate of scale buildup. Typical values for various services are given in the following table.

Typical fouling factors

Water	0.001 (sec water) to 0.003 river water
Hydrocarbons, no phase change	0.001 to 0.003
Gases	0.002 (air) to 0.01 (flue gas)
Condensing vapours	0.0005 to 0.002
Boiling organic liquid	0.002 to 0.001.

Problems on Heat Exchanger

1. One hundred thousand pounds per hour of water are to pass through a heat exchanger with a resulting temperature increase from 140°F to 200°F. Combustion gas to 800°F with specific heat of 0.24 Btu/lb°F is available at a rate of 100,000 lbs/hr to heat the water. The overall heat transfer coefficient for this system is 16 Btu/hr ft^2 °F. Determine (a) the exit temperature of the gas and (b) the required heat transfer area of a single-pass counter flow heat exchanger to accomplish the specified energy transfer.

2. An oil having a specific heat of 0.45 Btu/lb °F enters a single-pass counter flow heat exchanger at 260°F with a flow rate of 18000 lbs/hr. The oil must be cooled to 180°F, using water as the other fluid: water at 50°F is available at a rate of 15,000 lbs/hr. Find the required heat transfer area for an overall heat transfer coefficient of 42 Btu/hr ft^2 F.

3. Given the initial fluid temperature, flow rates, overall heat transfer coefficient and required heat transfer specified in the previous problem what heat transfer area will be required if the exchanger configuration is

(a) Shell-and-tube with oil in the tubes: one shell pass four tube passes.

(b) Shell-and-tube with oil flowing through the tubes: 2 two shell passes and four tube passes.

4. Determine the required heat transfer surface area for a heat exchanger constructed of an 1 inch 14 BWG mild steel tubing to cool 55000 lbs/hr of benzene from 150°F to 104°F using 50000 lbs/hr of water available at 50°F. The *U* based on the outer-tube area, is 110 Btu/hr ft^2 °F. Consider each of the following arrangements.

(i) Parallel flow, single pass

(ii) Counter flow single pass

(iii) Shell and tube with benzene in the shell two shell passes, 48 tube passes.

5. Hydraulic fluid (MIL-M-5606) flows in the tubes of a single-shell, two-tube-pass heat exchanger entering at 160°F and leaving at 100°F with a flow rate of 5000 lbs/hr. The oil is in turbulent flow. Water enters the shell at 60°F and leaves at 80°F. If the same exchanger is used with the same water flow rate and inlet temperature but with the oil flowing at 3500 lbs/hr and entering at 200°F, what will be the exit temperatures of the two fluids?

6. A heater is needed to heat water from 60°F to 160°F flowing at 6000 lbs/hr using 130 psi saturated steam condensing on the outside of 1¼ inch 16 BWG copper tubes. A steam side surface coefficient of 1400 Btu/hr ft^2 F may be assumed. The tubes are to be 10 feet long and the desired water velocity is 5 ft/sec.

Determine the required number of tubes per pass in the shell-and-tube heat exchanger the required heat transfer area and the arrangement which will accomplish the necessary heat transfer.

7. Solve for the required information in the previous problem when a fouling resistance of 0.002 hr/ft^2 °F/Btu is considered for the water side and one of 0.0005°F hr ft^2/Btu is considered on the steam side.

8. Aniline is cooled from 125°F to 95°F in the inner pipe of a double-pipe single-pass heat exchanger. Cooling water flows counter currently to the aniline entering at 60°F and leaving at 82°F. The exchanger is constructed of a 7/8 inch, 16 BWG copper tube jacketed with a 1 1/2 inch schedule 40 steel pipe. The velocity of the aniline is to be 5 ft/sec and that of the water is 4 ft/sec. Determine

(i) The '*U*' based on the outside area of the inner pipe.

(ii) The required length of the exchanger if the water flow rate is 3000 lbs/hr.

9. Hydraulic fluid is being cooled from 200°F to 110°F in the annulus of a double pipe heat exchanger by water which enters the inside pipe at 60°F and leaves at 95°F. The proposal is made that the double pipe unit be lengthened to cool the hydraulic fluid to a lower temperature.

If the fluid streams flow concurrently determine

(i) The minimum temperature to which the hydraulic fluid may be cooled.

(ii) The exiting hydraulic fluid temperature as a function of the fractional increase in heat exchanger length.

10. Given the same double pipe heat exchanger as in the previous problem with the same entering temperatures and flow rates of the two fluid streams except that the streams flow counter currently, determine

(i) The exit temperature of each stream.

(ii) The lowest temperature to which the oil might be cooled.

11. Forty thousand standard cubic feet per hour of nitrogen are to be heated from 75°F to 145°F by condensing saturated stream at 35 psia nitrogen is to flow through 1 inch 14 BWG tubes with a mass velocity of 8500 lbs/hr ft^2.

(i) Determine the required number of 8-foot tubes and the number of passes for this nitrogen preheating to be accomplished. One shell pass of condensing steam is to be used.

(ii) What flow rate of carbon dioxide in lbs/hr would be anticipated were this exchanger used to preheat co_2 rather than nitrogen? Steam at the same conditions would still be condensed in the shell.

12. A shell-and-tube heat exchanger with one tube pass is to be designed in which water in the tubes is to be heated by condensing steam in the shell. The steam is to condense film wise and the water is to be of turbulent flow in the smooth tubes. Specified quantities include water flow rate entering and exiting water temperatures, steam condensing temperature and available pressure drop (entering and exiting losses neglected). Assuming that the water flow remains turbulent and that the thermal resistance of the tube wall and condensate film are negligible, evaluate the effect of tube diameter on the total required heat transfer area in the exchanger.

13. Two identical heat exchangers are constructed, each 12 ft long. These units consist of a 2 inch schedule 40 steel pipe concentric with a 3 inch schedule 40 pipe. Water, at 65°F enters the inner pipe of one unit and flows through both exchanges in series at a rate of 50 gal/min. Two hot water streams are available for supplying heat to the cold stream one at 140°F and 30 gpm, the other at 210°F and 30 gpm. If the hot streams are mixed arbitrarily at any location before entering this system, specify the arrangement which will yield maximum heat transfer. For this optimum arrangement what will be the exiting temperature of the cold stream?

14. An oil cooler in an engine used jacket water at 150°F to cool lubricating oil from 210°F to 165°F, flowing at 2600 lbs/hr. A small shell-and-tube heat exchanger consiting of one shell pass and two tube passes with oil in the shell is to be used for this purpose. Fifty 1/2 inch 18 BWG copper tubes are involved 25 per pass. The oil has a specific heat of 0.42 Btu/lb °F an the oil-side heat transfer coefficient

may be taken as 45 Btu/hr ft^2 °F. If the cooling water temperature rise is 5°F, how long must the heat exchanger tubes be?

15. Exhaust steam from a turbine is to be used to heat 50,000 lbs/hr of water from 60°F to 115°F. Steam is to be in the shell of a 2 shell-pass 8-tube pass heat exchanger in which each tube pass employs 20, 1 1/4 inch 16 BWG copper tubes. The heat transfer coefficients on steam and water sides are 700 Btu/Hr ft^2 F and 250 Btu/hr ft^2 F respectively. What is the required length of the tubes in this exchanger? Neglect the fouling factors.

16. Solve for the length of tubes for conditions specified in the previous problem if fouling resistance of 0.001 hr. ft^2 °F/ Btu àre assumed for both the steam and water sides of the heat exchanger tubes.

17. A single-pass shell-and-tube heat exchanger with steam condensing in the shell is to be employed to heat 5000 gpm of water from 60°F to 100°F. The steam side heat transfer coefficient is 1800 Btu/hr ft^2 °F. Pumping capacity limits the pressure drop through the tubes to 14 psi. Calculate the length and number of 1-inch schedule 40 steel tubes which will accomplish the necessary heat transfer with the pumping capacity available.

18. A shell-and-tube heat exchanger, with 50 psi saturated steam condensing in the shell is to heat 550 lbs/min of normal butyl alcohol from 80°F to 180°F. The exchanger employs 3/4 inch 16 BWG copper tubes and the allowable pressure drop in the tubes is 1.5 psi. Neglecting the steam-side resistance as well as the conductive resistance through the thin-walled copper tubes, determine

(i) The velocity of alcohol in the tubes

(ii) The number of tubes in the bundle

(iii) The length of the tubes

19. A hot fluid at 160°F enters a double-pipe heat exchanger and is cooled to 80°F. A cool fluid at 50°F is warmed in the exchanger to a temperature of 70°F. Calculate the log-mean temperature difference for

(a) Counter flow

(b) Parallel flow.

20. A cool fluid is heated in a double-pipe heat exchanger from 15°C while a warmer fluid is cooled from 100°C to 90°C calculate the LMTD for

(a) Parallel flow

(b) Counter flow.

21. Calculate the LMTD for (a) counter flow (b) parallel flow for the following temperatures.

Hot fluid inlet temperature = 200°F

Hot fluid outlet temperature = 100°F

Cold fluid inlet temperature = 50°F

Cold fluid outlet temperature = 150°F.

22. Cutting oil is used in a machining operation. The oil acts as a lubricant and as a coolant so the oil temperature gets quite high. It is proposed to cool the oil in a double-pipe heat exchanger from an inlet temperature of 60°C with a mass flow rate of 2300 kg.hr water is used as the cooling medium and is available at a temperature of 20°C with a flow rate of 2500 kg/hr. An available exchanger is made of 2 × 1¼ type *M* copper tubing, 2 m long.

(a) Determine the overall heat transfer coefficient of the exchanger for this application.

(b) What is the outlet temperature of the oil?

Assume properties of the oil to be the same as those of unused engine oil, and use counter flow.

Problems on the Heat Exchangers/Condensers

1. What is the overal heat transfer coefficient for a condenser tube $k = 90$ k cal/ hr m °C having ID = 2.25 cm and OD = 2.5 cm. The steam and water side coefficients are 5000 and 3750 k cal/ hr m^2 C respectively.

2. In a multi-pass heat exchanger with 2 shell passes and 8 tube passes 45,000 kg/hr of water are heated from 80°C by means of hot exhaust gas ($c_p = 0.25$) entering at 340°C and leavinig at 170°C. The total surface based on the outer tube surface is 850 m^2. Determine

(i) LMTD considering the heat exchanger to be simple counter flow type.

(ii) Correction on factor for actual arrangement

(iii) Actual LMTD

(iv) Actual value of U [the units of c_p are k cal/ kgm °C.]

3. 850 kg/hr of oil $c_p = 0.5$ k cal/kg °C is to be cooled from 65 C to 30°C. Determine the minimum amount of water at 15°C, required if we employ.

(i) a counter flow exchanger

(ii) a parallel flow exchanger.

4. 110,000 kg/hr of water is heated from 40°C to 140°C by co_2 entering at 420°C and leaving at 200°C. If $U = 60$, determine the required area for

(i) a counter flow heat exchanger

(ii) a parallel flow heat exchanger.

5. A parallel flow heat exchanger is required to heat 6000 kg/hr If water from 20°C to 45°C with steam condensing at 100°C on the outside of the tubes. Tubes of 2.5 cm OD and 2.25 cm ID, 2.5 m long are available. The steam side film coefficient is 5000 k cal/hr m^2 C and $k_w = 100$ k cal/hr m °C. The velocity of water in each tube is 0.45 m. Calculate the required number of tubes per pass and the number of passes.

6. A two-pass condenser has 32 copper tubes of 3 cm ID and 4.5 cm OD and the tubes are 5 m long. Steam at 1 atm condenses on the outside of the tubes. If

10000 kg/hr of water are heated from 7°C to 85°C by condensation of 1800 kg/hr of steam, determine

(i) the LMTD

(ii) the U

(iii) the water side heat transfer coefficient.

7. A vapour condenser is required to remove 250,000 k cal/hr from a vapour at 150°C. The cooling fluid is water, which flows through 1.25 cm ID, 1.6 cm OD, 3 m long tubes at a flow N_{Re} of 10,000. The vapour side coefficient is 7500. The water enters at 15°C and leaves at 30°C. Determine the weight of water required per hour the number of tubes per pass and the number of passes. Ignore tube wall resistance.

8. A heat exchanger heats 25, 000 kg/hr of water entering at 30°C while cooling 20,000 kg/hr of water from 100°C to 85°C U = 1400. Determine the heat transfer area needed for

(i) a parallel flow arrangement

(ii) a counter flow configuration

(iii) 1–2 reversed current exchanger

If, in case (ii) above, the cold water enters at 20°C, what will be its outlet temperature when the inlet temperature and flow rate of the hot water remain unchanged.

9. In a counter flow exchanger oil is flowing through the tubes is cooled from 50°C to 35°C by water flowing on the shell side. The inlet and outlet temperature of water are 15°C and 32°C. The exchanger consists of an inner copper tube (3.25 cm OD), (0.32 cm thick) and an outer steel pipe (6 cm OD and 0.38 cm thick). The oil flow rate is 70 kg/min. Determine the length of the heat exchanger. The properties of oil are given below.

$\rho = 910$ kg/m^3

$c_p = 0.45$ k cal/kg °C

$k = 0.12$ k cal/hr m °C

$\mu = 0.042$ kg/m sec.

10. Calculate the length of 5 cm ID and 6 cm OD steel pipe ($k = 48$ k cal/hr m °C) required to heat 7800 kg/hr of oil from 30°C to 90°C, if the pipe is heated externally by steam condensing at 100°C. Steam side heat transfer coefficient is 8000 k cal/hr m^2 °C. The properties of oil are given below

$\rho = 850$ kg/m^3

$c_p = 0.47$ k cal/hr °C

$k = 0.115$ k cal/hr m °C

$\mu = 22$ kg/m hr.

11. A heat exchanger is required to heat 7000 kg/hr of water from 20°C to 93°C. Water flows through 90 tubes ($k = 90$ k cal/hr m °C) 1 cm O.D., 0.12 cm

thick and 4.5 m long. Steam at 100°C condenses on the outside of the tubes. If the value of h for the inner surface of tubes is 3000 kg k cal/hr m^2 °C, compute the heat transfer coefficient or the outer surface.

12. A water heater used for heating water from 20°C to 85°C consists of 10 horizontal tubes 3.75 cm OD and 3.15 cm ID. Water flows through the tubes at an average velocity of 1.2 m/sec. Steam at 150°C condenses on the outside of the tubes. If h_i = 7500 and h_o = 5000 k cal/hr m^2 C, calculate the outer surface area required.

13. A counter flow exchanger was constructed on the basis of U = 150. The area of the exchanger is 1 m^2. The water rate is 1300 kg/hr and the inlet water temperature is 15°C. If this exchanger is now used to cool 550 kg/hr of oil (c_p = 0.5 k cal/kg °C) at an inlet temperature of 94°C, what will be the outlet temperature of oil and water?

14. 9000 kg/hr of oil (c_p = 0.5 k cal/kg°C) are cooled from 80°C to 35°C by means of water entering at 25°C in a counter current exchanger. The exchanger area is 29 m^2 and U = 500. Determine the rate of flow of water.

15. Determine UA for a counter flow exchanger to cool 2000 kg/hr of oil (c_p = 0.5 k cal/kg °C) from 70°C to 45°C, if the cooling water enters at 15°C and leaves at 28°C. If the oil rate is reduced to 1000 kg/hr and the entering temperature is 75°C what will be the oil temperature at exit? The water rate and entering temperature remain the same.

16. The following data refers to a double pipe parallel flow heat exchanger having 2.25 cm ID and 2.6 cm OD inner pipe and 3.2 cm ID outer pipe of steel

High pressure steam entering inner tube: 400°C at the rate of 1000 kg/hr.

High pressure steam leaving inner tube: 385°C

Low pressure steam entering outer pipe: 220°C at the rate of 430 kg/hr.

Low pressure steam leaving outer pipe: 395°C

What will be the heat transfer rate if the steam flow in the inner tube is reversed: the entering temperatures remaining the same.

17. A counter flow heat exchanger is to be designed to cool 1000 kg/hr of oil (c_p = 0.5) from 100°C to 50°C by using water at 35°C. The maximum allowable exit temperature for water is 45°C. Determine the ratio $m_h\, c\, p_h / m_c\, c\, p_c$. Find the value of UA for given NTU = 1.5.

Problems on Heat Exchanger Performance and Design.

1. A shell and tube heat exchanger with two shell passes and four tube passes is employed to heat 10000 kg/hr of pressured water from 35 to 120°C with 5000 kg/hr water entering the exchanger at 300°C. If the overall heat transfer coefficient is 1500 w/m^2 °k determine the heat exchanger area that is required for this situation.

2. Consider the heat exchanger of the above problem. After several years of operation it is observed that the outlet temperature of the cold water reaches only

95°C rather than the desired 120°C for the same flow rates and inlet temperatures of the fluids.

Determine the cumulative (inner and outer surfaces) fouling factor that is the cause of the poorer performance.

3. A thin walled concentric tube heat exchanger is to be used to cool engine oil from 160°C to 60°C, water which is available at 25°C is to be used as the coolant. The water and oil flow rates are each 2 kg/sec and the diameter of the inner tube is 0.5 m. The corresponding value of the overall heat transfer coefficient is 250 w/m^2 k. What must be the length of the heat exchanger to accomplish the desired cooling?

4. A concentric tube heat exchanger for cooling lubricating oil consists of a thin walled inner tube of 25 mm diameter carrying water and an outer tube of 45 mm diameter carrying the oil. The exchanger operates in counter flow with an overall heat transfer coefficient of 60 w/m^2 k and the tabulated average properties

Properties	*Water*	*Oil*
ρ (kg/m^3)	1000	800
c_p (J/kg °k)	4200	1900
ν (m^2/sec)	7×10^{-7}	1×10^{-5}
k (w/m °k)	0.64	0.134
N_{pr}	4.7	140

Determine

(i) The total heat transfer and the outlet temperature of the water, if the outlet temperature of the oil is 60°C.

(ii) The length required for the heat exchanger.

5. Cold water at 20°C and 5000 kg/hr is to be heated by hot water supplied at 80°C and 10000 kg/hr. You select from a manufacturer's catalog a shell and tube heat exchanger (one shell and two tube passes) having a UA value of 11,600 w/°k. Determine the hot water outlet temperature.

6. A thin walled concentric tube heat exchanger of 0.19 m length is to be used to heat deionized water from 40 to 60°C at a flow rate of 5 kg/sec. The deionized water flows through the inner tube of 30 mm diameter while hot process water at 95°C flows in the annulus formed with the outer tube of 60 mm diameter. The thermo physical properties of the fluids are given below

Properties	*Deionized water*	*Process water*
ρ (kg/m^3)	982.3	967.1
c_p (J/kg °k)	4181	4197
k (w/m °k)	0.643	0.673
μ (sN/m^2) $\times 10^6$	548	324
N_{pr}	3.56	2.02

(i) Considering a parallel flow configuration of the heat exchanger, determine the minimum flow rate required for the hot process water.

(ii) Determine the overall heat transfer coefficient required for the conditions of part (i). Explain why it is not possible to achieve the conditions specified in part (i).

(iii) Considering a counter flow configuration, determine the minimum flow rate required for the hot process water. What is the effectiveness of the heat exchanger for this situation?

(iv) Saturated steam at 0.14 bar is condensed in a shell and tube heat exchanger with one shell pass and two tube passes consisting of 130 brass tubes each with a length of per pass of 2 m. The tubes have inner and outer diameters of 13.4 and 15.9 mm. respectively. Cooling water enters the tubes at 20°C with a mean velocity of 1.25 m/sec. The heat transfer coefficient for condensation on the outer surfaces of the tubes is 13,500 w/m^2 °k. Determine the overall heat transfer coefficient, the cooling water outlet temperature and the steam condensation rate?

8. A two fluid heat exchanger has inlet and outlet temperatures of 65°C and 40°C for the hot fluid and 15°C and 30° for the cold fluid. Can you tell whether this exchanger is operating under parallel flow or counter flow conditions. What is the effectiveness of the exchanger if the cold fluid has the minimum capacity ratio?

9. Water at 225 kg/hr is to be heated from 35°C to 95°C by means of a concentric tube heat exchanger, oil at 225 kg/hr and 210°C with a specific heat of 2095 j/kg °k is to be used as the hot fluid. If the overall heat transfer coefficient based on the outer diameter of the inner tube is 550 w/m^2 °k determine the length of the heat exchanger if the outer diameter is 100 mm.

10. A hot fluid enters a concentric pipe heat exchanger at 150°C and is to be cooled to 100°C by a cold fluid entering at 35°C and heated to 65°C. Would you use parallel flow or counter flow for the most effective design?

11. Consider a very long, concentric tube heat exchanger having hot and cold water inlet temperatures of 85 and 15°C. The flow rate of the cold water is 50 % of that of the hot water. Taking that the specific heats of the hot and cold water are equal determine the hot water outlet temperature for the following modes of operation (a) counter flow and (b) parallel flow.

12. A shell and tube heat exchanger is to heat 10000 kg/hr of water from 16 to 84°C by hot engine oil flowing through the shell. The oil makes a single shell pass, entering at 160°C and leaving at 94°C with an average heat transfer coefficient of 400 w/m^2 °k. The water flows through 11 brass tubes of 22.9 mm inside diameter and 25.4 mm outside diameter, with each tube making four passes through the shell. Assuming that the flow of water is fully developed, determine the required tube length per pass.

13. A shell and tube heat exchanger consists of 135 thin walled tubes in a double pass arrangement, each of 12.5 mm diameter with a total surface area 47.5 m^2. Water, which is the tube side fluid, enters the heat exchanger at 15°C. and 6.5 kg/sec and is heated by exhaust gas entering at 200°C and 5 kg/sec. The gas may be assumed to have the properties of atmospheric air and the overall heat transfer coefficient is approximately 200 w/m^2 °k. What are the gas and water outlet temperatures? Assuming fully developed flow what is the tube-side convection coefficient?

14. A shell and tube heat exchanger (one shell pass, multiple tube passes) is to be used to condense 2.73 kg/sec of saturated steam at 340°k. Condensation occurs on the outer tube surfaces and the corresponding convection coefficient is h_o = 12, 500 w/m^2 k. The condenser will be supplied with cooling water which enters the tubes at 291°k and is to exit the tubes at 300°k. Thin-walled tubes of 19 mm diameter are specified and the mean velocity of water flow through the tubes is to be maintained at 1.5 m/sec. How many tubes must be used? If the length of the heat exchanger is hot to exceed 1.5 m, how many tube passes should be made and what is the required length per pass?

15. Saturated process steam at 1 atm is condensed in a shell and tube heat exchanger (one shell, two tube passes). Cooling water enters the tubes at 15°C with an average velocity of 3.5 m/sec. The tubes are thin walled and made of copper with a diameter of 14 mm and length of 0.5 m. The convection heat transfer coefficient for condensation on the outer surface of the tubes is 21,800 w/m^2 °k.

(i) Find the number of tubes/pass required to condense 2.3 kg/sec of steam.

(ii) Find the outlet water temperature.

(iii) Determine the maximum possible condensation rate that could be accomplished with this heat exchanger using the same water flow rate and inlet temperature.

16. The feed water heater for a boiler supplies 10,000 kg/hr of water at 65°C. The feed water has an inlet temperature of 20°C and is to be heated in a single-shell two-tube pass heat exchanger by condensing steam at 1.30 bars. The overall heat transfer coefficient is 2000 w/m^2 k. Using both the LMTD and NTU methods, determine the required heat transfer area. What is the steam condensation rate?

17. A shell and tube with single shell and tube passes is used to cool the oil of a large marine engine. Lake water which is the shell side fluid enters the heat exchanger at 2 kg/sec and 15°C. While the oil enters at 1 kg/sec and 140°C. The oil flows through 100 copper tubes, each 500mm long and having inner and outer diameters of 6 and 8 mm. The shell side heat transfer coefficient if 500 w/m^2 k. What is the outlet temperature of the oil?

18. A shell and tube heat exchanger with one shell pass and 20 tube passes used hot water on the tube side to heat oil on the shell side. The single copper tube has inner and outer diameters of 20 and 24 mm and a length per pass of 3 m. The

water enters at 87°C and at the rate of 0.2 kg/sec and leaves at 27°C. Inlet and outlet temperatures of the oil are 7 and 37°C. What is the average convection coefficient for the tube outer surface?

19. A concentric tube heat exchanger used water which is available at 15°C. to cool ethylene glycol from 100 to 60°C. The water and glycol flow rates are each 0.5 kg/sec. What are the maximum possible heat transfer rate and effectiveness of the exchanger? Which is preferred, a parallel flow or counter flow mode of operations?

20. Saturated steam at 100°C condenses in a shell and tube exchanger (single shell two tube passes) with a surface area of 0.5 m^2 and overall heat transfer coefficient of 2000 w/m^2 k. Water enters at 0.5 kg/sec and 15°C. Determine the outlet temperature of the water and the rate of steam condensation.

21. Consider a concentric tube heat exchanger characterized by a uniform overall heat transfer coefficient and operating under the following conditions

	m kg/sec	c_p *J/kg °k*	T_i *°C*	T_o *°C*
Cold fluid	0.125	4200	40	95
Hot fluid	0.125	2100	210	–

What is the maximum possible heat transfer rate? What is the heat exchanger effectiveness? Should the heat exchanger be operated in parallel flower in counter flow mode? What is the ratio of the required areas for these two flow conditions?

22. Consider a concentric tube heat exchanger with hot and cold water inlet temperatures of 200 and 35°C, respectively. The flow rates of hot and cold fluids are 42 and 84 kg/hour respectively. Assume the overall heat transfer coefficient is 180 w/m^2 k.

(i) What is the maximum heat transfer rate that could be achieved for the prescribed inlet conditions?

(ii) If the exchanger is operated in counter flow with a heat transfer area of 0.33 m^2 determine the outlet fluid temperatures.

(iii) What is the largest possible heat transfer rate that could be achieved for the prescribed inlet conditions if the exchanger is operated in parallel flow and is very long. What is the effectiveness of the heat exchanger in this configuration?

23. Hot water at 2.5 kg/sec and 100°C enters a concentric tube counter flow heat exchanger having a total area of 23 m^2. Cold water at 20°C enters at 5.0 kg/sec and the overall heat transfer coefficient is 1000 w/m^2 °k. Determine the total heat transfer rate and the outlet temperatures of the hot and cold fluids.

24. Hot exhaust gases are used in a shell and tube exchanger to heat 2.5 kg/sec of water from 35 to 85°C. The gases assumed to have the properties of air, enter at 200°C and leave at 93°C. The overall heat transfer coefficient is 180 w/m^2 °k. Using the NTU-effectiveness method calculate the area of the exchanger.

25. Ethylene glycol and water at 60°C and 10°C respectively enter a shell and tube heat exchanger for which the total heat transfer area is 15 m^2. With ethylene glycol and water flow rates of 2 and 5 kg/sec, respectively the overall heat transfer coefficient is 800 w/m^2 °k. Determine the rate of heat transfer and the fluid outlet temperatures.

26. In a tubular heat exchanger with two shell passes and eight tube passes, 1000,000 lbs/hr of water are heated in the shell from 180°C to 300°F. Hot exhaust gases having roughly the same physical properties as air, enter the tubes at 650°F and leave at 350°F. The total surface based on the outer tube surface is 10,000 ft^2 Determine (a) log mean temperature if the heat exchanger were a simple counter flow type (b) the correction factor F for the actual arrangement, (c) the effectiveness of the heat exchanger (d) the average over all heat transfer coefficient.

27. Design (i.e determine the over all area and a suitable arrangement of shell and tube passes) for a tubular feed water heater capable of heating 5000 lbs/hr of water from 70 to 190°F. The following specification are given (a) Saturated steam at 134 psia is condensing on the outer tube surface (b) unit surface conductance on steam side is 1200 Btu/hr ft^2 F (c) tubes are of copper, 1 inch OD, 0.9 inch is ID 8 ft long and (d) water velocity is 3 ft/sec.

28. In a single pass counter flow heat exchanger 10,000 lbs/hr of water enter at 60°F and cool 20,000 lbs/hr of an oil having a specific heat of 0.50 Btu/lb F from 200°F to 150°F. If the overall heat transfer coefficient is 50 Btu/hr ft^2 F, determine the surface area required.

29. Determine the outlet temperature of the oil in the above problem for the same initial temperatures of the fluids if the flow arrangement is one shell pass and two tube passes but with the same total area and average overall heat transfer coefficient as the unit in the above problem.

30. Carbon dioxide at 800°F is to be used to heat 100,000 lbs/hr of water from 100°F to 300°F. While the gas temperature drops 400°F. For an overall heat transfer coefficient of 10 Btu/hr ft^2 °F, compute the required heat exchanger area in ft^2 for (a) parallel flow (b) counter flow and (c) a 2–4 reversed current exchanger.

31. A double pipe oil water heat exchanger is constructed of a 10 ft long brass tube BWG No. 18 (0.527 in ID, 0.625 in OD) concentric within a well insulated standard 3/4 in wrought iron pipe (0.824 in ID). Water flows in the annulus, entering at 60°F. A light oil flows in the tube, entering at 200°F. The flow arrangement is counter flow. For the velocities specified, determine (a) the exit temperature of the oil and (b) the rate of the heat flow.

The velocities specified are (i) water rate = 1 gpm, oil rate = 1 gpm.

(ii) water rate = 10 gpm, oil rate = 1 gpm

(iii) water rate = 1 gpm, oil rate = 20 gpm

and (iv) water rate = 10 gpm, oil rate = 20 gpm.

32. An ecomomizer is to be purchased for a power generation plant. The unit is to be large enough to heat 60000 lb/hr of water from 160 to 360°F. There are 100,000 lbs/hr of flue gases (c_p = 0.24 Btu/lb °F) available at 800°F. Estimate (a) the outlet temperature of the flue gases (b) the heat transfer area required for a counter flow configuration if the overall heat transfer coefficient is 10 Btu/hr ft² F.

33. Water is heated while flowing through a pipe by steam condensing on the outside of the pipe. For an overall heat transfer coefficient of 100 Btu/hr ft² F, based on the inside diameter of 2 in. a steam temperature of 220 F and a water flow rate of 500 lb/min calculate the length required to raise the water temperature from 60°F to 150°F.

34. An oil having a specific heat of 0.5 Btu/lb °F enters an oil cooler at 180°F at the rate of 20,000 lbs/hr. The cooler is counter flow unit with water as the coolant, the transfer area being 300 ft² and the overall heat transfer coefficient being 100 Btu/ft² F hr. The water enters the exchangers at 80°F. Determine the water rate required if the oil is to leave the cooler at 100°F.

35. Steam is to be condensed at atmospheric pressure in a shell and tube heat exchanger consisting of 72 eight foot lengths of standard 1 in 18 BWG copper condenser tubing (0.902 in ID, 0.049 in wall thickness). Water is available at a rate of 500,000 lb/hr and it flows inside the tubes, entering at 60°F. For the average and surface conductances (the heat transfer coefficients and the sealing factors) listed below, estimate the pounds of steam per hour condensed. The unit surface conductances (based on actual area) are

Water side 800 Btu/hr ft² °F

Steam side 2000 Btu/hr ft² °F

Scale on steam side 1800 Btu/hr ft² °F

Scale on water side 2000 Btu/hr ft² °F.

36. Show that the effectiveness of for a counter flow arrangement is

$$E = \frac{1 - e^{-[1-(c_{min}/c_{max})]\,NTU}}{1 - (c_{min}/c_{max})\, e^{-[1-(c_{min}/c_{max})]\,NTU}}\ .$$

37. Dry air is cooled from 150 to 100°F while flowing at the rate of 10000 lbs/hr in a simple counter flow heat exchanger by means of cold air which enters at 80°F and flows at a rate of 12, 500 lbs/hr. It is planned to lengthen the heat exchanger so that 10000 lbs/hr of air can be cooled from 150 to 80°F with a counter flow arrangement with air flowing at 12, 500 lbs/hr and entering at 60°F. Assuming that the specific heat of the air remains constant, calculate the ratio of the length of the new heat exchanger to the length of the original.

38. Water flowing at a rate of 100,000 lbs/hr is to be cooled from 200 to 150°F by means of an equal flow rate of cold water entering at 100°F. The water velocity will be such that the overall heat transfer coefficient U is 400 Btu/hr ft² °F. Calculate the ft² of heat exchanger surface required for each of the following

arrangements (a) parallel flow (b) counter flow (c) a multi pass heat exchanger with the hot water making one pass through a well-baffled shell and tube the cold water making two passes through the tubes.

39. A shell and tube counter flow heat exchanger is to be designed for heating an oil from 80 to 180°F. The oil is to pass through 1½ in schedule 40 pipes at a velocity of 200 ft/min and steam is to condense at 215°F on the outside of the pipes. The specific heat of the oil is 0.43 Btu/lb °F and its mass density is 58 lb/c ft. The steam side heat transfer coefficient is approximately 1800 Btu/hr ft^2 °F and the thermal conductivity of the metal of the tubes is 17 Btu/hr ft °F. The results of the previous experiments giving the oil side heat transfer coefficients for the same pipe size at the same oil velocity as those to be used in the exchanger are shown below

ΔT (°F)	135	115	95	75	35
T_{oil} (°F)	80	100	120	140	160
h_{oil} Btu/hr ft^2 °F	14	15	18	25	45

(a) Find the over all heat transfer coefficient U based on the outer surface area at the point where the oil is 100°F (b) Find the temperature of the inside surface of the pipe when the oil temperature is 100°F. (c) Find the required length of the tube bundle.

Problems on Heat Exchanger

1. 2.4 kg/sec of a fluid having a specific heat of 0.8 kj/kg °k enters a counter flow heat exchanger at 0°C and is heated to 400°C by 2 kg/sec of a fluid having a specific heat of 0.96 kj/kg °k entering the unit at 700°C. Show that to heat the cooler fluid to 500°C, all other conditions remaining unchanged, would require the surface area for heat transfer to be increased by 87.5 %.

2. Saturated non-oil bearing steam at 1 atm enters the shell pass of a two tube pass shell condenser. There are thirty 20 feet tubes in each tube pass. They are made of schedule 160, 3/4 in steel pipe (nominal diameter). A volume of flow rate of 0.01 ft^3/s of water entering at 60°F enters each tube. The condensing heat transfer coefficient is 2000 Btu/ft^2 hr °F and h = 1380 Btu/hr ft^2 °F for the water in the tubes. Estimate the exit temperature of the water and mass rate of condensate.

3. Consider a counter flow heat exchanger which must cool 3000 kg/hr of mercury from 150°C to 128°F. The coolant is 100 kg/hr of water supplied by at 70°F. If U is 300 w/m^2 °C. Complete the design by determining reasonable values for the area and the exit water temperature.

4. A single pass heat exchanger condenses steam at 1 atm on he shell side and heats water from 10 to 30°C on the tube side U = 2500 w/m^2 C. The tubing is thin walled 5 cm in diameter and 2 m in length (a) your boss asks whether the exchanger should be counter flow or parallel flow. How do you advise him? Evaluate (b) the LMTD (c) $m_{H_{2O}}$ and (d) the effectiveness.

5. 2 kg/sec of air at 27°C and 1.5 kg/sec of water at 60°C each enter a heat exchanger. Evaluate the exit temperature if $A = 12\ m^2$.

$U = 185\ w/m^2\ °c$ and

(i) the exchanger is parallel flow

(ii) the exchanger is counter flow ($T\,h_{out} = 54.0°C$).

6. Water at 15°C is supplied to a one-shell-pass two-tube-pass heat exchanger, to cool 10 kg/sec of liquid ammonia from 120°C to 40°C. You can anticipate a U on the order of 1500 w.m² °C when the water flows in the tubes. If the heat transfer area is to be 90 m², select the correct flow rate of water.

7

Thermal Energy Transfer by Radiation

In the case of heat transfer by conduction and convection, the transfer rate can be determined by integration of the usual rate equation and the size of the heat exchanging equipment can be obtained by combining the rate equation with appropriate material and energy balance equations.

Simultaneously and independently thermal energy transfer may occur by radiation.

q	$=$	q_c	$+$	q_k	$+$	q_r
$\uparrow$		$\uparrow$		$\uparrow$		$\uparrow$
total rate of thermal energy transfer		rate of heat transfer by convection mechanism		by conduction		by radiation

Though all these three mechanisms may be encountered in most of the situations of heat transfer, one or two will be more predominant than the other mechanisms. The transfer of radian energy my not be amenable for representation in terms of rate equations of the forms considered under conduction and convection.

7.1 NATURE OF RADIATION

It is emitted and received in discrete particles or pulses called photons. In transmission, radiation has all the properties of an electromagnetic wave that is a vibrating electrical field and a similarly vibrating magnetic field, the direction of vibration being perpendicular to the direction of radiant transmission. The speed of the beam is governed by the medium through which it travels.

$$c = \lambda n$$

where c = velocity of electromagnetic beam

λ = the wave length or distance from energy peak to energy peak

n = the frequency of radiation or the number of energy peaks passing any point per unit time.

On striking a receiver, the photons penetrate into the body before increasing its energy .... For liquids the penetration would be greater compared to solids in which absorption occurs close to the surface.

Thermal Radiation

All the solids, liquids and most of the gases receive and give off energy, in the form of electromagnetic waves which are similar whether they are perceived in the form of thermal energy, light or x-rays.

Radiation travels in straight-line paths from any emitter. It can be received *directly* only by bodies that can be 'seen' by the emitter. Receivers shade the area away from the emitter so that objects in this space receive no radiant energy directly from the emitter under consideration though objects so shaded receive radiation from all the other objects which they can see.

The Reception of Thermal Radiation

As in the case of optical waves thermal radiation falling on an object will be partly absorbed, partly reflected and partly transmitted; with the proportions of the incident energy that are absorbed. reflected and transmitted dependent primarily on the characteristics of the receiver though to some extent the wave length of the radiation and the temperature of the receiver have an effect on it.

most solids— absorb or reflect all the radiant energy

glass is transparent to radiation of short wave length while it is opaque for long wave length radiation

$$\rho + \alpha + T = 1$$

where ρ – reflectivity, the fraction of the total incident radiation that is reflected

α – absorbtivity

T = transmissivity— the fraction that is transmitted through the body.

The above equation indicates the possible means of disposal of the radiant energy arriving at a point.

For most of the solids and liquids, the above equation simplifies to

$$\rho + \alpha = 1$$

i.e. transmittivity is negligible

Penetration in metals is about 1 micron whereas in the case of electric non-conductors it may shoot up to 1000 microns.

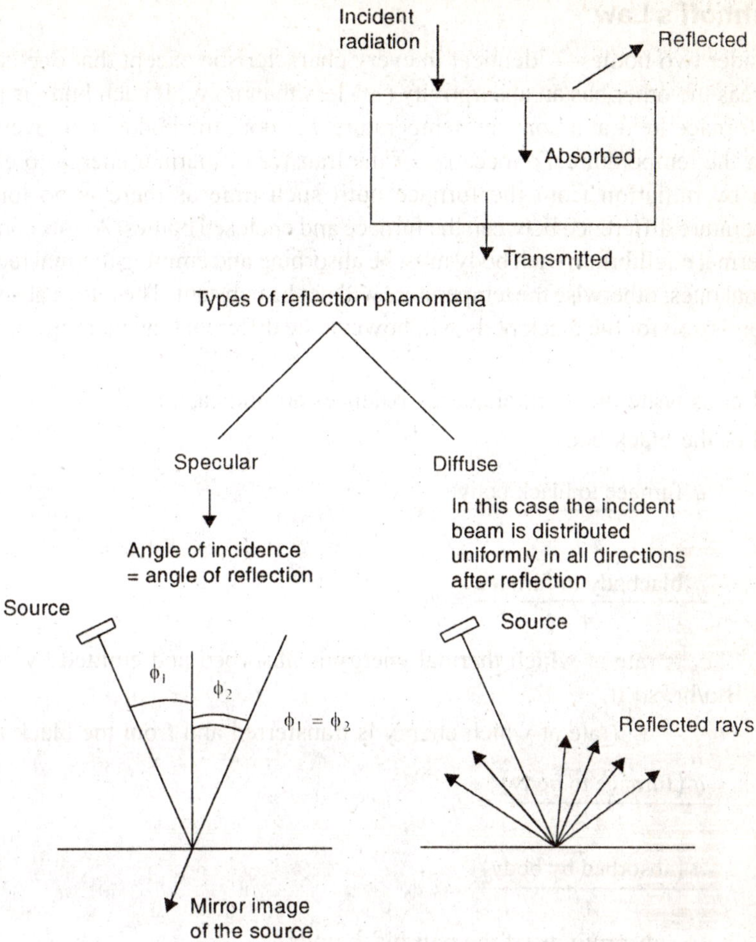

Fig. 7.1 *Types of reflection phenomena*

Normally specular is predominant in case of polished surface while rough surfaces would exhibit diffuse behaviour.

7.2 BLACK BODY—THE IDEAL RADIATION RECEIVER

A body which has $T = 0$, and $\rho = 0$, that is one in which all the incident radiation energy would be absorbed is termed black body (its surface need not be black in colour) Though such black bodies do not exist in nature with the only possibility being the materials approaching black behaviour. Black body *emits radiation* at the maximum rate possible of any body of equal size and shape at the same temperature.

Krichhoff's Law

Consider two bodies— identical in every characteristic except that one is black whereas the other has an absorptivity (α) less than unity. If each body is placed in a furnace held at a constant temperature *To*, both the bodies will eventually attain the temperature *To*, because of the transfer of thermal energy to each of them by radiation from the furnace until such time as there is no longer a temperature difference between the furnace and enclosed bodies. At this condition of thermal equilibrium each body must be absorbing and emitting thermal radiation at equal rates; otherwise the temperature will not be constant. The rate of absorption and emission for the black body will however be different than that the non-black body.

Let us write the thermal energy balances around each body.

For the black body

$$\frac{q \text{ furnace to black body}}{A} = e_b$$

$$\frac{q \text{ blacbody to furnace}}{A} = e_b$$

e_b = rate at which thermal energy is absorbed and emitted by a black body Btu/hr. sq ft.

(rate at which energy is transferred and from the black body)

$$\frac{q\ (\text{furnace to body})}{A} = e_b$$

$$\frac{q\ (\text{absorbed by body})}{A} = \alpha\ e_b$$

α – absorptivity of the non-black body

$$\frac{q\ (\text{reflected by body})}{A} = (1-\alpha)\ e_b \ \ \text{(there is no transmission)}$$

and $\quad e_b = \alpha\ e_b + (1-\alpha)\ e_b$

If e is defined as the actual rate at which radiant energy is absorbed and consequently emitted by the non-black body per unit of surface (Btu/hr ft^2)

$$e = \alpha\ e_b \left[\text{i.e. } \alpha = \frac{e}{e_b}\right]$$

we defined ε , the emissivity as

$$\varepsilon = \frac{\text{rate at which energy is radiated from a unit surface of non-black body}}{\text{rate at which it would be radiated from a black surface of equal area}}$$

i.e. $\varepsilon = \dfrac{e}{e_b}$

Therefore it could be seen that.

$\alpha = \varepsilon$

This is known as Krichhoff's law. Both α and ε are temperature sensitive.

Example of Body With Blackbody Characteristic

Approximated by a cavity in a solid susbstance which is open to atmosphere. Only through a pinhole in the solid surface. Any radiant beam arriving at this opening will pass into the cavity. After passing though this cavity it will bounce from wall to wall besides its being partly absorbed at each contact. Since the hole is very small and its chances of finding the, opening again are almost nil, the beam's energy will be completely absorbed eventually even though only a very small fraction of this energy is absorbed at each contact.

Under the condition that the pinhole and its surroundings are at thermal equilibrium, the emission from the pinhole must be at a rate equal to that at which radiant energy strikes it from the surroundings. Then since e_b is both absorbed and radiated, ε as well as α must be equal to one.

The cavity behaves approximately as a black body that is as an emitter as well as an absorber. The size of the cavity is very large compared with size of the opening.

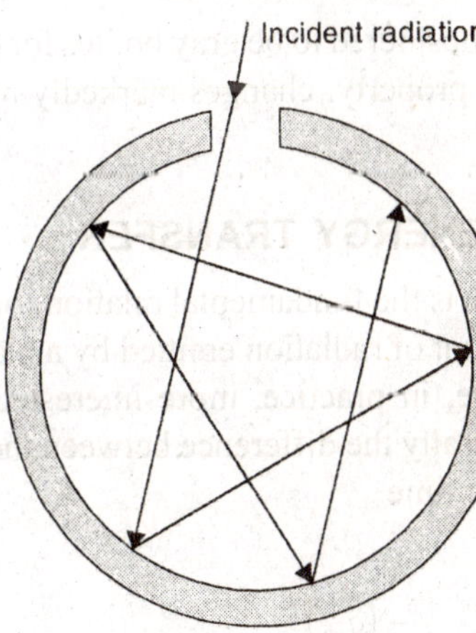

Fig. 7.2 *Example of a body with blackbody characteristic*

7.3 BLACK-BODY RADIATION LAWS

The emissive power of a black body depends on its temperature only and it could be expressed by the following relation.

$$\frac{q_{rb}}{A} = \sigma T^4$$

This relation is known as the Stefan-Bolizmann law and

σ = Stefan-Boltzmann constant 0.1713×10^{-8} Btu/(ft^2) (hr) (°R)^4

The Stefan-Boltzmann law is an exact consequence of the laws of thermodynamics and electromagnetism

The above equation may also be written as:

$$\frac{q_{rb}}{A} = 0.1713\left(\frac{T}{100}\right)^4$$

Most industrial surfaces of interest to chemical engineers give diffuse reflection and in treating practical cases the important simplifying assumption can be made that reflectivity and absorpitivity are independent of angle of incidence. The reflectivity may vary with the wave length of the incident radiation and the absorptivity of the entire beam is then a weighted average of the monochromatic absorptivities and it depends on the entire spectrum of the incident radiation.

For many industrial surfaces use of a second simplifying assumption can be made viz; the absorptivity of the surface is the same for all wavelengths. A body for which this *holds good* is called *gray* body. In diffuse radiation from a gray body the monochromatic absorptivity is independent of both angle of incidence and of wave length so the total absorptivity equals the monochromatic absorptivity and is also independent of angle of incidence.

Many materials are considered to be gray bodies for engineering calculations eventhough ε, a surface property, changes markedly by with small changes in surface conditions.

7.4 NET RADIANT-ENERGY TRANSFER

Stefan-Boltzman equation is the fundamental relation for radiant-energy transfer and it only gives the amount of radiation emitted by a black body without regard to that it absorbs. We are, in practice, more interested in finding out the net energy flow which is naturally the difference between the radiant energy emitted and that absorbed per unit time

For a black-body

$$(q_{rb})_{net} = (q_{rb})_{emitted} - (q_{rb})_{absorbed}$$

but since

$$(q_{rb})_{emitted} = \sigma A\ T_1^4$$

$$(q_{rb})_{absorbed} = \sigma A\ T_2^4$$

where T_1 = temperature of emitting body

T_2 = temperature of surrounding surfaces

A = surface area of body which is emitting and absorbing

The net rate of energy flow then is

$$(q_{rb})_{net} = \sigma A\, T_1^4 - \sigma A\, T_2^4 = \sigma A\left(T_1^4 - T_2^4\right)$$

The above relation is subject to the constraint that a black body must be emitting energy only to the surroundings and receiving black body radiation from its surroundings, i.e. the radiating body can not 'see' any of itself.

The net radiant- energy transfer case of imperfect radiator or a non black body will be

$$(q_r)_{net} = \sigma A\left[T_1^4\, \varepsilon - T_2^4\, \alpha\right]$$

where ε and α are the emissivity and absorptivity of the real body.

7.5 RADIANT HEAT TRANSFER COEFFICIENTS

In the case of heat transfer by convection and radiation, it may be convenient to reduce the calculations to the usual from of the '*h*' used in convection calculations.

Thus for interchange between black surface

$$q_{rh} = \sigma A\left(T_1^4 - T_2^4\right)$$

$$= h_{rh}\, A\,(T_1 - T_2)$$

from the above

$$h_{rh} = \sigma \frac{\left(T_1^4 - T_2^4\right)}{(T_1 - T_2)}$$

where h_{rb} = radiation "coefficient" for black body interchange.

Problem

Find out the amount of steam condensing per hour per foot of the pipe in the case of an uninsulated steam pipe running through a dark warehouse room. The o.d. of the steel pipe is 2" and the steam is saturated at 100 psia. The warehouse is kept at 40°F. The '*h*' by natural convection from outside pipe surface is 2.0 Btu/hr ft^2 °F and the pipe may be treated as a gray body.

Total heat loss

$$q = q_r + q_c \text{ Btu/hr (radiation)}$$

or $$q = \varepsilon A \sigma\left(T_p^4 - T_w^4\right) + h_c\, A\left(T_p - T_a\right)$$

T_p = temperature of pipe surface

T_w = temperature of the warehouse wall

T_a = temperature of warehouse air (°R)

ε = emissivity of the surface of pipe given as 0.95

A = area of outside surface of the pipe.

The pipe surface temperature (T_p) will equal the condensing steam temperature i.e. T_p = 328°F at 100 psia for saturated steam . Because the condensing film resistance and the pipe wall resistance are very small compared to the outside surface to warehouse resistance

i.e. ΔT (total) = ΔT from pipe surface to warehouse

Temperature of the warehouse wall and air will be identical that is 40°F.

$$q = 0.1713 \times A \times 0.95 \left[\left(\frac{T_p}{100}\right)^4 - \left(\frac{T_w}{100}\right)\right] + 2.0\, A\, (T_p - T_w)$$

A = 0.622 sq. ft

T_p = 328 + 460 = 788 °R

$T_w = T_a$ = 460 + 40 = 500 °R

q = 329 + 358

= 687 Btu/hr ft. of pipe

$$\text{Rate of condensation} = \frac{q}{\Delta H}$$

$$= \frac{687}{888.9} = 0.78 \text{ lb of steam per hr per foot of pipe.}$$

The Electromagnetic Wavelength Spectrum

Electromagnetic waves occur over on enormous range of wave, lengths, from the 10^{-16} m (cosmic rays) to 10^6 m (electric power waves). A schematic of the complete electomagnetic wave spectrum is shown in the figure given below. Thermal radiation is in the range $\lambda = 10^{7}$ to 10^{-3} m which is more conveniently described by small length unit called micron

$$1 \text{ micron} = 1\ \mu m = 10^{-6} \text{ m} = 10^4\ °A$$

In the figure given below the cross hatched thermal radiation region extends over wave lengths from $0.1\ \mu m$ to $1000\ \mu m$. Outside, this region thermal radiation, exists but usually has negligible energy.

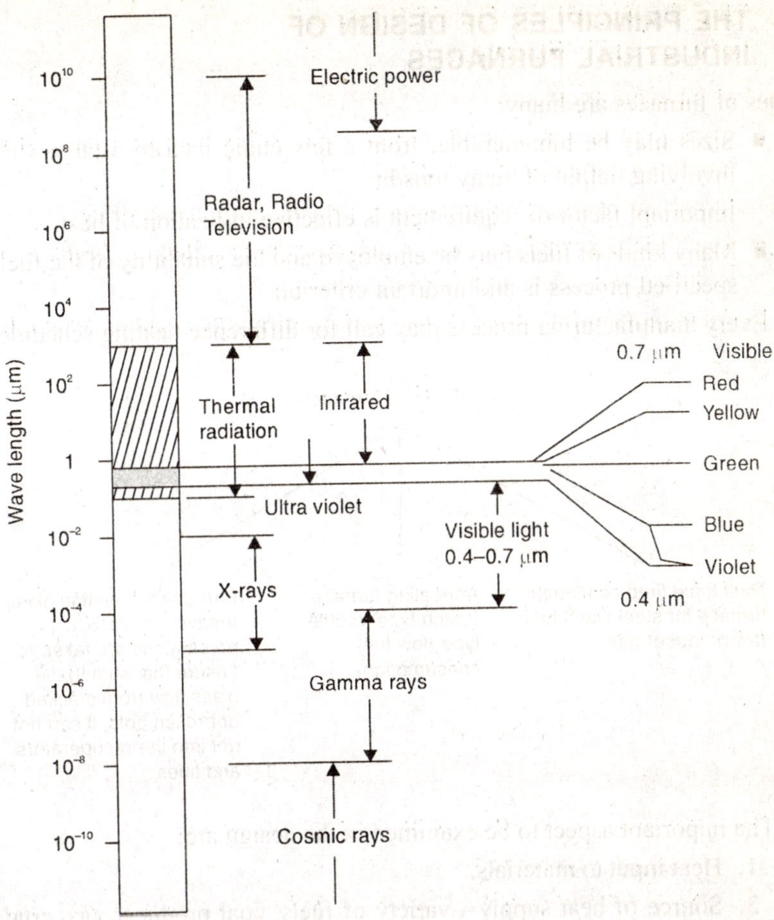

Fig. 7.3 *The electromagnetic wave spectrum*

Within the thermal radiation region in the above figure is a thin black region, $0.4 \leq \lambda \leq 0.7\ \mu m$, of visible radiation, where the human eye can detect radiation as colours of light. As we shall see, the eye can pick up radiation only from rather hot sources $T > 800° K$ such as the sun, lamp filaments or molton and metals. The 'colours' we see on ordinary objects are actually reflected radiation from these hot sources. Colder bodies, $T < 800° K$ emit radiation in the infrared range, $1 \leq \lambda \leq 1000\ \mu m$, which can be detected only by special optical means. The fact that the visible range is so narrow leads to some misleading beliefs about the effectiveness of a radiating surface, black surfaces being thought of as 'good' and the white as 'poor'. This holds good only in the visible range. Both white and black surfaces are excellent radiation emitters and absorbers in the infrared range where many practical engineering problems occur.

7.6 THE PRINCIPLES OF DESIGN OF INDUSTRIAL FURNACES

Types of furnaces are many:

- Sizes may be innumerable, from a few cubic inch to a large structure involving output of many tons/hr

  Important factor or requirement is effective utilization of heat.
- Many kinds of fuels may be employed and the suitability of the fuel for a specified process is an important criterion.

Every manufacturing process may call for difference heating schedule:

Types of furnaces

- Rust triple fired continuous furnace for steel slabs fuel: raw producer gas
- Annealing furnace (batch type) muffle type (low fuel consumption)
- Furnace for melting glass (heavily insulated) precautions are taken to ensure that should the glass flow from cracked or broken pots, it can not get into the recuperators and flues

The important aspect to be examined in the design are:

1. Heat input to materials.
2. Source of heat supply—variety of fuels, coal producer gas, crude oil tar or electrical heating.

   The rate at which the fuel can be burnt fixes the size of the combustion chamber.
3. Radiation and other heat losses.
4. Thermal balance calculations– main step in the design of a furnace.
5. Control Equipment—to monitor combustion conditions-temperature control, gas pressure and calorific value, flow meter for the quantity of gas and air used.

8

Evaporation

The objective of evaporation is to fortify or concentrate a solution consisting of a non-volatile solute and a volatile solvent which, in most of the situations, is water. Evaporation is carried out by vapourizing a portion of the solvent to produce a concentrated solution or thick liquor and it differs from drying which yields a solid product. It differs from distillation also in which the vapor is usually a single component. It differs from crystallization, in that, emphasis is placed of concentrating a solution rather than forming or growing crystals. But evaporation will be very close to crystallization in situations such as evaporation of brine.

Normally in evaporation the thick liquor is the valuable product and the vapour is condensed and discarded except in the production of potable wate from sea water.

8.1 FACTORS INFLUENCING THE EVAPORATION PROCESS

Characteristics of the Liquid

The practical solution of an evaporation problem is largely affected by the nature of the liquor to be concentrated. The important properties are:

Concentration

Dilute solution fed many have the same physical properties as those of water but as the concentration increases, the solution becomes more and more individualistic. Density and viscosity increase and the liquor may become too sluggish for heat transfer, the boiling point also increases with increase in concentration.

Foaming

Organic solutions foam during evaporation process resulting in heavy entrainment.

Temperature Sensitivity

Many fine chemicals, pharmaceuticals and food products are sensitive to heat. In these cases, reduced temperature and time of heating are needed.

Scale Formation

Some solutions deposit scales which impede the evaporation process culminating sometimes in the shutting down of the evaporator.

Material of Construction

Normally cast iron and steel are used. But corrosion and contamination may dictate the use of other materials Cu, Ni, Ss, Al, Pb and graphite etc are also used.

Type of Evaporators

Main types of steam-heated tubular evaporator in use presently are:

Vertical

1. Short-tube evaporator (stad.type)
2. Long-tube vertical evaporator
3. Horizontal tube evaporator- vapor inside tubes
4. Agitated film evaporator
5. Apparatus with heating medium in jackets
6. Apparatus heated by direct fire-steam boilers long.

Long-tube vertical evaporators may be categorized as

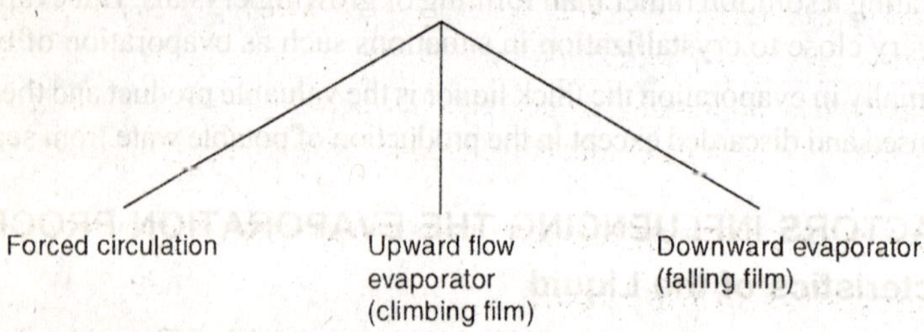

Fig. 8.1 *Classification of long-tube vertical evaporators*

8.2 ONCE THROUGH AND CIRCULATION EVAPORATOR

In once through operation the feed liquor passes through the tubes only once, releases the vapour and leaves the unit as thick liquor. Agitated film evaporators are always once through; falling and climbing film evaporators can also be operated in this way.

Once through evaporators are especially useful for heat sensitive materials. By operating under high vacuum the temperature of the liquid can be kept low. With a single rapid passage through the tubes the thick liquor is at the evaporation temperature and can be quickly cooled as soon as it leaves the evaporator.

In circulation evaporator a pool of liquid is held within the equipment. Incoming feed mixes with the liquid from the pool and the mixture passes through the tubes. Only part of the total evaporation occurs in one pass. All short tube and forced circulation evaporators are operated in this fashion.

We need to know the description and principle of working of the different types of evaporators and accessories like condensers vacuum pumps, traps etc.

Performance of an Evaporator

The principal measures of the performance of a steam-heated tubular evaporator are (i) the capacity (ii) the economy

Capacity is defined as the number of lbs of H_2O vapourized /hour

Economy is the number of lbs vapourized per lb of steam fed to the unit. In single effect evaporator, the economy is always less than one but in multiple effect equipment it may be considerably greater.

$$\text{Steam Consumption}\left(\text{lbs/hr}\right) = \frac{\text{capacity}}{\text{economy}}.$$

8.3 EVAPORATOR CAPACITY

The rate of heat flow 'q' through the heating surface of an evaporator is given by $q = UA\,\Delta T$.

If the feed to the evaporator is at the boiling point corresponding to the absolute pressure in the vapor space, all the heat transferred through the heated surface is available for evaporation and the capacity is proportional to q; If the feed is cold the heat required to bring its temperature to the boiling point may be quite large and the capacity for a given value of q is reduced accordingly. Conversely if the feed is at a temperature above the boiling point in the vapor space a portion of the feed evaporates spontaneously by a adiabatic equilibration with the vapor space pressure and the capacity is greater than the corresponding q. This process is called flash evaporation.

The actual temperature drop across the heating surface depends on the solution being evaporated, the difference in pressure between the steam chest and the vapour space above the boiling liquid and the depth of liquid over the heating surface. When the solution has the characteristics of pure water, its boiling point can be read from the steam tables, if the pressure is shown. In actual evaporators, however, the boiling point of a solution is affected by two factors: boiling point elevation and liquid head.

8.4 BOILING POINT ELEVATION AND DUHRING'S RULE

The vapor pressure of most of the aqueous solutions is less than that of water at the same temperature. Consequently for given pressure the boiling point of the solutions is higher than that of pure water.

The increase in boiling point of the solution over that of water is known as the boiling point elevation of the solution. The boiling point elevation must be subtracted from the temperature drop that is predicted from the steam tables.

For strong solutions the boiling point elevation is best found from an empirical rule known as Duhring's rule which states that the boiling point of a given solution is a linear function of the boiling point of pure water.

Effect of Liquid Head and Friction

The boiling point of the liquid layer at the surface corresponds to the pressure at the vapour space. When the depth of the liquid is appreciable the boiling point at the bottom layer is higher than that at the top. In addition when the velocity of the liquid is large frictional loss in the tubes further increases the average pressure of the liquid. The increase in boiling point decreases ΔT and hence the capacity.

Evaporator Calculations— Heat and Material Balance

Single Effect

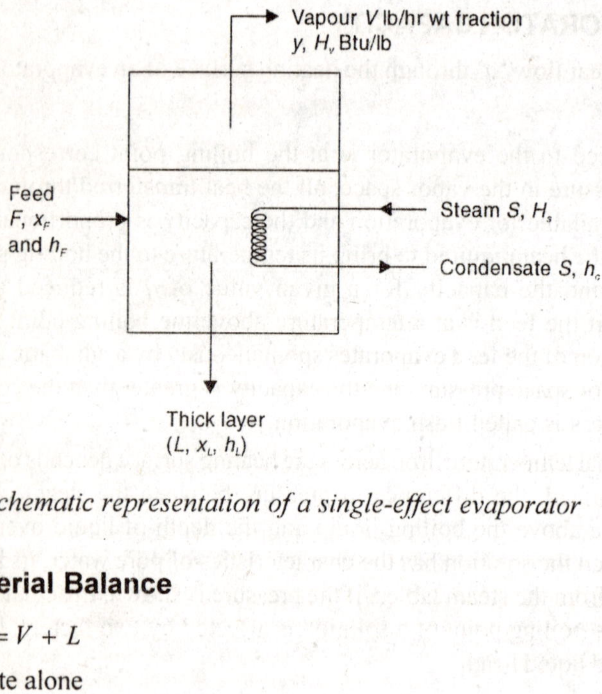

Fig. 8.2 *Schematic representation of a single-effect evaporator*

Total Material Balance

$$F = V + L$$

For the solute alone

$$Fx_F = Lx_L + Vy$$

Heat balance is given by:

(Heat in feed) + (heat in steam) = (Heat in thick liquor) + (heat in vapour) + (heat in condensate) + (heat lost by radiation)

(Heat lost by radiation is normally negligible.)

that is $\quad Fh_F + SH_S = VH_v + Lh_L + Sh_c$

or $$S(H_s - h_c) = VH_v + Lh_L - Fh_F$$

but $$V = F - L$$

i.e $$S\lambda_s = VH_v + Lh_L - Fh_F$$

$$\therefore q = S\lambda_s = VH_v + Lh_L - Fh_F$$

$$= (F - L)\, H_v + Lh_L - Fh_F$$

8.5 SINGLE AND MULTIPLE EFFECT EVAPORATION

Most evaporators are heated by steam condensing on metal tubes. Nearly always the materials to be evaporated flows inside the tubes. Usually steam is at low pressure say below 40 lbs/in^2 abs. Often the boiling liquid is under a moderate vacuum upto about 28″ Hg. Reducing the boiling point of the liquid increases the temperature difference between the steam and the boiling liquid and thereby increases the heat transfer rates in the evaporator.

When single evaporator is used, the vapour from the boiling liquid is condensed and discarded. The method is called single effect evaporation; it does not utilize the steam effectively. To evaporate 1 lb of water from a solution, 1000 to 1200. Btus is required and this calls for 1 to 1.3 lb of steam. If the vapour from one evaporator is fed into the steam chest of a second evaporator (the vapour from the 2nd is sent to the condenser), the operation becomes double-effect. The heat in the original steam is reused in the 2nd effect and the evaporation achieved by 1 lb of steam fed to the 1st effect is approximately doubled.

The general method of increasing the evaporation per lb. of steam by using a series of evaporators between the steam supply and the condenser is called multiple effect evaporation.

Illustration

A given evaporator is to be fed with 10000 lbs/hr of a solution containing 1% of solute by weight. The feed is at a temperature of 100°F. It is to be concentrated to a solution of 1.5% solute by weight. The pressure in the vapour space is 1 atm. The heating surface is supplied with steam at 5 psig (227°F). What is the weight of vapour produced ? What is the total weight of steam required? If *U* is 250, what is the heating surface required? Assume that the elevation of boiling point is nil.

Solution

$F = 10000$ lbs $\quad x_F = 0.01$

Total solids in feed = 10000 × 0.01 = 100 lbs

Total water in feed = 10000 – 100 = 9900 lbs

$t_F = 100\,°F \quad h_F = 68$ Btu / lb (from steam tables)

$x_L = 0.015$

Total solids in thick liquor = 100 lbs

$$L = \frac{100}{0.015} = 6670 \text{ lbs}$$

$V = 10000 - 6670 = 3330$ lbs

$t_L = 212\,°F \qquad h_L = 180$ Btu / lb (steam tables)

$t_v = 212\,°F \qquad H_v = 1150$ Btu / lb (steam tables)

$t_s = 227\,°F \qquad H_S = 1156$ Btu / lb (steam tables)

$t_c = 227\,°F \qquad h_c = 195$ Btu / lb (steam tables)

$\therefore\ 10000 \times 68 + 1156S = 3330 \times 1150 + 6670 \times 180 + 195S$

$\therefore\ S = 4530$

$\therefore$ Total heat to be transferred

= 4530 (1156 − 195)

= 4,350,000 Btu

The heating surface required

4,350,000 = 250 × A (227 − 212)

$$A = 1160 \text{ ft}^2.$$

Illustration

The above evaporator is fed with 10,000 lbs/hr of a 20% solution of NaoH at 100°F. This is to be concentrated to 50% solution. The evaporator is supplied with saturated steam at 5 psig and operates with the vapour space at a vaccum of 26″ Hg referred to a 30″ barometer. There is negligible entrainment. Condensate may be assumed to leave at the steam temperature and losses by radiation are to be ignored. What is the steam consumption? If U is 400, what the heating surface required?

The evaporator is provided with countercurrent contact condenser fed with H_2o at 60°F, The exit temperature of water is 120°F. How much cooling water is required?

Solution

$F = 10000$ lbs $\quad x_F = 0.2$

$t_F = 100°F \qquad x_L = 0.50 \qquad y = 0.0$

From steam tables

$H_s = 1156$ Btu / lb , $h_e = 196$ Btu/lb

$t_s = 228°F$

Boiling point of water at 4″ Hg = 125.4

Enthalpy of saturated steam at 125.4°F = 1116 Btu/lb (steam tables)

Boiling point of concentrated solution = 198.0 (from Duhring's chart)

$h_F = 56.5$ Btu / lb (enthalpy-concentration chart)

$h_L = 222$ Btu / lb (enthalpy-concentration chart).

Material Balance

$$F x_F = L x_L + V y$$

$$10000 \times 0.2 = 0.5 \times L + 0$$

$$\therefore L = 4000 \text{ lbs}, V = 6000 \text{ lbs.}$$

Enthalpy Balance

$$F h_F + S(H_s - h_c) = V H_v + L h_L$$

While calculating H_v, the enthalpy of vapour leaving the solution, it must be borne in mind that, this vapour is in equilibrium with the boiling solution at a pressure of 4" abs. and is superheated in comparison with vapour in equilibrium with H_2o at the same pressure. The specific heat of super-heated steam in this range may be taken as 0.46.

$$H_v = 1116 + (0.46)(198 - 125) = 1149 \text{ Btu/lb}$$

$$\therefore 10000 \times 56.5 + S(1156 - 196)$$

$$= 6000(1149) + 4000 \times 222$$

$$\therefore S = 7520 \text{ lbs / hr.}$$

Heating Surface

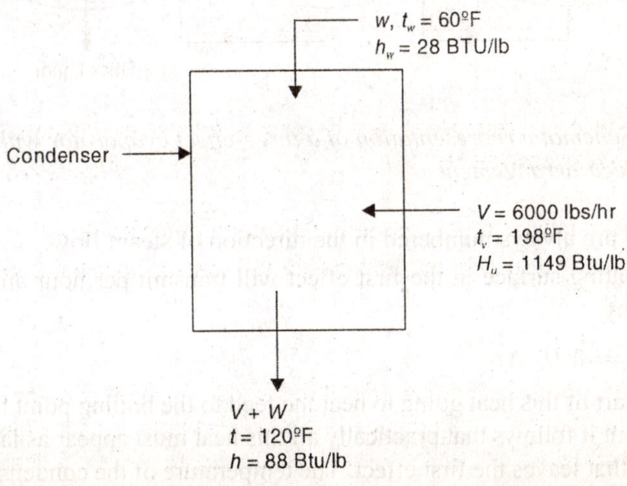

Fig. 8.3 *Condenser enthalpy balance*

$$7520\,(1156 - 196) = 400 \times A \times (228 - 198)$$

$$A = 600 \text{ sq. ft.}$$

Condenser Enthalpy Balance

$$VH_v + w\,h_W = (w + V)\,h$$

$$1149 \times 6000 + 28\,w = (w + 6000)\,88$$

$$w = 106{,}100 \text{ lbs/hr}$$

Multiple Effect Evaporation

The discussion on evaporation given above has been based on single effect evaporation. A modification of this system called multiple effect evaporation is commonly used in large scale operations—for the purpose of getting highest steam economy than those possible in single effect operations. The following sketch shows three-effect evaporator with forward feed arrangement.

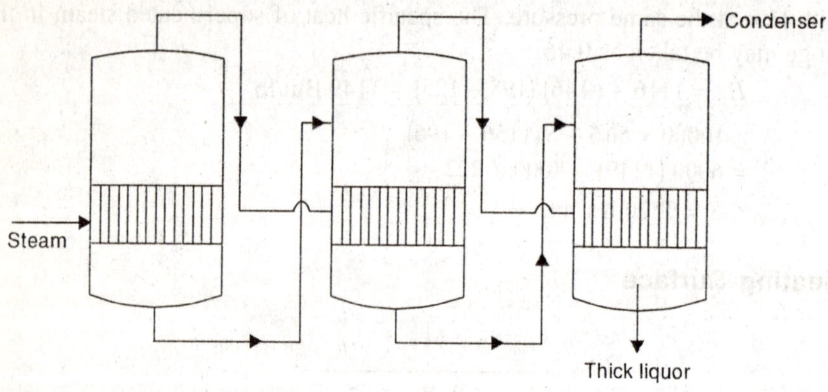

Fig. 8.4 *Schematic representation of a three-effect evaporator with forward feed arrangement*

Effects are always numbered in the direction of steam flow.

The heating surface in the first effect will transmit per hour an amount of heat given by

$$q_1 = A_1\,U_1\,\Delta t_1$$

If the part of this heat going to heat the feed to the boiling point is neglected for a moment it follows that practically all this heat must appear as latent heat in the vapour that leaves the first effect. The temperature of the condensate leaving 2nd effect very near temperature t_1 of the vapour from the boiling liquid in the first effect. Therefore all the heat that was expended in creating vapour in the first effect must be given up when same vapour condenses in the 2nd effect. The heat transmitted in the 2nd effect is given by the equation

$q_2 = U_2\ A_2\ \Delta t_2$: but q_1 and q_2 are nearly equal and

$A_1\ U_1\ \Delta t_1 = U_2\ A_2\ \Delta t_2$

The same reasoning when extended more effects shows that

$$A_1\ U_1\ \Delta t_1 = A_2\ U_2\ \Delta t_2 = A_3\ U_3\ \Delta t_3$$

The above two equations are only approximate equations and need to be corrected by inclusion of some more terms.

In ordinary practice, the heating surface in all effects of a multiple effect evaporator are equal for reasons of economy in construction.

Therefore, $U_1\ \Delta t_1 = U_2\ \Delta t_2 = U_3\ \Delta t_3$

From this it follows that the temperature drops in a multiple effect evaporator are approximately inversely proportional to the overall heat transfer coefficient.

Economy of Multiple Effect Evaporators

Since the latent heat necessary to evaporate a lb. of H_2o, at t is very nearly the same as the latent heat liberated in condensing a lb of steam at t_o the 1st effect of a multiple effect evaporator will evaporate approximately a lb. of H_2o for every lb of steam condensed. This is almost true if the feed to the 1st effect is nearly at the boiling point. It may be a bit erroneous if the feed is very cold. It follows that every lb. of vapour evaporated in the 1st effect and condensed in the 2nd effect, will evaporate approximately a lb. of H_2o in the 2nd effect and similarly for every lb. of H_2o evaporated in the 2nd effect, approximately a lb of water is evaporated in the 3rd effect. Since multiple effect evaporator may have any number of effects, it follows that in a N-effect evaporator there will be approximately N-lb of H_2o evaporated per lb of steam supplied. This relationship is only approximate and should not be used in actual calculations. Minor terms excluded in the above line of reasoning have an increasingly important effect as the number of effects increase. However the statement is approximately true and hence it follows that the reason for operating an evaporator in multiple effect is to secure increased steam economy.

Multiple Effect Calculations

The results usually required from calculations for a multiple effect evaporator are the amount of steam to be used, the amount of heating surface required, the approximate temperature in the various effects and the amount of vapour leaving the last effect. As in the case of single effect, the important relationship available are material balances, heat balances and the heat transfer equation, In the case of a multiple effect evaporation, a trial and error method must be employed rather than a direct algebraic solution.

Consider for instance a triple effect evaporator operation on a liquid whose boiling point elevations are negligible. There are seven equations that can be formulated. A heat balance for each effect, a rate equation for each effect, and the sum of evaporation in three effects. There are seven unknowns in these equations (i) steam to the 1st effect (ii), (iii) and (iv) evaporation in each effect (v) the temperature of boiling liquid in the first effect, (vi) temperature of boiling liquid in the 2nd effect and (vii) the heating surface in each effect. This is based on common situations where the heating surface of all the effects is to be equal. These even equations conceivably might be solved for the seven unknowns but this method is impracticable.

One has to go through the stepwise procedure for solving the above problem.

8.6 THE OCCURRENCE OF HEAT TRANSFER IN EVAPORATORS

As mentioned earlier, evaporation is the separation of water from a solution of a nonvolatile solutes. Since vapourization is usually effected by supplying heat to the solution, evaporation is sometimes considered a special case of heat transfer.

First it will be worthwhile to look at the various devices which are used to effect this separation. Several evaporator types are available. There are great differences in the performances characteristics of these units.

In the table given in the following section qualitative comparison between various steam-heated tubular evaporator (the most common type) are shown. This table shows the heat transfer characteristics of each type and indicates the kind of solution which can be used as a guide in the preliminary selection of equipment. Where two or more types, are indicated as satisfactory, the choice will be based on cost factors.

It is worthwhile to note that the short-tube natural circulation evaporators are gradually being displaced by long-tube evaporators having either forced circulation or natural circulation. The falling film and agitated film evaporators are essentially speciality evaporators for handling temperature sensitive materials.

In recent years fuel costs have steadily risen, and if this trend continues vapour-recompression systems may prove to be economical. In this type of evaporator the latent heat of vapourization is reused and good heat economy can be attained. However initial investment is high and the capacity rather low for present designs.

In this section attention will be given to steam heated evaporators. Feed handling capacity of an evaporator is the feed rate in lbs/hr which can be introduced and still effect a given change in concentration. This capacity may be expressed in terms of its ability to transfer heat. In turn, ability to transfer heat is dependent on (i) heat transfer coefficient attainable; (ii) area available; (iii) effective temperature difference available for heat transfer; (iv) physical properties of the solution such as latent heat, boiling point elevation and; (v) material balance.

Effective Temperature Difference

Potential for heat transfer in evaporation as in other heat transfer processes is a difference in temperature. In an evaporator heat is transferred to a liquid which is at or near the boiling point. The temperature drop which is useful for the transfer of heat is the difference in temperature of the source of heat (usually condensing steam) and the temperature of the boiling liquid.

Overall available temperature drop is made up of three parts: (i) that which is required to transfer heat (ii) that which is required to overcome the boiling point elevation caused by the presence of the solute: and (iii) that which is required to overcome the elevation in temperature super imposed by the hydrostatic head. In the evaporation process sufficient temperature differential must be provided to meet all these requirements.

The effects of boiling point elevation in reducing that portion of the available temperature difference which is effective in the transfer of heat are shown in the figure given below.

Overall Temperature Drop

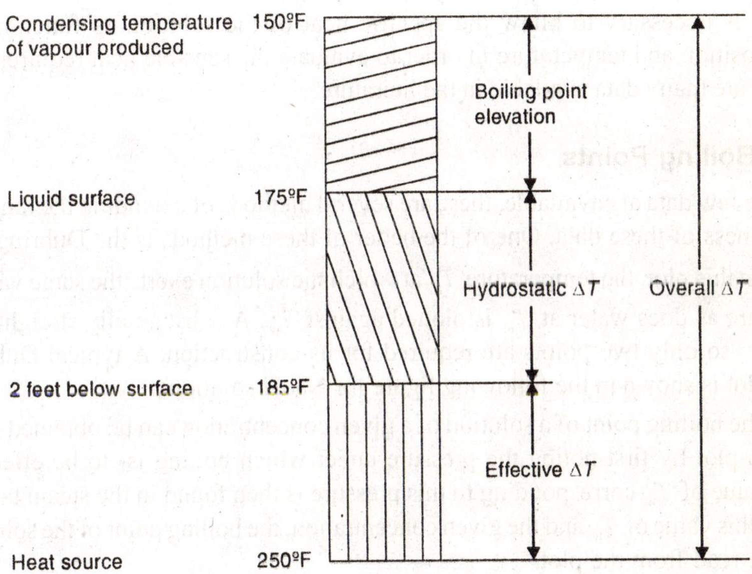

Fig. 8.5 *Effect of boiling point on heat transfer*

The temperatures shown here for heat being transferred to a 30% NaoH solution by steam at 250°F vapour is being produced at 7.6 in of mercury absolute which corresponds to a saturation temperature of 150°F. Average depth of the liquid below the surface is 2 feet.

With no agitation, the temperature of the liquid below the surface would be greater than that near the surface. Consequently the liquid below the surface is superheated with respect to that at the surface. However, boiling is usually accompanied by vigorous agitation and this superheated liquid is quickly brought to the surface where it loses its superheat by flashing. Vigorous agitation tends to make the temperature uniform and equal to surface temperature. Consequently the effect of hydrostatic head on boiling point elevation is more imaginary than real. In any case, the absolute magnitude of this effect cannot be evaluated until the final selection has been made.

For these reasons, in the remainder of this section, boiling point elevation due to hydrostatic head will not be considered. Thus the over-all ΔT will be treated as consisting of: (i) that necessary to transfer heat and (ii) that necessary to overcome boiling point elevation due to the presence of the solute.

The Physical Properties Involved

To effect vapourization, it is necessary that the heat of vapourization be supplied. Further since the solution is not always at its booling point, its sensible heat must be accounted for. These can be examined as follows.

It is necessary to know the specific heat of the solution as functions of composition and temperature in order to evaluate the sensible heat requirement. There are many data available in the literature.

The Boiling Points

Where few data are available, there are several methods of extending the range of usefulness of these data. One of the better of these methods is the Duhring line plot. In this plot, the temperature T_L at which the solution exerts the same vapour pressure as does water at T_w is plotted against T_w. A substantially straight line results; so only two points are required for its construction. A typical Duhring line-plot is shown in the following figure for NaoH solutions.

The boiling point of a solution of a given concentration can be obtained from such a plot by first noting the pressure under which boiling is- to be effected. The value of T_w corresponding to this pressure is then found in the steam tables. From this value of T_w and the given concentration, the boiling point of the solution can be read from the plot.

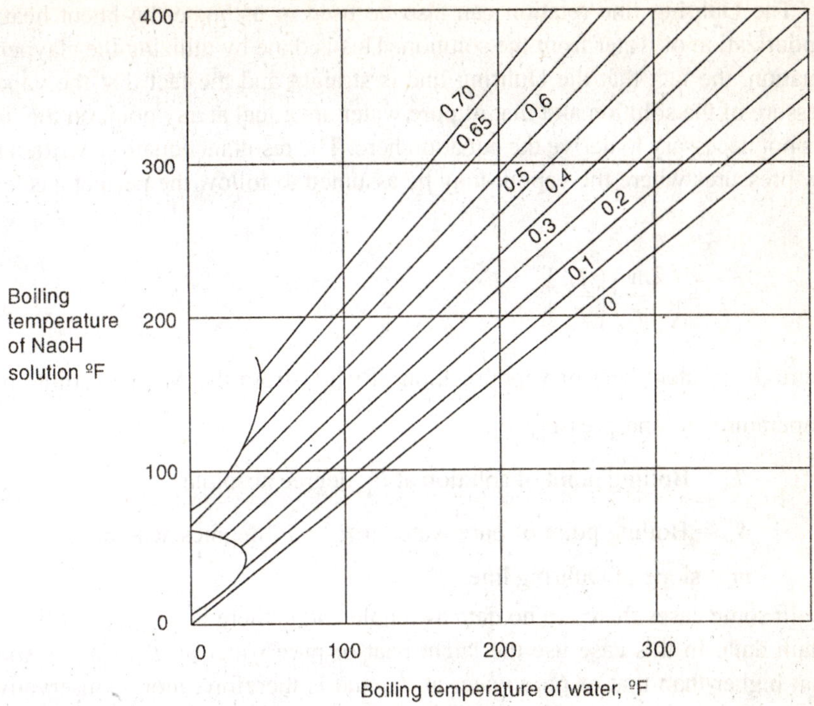

Fig. 8.6 *Duhring plot for NaoH solution*

Boiling Point Elevation

At any pressure, $(T_L - T_w)$ is the boiling point elevation and is designated here as B.P.E. This term is important in determining how much of the overall ΔT is effective for the transfer of heat.

Since the Duhring plot is a straight line, it may be represented by the equation

$$T_L = m\,T_w + b \qquad \text{...(i)}$$

where m is the slope of the line and b is a constant. It this fact is recognized., it becomes unnecessary to construct the Duhring line plot, since m and b can be easily obtained from the same points which would be used to construct the line.

By subtracting T_w, from both sides of Eqn (i), the boiling point elevation is obtained as

$$B.P.E. = (m-1)\,T_w + b \qquad \text{...(ii)}$$

Eqn (ii), shows that, even for a constant composition, the boiling point elevation is a function of pressure, because T_w is a function of pressure. This fact is often overlooked.

The Duhring line relation can also be used to estimate the latent heat of vapourization of water from the solution. This is done by utilizing the clayperon equation, the fact that the Duhring line is straight and the fact that the vapour pressure of the solution and that of pure water are equal at any point on the line. It is not necessary to derive the equation here. The resultant equation, written for low pressures where the vapour may be assumed to follow the perfect gas laws is

$$\lambda_L = \left(\frac{\lambda w}{m}\right)\left(\frac{T_L}{T_w}\right)^2 \quad ...(3)$$

where λ_L = latent heat of vapourization of water from the water in solution at temperature T_L and pressure P_s.

T_L = Boiling point of solution at P_s, degree absolute

T_w = Boiling point of pure water at $P_w = P_s$ degrees absolute .

m = slope of Duhring line.

In some cases there are no data available and is there is insufficient time to obtain data. In this case use the latent heat of pure water at T_w. This is some what higher than that of pure water at T_L and is therefore, more conservative. However, the true latent heat of water from solutions is higher than that of pure water at either T_L or T_w except for solutions which cool when the solute is dissolved.

Heat of Solution

Data on heat of solution of nonvolatile solutes in aqueous solutions are few and are not easily obtained with accuracy. Where such data are available, it is unnecessary to estimate the latent heat of vapourization. Latent of heat of vapourization of water from a solution of non volatile solute in water is

$$\lambda_L = H_g - H_L$$

where H_g is enthalpy of steam at T_L and P and H_L is the partial specific enthalpy of water in solution at the existing concentration and at the temperature T_L and pressure P.

Heat Transfer Coefficients

In steam-heated evaporators, as in other heat transfer-devices, the total resistance to heat transfer is the sum of five individual resistances

$$R_T = R_o + R_{Do} + R_w + R_{Di} + R_i \quad ...(iv)$$

where R_T = total resistance

R_o = Resistance of condensing steam

R_{Do} = Fouling resistance on steam side of tubes

R_m = Resistance of metal wall

R_{Di} = Fouling or scaling resistance boiling side of tubes

R_i = Resistance of boiling side film.

These resistances are related to the heat transfer coefficients thus:

$$R_T = \frac{1}{h_o\ A_o} + \frac{1}{h_{Do}\ A_o} + \frac{x_m}{k_m\ A_{av}} + \frac{1}{h_{Di}\ A_i} + \frac{1}{h_i\ A_i} \quad \text{...(v)}$$

$$R_i = \frac{1}{U_i\ A_i} = \frac{1}{U_o\ A_o}$$

Here h_0 – Heat transfer coefficient for condensing steam (usually on outside of tube area) Btu/hr ft^2 °F

h_{Do} – Fouling heat transfer coefficient on steam side of tubes Btu/hr ft^2 °F

h_{Di} – Fouling heat transfer coefficient on boiling side of tubes Btu/hr ft^2 °F

h_i – Heat transfer coefficient for boiling side (film,) Btu/hr ft^2 °F

x_m – Thickness of metal wall ft

k_m – Average thermal conductivity of metal wall, Btu/hr ft^2 °F

A_i – Inside area of tubes, ft^2.

A_0 – Outside area of tubes ft^2

A_{av} – Average area of tubes. ft^2.

U_i – Overall heat transfer coefficient based A_i Btu/hr ft^2 °F

U_o – Overall heat transfer coefficient based on A_0 Btu/hr ft^2 °F

Evaluation of these coefficients has been explained earlier in this book. By experimentation and with the aid of theory, we already have dependable methods of evaluating all film coefficients except the boiling side coefficient h_i . In the case of boiling, we do not have a completely general relation for prediction of h_i .

Most reliable relations for h_i are probably those for the forced circulation evaporator. For this type of unit most of the heat is transferred to liquids at high velocities on the order of 6 to 18 feet/sec and under sufficient pressure so that a minimum of boiling takes place within the tubes. Most of the vapourization occurs near the tubes exit by flashing. The following equation is found to be satisfactory

$$\frac{hD_i}{k} = 0.027\left(\frac{DG}{\mu}\right)^{0.8}\left(\frac{C_p\ \mu}{k}\right)^{0.4} \quad \text{....(vi)}$$

For long tube natural circulation evaporator with the liquid flowing upward through the tubes, the velocity is also probably quite high. Possibly an equation such as Equation (vi) may be satisfactory. However since the velocity is induced by vapourization, it is difficult to say just what velocity should be used or how it could be determined.

Short-tube natural circulation evaporators offer a more difficult problem than do forced circulation evaporators in the prediction of a boiling side coefficient. It has been shown that h_i is dependent on the temperature drop across the boiling film. In fact h_i seems to vary with the temperature drop raised to about the third power.

Viscosity of the liquid also has on important effect. The coefficient decreases with increasing viscosity. It is hard to separate the effects of Δt and viscosity on h_i. It may be that the effect of variations in these on the velocity of circulation is really important.

Other factors which are important are surface tension, liquid density and vapour density as well as specific heat and thermal conductivity. Attempts have been made to propose equations which show quantitatively the effects of these variables on the boiling side coefficient with little success. It may be that modern statistical design of experiments assisted by computers will crack this problem.

Because of the uncertainly in estimating the value of the boiling side coefficient, most data are reported in terms of the overall heat transfer coefficient which is determined experimentally. Examination of the following equation shows that the numerical value of the overall coefficient depends on the temperature drop as well as the area upon which it is based.

$$q = U_i\ A_i\ \Delta t = U_o\ A_0\ \Delta t \qquad \text{...(vii)}$$

As has already been pointed out, the overall temperature drop is not entirely available for transfer of heat. Some of it is lost as a result of boiling point elevation and hence is not effective in the transfer of heat. In Equation (vii) Δt is the difference between the condensing temperature on the steam side of the tubes and the boiling temperature of the solution at the existing concentration and under the pressure existing in the vapour space.

$$\Delta t = t_v - t_L$$

Where t_v is the temperature of condensing steam at saturation at the pressure existing in the steam chest and t_L is the temperature of the boiling liquid at the existing concentration and under the pressure existing in the vapour space.

The overall coefficient evaluated in this way is sometimes referred to as the coefficient corrected for boiling point elevation. On the other hand, Δt in Equation (vii) is sometimes used as the difference in the temperature of the condensing steam at saturation at the pressure existing in the steam chest and the temperature of saturated steam in the pressure existing in the vapour space of the evaporator. Coefficients defined in this way are called apparent heat transfer coefficients.

Use of apparent coefficients is not justified, but if they are used then the approximate Δt upon which they are based must also be used. Overall coefficients for boiling point elevation can be used in the analysis of the heat transfer equations for evaporators.

Typical range of data on heat transfer coefficients for various types of evaporators are shown in the table given below

Table 8.1 *Range of heat transfer coefficients in evaporators.*

	Overall coefficient Btu/hr ft² °F
Long tube vertical evaporators	
Natural circulation	200–600
Forced circulation	400–2000
Short-tube evaporators	
Horizontal type	200–400
Calendric type	150–500
Agitated-film evaporators	
Newtonian liquid	
1 centi poise	400
100 centi poise	300
10,000 centi poise	120

For natural circulation evaporators, the boiling side coefficients is greatly dependent on the temperature drop across the boiling side film. Therefore, it is not surprising to find that in equipment such as the short tube natural circulation evaporator, the temperature drop affects the vigour of boiling and the overall temperature coefficient. Also since viscosity is a function of temperature, the higher the boiling temperature at a given concentration of solution, the higher will be the coefficient.

The characteristics of some of the tubular evaporators are given in Table 8.2.

8.7 THE ANALYSIS AND PERFORMANCE CALCULATION OF MULTIPLE-EFFECT EVAPORATORS

Evaporators in processing operations may operate as single effect or multiple effect. Choice of arrangement reflects steam economy or evaporator capacity. In single operation, the heat supplied in the steam is used only once. Resulting economy, expressed lb vapour/lb. steam is very poor and is about 0.8 for a cold, feed. However, capacity is usually highest for single effect operation and is expressed as vapour weight per unit heat transfer area.

Table 8.2 *Characteristics of steam-heated tubular evaporators*

Evaporator type	*Relative cost*	*Heat Transfer characteristics*					*Suitability for use with various liquor characteristics*					
		Relative U in liquid	Relative U in visc. liquids	Entrance vel. of liquid	Residence time	Once through or recirculation	Thin	Visc.	Foaming	Scaling	Salting	Temp sensitive
Short tube natural circulation	Least	High	low	low 1 to 3 fps	High	Recirculation	Good	Poor	Poor	Fair	Fair	Poor
Forced circulation	Higher	High	High	High 6 to 18 fps	Mod. 1 to 3 sec	Either	Good	Good	Good	Good	Fair	Fair
Long tube, natural circulation (Upward flow)	Less than forced circulation	High	Mod. high	High	Short	Either	Good	Fair	Good	Fair	Poor	Fair
Falling film	Moderate	Fair	High	–	Short	Either	Good	Good	–	–	–	Good
Agitated film	High	High	High	–	–	Once through	Good	Good	–	–	–	Good

In multiple effect operation several evaporators are connected by appropriate piping so that vapour passes from one effect to the next in-series. Hence the heat supplied to the first effect is used to vapourize water in the first effect. This vapour in turn, passes to the next effect until finally the heat in the vapour supplied to the last effect passes to the condenser. The net result of this arrangement is multiple reuse of heat and a marked increase in the economy of the system. A rough approximation of the economy can be obtained by multiplying the number of effects by 0.8.

In addition to the saving in steam, there is also a saving in cooling water required to operate the last effect condenser. In the absence of detailed calculation a good rule of thumb is that about 30 lb of cooling water must be provided for each lb of steam supplied to the first effect. Sole advantage of multiple effect operation is the increased economy accompanied by the savings in steam and cooling water consumption for a given capacity.

We have already seen how the boiling point elevation reduces the effective temperature drop for heat transfer. In multiple effect operation this drop is cumulative. Thus, the effective total temperature is equal to the total overall temperature drop from the steam supplied to the first effect to the condensing temperature of the vapour from the last effect minus the sum of the boiling point elevations in all effects. This reduction in the effective temperature drop decreases the capacity.

In a multiple effect system the evaporators are connected in series. Consequently, any increase in the number of effects increases the resistance to heat transfer and decreases the capacity which is measured as lb vapour/ft^2. This means a increase investment to secure the same capacity for multiple effect as for single effect. Increased investment must be balanced against the savings in steam and condenser duty.

Analysis of Evaporator Calculations

In evaporator calculations, three relations must be satisfied (i) material balance (ii) heat balance and (iii) heat transfer rate equation. These relations must be applied to each evaporator in a multiple effect system and to the entire system as a whole.

Many multiple effect systems of operation are possible. The method of calculation is basically the same for each system. A multiple forward-feed triple effect system as shown in the sketch given below will be used to demonstrate the method. First the overall material balance on the entire unit and on the individual bodies must be made. The material balances are given in the figure.

Overall Material Balance

$$F - P = \sum E = E_1 + E_2 + E_3 + \cdots \qquad ...(1)$$

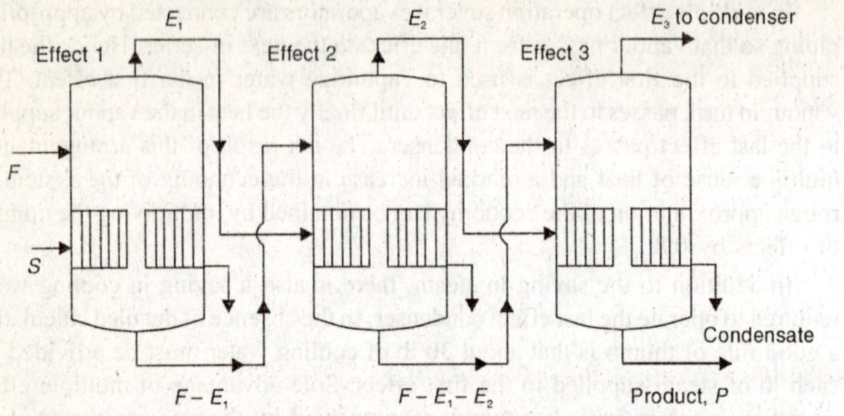

Fig. 8.7 *Triple effect evaporator—flow patterns for forward feed*

Product rate: $P = N_F \dfrac{F}{N_P}$ ...(2)

Total evaporation from all effects

$$\Sigma E = F\left(\frac{N_P - N_F}{N_P}\right) \quad ...(3)$$

Material balance for each effect are written to obtain the concentration of, the liquor leaving the effect.

$$N_1 = \frac{F N_F}{F - E_1} \quad ...(4)$$

$$N_2 = \frac{F N_F}{F - E_1 - E_2} \quad ...(5)$$

The equation similar to Equation (5) can be written for each proceeding effect.

The heat balance on each effect is written. These heat balances are best written in term of enthalpies if enthalpy-concentration diagram is available. However such charts are not always available. Since we have methods of calculating latent heats of water, the heat balance equations can be written without recourse to the enthalpy-concentration diagrams.

For Effect 1

$$S\lambda_S = F\,C_{p_F}\left(t_{L_1} - t_F\right) + E_1 \lambda_{L1} \quad ...(6)$$

For Effect 2

$$E_1\left[C_{pv_1}\left(t_{L_1}-t_{v_1}\right)+\lambda v_1\right]$$

$$= E_2\,\lambda_{L_2} - (F - E_1)\,C_{p_{L_1}}\left(t_{L_2} - t_{L_3}\right) \quad \text{...(7)}$$

For Effect 3

$$E_2\left[C_{pv_2}\left(t_{L_1}-t_{v_2}\right)+\lambda v_2\right]$$

$$= E_3\,\lambda_{L_3} - (F - E_1 - E_2)\,C_{p_{L_2}}\left(t_{L_2} - t_{L_3}\right) \quad \text{...(8)}$$

Also, the heat transfer equations are

$$q_1 = U_1\,A_1\,\Delta t_1 = U_1\,A_1\left(t_s - t_{L_1}\right)S\lambda_s \quad \text{...(9)}$$

$$q_2 = U_2\,A_2\,\Delta t_2 = U_2\,A_2\left(t_{v_1} - t_{L_2}\right)$$

$$= E_1\left[C_{pv_1}\left(t_{L_1}-t_{v_1}\right)+\lambda v_1\right] \quad \text{...(10)}$$

$$q_3 = U_3\,A_3\,\Delta t_3 = U_3\,A_3\left(t_{v_2} - t_{L_3}\right)$$

$$= E_2\left[C_{pv_2}\left(t_{L_2}-t_{v_2}\right)+\lambda v_2\right] \quad \text{...(11)}$$

In general the steps outlined in the following procedure can be used to solve these equations.

Step 1

An initial estimate of the temperatures in each effect is made. In making this estimate use the rough approximation of equal heat transfer rates in each effect.

$$U_1\,A_1\,\Delta t_1 = U_2\,A_2\,\Delta t_2 = U_3\,A_3\,\Delta t_3 \quad \text{...(12)}$$

$$\Sigma\Delta t = \Sigma\Delta t' - \Sigma PBE = \Delta t_1 + \Delta t_2 + \Delta t_3 \quad \text{...(13)}$$

$\Sigma\Delta t'$ = overall temperature difference from steam to condenser °F.

$$\frac{\Sigma\Delta t}{\Delta t_1} = 1 + \frac{U_1\,A_1}{U_2\,A_2} + \frac{U_1\,A_1}{U_3\,A_3} + \cdots$$

Note that only the ratios of the area need be known and not the actual areas. These can be defined as $R_{A_2} = \frac{A_2}{A_1}$ and $R_{A_3} = \frac{A_3}{A_1}$. In most cases these areas ratios are equal to unity but this is not necessarily so

$$\frac{\Sigma\Delta t}{\Delta t_1} = 1 + \frac{U_1}{U_2\,R_{A_2}} + \frac{U_1}{U_3\,R_{A_3}} \quad \text{...(14)}$$

use equation (14) to estimate Δt_1. From this $t_{L_1} = (t_s - t_1)$.

Temperature in the other effects are evaluated in a similar manner. Thus

$$t_{v_1} = t_{L_1} - (BPE)_1$$

$$\Delta t_2 = \frac{U_1 \Delta t_1}{U_2 R_{A_2}} = (t_{v_1} - t_{L_2}) \quad \text{...(15)}$$

$$t_{v_2} = t_{L_2} - (BPE)_2$$

$$\Delta t_3 = \frac{U_1 \Delta t_1}{U_3 R_{A_3}} = (t_{v_2} - t_{L_3}) \quad \text{...(16)}$$

Remember that these are rough estimates and as such serve as initial assumptions for solving Equations (6), (7) and (8). Having made an estimate of the temperatures and concentrations in each effect, then the latent heat of vapourization λ_L is evaluated by the methods already described. Heat balance equations and overall material balance equations are then solved simultaneously for $E_1, E_2, E_3 \cdots E_n$.

Step 2

Solve heat transfer Equations (9), (10) and (11) for the areas. Required areas ratios are then checked. In most common applications area ratios are unity.

Step 3

If the calculated area ratios do not check make new estimates of the values of Δt for each effect. This can be done by using the previously calculated areas as a guide. If the calculated area is too large, then Δt should be increased for that effect and vice versa.

Step 4

Using the values for E_1, E_2 and E_3 from step 1, make new estimates of the concentration in each effect. Using these new concentrations and values of Δt from step 3, get new values for boiling point elevations and temperature in each effect.

Step 5

Solve heat balance equations for more correct values of E_1, E_2 and E_3.

Even though poor initial assumptions were made, these new values will be very close to those obtained in the initial series of calculations.

Step 6

Solve the heat transfer equations for the areas in each effect and check the area ratios. If necessary, repeat the whole process but usually the second trial gives satisfactory results.

In summary, the usual method requires solving the various equations by successive approximations. This is not difficult but it would be time consuming and tedious.

Use of a Simplified Method

Much of the time required for making these calculations can be saved by recognizing that in a multiple effect evaporator heat is being transferred from the source through the evaporators in series. Finally, the heat discharged to the cooling water supplied to the condenser which receives the vapour from the last effect. Thus, the series resistance concepts can be employed here.

$$\Sigma q = \frac{\Sigma \Delta t}{R_{total}} \quad ...(17)$$

A convenient way for evaluating the terms in Equation (17) has been developed by investigators in this area. The key equations are

$$\Sigma q = \lambda_{av} \, \Sigma E \quad ...(18)$$

$$\lambda_{av} = \frac{F C_{PF} \left(t_{L_1} - t_F\right)}{\Sigma E} + \frac{\lambda_{L_1}}{b} \quad ...(19)$$

In equation (19), t_{L_1} is the temperature of the liquor in the effect the feed enters regardless of whether it is the first effect or not. The factor b equals $(1 + 0.1\, n)$ for the forward feed case and may be taken as unity for all other cases.

$$R_{total} = \frac{1}{U_e \, \Sigma A} \quad ...(20)$$

$$\lambda_{av} \, \Sigma E = U_e \, \Sigma A \, \Sigma \Delta t \quad ...(21)$$

$$\Sigma A = A_1 \left(1 + R_{A_2} + R_{A_3} + ...\right) \quad ...(22)$$

The effective coefficient is obtained from

$$\frac{U_1}{U_e} = \left(\frac{1 + R_{A_2} + R_{A_3} + \cdots}{n}\right)\left(1 + \frac{U_1}{U_2 R_{A_2}} + \frac{U_1}{U_3 R_{A_3}}\right) \quad ...(23)$$

where the area ratios are equal to unity, the term $\left(1 + R_{A_2} + R_{A_3} + \cdots\right)$ is equal to n and Equation (23) becomes.

$$\frac{1}{U_e} = \frac{1}{U_1} + \frac{1}{U_2} + \frac{1}{U_3} + \cdots \quad ...(24)$$

Steam consumption is estimated from the heat transferred in the first effect

$$q_1 = \Sigma E \left[\lambda_{av} - \lambda_{L_1} \frac{(n-1)}{n}\right] \quad ...(25)$$

Equation (25) gives the steam consumption in the first effect even though λ_{L_1} is evaluated at the conditions existing in the effect the feed enters. Having obtained q_1, steam consumption is calculated from

$$s = \frac{q_1}{\lambda_s} \qquad \text{...(26)}$$

The economy $\frac{\Sigma E}{s}$ is then easily obtained.

Illustration

A solution with a negligible boiling point elevation is to be concentrated from 10 to 50% solids in a triple effect evaporator. Steam is available at a gauge pressure of 15 psi and 249°F. Absolute pressure in the third effect is 4 in Hg. and corresponds to a boiling temperature of 125°F. Feed enters at a rate of 55,000 lbs/hr and 70°F.

Specific heat of the solution may be taken as 1.0 Btu/lb °F at all concentrations. Overall coefficients in Btu/hr ft^2 °F are estimated to be 550 in the first effect. 350 in the second and 200 in the third with forward feed. For backward feed, the corresponding coefficients are 450,350 and 275. Each effect is to have the same heating surface.

Calculate the amount of heating surface, the steam consumption and over-all economy (a) with forward feed and (b) with backward feed.

Solution

The overall evaporation rate E is found out by substituting Equation (3) which gives 44000 lbs/hr. Since boiling point elevations are negligible it is not necessary to calculate the concentrations of the liquor in each case. Since the areas are equal, it is found that R_{A_2} equals R_{A_3} and has a value of one.

From Equation (14), we estimate Δt_1, and then calculate.

$$\frac{\Sigma \Delta t}{\Delta t} = 1 + \frac{550}{350} + \frac{550}{200} = 5.32$$

Since the boiling point elevations are negligible

$$\Sigma \Delta t = (249 - 125) = 124° \text{ F}$$

$$t_{L_1} = 249 - \left(\frac{124}{5.32}\right) = 225.7° \ F$$

$$\lambda_{L_1} = 961.4 \text{ Btu/lb}$$

Substituting appropriate quantities in Equation (19) gives the average latent heat λ_{av} as 1127.5 Btu/lb. From Equation (23), we find the ratio of $\frac{U_1}{U_e}$ is 5.32.

Since the overall coefficient for the first effect is 550, U equals 103.0.

We can now find the total heating surface required for the evaporators by substituting appropriate quantities in Equation (21) to give an area of 3860 ft^2. Since each effect is to have the same heating surface, area of each effect is 1287 ft^2.

Substituting in Equation (25) gives 21.45 million Btu/lb as the heat transferred in the first effect. At 249°F the latent heat of steam is 946 Btu/lb. Hence steam consumption S is 22750 lbs/hr. Overall economy of the evaporator is 44000/22750 or 1.93 lb evaporated per lb steam. A better estimate of Δt_1 can now be obtained by using Equations (17) and (20).

$$\Delta t_1 = \frac{21{,}450{,}000}{550 \times 1287} = 30.3$$

Hence t_{L_3} equals (249 – 30.3) or 218.7°F.

Making several trials by the usual method employing simultaneous equations, we can get the following values

$$A_1 = A_2 = A_3 = 1202 \text{ ft}^2 \text{ / effect}$$

Overall economy = 1.96 lbs evaporated/lb steam

Steam consumption S = 22450 lb/hr

$$t_{L1} = 217° \text{ F}$$

The simplified method gives a somewhat high value for the area when the feed is cold but all answers compare well with those obtained by the usual method.

Backward Feed Case

In backward feed arrangement the feed enters the last effect and the concentrated product leaves the first effect. In Equation (19), t_{L_1} is 125°F and λ_{L_1} is 1022 Btu/lb. Hence average latent heat as computed from Equation (19) is 1097.1 Btu/lb.

From Equation (24), we calculate the ratio $\dfrac{U_1}{U}$ and find it to be 3.923.

Hence effective overall coefficient U_e equals $\dfrac{450}{3.923}$ or 114.8 Btu/hr ft^2 °F. Substituting in Equation (21) gives total heating surface of 3390 ft^2 or 1139 ft^2/effect.

Heat transferred in the first effect is obtained from Equation (25) even though λ_{L_1} and t_{L_1} are for last effect conditions. Substituting appropriate values in Equation (25) gives 18.35 million Btu/hr for the heat transferred. Dividing the heat transferred by latent heat of steam at 249°F gives steam consumption of 19,400 lb/hr. Overall economy is 44000/19400 or 2.26 lb evaporated/lb steam.

A better estimate for t_{L_1} now gives.

$$t_{L_1} = 249 - \frac{18{,}350.000}{450 \times 1130} = 213\ °F$$

By the exact method to the following values can be obtained.

$$A_1 = A_2 = A_3 = 1151\ \text{ft}^2\ /\ \text{effect}$$

$$s = 19150\ \text{lbs/hr}$$

Economy = 2.30 lbs evaporated/lbs steam

$$t_{L_1} = 214\ °F$$

Again the results check well with those obtained by the exact method. This problem is not intended to show the comparative performance of forward and backward feed. It shows only the method of calculation. For moderately warm feed the usual case the results differ from those obtained by the exact method by amount 2%. Maximum error found in any case would be about 7.5%.

This method should not be applied directly to systems in which steam is withdrawn from the vapour line between one or more of the evaporators of a multiple effect system. Radiation losses should be allowed for after application of this method.

9

Review of Heat Transfer Principles

The can be done by reviewing the elementary concepts of temperature and heat. These quantities are inseparables. According to the second law of thermodynamics heat will flow spontaneously only from points of higher temperature of those of relatively low temperature. Conversely temperature is an identifying characteristic of matter which is a measure of its ability to transfer heat.

For our purposes, we can visualize any molecule, whether in the fluid or solid state, as being in constant motion. However, in this case the motion is internal. It is vibration, rotation or oscillation of or within the molecule and need not result in displacement of one molecule relative to another. The higher the temperature of the molecule, the greater is its internal energy. Thus at higher temperatures one molecule collides more often with a neighbouring one. These collisions tend to equalize the terminal energy of the molecules. The faster ones slow down or cool while the slower ones accelerate or become hotter.

When this contact between molecules occur under conditions that the molecules do not or can not move from a fixed position such as in a solid, the mechanism of transmitting energy is termed conduction. In case after each collision, the molecule is free to move and circulate so that its next collision may exchange energy with a different molecule, the process is designated convection. This takes place in both gaseous and liquid fluids. These are the more significant methods of energy transfer at the temperature levels normally found in the chemical industries.

In a few cases, the internal motion of the molecules may result in the emission of electromagnetic waves which may be absorbed by a second molecule. Here the resulting mechanism of heat exchange is radiation. This mechanism is of great importance in boilers and furnaces but also takes place at lower temperature.

All the three mechanisms of heat transfer are shown pictorially in the figure given below

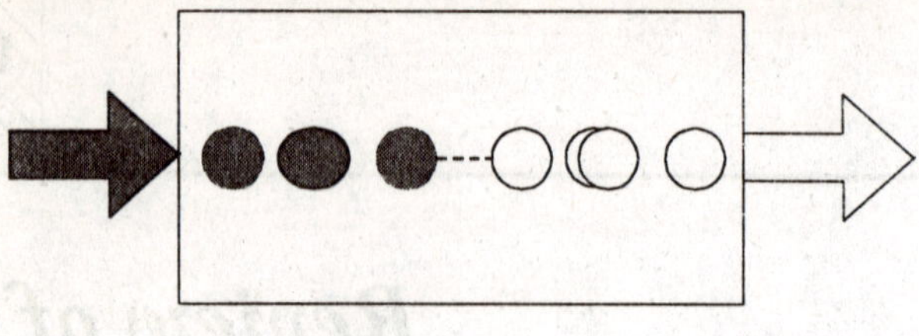

Fig. 9.1 *Heat transfer by conduction*

Conduction

Vibration of molecule results in collision with adjacent molecule and transfer energy

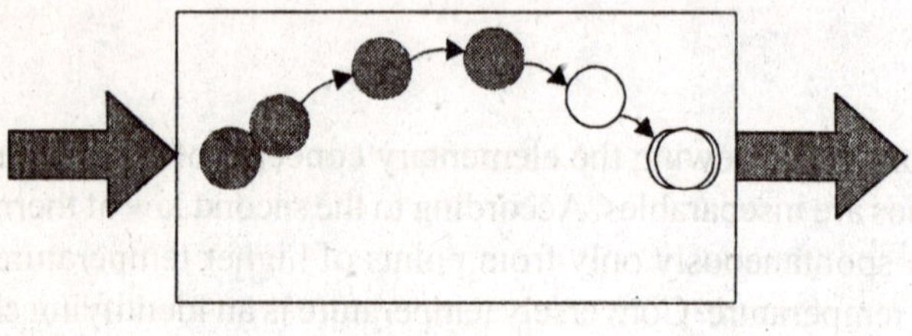

Fig. 9.2 *Heat transfer by convection*

Convection

Molecules receive energy by collision with hotter molecule and circulate to colder surface to transfer energy

Fig. 9.3 *Heat transfer by radiation*

Radiation

Energy as high frequency electromagnetic wave travels through space until it strikes another surface.

How Conduction Occurs

We will start with conduction for the following reasons:

1. Because of its widespread occurrence, it has been thoroughly studied and is described more completely and quantitatively than the other mechanisms
2. An understanding of several of the underlying concepts of conduction will facilitate the handling of convection and to a lesser extent radiation.

For conduction, the basic equation is Fourier's law which is written here in differential form

$$\frac{dQ}{d\theta} = -(k)(A)\frac{dT}{dx} \qquad ...(1)$$

The differential $\frac{dQ}{d\theta}$ is readily identified as the instantaneous rate of thermal energy transport through the body. Differential $\frac{dT}{dx}$ is the potential gradient measured along the path of heat flow, while k and A are physical factors which influence the thermal resistance of the body. As given by Equation (1), this differential from is a common starting point for all problems, whether steady or unsteady state, regular or irregularly shaped objects.

In most engineering applications we are concerned with the steady state where $\frac{dQ}{d\theta} = q =$ constant and with shapes that are readily described mathematically, with these restrictions, we can write Fourier's equation in integrated form

$$q = k_m\ A_m\ \frac{\Delta T}{X} \qquad ...(2)$$

Analogy to Ohm's law and to our potential resistance expressions in fluid flow is now apparent and defines the thermal resistance as $\frac{X}{k_m\ A_m}$. Reciprocal of thermal resistance is frequently used and is called the thermal conductance. It may be noted that electrical resistance of a wire or other solid object is given by $\frac{r\,L}{A}$. Resistivity r is the reciprocal of the conductivity. Thus the two systems are entirely analogous.

Working with finite objects, we must recognize that k and A can vary along the path of heat flow and that we must use appropriate average values. The problem of defining mean values does not exist with X and ΔT which are the finite differences between the terminal points as shown in the following figure.

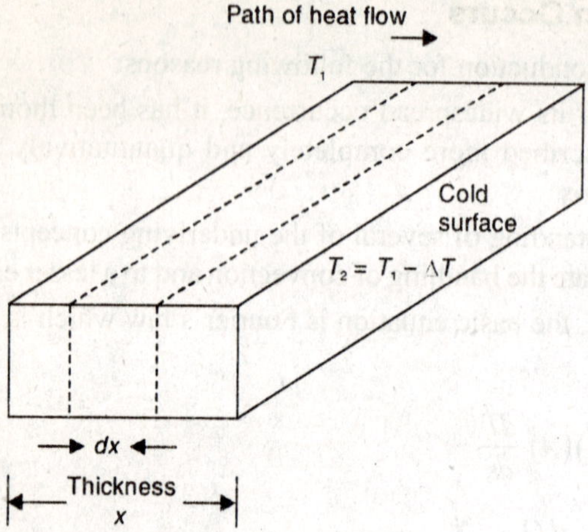

Fig. 9.4 *Heat conduction through a slab*

Recourse to the differential form enables to determine the correct mean value to use. If we rearrange Equation (1) grouping dependent quantities, we have two sets of terms to integrate

$$-k\,dT = q\,\frac{dx}{A} \qquad \text{...(3)}$$

By comparing Equation (3) with the integrated form of Equation (2), we can define each mean

$$-\int k\,dT = k_m\,\Delta T \text{ or } k_m = -\frac{1}{\Delta T}\int k\,dT \qquad \text{...(4)}$$

$$q\int \frac{dx}{A} = q\,\frac{X}{A_m} \text{ or } A_m = \frac{X}{\int \frac{dx}{A}} \qquad \text{...(5)}$$

What is Thermal Conductivity?

Thermal conductivity k is temperature dependent property of matter. Its value is the quantity of heat that would flow through a unit cube under unit temperature difference. Its magnitude varies greatly from one type of material to another.

For any substance, k can be expressed by the equation $k = a + bT + cT^2 + \cdots$. As shown by the following figure, the first two terms are adequate for most engineering materials. For other materials it is a good approximation over a limited temperature range under consideration.

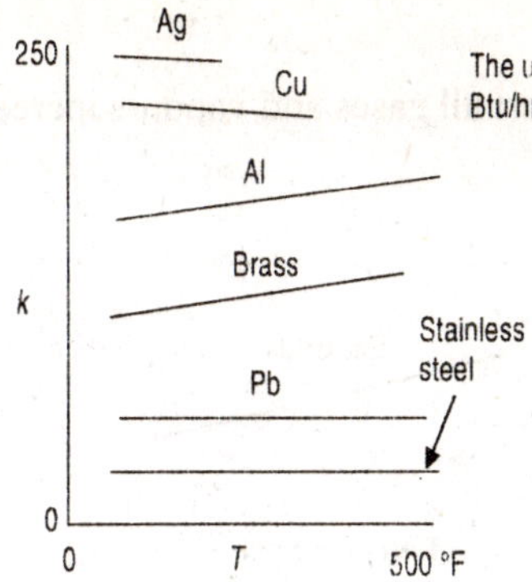

Fig. 9.5 *Thermal conductivity versus temperature for metals*

Metals

For most pure metals *k* decreases with rising temperatures with rising temperatures while for alloys *k* increases.

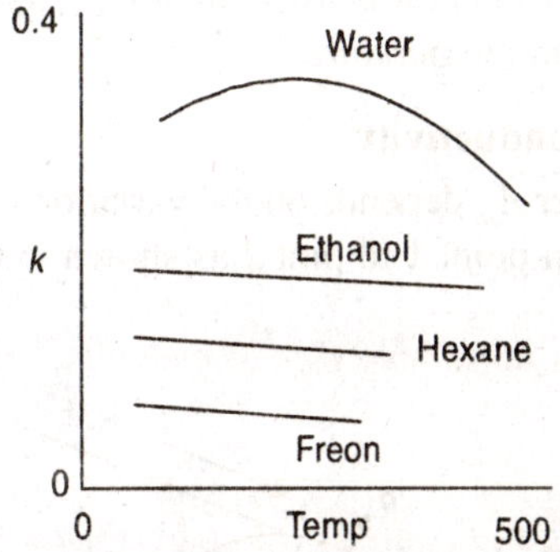

Fig. 9.6 *Thermal conductivity versus temperature for liquids*

Liquids

Increasing temperature yields decreasing *k* values for most liquids water is exception.

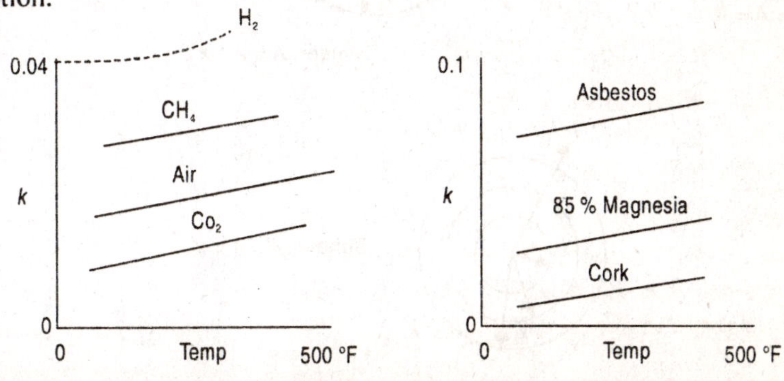

Fig. 9.7 *Thermal conductivity versus temperature for gases and insulating materials*

Gases and Vapours

Thermal conductivity of almost all gases and vapours increase with temperature.

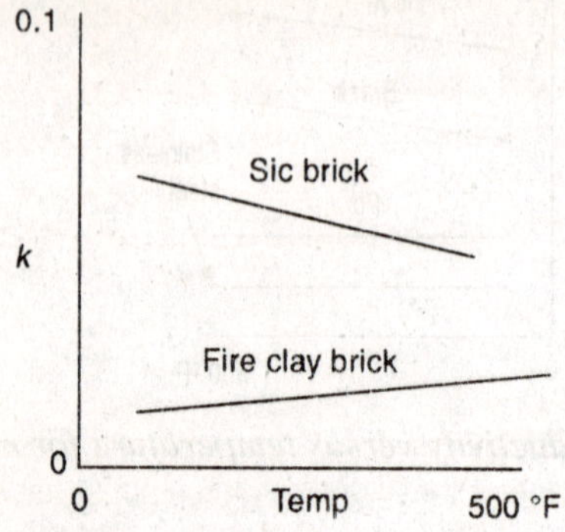

Fig. 9.8 *Thermal conductivity versus temperature for different types of brick*

When k varies linearly integration of Equation (4) shows that k_m is the value of k at the arithmetic mean temperature.

Shape Factor Effects Conductivity

Mean area for heat transfer A_m depends on the variation of A with position along the path of heat flow from point 1 to pint 2 as shown in the following figure.

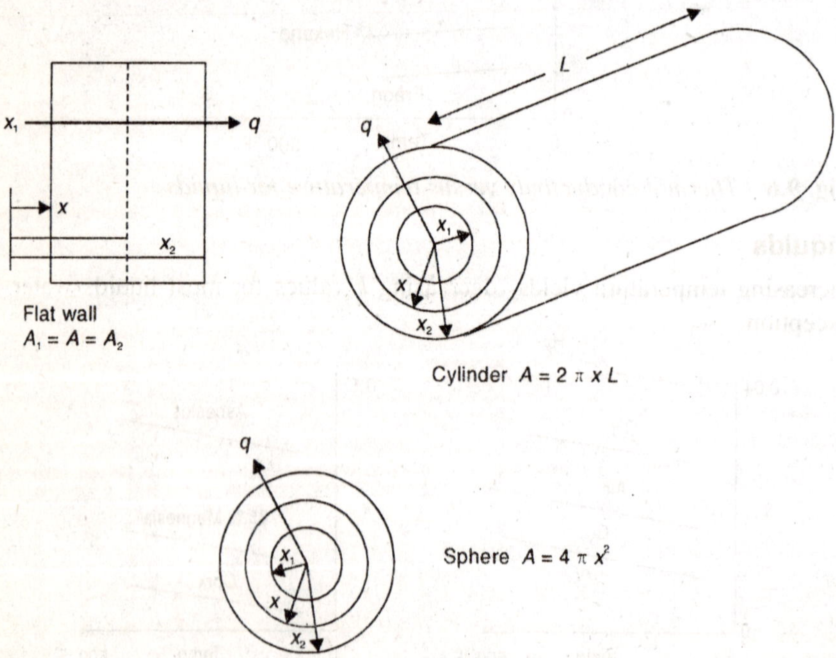

Fig. 9.9 *Heat flow through solids of different shapes*

There are three commonly occurring cases:

1. Area is constant and unvarying with position. This is true in a flat wall. For this case Equation (5) shows that $A_m = A_1 = A_2$.
2. Area varies linearly with position. This covers cylinders in which $A = 2\pi Lx$ or cylindrical sections. Substitution in Equation (5) and integration proves the logarithmic mean to be the correct mean.

$$A_m = \frac{A_1 - A_2}{ln \dfrac{A_1}{A_2}} = 2\pi L \frac{(x_1 - x_2)}{ln \dfrac{x_1}{x_2}}$$

$$= \pi L \frac{(D_1 - D_2)}{ln \dfrac{D_1}{D_2}}$$

3. Area varies as the square of the position along the path of heat flow. For example $A = 4\pi x^2$ in a sphere. The same treatment shows A_m in this case to be the geometric mean and equal to

$$(A_1 \; A_2)^{0.5} = 4\pi \, (x_1 \, x_2)^{0.5} = \pi \, (D_1 \, D_2)^{0.5}$$

Special techniques, mapping or relaxation are needed for irregularly shaped object such as rectangular ducts, insulation around certain items of equipment and buried pipe.

Problem 1

Calculate the heat loss expected from 100 ft of a 2 inch *OD* pipe that is covered with a 2 inch thickness of 85% magnesia insulation. Temperature conditions are such that the inside surface of the magnesia is 300°F and the outside surface at 100°F.

Since heat flow in this problem is steady state we can use the integrated form of the Fourier's equations Equation (2). As indicated in Figure 2, *k* for 85% magnesia is substantially linear with temperature. Hence the correct mean temperature to use is the arithmetic mean or $T_{am} = 200\,°\text{F}$.

The correct A_m is the logarithmic mean because heat is flowing radially and the area varies linearly with position. We calculate this A_m by using the relation given in case 2 above or $A_m = \pi \, L \, D_{lm} = 95 \text{ ft}^2$ we now substitute these quantities in Equation (2) using dimensionally consistent units and find $q = 4100 \text{ Btu/hr}$.

Series or Parallel Heat Flow in Materials

More frequently we must analyze the heat flow through composite bodies. These contain several different materials through which the heat flows in series or in parallel. When this occurs thermal resistances can be combined exactly the same as electrical or hydraulic resistances.

For series flow, the effective resistance equals the sum of the individual resistances or $q = \dfrac{\Delta T}{\Sigma R}$.

For parallel flow, the reciprocal of the effective resistance equals the sum of the reciprocals of the individual resistances or $q = \Sigma \dfrac{1}{R}$.

In applying these concepts, particularly in this case of series flow, the temperature dependence of k and consequently, of the thermal resistance makes analytical solution almost prohibitively complex for anything but a computer. Temperature at the boundary between adjacent materials is unknown and thus the mean conductivity is unknown. To avoid this difficu!ty the usual procedure is to assume all interface temperatures: base the thermal resistance on these and then check the interface temperatures by distributing the total temperature potential proportional to the resistances. For example, in series flow

$$q = \frac{\Delta T_{total}}{\Sigma R} = \frac{\Delta T_1}{\left(\dfrac{X}{k\,A}\right)_1} = \frac{\Delta T_2}{\left(\dfrac{X}{k\,A}\right)_2} = \cdots$$

Note the this problem does not occur in parallel flow because $\Delta T_1 = \Delta T_2 = \cdots \Delta T$. Thus the mean temperature of each substance is fixed and all values of k and R are readily determined.

Problem 2

A flat furnace wall is to be built with a layer of firebrick to withstand the furnace side temperature a layer of insulating brick to minimize the heat loss and a layer of building brick to protect the insulating brick from the weather or other damage. Characteristics of these bricks are from manufacturers bulletins.

The furnace side of the firebrick is expected to be at 3000°F and the outside of the building brick to be at 100°F. It is proposed to build the wall with 9 inch of firebrick 13.5 inch of insulating brick and 4.5 inch of building brick.

Calculate the expected heat loss per ft^2 of this wall and determine the maximum temperature of the building and the insulating bricks to assure that these temperatures are within their allowable values.

Since the wall is flat and the heat transfer area for all layers is the same

$$q = \frac{\Delta T}{\Sigma\left(\dfrac{X}{k\,A}\right)}$$

Since the temperature at the junction of the building and insulating bricks and of the insulating and firebricks is unknown, trial-and-error solution is necessary to establish k_m for each material. As a first attempt assume junction temperature between building and insulating bricks is 500°F and between insulating and fire-brick as 2400°F.

Hence for building brick, T_{am} = 300°F and k_m is 0.417.: insulating brick T_{am} is 1450°F and k_m is 0.246 and for firebrick T_{am} is 2700°F and k_m is 0.929.

Substituting values in equation for series heat flow:

$$\frac{q}{A} = \frac{3000 - 100}{\frac{0.375}{0.417} + \frac{1.125}{0.246} + \frac{0.75}{0.979}}$$

Next calculate temperature drop across the building brick and firebrick and compare to assumed values. For building brick $\Delta T = 463\left(\frac{0.375}{0.417}\right) = 415$ °F *vs* 400°F and for firebrick $\Delta T = 463\left(\frac{0.75}{0.979}\right) = 355$ °F *vs* 600°F.

As a second trial, assume function temperature to be 520°F and 2650°F. Repeating the calculations, we now find that $\frac{q}{A}$ = 477 Btu/(hr) (ft^2). Temperature drop across the building brick is now 429°F *vs* 420°F assumed and across the firebrick is 358°F *vs* 350°F assumed. This check is sufficiently close.

Therefore, the maximum temperature to which the building brick is subjected is 529°F *vs* 1000°F allowable and for the insulating brick is 2650°F *vs* 2500°F allowable. Hence more firebricks are needed.

Consideration of Unsteady State Heat Transfer

The unsteady state, as it occurs in batch operation is governed by the same principles as the steady state. However, these principles require special equation. because those given for the steady state apply only to differential time. As indicated in the figure given below part of the heat flows accumulates in the body and causes a temperature change. This alters the potential and to a lesser extent, the resistance of the body.

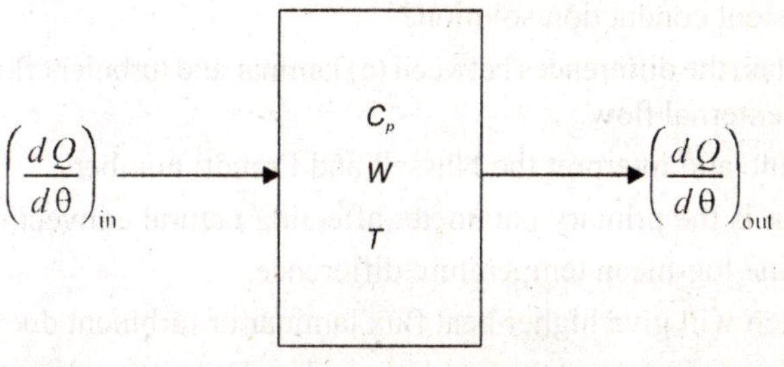

Fig. 9.10 *Unsteady state heat transfer*

$$\left(\frac{dQ}{d\theta}\right)_{in} - \left(\frac{dQ}{d\theta}\right)_{out} = wc_p \frac{d(T)}{d\theta}$$

Techniques for solving this basic equator for finite boundary conditions include (1) analytical solutions (2) Schmidt graphical method (3) numerical solutions. Where it is applicable, the first method is the most exact. The other two use infinite increments instead of true differentials. While these are not as rigorous, they permit handling of situations to which the first method can not be applied.

Review Questions

1. What is the difference between heat transfer and thermodynamics?
2. What basic laws govern heat transfer analyses?
3. Can control volume analyses be employed in heat transfer?
4. Define the three modes of heat transfer?
5. What is Fourier's law?
6. What property of a material determines (a) the amount of energy it can store per unit volume (b) the heat it can conduct under steady-state conditions: and (c) the rate at which it will react to transient temperature conditions.
7. Distinguish between homogeneous and heterogeneous materials.
8. What is Stefan-Boltzmann law?
9. What is a gray body?
10. Distinguish between forced and natural convection.
11. Explain the concept of critical radius of insulation. Why does it happen to cylinders but not to slabs?
12. Define the efficiency of a fin.
13. Give the essence of the finite-difference method.
14. What is the advantage of the finite-difference method compared to exact Fourier-series type analysis?
15. What dimensionless parameter basically determines the character of a transient conduction solution?
16. Explain the differences between (a) laminar and turbulent flow (b) internal and external flow.
17. Define and interpret the Nusselt and Prandtl numbers.
18. What is the primary parameter affecting natural convection?
19. Define log-mean temperature difference.
20. Which will give higher heat flux laminar or turbulent duct flow?
21. Define and interpret the Grashof number. Does it equal a ratio of physical effects?

22. Can natural convection develop in a gravity-free environment such as the inside of an orbiting space capsule?
23. What are the physical differences between the radiation and conduction heat transfer mechanisms?
24. How fast does radiation travel compared to conduction heat flux?
25. Define a black body and a gray body.
26. Define emissivity, absorptivity, transmissivity and reflectivity.
27. What is Kirchhoff's law?
28. Name six different types of phase change.
29. What dimensionless parameters affect the phase-change Nusselt number?
30. Define film type condensation and discuss its characteristics.
31. Can a film condensate achieve a turbulent flow condition.
32. Does it make sense to talk about Reynolds number of a film.
33. Discuss the characteristics of dropwise condensation.
34. Discuss the pool boiling curve regimes.
35. What is Leidenfrost point?
36. Name give different types of heat exchanges.
37. How does 'fouling' enter into a heat exchanger analysis?
38. Why do multipass heat exchangers need a correction factor for LMTD?
39. Define effectiveness and number of transfer units in a heat exchanger.
40. Why is the parallel-flow case less effective by its very nature than a counterflow design?

References

1. Holman, J.P., 'Heat Transfer' 5/e, McGraw-Hill Book Company, NY (1981).
2. Kreith, F., 'Principles of Heat Transfer' 3/e–Intext Educational Publishers, NY (9173).
3. Ozisik, M.N., 'Basic Heat Transfer', McGraw-Hill Book Company, NY (1977).
4. Mc Adams, W.H. 'Heat Transmission', McGraw-Hill Book Company, NY (1954).
5. Welty, J.R., 'Engineering Heat Transfer', John Wiley and Sons, NY (1974).
6. Wong, H.Y. 'Heat Transfer for Engineers', Longman, London (1978).
7. Kays, W.M. 'Convective Heat and Mass Transfer', Tata McGraw-Hill Company, New Delhi (1975).
8. Gebhart, B., 'Heat Transfer', McGraw-Hill Book Company, NY (1971).
9. Jakob, M., 'Heat Transfer', Wiley, NY (1971).
10. Rosenhow, W.M., 'Hand book of Heat Transfer', McGraw-Hill, NY (1973).
11. Kays, W.M. and London A.L., 'Compact Heat Exchangers', McGraw-Hill Book Company, NY (1968).
12. Faas, A.P. and M.N., Ozisik, 'Heat Exchanger Design', Wiley NY (1965).
13. Incropera, F.P. and D.P., Dewitt, 'Fundamentals of Heat Transfer', Wiley (1981).
14. Kern, D.Q, 'Process Heat Transfer', McGraw-Hill Book Company, NY (1957).
15. Tubular Exchanger Manufactures Associate, 'Standards of the Tubular Exchanger', Manufacturers Association 6/e, NY (1978).
16. Heat Transfer Engineering, 'Evolution of Heat Exchanger Design Techniques', Vol No.1 (1979).
17. Frass, A.P. and M.N,. Ozisik, 'Heat Exchanger Design', John Wiley and Sons Inc., NY (1985).

18. Lienhard, J.H., 'A Heat Transfer Textbook', Englewood Cliffs. Prentice Hall NJ (1981).
19. Karlekar, B.V. and R.M., Desmond, 'Heat Transfer' 2/e, St. Paul Minn. West (1982).
20. Sucec, J, 'Heat Transfer', Simon and Schuster, NY (1975).
21. Arpaei, V.S., Conduction Heat Transfer, Addison Wesley Publishing Company Inc., Mass. (1966).
22. Schneider, P.J., Conduction Heat Transfer, Addison Wesley Publishing Company, Mass. (1955).
23. Eckert, E.R.G and R.M., Drake, Heat and Mass Transfer, Tata McGraw-Hill, New Delhi (1979).
24. Knudsen I.D. and D.L., Katz, Fluid Dynamics Heat Transfer, McGraw-Hill Book Company, New York (1958).
25. Jakob, M and G.A., Hawkins, 'Elements of Heat Transfer', Wiley New York (1957).
26. Tongs, L.S., 'Boiling Heat Transfer and Two phase flow', Wiley New York (1965).
27. Perry, J.H., (ed) 'Chemical Engineers Handbook' (4th ed), McGraw-Hill Book Company, (1963).
28. Standards of the Tubular Exchanger Manufacturers Association, b/e Tubular Exchangers manufacturers Association, New York (1978).
29. Bosworth, R.C.L., 'Heat Transfer Phenomena', Wiley New York (9152).
30. Myers, G.E., 'Analytical Methods in Conduction Heat Transfer', McGraw-Hill, NY (1971).
31. Giedt, W.H., Principles of Engineering Heat Transfer Van Nostrand, NY (1957).
32. Karlekar, B.V. and R.M., Desmond, Engineering Heat Transfer, West Publishing Company, Ny (1977).
33. Kays, W.M., 'Convective Heat and Mass Transfer', McGraw-Hill, NY (1966).
34. Taborek, J, 'Evolution of Heat Exchanger Design Techniques', Heat Transfer Engineering, Vol. No.1 (1979).
35. Afgan, N.H. and E.V., Schlunder, 'Heat Exchanger', Design and Theory Source Book NY, McGraw-Hill (1974).
36. Brown, A.I. and S.M., Marco, 'Introduction to Heat Transfer' 3/e, McGraw-Hill, NY (1958).
37. Brown Aubrey I and Marco Salvative M., 'Introduction to Heat Transfer', McGraw-Hill Book Company, NY (1958).
38. John R. Simonson, 'An Introduction to Engineering Heat Transfer', McGraw-Hill, NY (1967).
39. Stoecker, W.F. Design of thermal systems, McGraw-Hill, NY (1980).

